CONCURRENT COMPUTATIONS

Algorithms, Architecture, and Technology

CONCURRENT COMPUTATIONS

Algorithms, Architecture, and Technology

Edited by
Stuart K. Tewksbury
AT&T Bell Laboratories
Holmdel, New Jersey

Bradley W. Dickinson and Stuart C. Schwartz
Princeton University
Princeton, New Jersey

PLENUM PRESS • NEW YORK AND LONDON

Library of Congress Cataloging in Publication Data

Princeton Workshop on Algorithm, Architecture, and Technology Issues for Models of
  Concurrent Computation (1987)
  Concurrent computations: algorithms, architecture, and technology / edited by Stuart
K. Tewksbury, Bradley W. Dickinson, Stuart C. Schwartz.
      p.   cm.
  "Proceedings of the 1987 Princeton Workshop on Algorithm, Architecture, and
Technology Issues for Models of Concurrent Computation, held September 30–October
1, 1987, at Princeton University, Princeton, New Jersey"—T.p. verso.
  Includes bibliographical references and index.
  ISBN 0-306-42939-X
  1. Parallel processing (Electronic computers)—Congresses. I. Tewksbury, Stuart K. II.
Dickinson, Bradley W. III. Schwartz, Stuart C. (Stuart Carl), 1939-   .IV. Title.
QA76.5.P728   1987   88-17660
004'.35—dc19   CIP

Proceedings of the 1987 Princeton Workshop on Algorithm, Architecture, and
Technology Issues for Models of Concurrent Computation, held September 30–October
1, 1987, at Princeton University, Princeton, New Jersey, Sponsored by Princeton
University Departments of Electrical Engineering and Computer Science, and AT&T
Bell Laboratories (Computer and Robotics Systems Research Laboratory)

© 1988 Plenum Press, New York
A Division of Plenum Publishing Corporation
233 Spring Street, New York, N.Y. 10013

Printed in the United States of America

1987 Princeton Workshop
Algorithm, Architecture and Technology Issues
for Models of Concurrent Computation.
Sept. 30 to Oct. 1, 1987.
Princeton University, Princeton, New Jersey.

PROGRAM COMMITTEE

David L. Carter	Bradley W. Dickinson	Mehdi Hatamian
Anis Husain	Thomas Kailath	Vijay Kumar
Sun-Yuan Kung	Adriaan Ligtenberg	James D. Meindl
Will Moore	Jack I. Raffel	Sailesh Rao
Sudhakar M. Reddy	John Reif	Stuart Schwartz
Kenneth Steiglitz	P. A. Subrahmanyam	James C. Sturm
Stuart K. Tewksbury	Leslie Valiant	Paul Vitanyi

SESSION	CHAIRPERSON
Technology Issues	S. K. Tewksbury
Algorithm Issues A	S.-Y. Kung
Real Time System Issues	M. Hatamian
Programming Issues	P. A. Subrahmanyam
Fault Tolerance/Reliability A	A. T. Dahbura
Network and Control Issues	A. Ligtenberg
Algorithm Issues B	A. L. Rosenberg
Fault Tolerance/Reliability B	V. Kumar

Foreword

The 1987 Princeton Workshop on Algorithm, Architecture and Technology Issues for Models of Concurrent Computation was organized as an interdisciplinary workshop emphasizing current research directions toward concurrent computing systems. With participants from several different fields of specialization, the workshop covered a wide variety of topics, though by no means a complete cross section of issues in this rapidly moving field. The papers included in this book were prepared for the workshop and, taken together, provide a view of the broad range of issues and alternative directions being explored. To organize the various papers, the book has been divided into five parts. Part I considers new technology directions. Part II emphasizes underlying theoretical issues. Communication issues, which are addressed in the majority of papers, are specifically highlighted in Part III. Part IV includes papers stressing the fault tolerance and reliability of systems. Finally, Part V includes systems-oriented papers, where the system ranges from VLSI circuits through powerful parallel computers.

Much of the initial planning of the workshop was completed through an informal AT&T Bell Laboratories group consisting of Mehdi Hatamian, Vijay Kumar, Adriaan Ligtenberg, Sailesh Rao, P. Subrahmanyam and myself. We are grateful to Stuart Schwartz, both for the support of Princeton University and for his organizing local arrangements for the workshop, and to the members of the organizing committee, whose recommendations for participants and discussion topics were particularly helpful. A. Rosenberg, and A. T. Dahbura joined with several members of the organizing committee to chair sessions at the workshop. Diane Griffiths served as workshop secretary and treasurer. I am also grateful to the authors both for providing the papers included in this proceedings and for their patience. Bradley Dickinson converted several papers to the LaTeX format. Miles Murdocca and Haw-Minn Lu of the Optical Computing Research Department at AT&T Bell Laboratories helped put together the tools used to produce this book. Finally, I thank D. O. Reudink and W. H. Ninke (Directors of the Computer and Robotics Systems Research Laboratory, AT&T Bell Laboratories) for their support of the workshop and the preparation of this proceedings.

Stuart Tewksbury: Program Coordinator and Senior Editor

Contents

I. TECHNOLOGY ISSUES 1

  1. An Ideology For Nanoelectronics 3
     G. Frazier

  2. Optical Digital Computers - Devices and Architecture 23
     A. Huang

  3. VLSI Implementations of Neural Network Models 33
     H. P. Graf and L. D. Jackel

  4. Piggyback WSI GaAs Systolic Engine 47
     H. Merchant, H. Greub, R. Philhower, N. Majid and J. F. McDonald

  5. Future Physical Environments and Concurrent Computation 65
     S. K. Tewksbury, L. A. Hornak and P. Franzon

  6. Understanding Clock Skew in Synchronous Systems 87
     M. Hatamian

II. THEORETICAL ISSUES 97

  7. On Validating Parallel Architectures via Graph Embeddings 99
     A. L. Rosenberg

  8. Fast Parallel Algorithms for Reducible Flow Graphs 117
     V. Ramachandran

  9. Optimal Tree Contraction in the EREW Model 139
     H. Gazit, G. L. Miller and S.-H. Teng

 10. The Dynamic Tree Expression Problem 157
     E. W. Mayr

 11. Randomized Parallel Computation 181
     S. Rajasekaran, J. H. Reif

 12. A Modest Proposal for Communication Costs in Multicomputers 203
     P. M. B. Vitányi

13. Processes, Objects and Finite Events: On a formal model of
concurrent (hardware) systems 217
P. A. Subrahmanyam

14. Timeless Truths about Sequential Circuits 245
G. Jones and M. Sheeran

III: COMMUNICATION ISSUES 261

15. The SDEF Systolic Programming System 263
B. R. Engstrom and P. R. Cappello

16. Cyclo-Static Realizations, Loop Unrolling and CPM:
Optimum Multiprocessor Scheduling 303
D. A. Schwartz

17. Network Traffic Scheduling Algorithm for
Application-Specific Architectures 325
R. P. Bianchini and J. P. Shen

18. Implementations of Load Balanced Active-Data Models
of Parallel Computation 353
C. Jesshope

19. A Fine-Grain, Message-Passing Processing Node 375
W. J. Dally

20. Unifying Programming Support for Parallel Computers 391
F. Berman, J. Cuny and L. Snyder

IV: FAULT TOLERANCE AND RELIABILITY 409

21. System-Level Diagnosis: A Perspective for the Third Decade 411
A. T. Dahbura

22. Self-Diagnosable and Self-Reconfigurable VLSI Array Structures 435
A. Rucinski and John L. Pokoski

23. Hierarchical Modeling for Reliability and Performance Measures 449
M. Veeraraghavan and K. Trivedi

24. Applicative Architectures for Fault-Tolerant Multiprocessors 475
M. Sharma and W. K. Fuchs

25. Fault-Tolerant Multistage Interconnection Networks
for Multiprocessor Systems 495
V. P. Kumar and S. M. Reddy

26. Analyzing the Connectivity and Bandwidth of Multiprocessors
with Multi-stage Interconnection Networks 525
I. Koren and Z. Koren

27. Partially Augmented Data Manipulator Networks:
Minimal Designs and Fault Tolerance 541
D. Rau and J. A. B. Fortes

28. The Design of Inherently Fault-Tolerant Systems 565
L. A. Belfore, B. W. Johnson and J. H. Aylor

29. Fault-Tolerant LU-Decomposition in a Two-Dimensional
Systolic Array 585
J.H. Kim and S.M. Reddy

V: SYSTEM ISSUES 597

30. Programming Environments for Highly Parallel Scientific Computers ...599
A. P. Reeves

31. Systolic Designs for State Space Models: Kalman Filtering and
Neural Network 619
S-Y Kung and J. N. Huang

32. The Gated Interconnection Network for Dynamic Programming 645
D. B. Shu and G. Nash

33. Decoding of Rate k/n Convolutional Codes in VLSI 659
V. P. Roychowdhury, P. G. Gulak, A. Montalvo and T. Kailath

34. IC* Supercomputing Environment 675
E. J. Cameron, D. M. Cohen, B. Gopinath, W. M. Keese, L. Ness,
P. Uppaluru and J. R. Vollaro

35. The Distributed Macro Controller for GSIMD Machines 689
W. Holsztynski and R. Raghavan

36. The Linda Machine 697
V. Krishnaswamy, S. Ahuja, N. Carriero and D. Galernter

Index 719

PART I
Technology Issues

Introduction

The papers in this section address technology issues. The emphasis is on emerging or future technologies likely to impact the successful development of high performance parallel computing systems. The emphasis is also on new technologies, rather than the evolution of entrenched technologies such as silicon VLSI.

Frazier looks beyond the scaling limits of classical logic devices to devices with nanometer scale dimensions (i.e. dimensions below 100 nm = 1000 Å). In a semiconductor, doped at 10^{15} dopants/cm^3, the individual dopant atoms have an average separation of about 100 nm, providing a view of the dimensions considered here. At such small dimensions, the behavior of neighboring devices is coupled by the quantum mechanical interactions between them. In this environment of very small devices, Frazier considers several conventional system issues, providing an intriguing look into the possibilities of future system physical environments.

Huang also looks beyond classical electronics, in this case to photon-based computing (i.e. optical computing). The very high speed of optical switching elements combined with the use of highly parallel beams of non-interacting optical interconnections provides a computing environment with unique capabilities and constraints. Huang's paper reviews the work of several Bell Labs researchers pursuing a variety of directions in this active field.

Neural networks have received considerable attention. At the workshop, **D. Psaltis** reviewed optical neural computing directions. A very readable description of his work appeared recently in Scientific American. Much of the research on neural networks emphasizes algorithm development and network simulations. Electronic realizations of neural networks, with small dimension analog "neuron" circuits are discussed by **Hans Graf et al.** This work draws on submicron structures fabrication research at Bell Labs. Working neural network chips are being used for studies on pattern recognition.

McDonald et al. discuss two major directions in advanced technologies for high speed integrated electronics. The first is use of GaAs IC's to achieve higher speeds than possible with silicon IC's. The second direction is developing a packaging approach which permits the high internal intrinsic speed of such circuits to be extended beyond the perimeter of the IC. Silicon circuit boards with high bandwidth transmission lines are described. The architecture of a systolic array processor to achieve 1000 GFLOPS is described.

Tewksbury et al. consider the specific issue of high performance communication environments for fine grained computation and communication in multi-purpose, highly parallel computing environments. From this perspective, technologies able to achieve very high performance communications are considered. Wafer-scale integration, hybrid wafer-scale integration, optical interconnections and high transition temperature superconducting transmission lines are described.

The highest data rates are obtained between devices separated by the smallest distances (e.g. neighboring gates within a monolithic IC). A high local data rate can be propagated by pipelining computations through successive cells of a more complex function, retaining minimum physical distance between successive cells of the pipeline. However, that high data rate is managed, typically, by a global clock signal, distributed over long distance and complex paths to all retiming elements in the circuit. **Hatamian** addresses the issue of preserving the high local data rates under such conditions, emphasizing the issue of clock skews between retiming circuits in different areas of a circuit.

Chapter 1

An Ideology For Nanoelectronics

Gary Frazier [1]

Abstract

The performance limits of conventional integrated circuits will be reached within twenty years. Avoiding these limits requires revolutionary approaches to both devices and architectures that exploit the unique properties of nanometer- sized electronic structures. The casualties of this revolution will include high connectivity architectures, transistors, and classical circuit concepts. One approach, Nanoelectronics, combines quantum coupled devices and cellular automata architectures to provide computing functions that are downscalable to fundamental physical limits. This paper reviews the motivations for nanoelectronics are reviewed, and presents a framework for developing this technology. Some of the issues relevant to nanoelectronic computation are also addressed.

1.1 Introduction

The information economy is a direct offspring of semiconductor industry. In large part because of the invention and improvement of the integrated circuit (IC), information management has become decentralized, convenient, and cost effective. The further development of IC technology is fueled by an ever-increasing demand for information processing in science, medicine, education, industry, and national defense. Certainly, the motivation to improve the performance of computing systems will intensify as we move into the age of widespread robotics, desktop supercomputers, man-machine interfaces, and hardwired artificial intelligence [1].

---

[1] Texas Instruments Incorporated, Dallas, TX

There are two approaches to providing the computational resources that will meet the information processing requirements of the 1990s and beyond. First, new algorithms can be developed that give exponential improvements in the mapping of a problem solution onto present technology. The Fast Fourier Transform is an archetypical example of a performance breakthrough achieved by algorithmic optimization alone. Embodiments of optimized algorithms range from systolic digital signal processor arrays to silicon neural networks. Second, processor performance can be enhanced by increasing the level of chip integration and functional density. Function scaledown, which also provides increases in speed as well as reductions in power and weight, has been the dominant system enabler to date. In fact, the growth of the semiconductor industry has paralleled the physical downscaling of computer components.

Until recently, the potential for advancing both algorithm science and chip technology was open ended. There are now clear indications that chip-level functional density will saturate within twenty years. When this maturation of technology occurs, the return on the investment in new algorithms will also decline. The resulting slowdown of chip enhancement will impact the full spectrum of semiconductor computer technology. The impending catastrophe cannot be avoided by anything short of a revolution in chip technology [2].

This paper presents an ideology for post–microelectronics that could avoid projected limits. First, we will review some of the problems with current practice and discuss the essential features of alternative approaches. Second, candidate devices and architectures will be outlined. Finally, we will speculate on the possible impact of the proposed technology on future computing resources.

1.2   The Scaling Limits of Conventional Technology

An important factor responsible for the pervasiveness of the integrated circuit has been the sustained exponential decrease over time of minimum lateral circuit geometries. However, there are limits to the use of geometry scaledown as the means to further improve the performance of conventional integrated circuits. These limits are so fundamental as to force the ultimate abandonment of high connectivity architectures, transistors, and the classical circuit concept. We can review briefly a few of the basic problems.

1.2.1   Transistor Scaling Limits

The linchpin of solid state electronics is the p-n junction. The depletion layers between p- and n- regions of a circuit serve as the potential barriers required to guide the flow of electronic charge. A transistor is simply a device that can modulate the effectiveness with which p-n structures electrically isolate two points in a circuit. Any integrated technology must possess a similar ability to control the direction and magnitude of charge transport. Unfortunately, depletion layers begin to lose their ability to confine electrons as p-n widths are scaled below 0.2 micrometers in size. Based in part upon the failure of depletion isolation, it has been estimated that junction-based silicon switching devices cannot be shrunk appreciably below 0.1 micrometers [3]. The classical transistor must eventually lose its ubiquity.

1.2.2 Connectivity Scaling Limits

The overall functional density (functions/area-time) of a computing machine is controlled by the available interdevice communication bandwidth and device density at the board, chip, function, and gate levels. Since the speed of light sets an upper limit on communication rate, system performance can be improved if more functions are integrated on-chip. However, current models of circuit parasitics show that with scaling, RC delays and interline coupling rapidly degrade the speed and noise performance of submicrometer interconnections [4,5]. Moreover, device size and electromigration factors set upper limits on the available current density at the device port. For this reason, even well-isolated long interconnects will have self-capacitances that lead to unacceptable effective device switching speeds. In the near future, the computational advantage of long, multi level interconnects will be negated by the saturation and/or reduction of the effective communication bandwidth per wire. Therefore, merely combining submicrometer active devices with conventional interconnect networks will not significantly improve the performance of integrated systems.

1.2.3 Yield Scaling Limits

At present, the unavoidable errors induced by substrate defects, cosmic rays, and thermal fluctuations are second-order considerations in VLSI design. A low density of fixed and transient errors can be handled by production culling and error control coding, respectively. Further component scaling will make these *ad hoc* fault management schemes obsolete. Scaledown reduces the number of electrons that can participate in each computation. This reduction translates into an increase in both the informational impedance and the noise sensitivity of the switching event. In the future, media noise and transient upset events will affect entire groups of submicrometer devices. Hard and soft faults will become inherent technological characteristics of ultra-scaled structures. The age of the 100% functional, fault-free integrated circuit is fading fast.

1.2.4   Addressability Scaling Limits

An inherent problem with ultra-integration will be the further decrease in our ability to access directly a given functional resource. For a minimum feature size $1/S$, resources may grow as fast as S^2 and S^3 for two- and three- dimensional circuits respectively. However, conventional access methods are essentially peripheral, so that I/O accessibility may always be one dimension behind component density. With scaledown, it will become increasingly difficult to address directly either single devices or even entire device groups. This growing inaccessibility presents the potential for severe I/O bottlenecking at function boundaries.

1.2.5   Superposition Limits

The issue of intercomponent crosstalk points out a subtle influence scaling will have on the circuit paradigm. Traditionally, a *Principle of Superposition* has been assumed in the design of complex information processing structures. This principle, quite valid in the 1970s, asserted that the properties of active elements such as transistors and logic gates were independent of the physical environment of the device. In addition, the superposition argument maintained that the behavior of the computing system at any level of complexity could be expressed as a piecewise-linear sum of the characteristics of more basic functions. Many of the impediments to the further enhancement of conventional computing structures can be traced to the invalidation of superposition at one or more levels of functionality. For example, it is clear that successfully avoiding the problems associated with interconnect scaling would allow device density to control the limiting on-chip functional density. However, as interdevice geometries decrease, the coupling between these active, nonlinear agents will also increase. Device densities are now reaching the point where classical notions of isolated, functionally independent active elements are less appropriate than are distributed models that include the possibility of collective modes of device interaction. Conventional device simulation models already recognize the need to account for interdevice parasitics [6]. Scaling into the nanometer size regime only exacerbates the problem of isolating device function from the local environment. Further scaling will make it impossible to partition a circuit into active and passive components. Lumped-constant rules will be of little use in nanometer design.

In addition to architectural problems at the device level, certain chip applications cannot employ the principle of superposition in the design of functions. For even though traditional sequential computers are computationally universal, the complexity of NP-complete optimization problems requires prohibitively large serial processing power to reach good solutions within an acceptable time. Important tasks such as artificial intelligence, image understanding, adaptive learning, and general speech recognition demand massively parallel, global computations. It is possible that true machine intelligence can evolve only if the behavior of the system as a whole exceeds the sum of its separate parts [7]. If that is true, even highly concurrent architectures that can be decomposed into modules of more primitive functionality may not exhibit the emergent collective properties required to solve important compute-bound problems.

1.2.6 Limits to Static Design

Interpretation and Compilation are popular approaches to hardware and software development. The two methods trade speed for flexibility. In program compilation, the computer programmer expresses an algorithm in terms of a language, then converts this logical literature into a fixed, machine-dependent set of basic operations. Similarly, the computer architect may employ a library of hardware primitives that can be used with a "silicon" compiler to map algorithms into an application-specific integrated circuit (ASIC). In contrast, program interpreters use a core routine to execute arbitrary algorithms phrase by phrase, while machine interpreters utilize a microprocessor to emulate arbitrary hardware at reduced throughput.

These methodologies suggest that you cannot have simultaneously algorithmic flexibility and optimum throughput. The more optimized the hardware, the lower its adaptability to new computing environments. For this reason, ASIC may be destined to accelerate, but never displace, general-purpose, possibly dinosauric architectures. Our inability to erase and "re-write" transistors and wires as easily as we modify code warns of an ultimate saturation of the scope-performance product of static architectures.

1.3 The Law of Lilliput

The aforementioned limits are compelling, but are for the most part technological in origin. There is no fundamental reason why physical scaledown cannot be used to improve the functional density of integrated circuits indefinitely. However, dramatic shifts from convention are necessary if the advantages of scaling are to be realized beyond the 1990s. To avoid the limits of current practice, we propose the following set of six essential requirements for a successful post microelectronic IC technology.

- Devices must be scalable to fundamental physical limits.

- The most basic functions of a computing system must employ only local connection schemes.

- Device function must transcend simple switch operations.

- Intercomponent coupling must be exploited as a method of communication and control.

- Fault management must be incorporated at the lower levels of functionality and feature size.

- Functionality must be reconfigurable.

1.3.1   Discussion

Further increases in the density of on-chip computational resources can be obtained only through the combined development of revolutionary devices that are based upon nanometer physical phenomena and chip architectures that avoid interconnect saturation. This implies an absolute minimization of device connectivity within low-level functions. Therefore, it will be necessary to recast existing random logic networks into low-connectivity equivalents. To avoid I/O bottlenecks in the reorganized functions, a high degree of functional concurrency and pipelining will be necessary. We expect that the advantages that accrue from regular VLSI layout will carry over to regularized nanometer functions. The combination of layout regularity and pipelining at several levels of complexity suggests that these architectures will have a measure of topological scale invariance. In short, the future is fractal.

It will be necessary to construct functions holistically through an appreciation of the impact of environment upon active and passive structures. Integrated circuit models in the late 1990s may abandon entirely the idea of point-in- space design in favor of the construction and solution of the continuous partial differential equation that describes the effective computational volume.

To compensate for the penalties incurred by reducing interconnect complexity, device behavior must transcend simple transistor-like functions. By "device" we mean a set of physical structures that collectively perform a computing function without the use of wiring. Since reduced connectivity translates into communication delays, these most basic computing elements must execute higher level operations when the data finally arrive. Increasing device complexity amounts to extracting greater functionality from known physical phenomena. For example, devices based on superconductivity effects demonstrate how function can be wrought from phenomena. Submicrometer Josephson Junction (JJ) technology has been used to construct an A/D convertor using only one JJ device per digitized bit [8]. This significant improvement in functional density arises not from the cleverness of the device architecture, but directly from the behavior of very small, very cold things. A similar strategy must be used to leverage nanometer physics into complex device function.

The negative effect of scaledown on signal-to-noise ratios emphasizes the need for a generic method for managing soft and hard failures within ultra-scaled circuits.

Current efforts to develop fault-tolerant architectures at the system and processor levels must be extended to include fault management at the most basic levels of functionality. Finally, to route around defects, increase adaptability and avoid early obsolescence, chip functions must be explicitly reconfigurable. Functional elasticity will also eliminate many design turnaround problems by softening the distinction between software and hardware.

In summary, successful next-generation chip architectures must be based upon minimal connectivity and sophisticated active devices. Chip functions must be pliable and fault tolerant.

1.4 Nanoelectronics

The complete approach to meeting the post- microelectronic challenge of next-generation IC technology is called *Nanoelectronics* [9]. This section examines candidate nanoelectronic devices and architectures.

1.4.1 Quantum Coupled Devices

Conventional electronic devices are designed from a set of approximations that allow explicit quantum mechanical considerations to be ignored. As dimensions scale below 0.1 micrometer, new effects associated with the wave-like nature of the electron begin to dominate the mechanisms of charge transport. Recent advances in materials science have made it possible to construct devices that demonstrate strong quantum effects. These nonclassical phenomena, including quantum size effects and resonant tunneling, can be used to build a new technology base for nanometer electronics.

Semiconductor technology now provides the means to construct electronic nanostructures. Heterostructure techniques combine dissimilar semiconductors layer-by-layer to tailor the effective electronic band structure of the resulting composite on an atomic scale. The energies and density of electron states can be engineered into a heterostructure by controlling the stoichiometry of the final chemical compound. Popular fabrication methods include molecular beam epitaxy and metal-organic chemical vapor deposition [10]. As a heterostructure is grown, abrupt changes in composition can be used to shift electronic material properties. Unlike broad, fuzzy p-n junctions, the effective potential variations across a heterostructure are sharply defined.

To discuss device possibilities, we will take nanofabrication as a given and consider a nanoelectronic structure that incorporates most of the relevant physical phenomena required to build a useful device. We consider two nanometer-sized cubes of

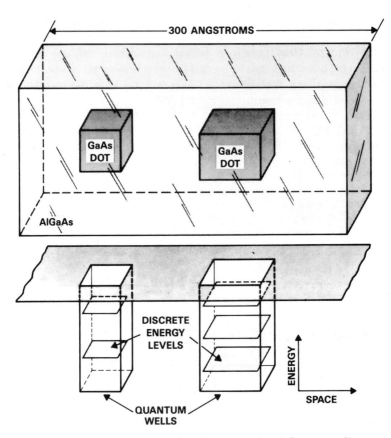

Figure 1.1: GaAs quantum dots and their potential energy diagram

semiconductor that are embedded in a sea of another material as shown in Figure 1.1a. For clarity, we will use the aluminum gallium arsenide (AlGaAs) material system in our discussion. The energy diagram for the possible electron energies in this structure is shown in Figure 1.1b. Due to the difference in the electronic properties of GaAs and AlGaAs, the conduction electrons within the GaAs material are surrounded by an electrostatic potential barrier. In addition, the small size of each cube squeezes out most of the normal energy states of conduction band GaAs, leaving only a few sharply defined energy levels, as shown in the potential well diagrams. State quantization due to dimensional scaling is known as the Quantum Size Effect (QSE). To first order, the number of distinct electron energy levels within a potential well is inversely proportional to the square of the well width. Using the dimensions of Figure 1.1a, our hypothetical GaAs structures would have only a few bound states as shown. The QSE can have a controlling influence on device properties even if only one or two physical dimensions are quantized [10]. As we will see below, the exploitation of the QSE is basic to nanoelectronics. Physical structures that display size quantization effects are called *quantum wells*. Three-dimensionally quantized structures, like our quantum cubes above, are called *quantum dots* [11].

Classically, electrons could be trapped forever inside potential cages. However, the wave properties of electrons enable them to escape from quantum wells by the process of *resonant tunneling* [10].

The resonant nature of electronic tunneling between quantum wells also provides the mechanism with which to control charge transport. Charge can be exchanged between quantum wells if

1. they are in close proximity so that tunneling effects are strong,

2. energy is conserved,

3. momentum is conserved, and

4. the destination state of the transported charge is unoccupied.

These conditions are satisfied when quantum wells are spaced by less than a few hundred angstroms, and a compatible occupied state in one well is in energy resonance with an empty state in the other. A schematic of the charge transfer process between two quantum dots is shown in Figure 1.2. As shown in Figure 1.2, energy is easy to conserve in tunneling between identical quantum wells, since symmetry requires the allowed energies to be the same in either well. The tunneling probability between dissimilar wells is determined by the relative position of the energy levels between the two wells. In principle, the communication between nonresonant quantum wells can be reduced to an arbitrarily small value.

The quantum size effect and resonant tunneling provide all the essential phenomena required to control the direction and amplitude of charge transport in nanostructures. Devices that operate ostensibly through the use of the QSE and resonant tunneling effects are called *quantum coupled devices*. Preliminary ideas for devices based upon resonant tunneling have been conceived [9]. In addition, the empirical and theoretical understanding of transport in quantum devices has increased considerably in recent years [12,13].

1.4.2 Quantum Architectures

The effective utilization of the scaling advantages of quantum coupled devices requires the development of equally scalable device architectures. This requirement is

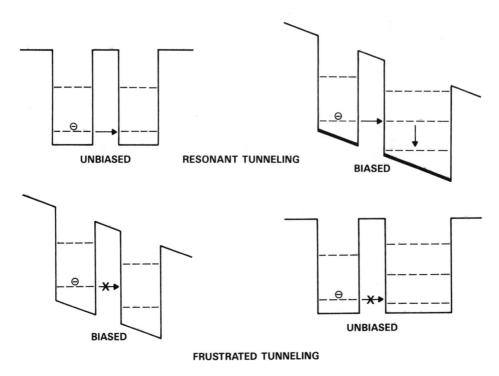

Figure 1.2: Elastic (strong) and Inelastic (weak) tunneling between quantum wells

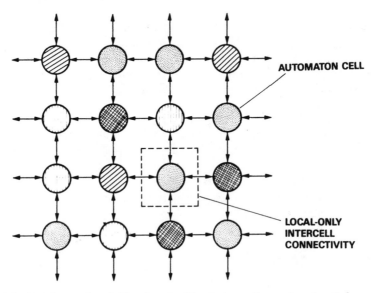

Figure 1.3: Lattice and neighborhood of basic two-dimensional cellular automaton

best satisfied by the cellular automaton [14]. Basically, a cellular automaton (CA) is a collection of simple active devices that interact in discrete space and time. Each active device, or cell, is placed at one of the vertices of a regular lattice as shown in Figure 1.3. The function of each cell is specified by a rule of interaction analogous to a state transition lookup table. Cell rules are usually deterministic functions of the instantaneous value of the states of neighboring cells. However, cell types may be defined that include memory. More elaborate models can test the impact of clock skew and fault tolerance by allowing cells to interact asynchronously and obey time-dependent rules. Spatio-temporal "snapshots" of the dynamics of several one-dimensional cellular automata are shown in Figure 1.4. All these examples demonstrate nearest-neighbor interactions between two-state cells. Light areas (paper) represent state "0" cells, while dark areas (print) correspond to state "1" cells. Even this simple class of CA displays a variety of behaviors ranging from time-independent to nearly chaotic time evolution.

The most basic characteristic of a CA is the formal limit upon the range of direct coupling between lattice cells. Typically, cells are only allowed to influence the dynamics of nearest and perhaps next-nearest neighbors. Thus, CA embody the precepts of eliminating long interconnects while also explicitly defining the behavior of a cell in terms of a tightly coupled local environment.

Computation in a CA results from the adjustments each cell makes to its internal state in response to changes in the state of its local environment. During a computa-

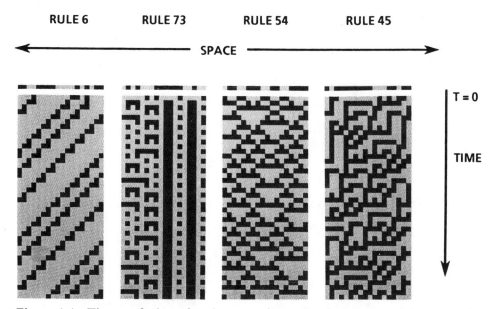

Figure 1.4: Time evolution of various one-dimensional, nearest neighbor coupled cellular automata.

tion, all cells act in parallel, using their initial state and individual interaction rules to calculate future lattice states. Despite minimal connectivity, many CA are known that can perform all the general logic operations required to build a computer. The Game of Life is a popular example of a two-dimensional CA that can be set up to perform general-purpose computation [15]. A natural mode of computation in rule-homogeneous CA can be based upon interactions between spatio-temporal patterns in the cell dynamics. The basic events necessary for interesting computation arise out of a purely homogeneous active medium. As seen in Figure 1.5, spatially static (vertical) and propagating (slant) patterns are supported in this one-dimensional automaton. Space-time events include pattern creation, annihilation, distortion-less crossover, and phase shift. We can define a computational event in this CA as a nonlinear collision between patterns. Pattern annihilations and creations can be translated into logically equivalent Boolean functions such as AND, NOR, and EXOR, etc., by assigning logical states to distinct space-time patterns. It is interesting that in this automaton, activities are more appropriate descriptors of system state than are instantaneous cell patterns. Many other examples of computation in homogeneous CA are known.

1.4.3   Combining Quantum Coupled Devices and Cellular Automata

The close analogy between quantum coupled device arrays and cellular automata can be seen by considering the one-to- one mapping of quantum devices with CA lattice cells. An automaton cell is completely defined by its interaction rule, which determines how information flows through the network. In an analogous fashion,

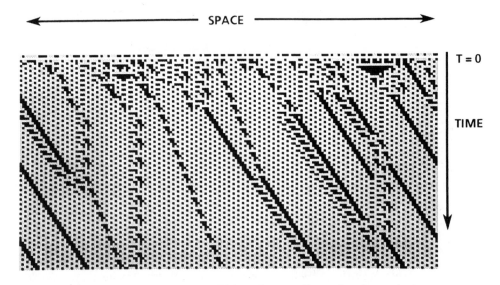

Figure 1.5: Computing collisions in one-dimensional automata.

the electrostatic coupling between quantum wells provides a mechanism for tailoring charge flow between wells. Conditions of resonant and nonresonant tunneling between quantum structures are equivalent to connections and isolations between cells, respectively. The strength of interwell connections can be modulated by the electric fields that are broadcast from each well to its neighbors. Lithographic tuning of quantum well energy levels can be used to express the variations in cell type and connectivity. We call the combination of quantum devices and CA a *quantum cellular automaton.*

1.4.4 Increasing Device Sophistication in Nanoelectronics

The quantum size effect can be used to design quantum wells that have several well defined states. One approach to exploiting this additional complexity is to use quantum devices to perform multivalued logic operations. The most important benefit offered by multivalued logic is the potential to reduce interconnect complexity by embedding more states, and more functionality, within the same number of switching elements [16]. For example, ternary (three-valued) multipliers have been conceived that contain 60% fewer interconnects and 20% fewer devices than do equivalent binary circuits [17]. Quantum coupled devices are ideally suited to multivalued representations, since the number and spacing between energy (logic) levels within a device is adjustable. We have demonstrated in principle that individual quantum coupled devices can perform general-purpose, multivalued logic [18]. The ability to freely mix binary and n-ary logic within the same technology opens exciting new possibilities for function design.

1.4.5 Fault Management in Nanoelectronics

Several approaches can be used to design a fault-tolerant CA capable of reliably storing and processing information. Majority voting from redundant cells can be used to correct local errors. In this method, local variables may be defined in terms of the group properties of small cell arrays so that the upset of one member of the array does not significantly alter the group dynamics. Multivalued cells can be used to perform the equivalent of binary logic within a redundant, super-binary representation. Tolerance of transient errors can be obtained by combining logically reversible CA functions with a novel form of garbage collection. In addition to its mathematical result, a reversible logic sequence also provides the information required to undo the computation. The extra computational "garbage" acts as a tag on the trajectory of the computation through state space. At the end of a series of reversible logic operations, this normally superfluous trajectory information can be used to check for computational errors. Detecting errors in a reversible logic system is straightforward. First, the system is initialized with the input variables.

Second, the logic operations are carried forward, generating the desired function of the input variables as well as the trajectory code. To check for errors, an identical but time reversed sequence of logic operations is applied to the tentative output result along with the trajectory "garbage." If no errors occur during the round-trip, the forward and reversed logic operations will precisely cancel, leaving the system in its original state. Any irreversible CA can be made reversible by adding states to each cell and performing a simple rule transformation [19]. In principle, then, long sequences of complex CA logic operations can be verified by a postcomputational test. Most importantly, a reversible logic approach would not need to rely upon local mechanisms to manage errors. Of course, this technique has the disadvantage of lower throughput than real-time methods.

1.4.6   The Role of Optics in Nanoelectronics

The novel optical properties of quantum well structures are suggesting a plethora of device applications [10]. Quantum well lasers, nonlinear electro-optic switches, and light modulators have been demonstrated. These successes suggest that nanoelectronics could employ optical buses to avoid interconnect crosstalk and use optically active quantum devices to perform logic operations. Three points are important in this regard. First, all other parameters being equal, the speed and low-crosstalk properties of light are more than offset by the energy and noise figure spent in converting between the electrical and optical domains. Second, photons are much larger than electrons. This implies that those nanoelectronic structures that are designed to collect and process light will always be much larger than their all-electronic equivalents. Thirdly, we note that light does have the unique property of being easily broadcast over large areas. A natural application of light will be in providing optical clocks and energy to large arrays of nanoelectronic devices. Low resolution optical interactions have great potential to synchronize the movement of information and supply contactless operating power.

1.5   The Shape of Things to Come

Nanoelectronics should permit packing densities to exceed 10 billion devices/cm^2 per layer. As an example in contrast, a conventional VLSI bondpad consumes an area equivalent to more than 400,000 laterally quantized quantum devices. Such a dramatic increase in device density will lead to novel functions and system concepts.

1.5.1   High Density Memory

It is estimated that conventional memory capacity will not exceed 64 megabits/chip. Capacity limits are due in no small measure to the saturation of the density of the

X-Y matrix crosspoints used to access data within the memory array. One approach to avoiding interconnect saturation would use the dynamics of a quantum cellular automaton to store and retrieve information through the propagation of virtual address wavefronts across the cell array. No internal addressing structure would be used. Rather, information would be stored as points in the large state space of the network. The memory would act much like a data-driven systolic processor array, except that the processing nodes would serve as memories for single items in the manner of a random access memory. The random access of information from the memory would be necessarily delayed by the time required to

1. propagate addresses across many millions of cells,

2. allow the dynamics of the system to retrieve the desired memorized item from state space, and

3. propagate the readout to the array periphery.

A clear application of such a system would be in the replacement of large memories including magnetic and optical disk machines.

1.5.2 Ultra-Scaled Functional modules

As shown in Figure 1.6, one approach to the early application of nanoelectronics would be the implementation of low-level functional modules that can be embedded in evolutionary VLSI architectures. Often used functions such as memory, gate arrays, and ALUs would be embodied by special-purpose minicells. These ultra-scaled functions would be integrated into more conventional connection-intensive architectures where the advantages of buses, I/O drivers, etc. can be exploited at higher levels of function interconnection. It should also be possible to construct very large, all- systolic processor arrays on-chip. Complete pipelining of digital signal processing functions has been suggested [20]. Systolic methods have the same structural regularity as the proposed cellular minicell. Therefore, nanoelectronics should merge naturally with existing iterative architectures and provide orders of magnitude of additional processing parallelism per chip. Embodiments of generic tissue operations would add both fault tolerance and scalability at the level of function where systems are currently most vulnerable [21].

1.5.3 Reconfigurable 3-D Superchips

We pointed out earlier that local field effects can be used to isolate or interconnect quantum coupled devices. The ability to reconfigure connections at the device level

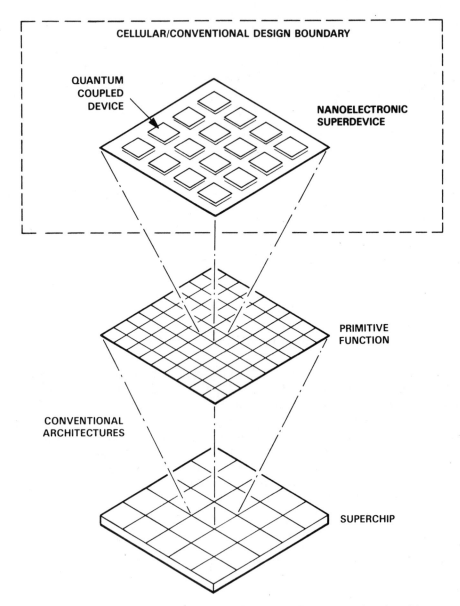

Figure 1.6: Providing ultra-dense functions by combining conventional architectures with quantum cellular automata.

provides new opportunities for dynamic resource allocation on-chip. In principle, hard errors can be routed around optimally, since entire sublattices can be functionally restructured. The physical topology of such a chip would be fixed, but the logical machine description would be programmable. In particular, gates, functions, variable grain size processors, and even buses and bondpads could be configured by the proper initialization of the array into regions of control, connection, and computation.

Full functional reconfigurability of a quantum CA also portends other futuristic system functions such as machine self-organization. Functionally complete, dynamically reconfigurable CA are known that can construct, through an internal program, new CA functions, and even replicas of the parent construction automaton itself. Although a long-term prospect, nanoelectronics may be the first realistic technology with which to develop a self-repairing and self- organizing superchip.

Finally, it is noteworthy that the formal lack of a connection infrastructure should allow nanoelectronic devices to be stackable in the vertical dimension. As suggested by Figure 1.7, computations might be carried out along the vertical dimension, with interfunction communication taking place in the chip plane. The construction of Teradevice chips will not be physiologically impossible.

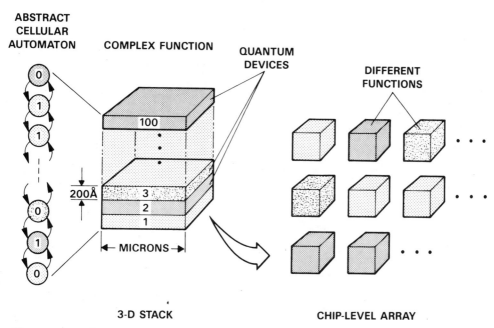

Figure 1.7: Three-dimensional superchip. Computation proceeds in the vertical dimension while communication is managed in the chip plane.

1.6   Conclusions

The downscaling of the minimum geometries of transistor- based integrated circuits will eventually be brought to an end by a combination of problems related to devices, interconnections, noise, and reliability. The resulting saturation of circuit densities will end the exponentially downward trend in the cost effectiveness of integrated circuits. Avoiding this catastrophe may involve a blurring of the traditional division between device and circuit design, and even the abandonment of circuits in favor of super-devices that perform relatively sophisticated logic and memory functions. The potential benefits of a wholly revitalized semiconductor technology are great, but true paradigm shifts are required to fully exploit the scaling advantage of nanometer scaled structures. In our view, the most likely solution will evolve from the combination of quantum coupled devices and cellular automata architectures.

Acknowledgements

I would like to thank Bob Bate, Bill Frensley, and Mark Reed for useful discussions. This work was supported in part by the Office of Naval Research.

References

[1] Edward A. Torrero ed., *Next-generation Computers,* IEEE Press, (1985).

[2] R.T.Bate, *Limits to the Performance of VLSI Circuits,* VLSI Handbook, Academic Press, (1985).

[3] P.K.Chatterjee, P.Yang, and H.Shichijo, *Modeling of Small MOS Devices and Device Limits,* Porc.IEEE, Vol. 130 (Part 1), No.3, p. 105 (1983).

[4] C.P.Yuan, T.N.Trick, *Calculation of Capacitance in VLSI Circuits,* Proc.ICCD, p.263, (1982).

[5] K.C.Saraswat and F.Mohammed, *Effect of scaling interconnects on time delay of VLSI circuits,* IEEE J.Solid-State circuits, vol. SC-17, p 275, April (1982).

[6] M.Uenohara et al., *VLSI microstructure science,* vol. 9, chapter 11, Academic Press, (1985).

[7] G.A.Frazier, *Simulation of Neural Networks,* International Conference on Neural Networks, Santa Barbara Calif. (1984).

[8] C.A.Hamilton et al., *A High Speed Superconducting A/D Convertor,* 37th Annual Device Research Conference, June 25-27, (1979).

[9] R.T.Bate et al., *Prospects for Quantum Integrated Circuits*, Baypoint Conference on Quantum Well and Superlattice Physics, Proc. of the SPIE, March (1987).

[10] IEEE Journal of Quantum Electronics, Vol QE-22, No.9, September (1986).

[11] M.A.Reed et al., *Spatial Quantization in GaAs-AlGaAs Multiple Quantum Dots*, J. Vac. Sci. Technol. B, Vol 4, No.1, p. 358, (1986).

[12] M.A. Reed, Superlattices and Microstructures, Vol. 2, p. 65, (1986).

[13] William R. Frensley, *Transient Response of a Tunneling Device Obtained From the Wigner Function*, Phys. Rev. Lett., Vol 57, No.22, p2853, (1986).

[14] Stephen Wolfram ed., *Theory and applications of cellular automata*, Singapore Pub. (1986).

[15] E.R.Berlekamp et al., *Winning Ways For Your Mathematics Plays*, vol.2, Academic Press (1982).

[16] IEEE Transactions on Computers, Vol C-35, No.2, February (1986).

[17] Z.G.Vranesic and V.C.Hamacher, *Ternary Logic in Parallel Multipliers*, Journal of Comp., Vol.15, No.3, p.254, November (1972).

[18] G.A.Frazier, To be published.

[19] Gerard Y. Vichniac, *Simulating Physics With Cellular Automata*, Physica D, vol.10D, No.1 & 2, p.96 (1984).

[20] J.V.McCanny et al., *Completely Iterative, Pipelined Multiplier Array Suitable for VLSI*, Proc. IEEE, vol.129, No.2, p.40, (1982).

[21] C-W Wu, P.R.Cappello and M.Saboff, *An FIR filter tissue*, Proc. 19th Asilomar conference on circuit systems, and computers, Pacific Grove, Ca, November, 1985.

Chapter 2

Optical Digital Computers: Devices and Architecture

Alan Huang [1]

2.1 Introduction

Advances in computation are being limited by communication considerations. The fastest transistors switch in 5 picoseconds whereas the fastest computer runs with a 5 nanosecond clock. This three orders of magnitude disparity in speed can be traced to communication constraints such as connectivity and bandwidth.

2.2 Optical Interconnects

Optical digital computing is a natural extension of optical interconnections. Optics now connect city to city, computer to computer, computer to peripheral, and board to board. Optics will eventually connect chip to chip and gate to gate. At this point, the computer will be as much optical as electronic [1]. This infiltration is being driven by the need for increased interconnection bandwidth, ground loop isolation, and freedom from electrical interference. It is being limited economically by the costs of comparable coaxial interconnects. In terms of systems, this advance is being limited by protocols and hardware handshakes originally intended to protect against the communications inadequacies associated with electronic interconnects. In terms of technology, this advance is being limited by difficulties associated with integrating and packaging lasers and light emitting diodes with electronics.

---

[1] AT&T Bell Laboratories, Holmdel, NJ

One problem with this evolution from using optics to connect city to city and eventually gate to gate is that it fails to take advantage of the connectivity of optics. A lens can easily convey a 100 by 100 array of spots. This can be considered a 10,000 pin connector. The problem is that since electronics can not support this connectivity there are no architectures around that utilize this parallelism [2].

2.3   Using the Parallelism of Optics

As mentioned previously, optics can easily provide a 100 by 100 very high bandwidth interconnect. The problem is that this only true for regular interconnects. Arbitrary interconnects quickly run into space-bandwidth limitations. The question arises, "are regular interconnects sufficient?"

There are two approaches to this problem. One is to take an arbitrary interconnection and try to implement it via a regular interconnect such as a perfect shuffle [3] as shown in Figure 2.1. Thirty two, 32 point optical perfect shuffles have been demonstrated experimentally. See Figure 2.2. There should be no difficulty extending this to one hundred twenty eight, 128 point perfect shuffles. Techniques have also been developed to combine these perfect shuffles into even larger perfect shuffles.

Another approach to dealing with arbitrary interconnects is to avoid using them at the very beginning. This can be accomplished by building an arbitrary circuit out of uniform modules interconnected in a regular manner. The technique involves trying to find the smallest "atom" of computation. Computation is traditionally divided into logic and communication. The difficulty with this partitioning is that if you try to regularize the logic you push your complexity into the communications and vice

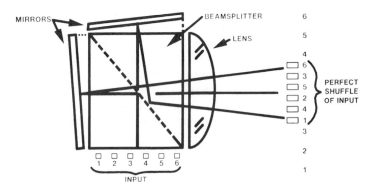

Figure 2.1: Diagram of an optical perfect shuffle.

test pattern (computer generated, inversely shuffled):

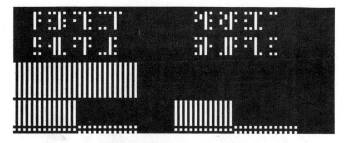

optical perfect shuffle:

Figure 2.2: Experimental demonstration of an optical perfect shuffle unit.

versa. The linking of the logic and communications together in a computational atom allows the construction of computational structures which are regular both in terms of logic and communications. The regularity of these structures have no bearing on the computational complexity which can be achieved. The use of geodesic and tessellated structures proposed by Buckminster Fuller bear witness to this[4]. We call this technique based on computational atoms, *Symbolic Substitution*. It facilitates the design of circuits with constant fan-in, constant fan-out and regular interconnections. The constant fan-in and constant fan-out simplify the demands on the optical logic gates while the regular interconnections match the space invariant connectivity of optics. The basic mechanism involves recognizing all the occurrences of a particular pattern within an image and replacing these occurrences with another pattern. See Figures 2.3 and 2.4. The technique can be used to emulate Boolean logic, implement binary arithmetic, perform cellular logic operations, and build Turing machines.

There is another problem associated with using parallelism that has nothing to do with optics. The problem involves how to partition an arbitrary computational into a several processors. If one is not careful the partitioned computation can quickly become dominated by communication or even deadlocked. We have developed a technique which we call *Computational Origami* which takes advantage of the regularity imposed by symbolic substitution to partition and schedule a computation

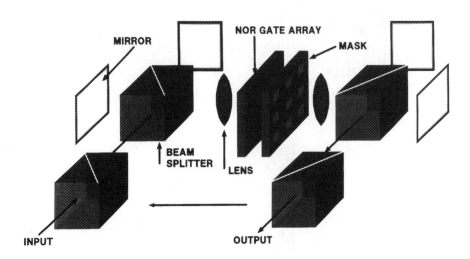

Figure 2.3: A diagram of an optical symbolic substitution unit.

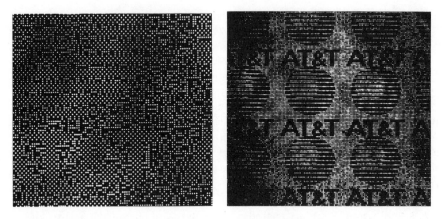

Figure 2.4: Experimental demonstration of the recognizer portion of an optical symbolic substitution unit.

to achieve an arbitrary hardware/time trade-off. The technique works by forcing an algorithm into special regular array in which all the data dependencies can be broken. This allows the algorithm to be folded upon itself to an arbitrary degree. The technique essentially allows a hardware window to be scanned over a problem. This ability to dial the amount of parallelism allows a computation to be matched to the parallelism of optics and should also resolve the age old debate between serial and parallel approaches.

2.4 Optical Logic Gates

The search for an optical logic has been compared to the quest for the Holly Grail. In retrospect, it is not a question of existence but rather engineering. Such a gate can be implemented by integrating some photodiodes, an electronic logic gate, and some light emitting diodes in GaAs and would have optical inputs, optical outputs, be a three port device, cascadable, with fanout, etc. This might seem to be cheating but all that has really been done is to selectively dope a crystal with various impurities. The important questions are really ones of practicality and advantage.

We are exploring these questions with several different approaches. One approach, the Self-Electrooptic Effect Device (SEED) implemented via molecular beam epitaxy relies on combining a multiple quantum well based optical modulator, a photodiode, and various electronics to implement a latching optical NOR gate [6]. See Figure 2.5. A 6 by 6 array of these devices have been fabricated and utilized

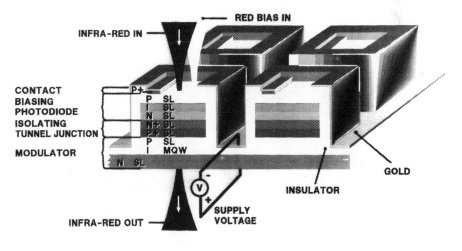

Figure 2.5: A diagram of a SEED device.

experimentally. Subsequent devices are expected to have powers and and speeds comparable to the fastest transistors.

A second approach, the Optical Logic Etalon (OLE') device implemented via molecular beam epitaxy relyes on a light-sensitive semiconductor optical resonator [6]. See Figure 2.6. Repetition times as fast as 50 - 60 ps. have been demonstrated with powers comparable to the SEED device.

A third device, based on the Quantum-Well Envelope State Transition (*QWEST*) effect uses molecular beam epitaxy to quantum engineer a new light sensitive material by enabling a transition between the two lowest energy levels in the conduction band of a quantum well [7]. See Figure 2.7. The material has been fabricated and the effect demonstrated.

It is important to note that the three previous devices all rely on molecular beam epitaxy. The SEED device takes more of an electronics approach, while the OLE' takes more of an optics approach, whereas the QWEST takes more of a materials approach.

A fourth type of optical gate is based on optical nonlinearities in organic molecules. Various organic molecules are being investigated for their χ^3 effects [8]. The hope is to find stronger and faster nonlinearities than are possible in semiconductors.

2.5 Limits

The question arises as to the ultimate speed of optical logic. One of the fastest optical effects which could be used for optical logic is frequency mixing. When two

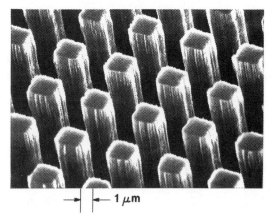

Figure 2.6: A picture of an array of OLE' devices.

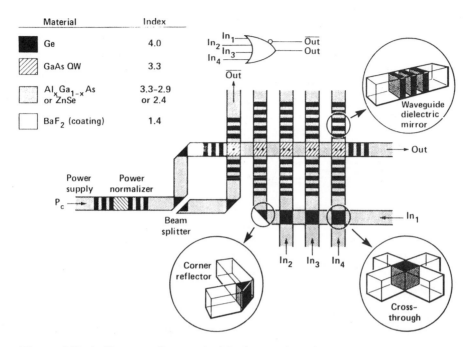

Figure 2.7: A diagram of an optical logic gate based on the QWEST effect.

frequencies are mixed the sum and difference frequencies will not appear unless both input frequencies are present. This effect can be viewed as an AND gate and this interaction occurs within femtoseconds.

Philosophically, the speed of a phenomena is inversely proportional to the distance that the electron has to travel. In electronics the electrons have to effectively traverse wires. This speed limitation can be seen in the SEED devices. The repetition rate of the OLE' device is limited by the diffusion time of the carriers in the etaleon. The speed of the QWEST effect is limited by the transition between two energy levels in the conduction band of the quantum well. The speed of the organic nonlinearities is limited by the traverse of the electrons across the bonds of a molecule. The speed of frequency mixing is limited by the motion of the electron within their atomic energy levels. The shorter the distance that the electron has to move the faster the phenomena.

There is also the question of how much power will an optical logic gate require. The semiconductor based optical logic gates will probably require about the same amount of power as a transistor. The power of the gates based on χ^3 effects are still being investigated. Their power will depend on the strength of the effect and the instantaneous E-field which can be achieved.

Philosophically, computing is the representation of information by energy, moving this energy around, and interacting it. How much energy it takes to interact is a function of how sensitive the material is and how well the energy is concentrated. Electronics concentrates energy by using very small junctions to generate very large E-fields. Optics can concentrate energy in a comparable manner via focusing and waveguiding. Optics can also concentrate energy and generate very large instantaneous E-fields via pulse compression techniques. Electronics is at a disadvantage since it does not have the communications bandwidth necessary to propagate such signals. This give optics another degree of freedom in compressing energy. It is particularly relevant in the case of χ^3 nonlinearities which are quite weak and yet a function of the cube of the instantaneous E-field.

There is a tendency to try and minimize the power of optical logic devices since this is the approach used by electronics. It should be noted however that there are some applications in which you would instead like to maximize the speed. In these situations the only relevant parameters are how much power can be generated, how much power can be dissipated, and how much will it cost per computational operation.

2.6   At What Point Does Optics Become Competitive?

The answer to this questions depends a large part on the application involved. The more communications intensive a problem is the better it will be for optics. The most likely candidates will be switching computers, super computers, and specialized signal processors.

Some of the quoted strengths of electronics are a bit misleading. The powers and speeds of optical and electronic gates are constantly being compared. This is not a fair comparison since communication costs are not really factored in. It is relative easy of an electronic gate to talk to another gate on the same chip however the problem gets considerable more difficult when the gate wants to talk with a gate on another chip or on another board. The difference between an electronic and an optical gate is that the optical logic gate already has this communications capability built in and should be be competitive in the previously mentioned applications in the relatively near future.

Summary

Computation is essentially the representation of information by energy, communicating it, and interacting it. Optics has a greater connectivity and a higher bandwidth than electronics and it should be able to interact the signals with speeds and energies comparable to electronics. Optics also has the potential of communicating and interacting signals with speeds far faster than electronics.

Acknowledgments

I would like to acknowledge the contributions and helpful discussions with K-H Brenner, L. Chirovsky, M. Downs, C. Gabriel, J. Henry, J. Jahn, J. Jewell, S. Knauer, Y Lee, A. W. Lohmann, H-M Lu, S. McCall, D. A. Miller, M. Murdocca, M. Prise, C. Roberts, T. Sizer, N. Streibl, S. Walker, L. West, and N. Whitaker.

References

[1] Trudy E. Bell, *Optical Computing: A Field in Flux*, IEEE Spectrum, vol. 23, no. 8, pp. 34-57, August 1986.

[2] A. Huang, *Architectural Considerations Involved in the Design of an Optical Digital Computer*, Proceedings of the IEEE, vol. 72, no. 7, pp. 780-786, July 1984.

[3] K-H Brenner and A. Huang, *Optical Implementations of the Perfect Shuffle*, to be published in Applied Optics.

[4] B. Fuller and F. J. Applewhite, Synergetics: Explorations in the Geometry of Thinking, Macmillan, 1975.

[5] K-H Brenner, A. Huang, and N. Streibl, *Digital Optical Computing With Symbolic Substitution*, Applied Optics, vol. 25, no. 18, pp. 3054-3060, September 1986

[6] A. L. Robinson, *Multiple Quantum Wells for Optical Logic*, Science, vol. 225, pp. 822-824, August, 1984.

[7] L. C. West, *Spectroscopy of GaAs Quantum Wells*, Department of Applied Physics (dissertation), Stanford University, July 1985.

[8] M. C. Gabriel, *Transparent Nonlinear Optical Devices*, Proceedings of the SPIE O-E/Fiber Conference, San Diego, August 1987.

Chapter 3

VLSI Implementations of Neural Network Models

H. P. Graf [1]
L. D. Jackel [1]

Abstract

Three experimental CMOS VLSI circuits implementing connectionist neural network models are discussed in this paper. These chips contain networks of highly interconnected simple processing elements that execute a task distributed over the whole network. A combination of analog and digital computation allows us to build compact circuits so that large networks can be packed on a single chip. Such networks are well-suited for pattern matching and classification tasks, operations that are hard to solve efficiently on a serial architecture.

3.1 Introduction

Models of neural networks are receiving widespread attention as potential new architectures for computing systems. The models we are considering here consist of highly interconnected networks of simple computing elements. A computation is performed collectively by the whole network with the activity distributed over all the computing elements. This collective operation results in a high degree of parallel computation and gives the network the potential to solve complex problems very fast.

---

[1] AT&T Bell Laboratories, Holmdel, NJ

These models are inspired by biological neural networks but the electronic circuits we are implementing are not an attempt to imitate precisely real neurons. It is clear that the models used are gross simplifications of the biological networks. But even these relatively simple models have very complex dynamics and they show interesting collective computing properties. It is this collective computation we try to exploit by building such networks.

To date, the research concerning neural network models has focused mainly on theoretical studies and computer simulations. The real promise for applications of the models, however, lies in specialized hardware, in particular in specialized micro-electronic circuits. Simulations of large networks on standard computers are painfully slow, and only with customized hardware, can one hope to realize neural network models with speeds fast enough to be of interest for applications. Digital accelerators to simulate neural networks are now commercially available, but they are still orders of magnitude slower than what can be achieved by directly fabricating a network with hardware. The most promising approach for implementing electronic neural nets is certainly to fabricate special purpose VLSI chips. With today's integration density a large number of simple processors can be packed on a single chip together with the necessary interconnections to make a collective computing network. Experiments with VLSI implementations have been initiated by several groups and a few functioning circuits have been demonstrated [1-5].

3.2 Model Neurons

The circuits described here are known as *connectionist models*. An individual processor or "neuron" does very little computation, often just a thresholding of its input signal. The kind of computation performed by the whole network depends on the interconnections between the processors. Figure 3.1 shows a possible electronic circuit for one such simple "neuron". An amplifier models the cell body, wires replace the input structure (dendrite) and the output structure (axon) of a neuron and the synaptic connections between neurons are modeled by resistors. Each of

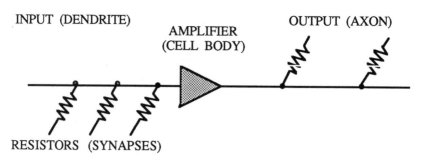

Figure 3.1: An electronic "neuron".

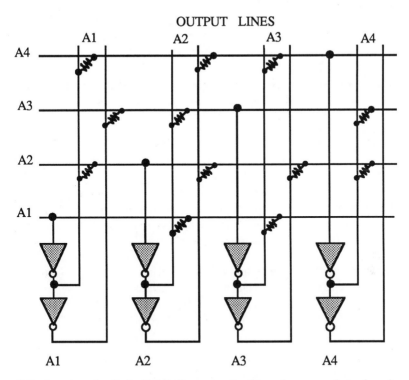

Figure 3.2: A network of electronic "neurons". At each crosspoint of an input and an output line a resistive connection may be set. The configuration of resistors shown here is just one example. Two inverters are connected in series and work as one amplifier unit. Inverted and noninverted outputs are needed to make excitatory and inhibitory connections.

the resistors connects the input line of the amplifier to the output line of an other amplifier. In the case where the amplifier measures the current flowing into the input line the output voltage of the amplifier is given as:

$$V out_j = f(\sum_i I_i) = f(\sum_i (V out_i - V in_j)T_{ij})$$ (3.1)

I_i: currents through the resistors.

$V in, V out$: input, output voltage of an amplifier.

T_{ij}: conductance of the resistor connecting amplifier i with amplifier j.

$f()$: transfer function of the amplifier.

This circuit can be used to compute a sum of products in an analog way. The state of the amplifier is determined by the output voltages of all the other amplifiers multiplied by the conductances of the connecting resistors. If the conductances of the resistors represent the components of a vector and the output voltages of the amplifiers represent the components of an other vector the current is proportional to the inner product of the two vectors (for simplicity we set the input voltage $Vin = 0$; eg. in a virtual ground configuration). For the multiplication Ohm's law is used and the summing of the currents is done 'for free' on the input wire. Exploiting these physical properties of the components it is possible to build very compact circuits computing sums of products.

Figure 3.2 shows a network of such elements; an array of amplifiers is interconnected through a matrix of resistive elements. In the case of a fully interconnected network each amplifier is connected to every other amplifier resulting in N^2 interconnection elements, where N is the number of the amplifiers. In many cases the network does not have to be fully interconnected but a high degree of interconnectivity is typical for these models. The requirements for the components (amplifiers and interconnections) depend on the task the circuit is intended to do; e.g. for learning tasks fine changes of the resistor values are needed as well as a control of the transfer characteristic of the amplifiers.

3.3   Computing with Neural Network Models

Connectionist neural network models have been used to demonstrate many different functions such as associative memory, learning from example and optimization [6,7]. Theoretical work has focussed strongly on learning aspects and there have been interesting developments in the area of training and self-organization of connectionist networks in the last few years [7-10]. However, there is still strong debate about the efficacy of neural learning rules [7,11,12] and it remains to be seen for which tasks the neural algorithms offer advantages over conventional statistical methods.

Regardless of the outcome of the current debate on learning rules it is clear that neural network hardware can perform tasks of pattern matching in a fast and efficient manner. The implementations we describe here have been developed for this type of application. Equation 3.1 is well suited as the basis for operations such as feature extraction and classification. For a feature extraction operation the network is configured as shown in Figure 3.3. A set of vectors (referred to as features) is stored in the network, each feature along the input line of an amplifier (referred to as 'label' unit). The conductances of the resistors represent the components of the features. A vector is given at the input of the network and is compared with all the stored features in parallel. The inner product between the input vector and each feature is evaluated and if the result exceeds a certain threshold the 'label'

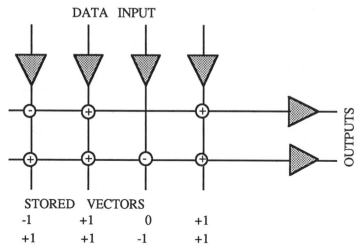

DATA INPUT

OUTPUTS

STORED   VECTORS
-1   +1   0   +1
+1   +1   -1   +1

Figure 3.3: A network set up for pattern classification or feature extraction. The '+' and the '-' mark excitatory and inhibitory connections respectively (an excitatory connection inputs current into the line; an inhibitory connection draws current from the line). Below the schematic the stored vectors are indicated.

unit connected to that feature goes high indicating the presence of the feature in the input.

For a classification task one wants to know which feature is the closest match to the input. Therefore, we have to find the 'label' unit with the maximum value. In a network this can be accomplished by adding mutually inhibitory connections between the 'label' units as shown in Figure 3.4. These inhibitory interconnections form a "winner take all" circuit. If the inhibitory connections among the 'label' units are much stronger than than the excitatory interconnections, the network will evolve to a stable state in which the 'label' unit connected to the feature with the largest result will be on, and all other 'label' units will be off. The whole operation corresponds to a nearest neighbor classification.

In the hardware implementations described below the components of the vectors are binary (1,0) or ternary (+1, 0, -1). This simplifies the design considerably and it turns out to be adequate for many applications. However, the same architecture can also be used with analog input signals and analog values for the resistors.

3.4   Electronic Implementations

For an electronic implementation the large number of interconnections and wires needed in these highly interconnected networks represent the main challenge. The

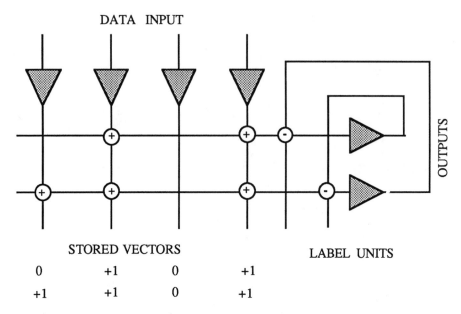

Figure 3.4: Interconnections set for a nearest neighbor classifier (associative memory).

whole network could be built with digital electronics, but this would require a rather complex and bulky adder circuit at the input of each processor. Using analog computation this can be built in a much smaller area and a larger network can be packed on a single chip. The general trend in signal processing today is going away from analog computation to digital computation wherever possible. But the architecture of the neural network models and the modest precision required for the signals make an analog solution attractive. Many of the circuits built or proposed so far use either full analog computation or a mixture of analog and digital electronics [1-5].

3.4.1 Mask programmable network chip

For the implementation of large analog networks resistors with relatively high values are required to keep the power dissipation at an acceptable level. There are no resistors available in standard VLSI technology, in particular in CMOS technology, that are high enough in value with a size small enough to integrate networks with hundreds or thousands of amplifiers (requiring millions of interconnections) on a single chip. We developed a fabrication process to make high density interconnection matrices using amorphous silicon resistors. By stacking amorphous silicon vertically between crossing metal wires it was possible to build resistors with submicron dimensions and pack as many as 4 resistors per square micron [13]. A similar fabrication process was used to add amorphous silicon resistors to a CMOS VLSI chip containing 256 amplifiers [14]. This technology requires programming the circuit during the fabrication process; once the fabrication is done all the connections remain fixed and the content of the circuit can not be changed. But these mask-programmable interconnections are much smaller than programmable interconnections and in applications where the connections can be determined ahead of time many more neurons can be packed on a chip.

3.4.2   Network with programmable interconnections

Programmable interconnections are required where the matrix configuration can not be determined beforehand. We built a network chip containing 54 amplifiers fully interconnected through a network of programmable resistive interconnections [2]. The architecture of this chip corresponds basically to the one shown in Figure 3.2. The circuit can be set up to perform several different functions simply by programming the interconnections. Among others, the configurations described in section 3.3 were tested. When set up to extract features, the circuit can store up to 50 features, each 50 bits long. The whole operation of determining good matches between an input vector and all the features requires less than 1 μs. When the chip is configured as a nearest neighbor classifier it stores up to 20 features, each 30 bits long. An input vector is compared to these features and the best match is picked. This whole operation is performed without any external synchronization among the amplifiers or any external control (except for the start of the operation). Once started the circuit evolves by itself to a stable configuration within 50 ns and 600 ns. This convergence time depends strongly on the data.

This chip has been interfaced to a minicomputer and is used as a coprocessor in character recognition experiments. In this application it performs the computationally intensive tasks of line thinning and feature extraction [15].

3.4.3   Network for feature extraction

A new version of a programmable network chip, specialized for the feature extraction operation has been built. We concentrated on this operation since it is very versatile and can be used for many applications in pattern recognition and image processing. Also, an implementation of this function with a connectionist network has clear advantages in terms of size compared with a digital circuit. Figure 3.5 shows a schematic of this circuit.   It resembles the circuit of Figure 3.3 but now the architecture of the chip as well as each component is optimized for one particular function. This results in a computing speed roughly 10 times faster than in the first programmable network chip and also in higher storage density.

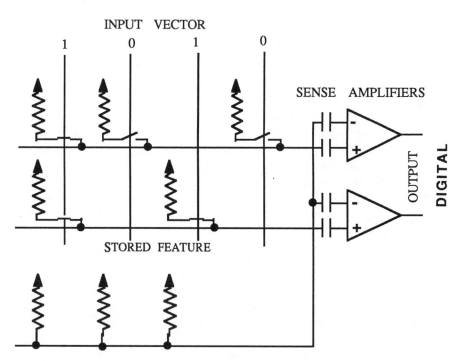

PROGRAMMABLE REFERENCE CURRENT

Figure 3.5: Schematic of the circuit implemented with the network specialized for feature extraction. The resistive interconnections are actually the circuits shown in Figure 3.8 and are present at each crosspoint of a data line and a input line of a sense amplifier. Drawn here are only a few of the resistive interconnections to indicate where a weight bit has been set to '1'.

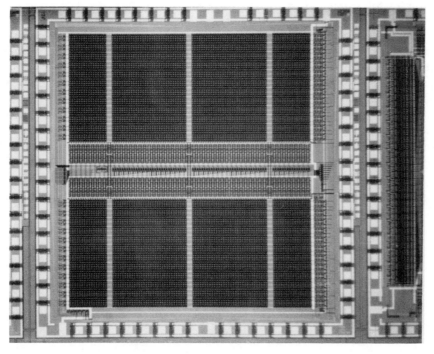

Figure 3.6: Photomicrograph of the "feature extraction" chip.

A feature is stored along a wire with each resistor representing one component of the feature. During a computation a current proportional to the inner product between the input vector and the feature is injected into the wire. This current is compared to a programmable reference current in a sense amplifier. If the reference current is larger than the current representing the result the amplifier's output stays low, otherwise it goes high indicating a close match between the input vector and a feature.

Figure 3.6 shows a photomicrograph of this chip. It has been fabricated in CMOS technology with 2.5 μm design rules and contains around 70,000 transistors in an area of 6.7 x 6.7 mm. Its block diagram is shown in Figure 3.7. The whole circuit is split in half. In each half-matrix 23 features can be stored, each 96 bits long. A shift register loads the input vector into the network. When an image is analyzed for features, a window (size up to 96 pixels) is scanned over the picture and at each pixel position the window is checked for the presence of a number of features. By using a shift register, a new input vector, representing the pixels in the window, is ready for a run at each clock cycle. Only the new pixels in the window have to be loaded onto the chip from run to run, the rest of the pixels are just shifted by a number of positions. From the shift register the input vector is loaded into static registers (data registers) from where the data bits control the connections in the resistor matrix. With mask registers, parts of the input vector can be blocked from

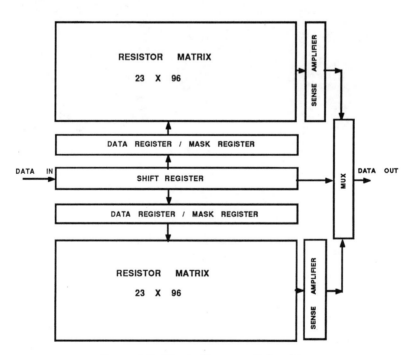

Figure 3.7: Block diagram of the chip.

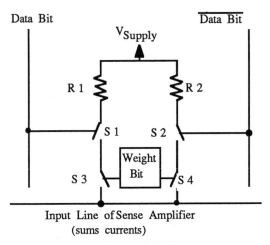

Figure 3.8: Schematic of one of the interconnection elements.

the computation ("don't care" pixels). The mask registers can also be used to set the threshold for a match individually for each feature.

Figure 3.8 shows a schematic of one of the programmable interconnections. Four transistor switches form an NXOR gate controlled by a bit from the input vector and a weight bit. When two switches connected in series are enabled current flows through a resistor into the input line of a sense amplifier where all the contributions from the different connections are summed up.

The circuit is designed to complete a full operation in less than 100 ns. This time includes resetting the network and the execution of the analog matching operation. The analog computation is actually completed in 20ns - 30ns. With a clock frequency of 10 MHz the chip can evaluate close to 500 Million inner products of two 96 bit vectors per second. This chip is now being tested and all the modules have been determined to work properly.

3.5 Discusssion

The main issue is certainly whether such neural network models provide any advantage compared with more conventional circuits. The networks we implement are special purpose chips and their performance has to be compared with application-specific digital circuits. The algorithms implemented are not principally different from those used in other pattern recognition systems, the main function performed by the networks is computing inner products of vectors. High speed and compact size is accomplished by exploiting the physics of the devices for analog computation. Digital circuits built for similar functions are certainly considerably larger in size (see for comparison e.g. [16] for a digital pixel matching circuit using a parallel counter). We estimate that compared with parallel digital counters an improvement of a factor of 3 to 10 in density can be achieved. This results in considerably higher performance of the analog circuit.

On the other hand there is always the problem of limited precision with analog computation and for a general-purpose processor this represents a serious handicap. For pattern recognition applications, however, this is less of a problem. Very often large amounts of noisy data with low information-content have to be searched through for the relevant information. High speed is crucial for such an operation and limited precision can be tolerated.

The results obtained in the few applications tried so far with our networks are very encouraging but more research is needed to determine how well the properties of these networks are suited for a wide range of recognition tasks.

References

[1] M.A. Sivilotti, M.R. Emerling, and C.A. Mead, *VLSI Architectures for Implementation of Neural Networks,* Proc. Conf. Neural Networks for Computing, Snowbird, Utah, 1986, J.S. Denker ed., p.408, AIP Proc. 151.

[2] H.P. Graf and P. deVegvar, *A CMOS Implementation of a Neural Network Model,* in Advanced Research in VLSI, Proc. Stanford Conf. 1987, P.Losleben ed., MIT Press, p. 351.

[3] J.P. Sage, K. Thompson, and R.S. Withers, *An Artificial Neural Network Integrated Circuit Based on MNOS/CCD Principles,* Proc. Conf. Neural Networks for Computing, Snowbird, Utah, 1986, J.S. Denker (ed.), p. 381, AIP Proc. 151.

[4] J. Raffel, J. Mann, R. Berger, A. Soares, S. Gilbert, *A Generic Architecture for Wafer-Scale Neuromorphic Systems,* Proc. IEEE Int. Conf. Neural Networks, San Diego, 1987, p. III-501, M. Caudill and C. Butler (eds.), IEEE 87TH0191-7.

[5] A. P. Thakoor, A. Moopen, J. Lambe and S. K. Khanna, *Electronic Hardware Implementations of Neural Networks,* Applied Optics, **26,** p. 5085 (1987).

[6] D. W. Tank and J. J. Hopfield, *Simple "Neural" Optimization Networks: An A/D Converter, Signal decision Circuit and a Linear Programming Circuit,* IEEE Trans. Circuits and Systems, **CAS-33,** p. 533 (1986).

[7] For an up-to-date picture of the status of neural network modeling see e.g.: Proc. IEEE Conf. *Neural Information Processing Systems - Natural and Synthetic* in Denver, Colorado,1987, D.Anderson (ed.), American Institute of Physics (publisher), (1988).

[8] D.E. Rumelhart, J.L. McClelland, and the PDP Research Group, *Parallel Distributed Processing*, MIT Press, (1986).

[9] T. Kohonen, *Self-Organization and Associative Memory*, Springer Verlag, New York, (1984).

[10] G. Carpenter and S. Grossberg, *ART 2: Self-Organization of Stable Category Recognition Code for Analog Input Patterns*, Applied Optics, **26**, p. 4919, (1987).

[11] R.P Lippmann, *An Introduction to Computing with Neural Nets*, IEEE ASSP Magazine, **4/2**, p. 4, (1987).

[12] J.S. Denker, D. Schwartz, B. Wittner, S. Solla, R. Howard, L.Jackel and J. Hopfield, *Large Automatic Learning, Rule Extraction, and Generalization*, Complex Systems, **1**, p. 877, (1987).

[13] L.D. Jackel, R.E. Howard, H.P. Graf, B. Straughn, and J.S. Denker, *Artificial Neural Networks for Computing*, J. Vac. Sci. Tech., **B4**, p. 61 (1986).

[14] H.P. Graf, L.D. Jackel, R.E. Howard, B. Straughn, J.S. Denker, W. Hubbard, D.M. Tennant, and D. Schwartz, *VLSI Implementation of a Neural Network Memory with Several Hundred Neurons*, Proc. Conf. Neural Networks for Computing, 1986, J.S. Denker (ed.), p.182, AIP Proc 151.

[15] H.P. Graf, L.D. Jackel, and W. Hubbard, *VLSI Implementation of a Neural Network Model*, IEEE COMPUTER, March 1988.

[16] M.J.M. Pelgrom, H.E.J. Wulms, P. Van Der Stokker, and R.A. Bergamaschi, *FEBRIS: A Chip for Pattern Recognition*, IEEE J. Solid-State Circuits, **SC-22**, p.423, (1987).

Chapter 4

Piggyback WSI GaAs Systolic Engine

H. Merchant [1]
H. Greub [1]
R. Philhower [1]
N. Majid [1]
J. F. McDonald [1]

Abstract

GaAs digital circuits have been much heralded as a means for achieving high computational throughput rates. For example, recently Hughes has published accounts of a flip-flop exhibiting toggle rates approaching 20 GHz using 0.2 micron BFL-CEL MESFET logic operating at room temperature. This impressive performance must, however, be taken in the context of the severe yield problems which inhibit the use of this technology in a cost effective manner in large systems. This leads to the fabrication of big systems using a large number of relatively small dies. Use of conventional packaging for such small dies then introduces parasitics which largely negate the performance improvements promised by the underlying GaAs technology. In this paper we examine one approach to packaging a large number of small GaAs circuits to implement a system of 1000 heavily pipelined systolic processors each operating at a rate of 1 billion floating point operations a second resulting in a sustained throughput of 1000 GFLOPS. This stunning performance could be accomplished using a package with a volume of a few cubic feet, and dissipating only about 10 KW of power. The impact on aerospace tactical and strategic signal processing applications of such a technology could be substantial.

---

[1] Rensselaer Polytechnic Institute, Troy, NY

4.1 Introduction

Commercial ventures such as Tektronix/TRIQUINT, Vitesse, and Gigabit Logic have provided foundry service for GaAs digital circuits capable of roughly 4 GHz counter operation using 1 micron design rules. Conventional packages such as the ceramic DIP or PGA are inadequate for building large scale systems such as supercomputers. These packages exhibit large interlead coupling [1], large self parasitics and large lead dimensions resulting in lengthened interdie signal path lengths and delays. Therefore, some advanced packaging concepts will be required to fully exploit the potential performance advantages which might be offered by GaAs.

Wafer Scale Integration (WSI) [2] and its companion technology of Wafer Scale Hybrid Packaging (WSHP) [3] are experimental methods for packaging large systems on a single wafer substrate [4]. In WSI the active logic circuits are fabricated on the wafer substrate, tested under burn-in conditions, and the circuits found to operate correctly are connected either by programmable switches or discretionary wiring involving subsequent fabrication steps. In the WSHP approach, the wafer substrate is fabricated as a hybrid or "silicon wiring board" containing a multilayer structure consisting of interconnections, insulators, and possibly other items such as bypass capacitors, heat pillars, or terminators [5] . The chips containing the active devices such as logic gates are fabricated on separate wafers and after testing and dicing are simply mounted on top of this structure. Since the active devices are not actually in the hybrid substrate, it can be made of other materials such as ceramics or metals. However, the greatest compatibility with semiconductor processing is maintained by making the substrate from silicon. It is this form of wafer integration which seems to be receiving the greatest attention by industry today. Nevertheless, if the WSHP technology can be made successful, a cautious reexamination of the fully monolithic WSI approach is probably likely [6,7], although a GaAs epitaxial growth on a silicon substrate might be necessary.

In this paper we consider the construction of an extremely large system consisting of 1000 32 bit floating point GaAs systolic processors. We desire to use semiconductor fabrication techniques as much as possible in order to keep the spatial density of the processors as high as feasible, and to keep the cost low. The low cost consideration indicates that a small die size be employed for the GaAs building blocks. The net impact of this will be that a great deal of inter die interconnect will be required just to package one systolic processor. In addition, a fairly substantial amount of interconnect will be required to provide busing and control signals between the systolic processors and to provide data communication to the external environment. To cope with the extremely large number of interconnections we propose a style of WSHP which we call "piggyback" wafer scale integration. In this scheme small wafer scale hybrids containing the large number of interconnections within a systolic cell are mounted on larger wafer scale hybrids which support the large number of intercell communication lines. This is illustrated in Figure 4.1. In this way one avoids the yield difficulties associated with having to fabricate all of the interconnections

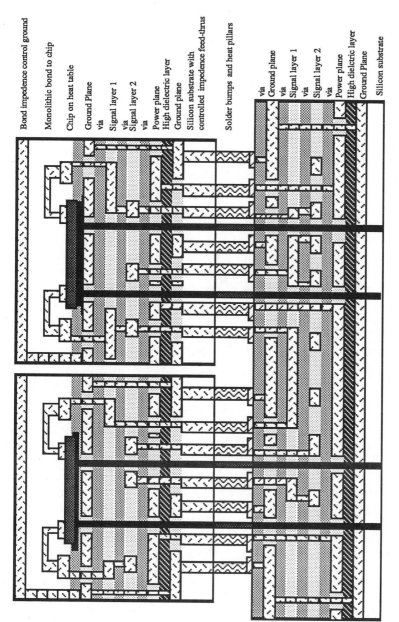

Figure 4.1: Schematic representation of a cross section of "Piggyback" Wafer Scale Hybrid Packaging.

on one substrate. Yet the surface die density can be extremely high. In effect one considers a multilayer interconnection structure in which a hierarchy exists in the wiring thus permitting it to be partitioned into sections which can be fabricated and tested separately.

4.2   The Performance Goals

In 1985 the Strategic Defense Initiative (SDI) announced in a white paper solicitation sent to universities that the throughput goal of their program required 1000 GFLOPS (one thousand billion floating point operations per second) in a form suitable for solving signal processing, linear equation, or matrix manipulation problems. In formulating a response to this solicitation it was noted that systolic architectures have been shown to be appropriate for many of these applications [8-11]. However, in order to meet or exceed the prodigious throughput requirements of the white paper solicitation with a modest number of processors and with volume and heat dissipation demands compatible with aerospace environments, a GaAs implementation was indicated. Use of GaAs circuitry could also satisfy readily the radiation hardness requirements with existing technology, although Silicon On Insulator (SOI) CMOS is a possible future alternative at the appropriate lithographic dimensions.

The throughput requirements of SDI and related programs would be met by 1000 GaAs systolic processors operating at a rate of one billion floating point operations per second. Any substantially larger number of systolic processors would create a difficult maintenance situation, despite the use of on-line spares. Unfortunately, even with the best commercially available GaAs technology, a throughput of 1 GFLOPS per systolic cell could not be achieved today without heavily pipelining the cell. This increases the latency time of the cell (the time between the point at which data appears at the input of the cell and the point at which the corresponding answer cycles to the output.) The practical implication of this limitation is that signal processing schemes demanding real time feedback will suffer. One example of this is the so called IIR or recursive filter. Some of the performance limitations resulting from this long latency time can be overcome by block processing [12], in which several input samples are handled concurrently. However, until the very best GaAs circuit technologies are available in the commercial market the latency problem will persist.

In the initial study performed in our group, we obtained access to the TRIQUINT performance data for their E/D MESFET logic circuit fabrication line. The Enhancement/Depletion mode power dissipation can be as low as 0.25 milliwatts per gate. This compares very favorably with D mode only circuits which are often in the range of 1-10 milliwatts per gate. The E/D MESFET logic is also more similar to NMOS circuits than D mode only circuits, thereby making it easier for designers to master. The published characteristics of the TRIQUINT E/D MESFET logic

[13] indicate that a lightly loaded gate delay of 200ps can be expected, with 66ps being feasible for an essentially unloaded gate dissipating 1 milliwatt. These speeds are only feasible on one input (called the FS or fast speed) input of each gate. Clearly critical paths should be routed on the FS inputs of lightly loaded gates wherever possible to achieve the best performance. In typical loading situations this implies about three levels of gates totaling 600ps, followed by a 400ps fast flip flop to achieve a 1000ps or 1 GFLOPS throughput pipeline. Additional constraints are a maximum fanin of 3 (due to the turn-off characteristics of the E mode device) and a maximum fanout of 5. Fortunately, because the interior systolic cell structure can be viewed as a series of cells with relatively short intercell connections, in most cases the pipeline model works quite well. The only serious problem is the tuning of the clock distribution network so that all of the flip flops in the pipeline registers receive their stimuli simultaneously, and that the spacing between clock pulses be adequate for proper functioning of each layer of the processor. It has been shown that it is possible to distribute a 1 GHz clock properly deskewed within the postage-stamp (4 square cm.) surface area of one processor cell [14,15]. However, good synchronous clock distribution in a system of 1000 cells at this rate is not feasible.

In fact, interprocessor communication is a significant consideration for this problem, not only from the point of view of synchronization, but also from the point of view of yield. Each systolic cell would compute a function at least as complex as Z = X + A*B. This would require at least 96 lines for the input data items X, A, and B and 32 lines for the output data item C, in addition to timing or control signals. For synchronous operation tight control on hundreds of thousands of data path length delays would be required. This degree of skew control would be infeasible at synchronous clocking rates of 1 GHz due to variations in dielectric constants of materials, length variations, linewidth and thickness or separation variations, and thermal variations. The clock distribution problem would be aggravated in the presence of certain fault tolerance schemes where wiring of differing lengths to remote spares might become necessary. The only alternatives consistent with maintaining a 1 GHz processing rate for the output of the systolic cell would be to have multiple, parallel, simultaneous, synchronous word transfers for each data item operating at a lower clock rate than 1 GHz where the sensitivity to time delays would be less severe, or to have asynchronous data transfers. The latter approach might be able to operate at significantly higher bit rates than 1 GHz thereby actually reducing the number of data lines. Table 4.1 partially summarizes the data communication rate tradeoffs between synchronous very wide multiword transfers and asynchronous transfers for transferring a single 32 bit data item for various systolic processor internal clock rates (an internal clock pulse being generated when a complete set of input data items is assembled at the cell for processing).

For a system of the projected size a synchronous transmission bandwidth of 125 MHz (8ns per transfer) is probably the highest one can tolerate since a universal global clock would have to be provided with some degree of accuracy at this rate.

Table 4.1: Communication bandwidth requirements for parallel synchronous and serial asynchronous transmission (All bandwidths in GHz).

Systolic Internal Clock Rate	Synchronous Number of 32 bit words transfered in parallel per 32 bit data item				Asynchronous Number of serial channels used to transfer one 32 bit data item			
	1	2	4	8	1	2	4	8
1.0 GHz	1.0	0.5	0.25	0.125	32	16	8	4
0.5 GHz	0.5	0.25	0.125	0.063	16	8	4	2
0.25 GHz	0.25	0.125	0.063	0.034	8	4	2	1
0.125 GHz	0.126	0.063	0.034	0.017	4	2	1	0.5

Hence, to maintain a 1 GHz systolic cell rate each data item would have to be fetched as 8 parallel words creating a severe inter-processor wiring yield problem. On the other hand by operating asynchronously a wiring yield improvement might actually be achieved. For example, a wiring bandwidth of 4 GHz between systolic cells might be feasible. Transmission above 8 GHz across the 8 inch substrate is extremely difficult due to skin effect attenuation. A 4 GHz serial asynchronous communication link is also feasible using GaAs. Therefore, the total number of data wires might be reduced by a factor of 4. However, operation of the interconnect at this high frquency demands controlled impedance leads for the die-to- substrate interconnections, and other feed through fixtures.

While we are considering these same numbers, we can see that to achieve comparable performance using SOI CMOS at VHSIC-II clock rates of 125 MHz would require eight times as many processors and correspondingly eight times the number of I/O lines. A wiring yield comparable to the worst case considered for the GaAs implementation in Table 4.1 would result at essentially equal power dissipation levels.

The picture which is beginning to emerge from these discussions is that the packaging of the architecture might consist of roughly 70 postage-stamp sized WSHP processors mounted "piggyback" on each of 16 eight inch WSHP substrates. This is illustrated in Figure 4.2. Each processor would have 24 4 GHz serial input links and 8 similar output links to carry the 32 data items between processors. This would permit approximately 8% of the cells to be allocated as spares. The 16 eight inch diameter substrates might be mounted in a vertical stack in a cylindrical enclosure (Figure 4.3) or "silo" similar to an early IBM WSI project. Wiring in between the eight inch diameter substrates might employ vertical feed through vias located

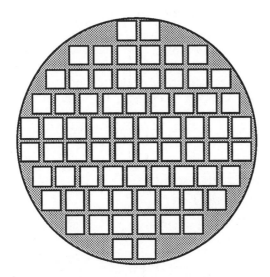

Figure 4.2: An eight inch diameter wafer with 70 2 cm × 2 cm systolic floating point processors mounted in "piggyback" fashion on its surface. The eight inch substrate supports the interprocessor connections.

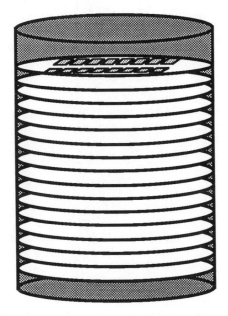

Figure 4.3: Conceptual representation of a cylindrical multiwafer package or "silo" housing 16 eight inch wafers

across the face of the wafer, or along the surface of the silo enclosure. Careful control of transmission line impedance over these lines and contacts is essential. In some cases optical communication links operating at 4 GHz might be attractive for these lines [16,17]. Cooling of the system would require either liquid fed heat sinks under each wafer (such as a heat pipe or thermosyphon) or immersion cooling. Management of the data flow, interconnections, fault tolerant cell assignment, and overall input-output operation of the system would require one or more conventional processors operating at high MIPS rates and performing in a supervisory mode.

4.3   GaAs Systolic Cell Design

The floating point processor for the systolic cell is one of conventional design except for the high degree of pipelining necessary to achieve the 1 GHz throughput rate. Each processor requires approximately 43,200 gates using the design limitations of the TRIQUINT E/D MESFET process as the model. A working assumption in the design has been that a standard cell approach would be employed using scalable transistors for each cell's outputs. Another assumption has been that attractive GaAs yields would be available for circuits of about 1200 gate equivalents. With 100 micron bond pads these circuits fill a 3 mm × 3 mm die. Hence, 36 of these mounted on the top "piggyback" level substrate would fill an area of about 4 square centimeters per processor depending on the type of die and wire bonding used. An architectural overview of the processor is shown in Figure 4.4. Figure 4.5 shows the partitioning of this diagram into the dies actually needed to implement the architecture. The subcellular structure of the 24 bit mantissa multiplier portion is shown in Figure 4.6 and illustrates the fine grain of pipelining. The total pipeline depth is 113 stages for the full processor.

The placement of the 36 dies within a 2.03 cm × 2.03 cm layout is shown in Figure 4.7. A total of 12 different dies are required which are identified as 12 reticle masks in Figure 4.8.

Finally, the inter die wiring on this "postage stamp" sized substrate is illustrated in Figure 4.9. The computer aided design programs used to compute this routing also provide statistics of the wiring within the systolic processor. Figure 4.10 shows the histogram of total wire lengths per interconnection. The longest lines are clock distribution lines and these tend to be in the 1 to 2 cm range. Since none of the other lines are bus lines, they are generally much shorter and the line drivers for these signals can be source terminated. The distribution of numbers of wire segments per line is shown in Figure 4.11. Most of the connections consist of 2-3 segments leading to good transmission line performance since only a few right angle turns arise. However, a small number of these connections unavoidably contain as many as 20 segments. However, as can be seen from Figure 4.12 only a very small number of line segments are much longer than 250 microns. Since the longest lines

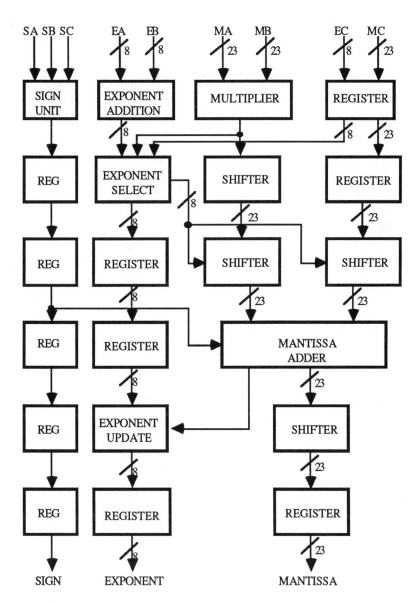

Figure 4.4: Architectural overview of Systolic Processor Cell.

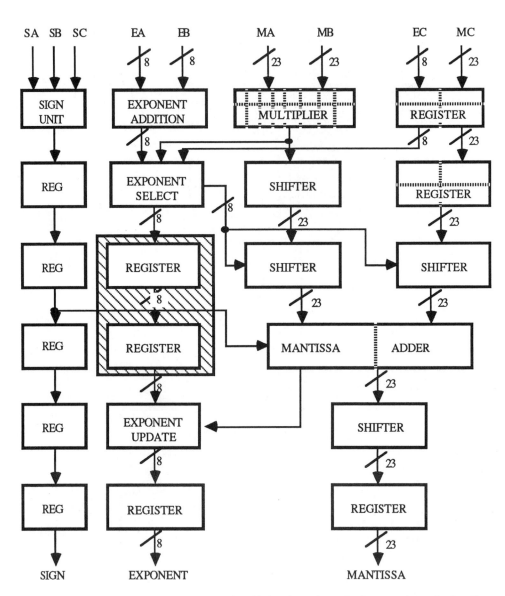

Figure 4.5: Partition of architecture identifying location of where various GaAs dies would be used.

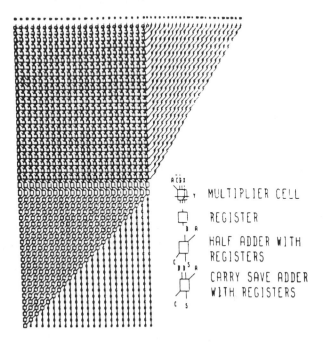

Figure 4.6: Fine grained pipeline structure for the 24 bit mantissa multiplier.

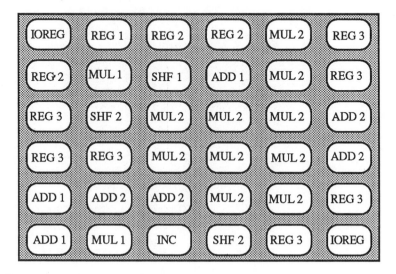

Figure 4.7: Placement of 36 small GaAs dies on the upper substrate.

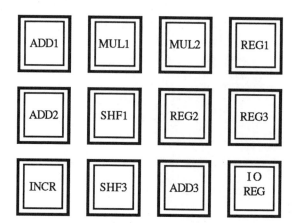

Figure 4.8: Twelve reticle masks required for the different dies used in the architecture.

Figure 4.9: Routed interdie interconnections on each upper substrate.

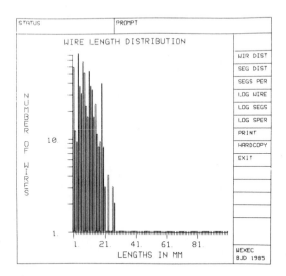

Figure 4.10: Histogram of wire lengths for interconnections in Figure 4.9.

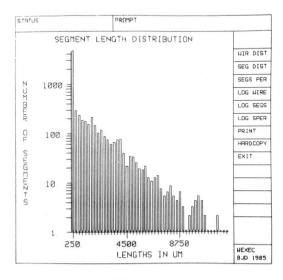

Figure 4.11: Histogram of line segment lengths of Systolic Cell.

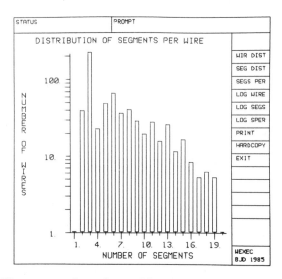

Figure 4.12: Histogram of numbers of line segments in an interconnection.

tend to get routed first, these generally contain fewer segments thus providing a performance compensation for their length.

The most important number which can be extracted from this routing infromation is that the total wiring length on the top "piggyback" substrate for each systolic processor is 6 meters. This is in a range which is small enough that a high yield can be expected for this wiring provided a suitably clean environment is employed for fabrication. Indeed this is comparable to the length of interconnect in a four megabit memory. Furthermore, the pitch of the wafer wiring is 5 or 10 times as large so the yield should be quite high. Clearly, however, an attempt to fabricate the wiring for 70 such processors on one substrate would result in a much lower yield. In addition, one must produce the inter cell communication links. The total length of these interconnections depends on just how the cells are to be networked. If only immediately adjacent cells are connected then the total wiring length would be quite short. On the other hand, if every processor needed links to a remote spare for fault tolerance then the total wiring length required might be as much as several hundred meters. The capacity of an 8 inch wafer for 20 to 40 micron pitch wire in two layers is more than a kilometer. Hence, sufficient capacity exists, but the defect density necessary to achieve satisfactory yield for these lengths might have to be exceptionally low. Therefore, some form of wafer wiring repair will probably be necessary even for a wafer which just contains this wire.

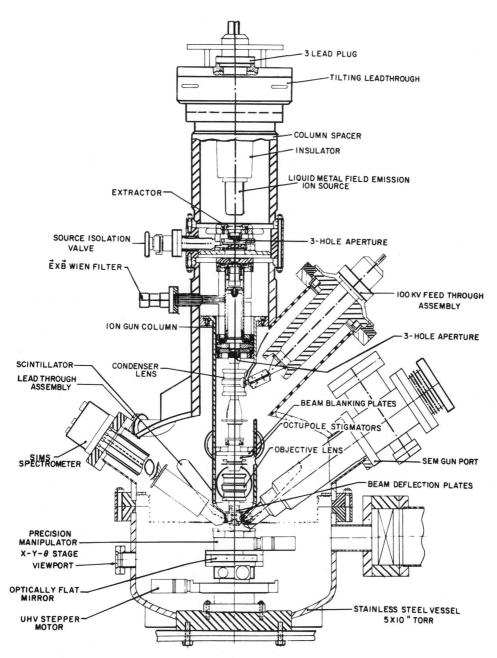

Figure 4.13: Focused Ion Beam (FIB) system suitable for repair of WSI interconnections.

4.4   Focused Ion Beam Repair for Thick Film Interconnections

A key conclusion from the aforementioned yield problem is the need for some systematic means of repair of interconnections. These interconnections must be capable of operation as microtransmission lines out to a frequency of at least 4 GHz. Such high bandwidths require metal and dielectric thicknesses which are larger than used in conventional integrated circuits but not so thick for the metal that the skin effect will limit its high frequency cutoff. Cutting and welding of such thick lines for repair purposes requires a special kind of tool. One possible system which shows promise in this application is the focused ion beam. Research is currently underway at Rensselaer on an experimental system shown in Figure 4.13 for use in WSI repair [18,19].

4.5   Conclusions

In this paper we have considered a packaging approach which we call "piggyback" wafer scale hybrid integration. We have been led to consider this form of packaging during the evolution of an architecture designed to meet the stated SDI computational throughput goal of 1000 GFLOPS. We have seen that these goals could be met by a system consisting of 16 eight inch diameter wafer hybrids providing interprocessor wiring on which are mounted 70 floating point systolic processors. These systolic processors themselves use a smaller wafer hybrid for supporting the 36 GaAs chips necessary to implement the 43,000 gate processor. The limiting consideration on pursuing even further hierarchical packaging is the number of solder points between the substrates, and the need to conduct heat through the assembly.

Acknowledgements

The authors wish to express their gratitude to the General Electric Corporate Research and Development Center in Niskayuna, New York for their continued support of the Wafer Scale Integration project at the Rensselaer Center for Integrated Electronics. Thanks are also due to the Corporate Sponsors of the Center and the Semiconductor Research Corporation for their partial support of various aspects of this work.

References

[1] McDonald, J.F., Donlan, Capt. B.J., Steinvorth, R.H., Greub, H., Dhodhi, M., Kim, J.S., and Bergendahl, A.S., *Yield of Wafer Scale Interconnections,* VLSI Systems Design, December, 1986, pp. 62-66.

[2] Lathrop, J.W., Clark, R.S., Hull, J.E., and Jennings, R.M., *A Discretionary Wiring System as the Interface between Design Automation and Semiconductor Array Manufacture,* Proc. of the IEEE, Vol. 55(1), Nov. 1967, pp. 1988-1997.

[3] Huang, G., Nunne, W., Spielberg, D., Mones, A., Fett, D., and Hampton, F., *Silicon Packaging - A New Packaging Technique,* Proc. Custom Int. Circ. Conf., Rochester, N.Y., May, 1983, pp. 142-146.

[4] Johnson, R.R., *The Significance of Wafer Scale Integration in Computer Design,* Proc. of the IEEE Int. Conf. on Comp. Des., Oct. 1984, pp.

[5] Donlan, Lt. B.J., McDonald, J.F., Taylor, G.F., Steinvorth, R.H., and Bergendahl A.S., *Computer-Aided Design and Fabrication for Wafer Scale Integration,* VLSI Design, April 1985, pp. 34-42.

[6] Raffel, J.I., Anderson, A.H., Chapman, O.H., Konkle, K.H., Mathur, B., Soares, A.M., and Wyatt, P.W., *A Wafer Scale Digital Integrator,* Proc. of the I.E.E.E. Int. Conf. on Comp. Des., Oct. 1984, pp. 121-126.

[7] McDonald, J.F., Greub, H.J., Steinvorth, R.H., Donlan, Capt B.J., and Bergendahl, A.S., *Wafer Scale Interconnections for GaAs Packaging - Applications to RISC Architecture,* IEEE Computer, April 1987, pp. 21-35.

[8] Kung, H.T., Sproull, B., and Steele, G., Eds. *VLSI Systems and Computations,* Computer Science Press, Rockville, MD, 20850, pp 235-284.

[9] Kailath, T., *Modern Signal Processing,* Hemisphere Publishing Corporation, New York, 1985.

[10] Kung, S.Y., Whitehouse, H.J., and Kailath, T., Eds., *VLSI and Modern Signal Processing,* Prentice Hall, 1985.

[11] Swartzlander, E.E., *VLSI Signal Processing Systems,* Kluwer Academic Publishers, Boston, 1986.

[12] Cohen, R., McDonald, J.F. , Sanya, M. , and Woods, J.W., *A Wafer Scale Integration Video Rate fully Recursive Two- Dimensional Filter,* Proc. IEEE Conf. on Comp. Des., Oct. 1985, pp. 234-239.

[13] Product Documentation, *GaAs E/D LSI Programmable Cell Array - Preliminary TQ3000 Product Description,* TriQuint Semiconductor, Inc., Tektronix Industrial Park, Group 700, P.O. Box 4935, Beaverton, OR, 97075.

[14] Noll, T.G., Schmitt-Landsiedel, D., Klar, H., and Enders, G., *A Pipelined 8×8 330 MHz NMOS Multiplier,* IEEE J. Sol. State Ckts. (JSSC), Vol. SC-21(3), June 1986, pp. 411-416.

[15] Bakoglu, H.B., Walker, J.T., and Meindl, J.D., *A Symmetric Clock-Distribution Tree, and Optimized High-Speed Inter- connections for Reduced Clock Skew in ULSI and WSI Circuits,* Proc. IEEE Conf. on Comp. Des., Oct. 1986, pp. 118-122.

[16] Hornak, L.A., and Tewksbury, S.K., *On the Feasibility of Through-Wafer Optical Interconnects for Hybrid Wafer-Scale Integrated Architectures,* IEEE Trans. on Elec. Dev., Vol. ED-343(7), July 1987,pp. 1557-1563.

[17] Grinberg, J., Nudd, G.R., Etchells, R.D., *A Cellular VLSI Architecture,* Computer, vol. 17, no. 1, pp. 69-81, Jan. 1984.

[18] McDonald, J.F., Stanton, M., Rajapakse, R. Lin H., Selvaraj, R., King, N., King, D., and Haslam, M., *Adaptive Discretionary Wiring for Wafer Scale Integration using Electron Beam Lithography,* Proc. SPIE, Vol 773 (Electron-Beam, X-Ray, and Ion-Beam Lithographies VI), P. D. Blais, Ed., Mar. 1987, pp. 140-149.

[19] McDonald, J.F., Rajapakse, R.U., Lin, H.T., Selvaraj, R., Corelli, J.C., Jin, H.S., Balakrishnan,S., and Steckl, A.J., *Optimized Focused Ion Beam Inspection and Repair of Wafer Scale Interconnections,* Proc. SPIE, Vol 773, P. D. Blais, Ed., Mar. 1987, pp.206-215.

Chapter 5

Future Physical Environments and Concurrent Computation

S. K. Tewksbury [1]
L. A. Hornak [1]
P. Franzon [1]

5.1 Introduction

Using graph-based representations of computation problems [1]-[3], the communication function of a "pseudo-general purpose," massively parallel computing environment is discussed to help define technology-focussed realizations of that communication function. Compatible computation problems are neither constrained to highly regular structures (such as systolic arrays and their generalizations [4]) nor extended to the globally non-deterministic behavior of many general purpose problems [5]. A fully distributed [6], data driven [7] computing environment is assumed, emphasizing the impact of communications on algorithm execution [8]. Evolution of such massively concurrent computing environments is necessary to sustain the growth of computing power as device technologies approach fundamental limits on dimensional scaling and higher device performance [9],[10].

Decomposition of the overall problem into fine-grain tasks maximizes available concurrency but emphasizes computation and communication delay variations and delay fluctuations. Section 5.2 considers the structure of computation algorithms, emphasizing elasticity within the graph, locally absorbing delay uncertainties (a cause of non-determinism in otherwise deterministic algorithms). Section 5.3

---

[1] AT&T Bell Laboratories, Holmdel, NJ

considers the related impact of communication delay uncertainties. Physical implementations that might evolve in the future are considered in Section 5.4. A compact computer implemented using stacked wafer-level components models a future, high-performance computing site. Two future technology directions for communication fabrics are considered. The first direction considers a mesh network, using high density inter-wafer interconnections to minimize delay uncertainties. The second direction speculates on a parallel data, high- T_c superconducting bus operating at multi-Gbit/sec line rates and providing guaranteed time slot access to each of thousands of attached processors.

5.2   Simple Model

The mapping of complex computation problems onto a concurrent computing environment (with N_{cs} local computing sites interacting through a distributed communication environment as in Figure 5.1) helps define the spatial and temporal characteristics of the communication environment. The heuristic description given here is not applicable to all problem classes. However, it does provide a basis for introducing performance issues that might be addressed by advanced technologies for communications in massively parallel computing systems.

5.2.1   Spatially Expanded Graph Representation

A "fully expanded," acyclic graph model of the computation directly displays temporal issues. The representation used here and illustrated in Figure 5.2 fully expands all iterations external to the individual tasks T_J along the time axis of the space-time task distribution and satisfies the following conditions.

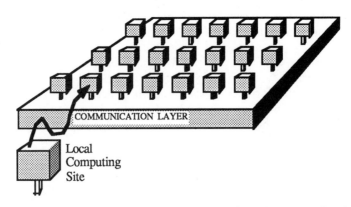

Figure 5.1: Local computation and distributed communication

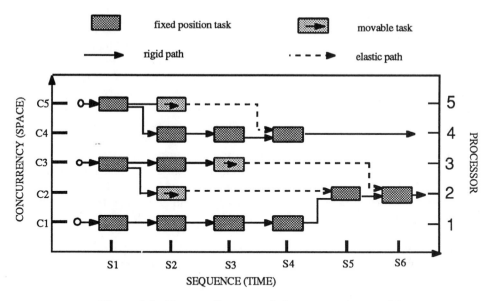

Figure 5.2: Temporally expanded computation model

1. The graph is unbounded for positive space and time.

2. All (directed) arcs point from left-to-right (i.e. successor nodes of T_J lie to the right of T_j).

3. Each computing task is placed at a distinct grid points (C_J, S_J). The nodes of the intrinsic problem are "point tasks" with no measure of their internal computation complexity. Tasks at a given S_J can be executed simultaneously in different computing elements. The sequence axis represents the order in which tasks are executed.

4. The graph is "left-packed", i.e. the sequence point S_J of T_J is the earliest point along the sequence axis at which all inputs to T_j are available.

5. It is assumed that such a graph model exists, either independent of or depending on the specific problem instance.

In Figure 5.2, each task T_J may be labeled by its execution time $\tau_{comp}(J)$ and each directed arc $a_{J,K}$ from T_J to T_K may be labeled by its communication delay $\tau_{comm}(J, K)$. The intrinsic algorithm model can assume $\tau_{comp}(J) \equiv 0$ for all tasks and $\tau_{comm}(J, K) \equiv 0$ for all arcs, with time flow given by successor/ancestor relations in the graph. The graph model is not conceptually changed by taking $\tau_{comp}(J) \equiv \tau_{comp}^{(0)}$ for all T_J and taking $\tau_{comm}(J, K) \equiv \tau_{comp}^{(0)}$ for all arcs $a_{J,K}$ between T_J and T_K. The sequence axis can then be labeled by an integer L_t giving task execution times (starting times) $t = L_t(\tau_{comp}^{(0)} + \tau_{comm}^{(0)})$. Let $L_t^{(max)}$ be the latest discrete time at which at least one task is executed in the "left-packed graph". A *globally rigid temporal path* originates on a task with no predecessor, terminates on a task with no successor and intersectes a task at each time point between the originating and terminating tasks. Delay uncertainties on such global critical paths induce uncertainties in the problem execution time. Internal to the graph there may be *internal rigid temporal paths,* coupling local time uncertainties over the limited range $K_{\min} \le L_t \le K_{\max}$. In contrast to these rigid paths, an *elastic temporal path* traverses time points without intersecting a task, locally absorbing time uncertainties up to some maximum without disturbing uncoupled neighboring rigid paths. Rigid and elastic paths are contrasted in Figure 5.2. The elasticity of arcs and the distribution of such elasticity in the algorithm structure provides a mechanism allowing introduction of limited time uncertainties (e.g. recursions) while preserving a deterministic mapping of the overall problem onto the time axis.

The concurrency $C(L_t)$ is the number of tasks at time point L_t. Temporal elasticity may allow reduction of local concurrency by moving tasks from heavily populated time points into neighboring, less populated time points without increasing $L_{\max}$. Such manipulations help map a problem onto a number of computing sites less than the maximum $C(L_t)$.

5.2.2 Computation Time Enhanced Model

Above, $\tau_{comp}(J)$ and $\tau_{comm}(J, K)$ were assumed equal to constants $\tau_{comp}^{(0)}$ and $\tau_{comm}^{(0)}$. In practice, different tasks may have different execution times and different arcs may impose different communication delays. The introduction of task-dependent computation and arc-dependent communication delays can change the problem's intrinsic space-time behavior. This section considers task-dependent execution times (assuming negligible communication delays). The next section considers arc-dependent communication delays.

Let $\tau_{comm}(J, K) = 0$ and, using a time step t_{inst}, let $\tau_{comp}(J) = M_j t_{inst}$ where $M_J = M_j^{(0)} + \delta M_j$. Here, $M_j^{(0)}$ is the minimum, instance independent execution time of T_J and δM_j is the specific instance fluctuation in $\tau_{comp}(J)$. We take $0 \le \delta M_j \le \Delta M_j$, assuming bounded fluctuations. Consider first task-dependent computation times independent of the specific problem instance, i.e. $\delta M_j \equiv 0$ for all T_J.

Task-dependent execution times induce a new time-mapping on the problem graph, as illustrated in Figure 5.3. The temporal extent $M_j^{(0)}$ of T_J is taken as proportional to its computational complexity (i.e. granularity), since we assume each task is sequentially executed at a single computing site. The extent to which the original graph is distorted depends on the granularity variations among tasks. Let $M_{\min}$ ($M_{\max}$) be the smallest (largest) and $\hat{M}$ be the average value of M_J over the

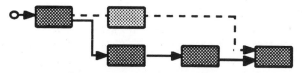

(A)   EQUAL EXECUTION TIMES (ALL TASKS)

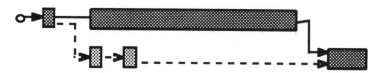

(B)   TASK-DEPENDENT EXECUTION TIMES

Figure 5.3: Equal vs. unequal task times

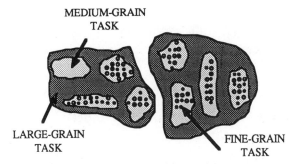

Figure 5.4: Granularity hierarchy

tasks T_J of the problem. The overall *granularity variation* of tasks, normalized to $\hat{M}$, is defined as $\zeta_{comm} = (M_{\max} - M_{\min})/\hat{M}$. Large ζ_m distort the $\zeta_m = 0$ time-space model, e.g. (1) the critical paths may change, (2) the temporal elasticity may change, (3) the problem concurrency generally decreases and (4) internal synchronization becomes a more serious issue. Small ζ_m introduces less distortion and supports "machine-independent" partitioning of a problem onto computing sites. Such small ζ_M may be achieved in some cases with large $\hat{M}$. However, a more general approach is to partition the problem at the finest grain level (i.e. approaching the single instruction level of detail). A large-grain task T_J may be expanded into its finer grain internal structure of subtasks $\tilde{T}_l(J)$, as suggested in Figure 5.4. These fine-grain internal tasks of T_J may then be distributed among computing sites, possibly decreasing the granularity variations and perhaps increasing usable concurrency. However, partitioning a problem at a finer grained level typically increases the communications among computing sites, requiring higher performance communication networks, as considered later. Despite the distortion of the intrinsic problem by large ζ_{comm}, a deterministic mapping of the problem onto computing sites is still possible since (having assumed $\delta M_j = 0$) the time behavior is deterministic (does not depend on problem execution).

Instance-dependent execution times (i.e. $\delta M_j \neq 0$) impose more serious problems. If T_J is on a global rigid path of the problem graph, both the overall problem execution time and timing relationships among internal rigid paths referenced to that global critical path vary with δM_J. Deterministic partitioning of the problem therefore becomes more difficult as $\delta M_j/M_j^{(0)}$ increases. However, if T_J with execution time fluctuation $0 \leq \delta M_J \leq \Delta M_J$ appears on an elastic path with temporal elasticity $t_e > \Delta M_J$, then the fluctuation may be locally absorbed by that local elasticity, leaving other portions of the problem structure unaffected and preserving global partitions.

The discussion above emphasized locally executed computations. Next, the impact of communications among computation tasks is considered.

5.2.3   Communication-Delay Enhanced Model

Here, we take $\tau_{comp}(J) = \tau_{comp}^{(0)}$ and represent the communication delay $\tau_{comm}(J, K)$ of the arc from T_J to T_K as $\tau_{comm}(J, K) = N_{J,K} t_{trans}$, $N_{J,K} \geq 1$ an integer. We express $N_{J,K}$ as $N_{J,K} = N_{J,K}^{(0)} + \delta N_{J,K}$, with an execution-instance independent part $N_{J,K}$ and an execution-instance dependent part $\delta N_{J,K}$, mimicking the earlier representation $N_J = N_J^{((0)} + \delta N_J$ for computation delays. Again, we assume bounded fluctuations, i.e. $0 \leq \delta N_{J,K} \leq \Delta N_{J,K}$. fluctuations in communication delays (i.e. ζ_{comm}) is similar to the definition of ζ_{comp}. Communication delay issues are qualitatively similar to the computation delay issues considered above.

If $\tau_{comm}(J, K) \ll \tau_{comp}^{(0)}$, then the communication delays do not severely distort the intrinsic problem structure. This condition $\tau_{comm}(J, K) \ll \tau_{comp}^{(0)} \equiv M^{(0)}\tau_{inst}$ can be achieved by using large grain tasks (i.e. choosing $M^{(0)}$ large) or by using slow computing sites (i.e. choosing τ_{inst} large). However, fine-grained tasks were suggested earlier to increase available concurrency and to decrease effects of execution time variations and fluctuations. Slow computing sites are contrary to achieving the highest performance general purpose computing machine recognizing that computational problems do not exhibit unbounded concurrency.

If, on the other hand, $\tau_{comm}(J, K) \gg \tau_{comp}^{(0)}$, the communication environment may completely dominate the execution of the problem. The effects of large communication delay variations (large ζ_{comm}) on critical paths, the distribution of temporal elasticity and the degree of concurrency are similar to the effects caused by computation time variations ζ_{comp}. Similarly, the effects of delay fluctuations $\delta N_{J,K}$ are similar to those of computation time fluctuations δM_J. This suggests the simple model illustrated in Figure 5.5. It is the total delay variation (computation plus communication) and the total delay fluctuation that defines the execution's space-time behavior. This suggests that seeking computation algorithms with well behaved computation times (i.e. low computation delay fluctuations to allow problem partitioning and perhaps low computation delay variations to avoid partitions sensitive to the realization of local computing sites) has little value if the physical communication environment does not provide low communication delay fluctuations

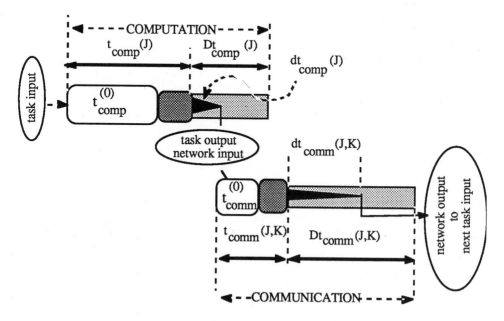

Figure 5.5: Computation and communication delays

and low communication delay variations. Here, we seek physical communication environments which minimize both $N_{J,K}^{(0)}$ and $\delta N_{J,K}$.

5.3   Architectural Issues

Here, we consider the communication function implied by the computation problem's graph and communication delays imposed by physical implementations of that function. The fully extended graph model can display the assignment of tasks to computing sites by associating each concurrency ordinate C_j with a distinct computing site. Then, the sequence of tasks with the same concurrency ordinate is sequentially executed in the same computing site, as indicated in Figure 5.2. The arcs connecting tasks with the same concurrency ordinate are implemented within the computing site and represent *local communication functions*. Arcs connecting between different concurrency ordinates extend between computing sites and represent *non-local communication functions*. We assume that the problem has been fit onto the finite number of computing sites (e.g by moving tasks in elastic regions of the graph to idle computing sites when possible and otherwise folding the graph onto the finite number of computing sites) and consider below some general issues impacting the architecture of the communications function.

5.3.1   Classes of Communication Function

In general, a given task can spawn $n_1 \geq 1$ output arcs and each such output arc can fan out to $n_2 \geq 1$ successor tasks. To simplify discussions below, we take $n_1 = n_2 \equiv 1$ and define the communication function according to the sequence of arcs spawned by a given computing site. In particular, the communication function $\Gamma_l(k)$ for the lth task executed along C_k is specified by its intrinsic spatial distance $\delta x_l(k)$ and by its intrinsic arc delay $\delta \tau_l(k)$. Using the notation $J_p = (l_p, k_p)$, T_{J_1} spawns $\Gamma_{l_1}(k_1)$ and its output arc terminates on T_{J_2}. The spatial distance of $\Gamma_{l_1}(k_1)$ is then $\delta x_{l_1}(k_1) = k_2 - k_1$. Letting $t_J^{(in)}$ and $t_J^{(out)}$ be the start and completion time of T_J, the intrinsic arc delay is $\delta \tau_{l_1}(k_1) = t_{J_2}^{(in)} - t_{J_1}^{(out)}$, with non-zero positive values allowed even if physical communication delays are taken as zero. These "space-time" displacements then define the communication function spawned by $T_{J(l,k)}$, i.e. $\Gamma_l(k) = [\delta x_l(k), \delta \tau_l(k)]$. The communication function $\Gamma(k)$ associated with a given computing site C_k is the set of $\Gamma_l(k)$ for that site. Finally, the overall communication function for the problem is the set of $\Gamma(k)$ over all computing sites.

The graph model of the problem largely defines the $\delta \tau_l(k)$ and their statistical distribution. However, the $\delta x_l(k)$ depend on the particular mapping of tasks onto computing sites, allowing adjustment of the distribution of δx during mapping to match the distribution favored by a particular communication network.

The communication functions discussed above are defined for a specific problem. Often, a simplified representation for $\Gamma(k)$ and Γ for representative problems is useful. Two extreme cases are common.

- Class R: The values of $\delta x_l(k)$ for each computing site C_k are assumed randomly distributed over all computing sites. Typically, $\delta \tau_l(k)$ is taken as zero (more precisely, ignored). This corresponds to the familiar model of randomly distributed destinations for data transfers originating from each computing site.

- Class F: Here, $\delta x_l(k) = \delta x(k)$, independent of l. In this case, the connections are time-independent and may be implemented with hard-wired, point-to-point connections for efficient, special purpose algorithm realizations.

A third, intermediate, case represents local communications, i.e.

- Class L, with the distribution of δx peaked at $\delta x = 0$ and decreasing with larger $|\delta x|$.

This "local communication model" may evolve naturally in top-down, hierarchically synthesized algorithms as illustrated in Figure 5.2. If the exhibited communication activity (i.e. number of non-local arcs) increases at the finer granularity level of description of a problem segment, then the distribution of δx for a given computing site will tend to be locally peaked in the region of fine-grained tasks of that segment.

5.3.2 Delay Issues in Physical Networks

Communication delays through a physical network are considered here for class L and class R communication functions. (Class F, important for special purpose systems, is not considered here.) The distributed nature of communications and the localized execution of tasks suggest the general model of the physical computing environment shown earlier in Figure 5.1. The communication layer includes several functions, each contributing to the total end-to-end delay $\tau_{comm}(J, K)$. Arcs internal to computing sites achieve values $\tau_{comm}(J, K) = 0$ whereas the minimum delay $\tau_{comm}(\min)$ of non-local arcs is non-zero, imposing an essential communication delay variation between local and non-local communication arcs. Both to minimize distortion of the intrinsic problem structure and to avoid high sensitivity to alternative local/non-local partitioning decisions during mapping, a physical network should provide a small, minimum non-local delay $\tau_{comm}(\min)$. The major network functions and their delay contributions are as follows.

- Assembling multi-data packets before entering data into the network induces an unnecessary delay $\tau_{pck} = N_{pck}\tau_{inst}$, where $N_{pck} \geq N_w$, the number of words per packet. Construction of large packets conflicts with achieving small τ_{comm}(min). The network should preferably allow direct entry of single word data into the network as soon as it is generated.

- The network access function determines whether the local network switch can accept data from the computing site. The corresponding delay $\tau_{acs} = N_{acs}\tau_{inst}$, N_{acs} an integer, depends on the software/hardware protocols for network control. However, ideally the network should, with high probability, accept data on request. Operation of a distributed network near saturation conflicts with this objective, suggesting that such networks be operated well below saturation.

- A non-local graph arc is implemented as one or more links between switching sites of the network. Minimum delay τ_{lnk} across a single link favors parallel data transfers (assuming wire bandwidth can be maintained independent of the width of the connection). The link delay is then $\tau_{lk} = M_{lk}\tau_{trans}$, where M_{lk} represents the number of periods τ_{trans} to transfer data across a link.

- Distributed networks connect individual links through routing circuits connecting an incident link to one of its outgoing links. Assuming the data is accompanied by its destination information, used by the router to select the appropriate outgong link, the delay τ_{rte} through the routing function is taken as $\tau_{rte} = M_{rte}\tau_{trans}$, allowing for pipelined routing algorithms operating at the clock period τ_{trans}.

- Finally, the data extraction function enters data from the network into the destination computing site. The delay $\tau_{ext} = M_{ext}\tau_{inst}$ can become appreciable if multiple sources simultaneously access the same destination node. For the quasi-deterministic mapping assumed here, such conflicts can be minimized if communication delays are well defined during mapping of the problem onto computing sites.

Combining the delays above, the end-to-end delay $\tau_{comm}(J, K)$ is roughly

$$\tau_{comm}(J, K) \approx (M_{pkt} + M_{acs} + M_{ext})\tau_{inst} + K_{link}(M_{lk} + M_{rte})\tau_{trans},$$

with K_{link} the number of network link segments traversed between T_J and T_K. Taking the delay per link as $\tilde{\tau}_{link} = (M_{lk} + M_{rte})\tau_{trans}$ and $K_{acs} = M_{pck} + M_{acs} + M_{ext}$, $\tau_{comm}(J, K) \approx K_{acs}\tau_{inst} + K_{link}\tilde{\tau}_{link}$. If the network entry delays are negligible (i.e. $K_{acs}\tau_{inst} \ll K_{link}\tilde{\tau}_{link}$), then the communication delay is proportional to the number of links traversed. Several networks emphasize minimum K_{link} despite large packet formation and access delays because link delays are comparable or even longer. Here, absolute minimum delays into and through the network are sought, suggesting somewhat different optimum network structures given physical limits on minimum delays.

Also important is the communication rate across links of the network. The communication rate is not necessarily the reciprical of the communication delay since the network can be pipelined (e.g. store-and-forward) and the individual routing sites can have pipelined routing logic. Communication rates are considered separately in the examples of the next section where future technologies for high-performance communciation fabrics are considered.

5.4 Network Examples

Assuming fine-grained tasks, quasi-deterministic mapping of problems onto the physical computing environment and both control and minimization of delays and delay uncertainties as discussed above, we **speculate** here on some technological directions to implement compatible concurrent computing systems. First, a model of a high performance computing site using silicon MOS wafer-level components is described. Then, assuming this (or a comparable) implementation of the local computing sites, a mesh network using high density, moderate rate interconnections and a very high bandwidth superconducting bus network are considered. These examples highlight technology directions that may lead to communication networks with performance well beyond the capabilities now achieved.

5.4.1 Compact, High Performance Computing Sites

The local computing sites are here assumed to be high performance, parallel data, sequential computers. Figure 5.6 illustrates schematically a multi-wafer, silicon MOS realization of such a computing site (e.g. 32-bit processor and associated user/disk memory). The individual wafer levels are either monolithic wafer-scale integrated (WSI) circuits or hybrid wafer-scale circuits (H-wsi). H-wsi directly mounts unpackaged VLSI IC's on a silicon circuit board on which line drivers, multiplexers, test circuitry and other "glue" circuitry are monolithically integrated.

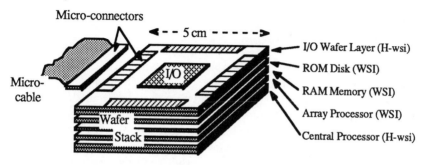

Figure 5.6: Stacked wafer-level computing site.

Such wafer-level components are stacked to achieve a compact structure (each wafer about 20-30 mils thick). The wafer-level layers shown are (1) a H-wsi implemented processing layer with the CPU, coprocessor, MMU and other associated processor functions, (2) possibly a WSI implemented array of fine-grained computing elements for matrix calculations, signal processing, etc. and used as a co-processor, (3) a WSI implemented user RAM, providing >10 Mbytes of user memory per site, (4) a WSI implemented "solid state disk" using EEPROM's and PROMS, providing > 10 Mbyte of local disk storage for operating systems, etc., and (5) a H-wsi implemented layer for connection to the network (and perhaps for implementation of the routing function associated with the network).

Wafer-scale integration, originally proposed several years ago [11], is receiving increased attention as limits on scaling devices to smaller sizes are approached and as requirements for yield enhancement (necessary for WSI) extend to chip-level components. Faults on WSI circuits are virtually certain [12] and explicit schemes [13]-[16] to tolerate such faults must be provided. For computational logic, fault tolerance must typically be integrated directly on the wafer, favoring regular arrays of fine-grained cells and serial data communications among cells. Furthermore, mixtures of digital functions monolithically integrated in WSI are limited by the variety of their device structures. Logic gates, dynamic RAM cells and EEPROM cells each requires distinct device structures with different fabrication processes; if combined on a single wafer, more fabrication steps and lower yield results. Implementing a complete, monolithic WSI high-performance computer on a single wafer will be difficult. Therefore, several monolithic components are assumed necessary for each computing site.

Faults in memory cells can be handled by avoiding defective cells much as faulty regions of a hard disk are avoided using "bad block" lists. Externally providing fault tolerance using such lists, WSI integration of memory arrays becomes efficent [17]. Distinct device structures may still require separate monolithic circuits for RAM, ROM, EEPROM, etc. However, a high performance, general purpose 32-bit local computing site can probably consume sufficient memory that separation of the memory onto different wafers is acceptable. With such WSI memory, not only large user memory but also copies of the operating system can be provided locally at each site (avoiding access to a central facility for system software facilities). The "solid-state" disk memory wafers in Figure 5.6 also can be randomly accessed, eliminating the slow access of real disk storage systems.

The processor layer in Figure 5.6 is implemented with H-wsi, directly mounting unpackaged VLSI IC's on a silicon substrate. The term "hybrid-WSI" signifies that the silicon circuit board is not the passive interconnection circuit board of present advanced packaging schemes [18] but instead contains monolithically integrated circuitry [19]-[22] such as drivers for long lines (relaxing requirements for power-hungry drivers on the VLSI IC's), multiplex/demultiplex circuitry, bus-control/access circuitry and other small-scale system functions that comprise the inefficiently implemented "system glue" circuitry of present systems. Issues of fault tolerance and yield for such active circuit boards are less severe than for VLSI WSI since the circuitry is less dense (with available space for redundancy) and since less aggressive technologies with higher yields might be used. A H-wsi module is suggested in [19], using a second silicon wafer with precision etched cavities to "imbed" and align the flip-chip mounted VLSI IC's. Flip-chip mounting of IC's significantly increases the possible number of I/O connections to IC's (using area distributed rather than edge-distributed I/O pads) and the fine lines on the silicon substrate support a high inter-IC wiring density. The processor board, with high power dissipation, is located at the bottom of the stack in Figure 5.6, with the lower dissipation memory wafers in the center of the stack, to minimize heating of the center of the stack. This partitioning of computer functions onto separate wafers also tends to minimize the number of vertical interconnections.

The network interconnection layer at the top of the wafer stack provides intermodule (e.g. the mesh network example in Section 5.4.2) or module/network (e.g. the superconducting system bus example in Section 5.4.3) interconnections. This H-wsi layer allows direct mounting of unpackaged GaAs IC's and optoelectronic devices on the silicon substrate, providing efficient implementation of a wide variety of inter-module and/or network access connection schemes.

Finally, some brief comments concerning the vertical connections through the stacked wafer layers are appropriate. Thermally migrated aluminum interconnections through silicon wafers have been demonstrated in a high-performance, stacked wafer module [23]. Compact, innovative "interwafer" packaging schemes were developed for superconducting, Josephson junction computers [24]. Optical interconnections have also received considerable attention [25]-[27]. Infrared, thru-wafer (since silicon is transparent to infrared) optical connections using microintegrated Fresnel zone-plate lenses to focus optical beams on detectors are described in [27]. Recent advances in optoelectronic components [28] are likely to stimulate further efforts toward exploiting optical interconnections at low levels of the system interconnection hierarchy. Optical, free space communication links may allow development of large-scale 3-dimensional architectures, with many of the same advantages for communications suggested for 3-D VLSI layouts [29]-[30].

With this model of a future, compactly integrated high performance computing site, we consider next realization of the communication network function, first using a mesh network example supporting Class L communication functions and then considering a bus network supporting Class R communication functions.

5.4.2 Mesh Network Example

The high density electrical and free space optical connections considered here favor short physical interconnection lengths, suggesting the nearest-neighbor mesh-

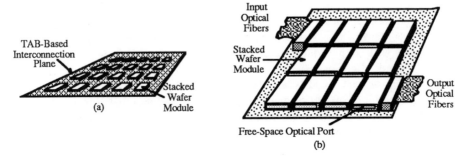

Figure 5.7: Mesh arrays (electrical plane and free space optics interconnections)

connected arrays assumed here and shown in Figures 5.6 and 5.7 with N_{cs} attached computing sites. Large mesh networks favor Class L communication to avoid saturation of network links. The limitations for Class R communications are easily seen. Bisecting the mesh vertically into two equal halves, the cut intersects $\sqrt{N_{cs}}$ links. Let R_a be the average rate at which each computing site enters data into the network and assume destinations are uniformly distributed (Class R) over all sites. Each site to the left of the cut communicates across the cut at an average rate $R_a/2$. The net transfer rate across the cut is $N_{cs}R_a/4$ while the rate per cut link is $R_{lnk} \approx \sqrt{N_{cs}}R_a/4$. R_{lnk} becomes large relative to t_{inst} and eventually increases beyond the limits of electronic interconnections for larger arrays (i.e. larger N_{cs}) and finer grain tasks (i.e. larger R_a). Class L communication, with locality diameter L_d, limits communications across the cut to sources within a distance L_d of the cut, giving $R_{lnk} \propto L_d R_a$, independent of network size for $N_{cs} > L_d$. We shall assume locality of communications when using mesh networks. To obtain minimum network access delays for fine grain tasks, computing sites must be able (with high probability) to enter data into the network within a time $\approx t_{inst}$ while not impeding the flow of other data propagating through the attached network node. In particular, the network must be operated well below local bandwidth saturation of its links and nodes. Let R_{link} be the maximum link bandwidth, αR_{link} with α small (e.g. < 0.1) be the maximum link bandwidth usable to avoid saturation, f_{clk} be the processor clock rate, $R_{inst} \approx \beta f_{clk}$ be the instruction execution rate and N_{grn} be the task granularity (average number of instructions per non-local output transfer). Then the link's communication bandwidth is related to the processor clock rate roughly according to

$$R_{link} \geq \frac{L_d \beta}{\alpha N_{grn}} f_{clk}.$$

Taking $f_{clk} = 20MHz$, $\beta = 10$, $\alpha = 0.1$, $N_{grn} = 10$ and $R_{link} = 100MHz$, then $L_d \leq 50$, i.e. a computing site's "local communication" region contains about 2500 nearest sites. The link communication rate R_{link} is the rate at which full data words and header information associated with each word is transferred. For 32-bit data and about 32-bits of header, this requires transfer of 64 data bits in parallel at

R_{link} tranfers/sec. Smaller data path widths (K-bits at a time) reduce the locality distance by a factor 64/K, e.g. 8-bit transfers reduce L_d in the numerical example from 50 to about 6 (and the local communication region from about 2500 to about 40 computing sites).

Smaller data paths and constant R_{link} are achieved by increasing the physical data rate on each wire above R_{link}. 1 Gbit/sec optical interconnects provide $R_{link} = 100$ MHz with a ten-fold reduction in path width in the example above. Since each mesh switching site has 4 input and 4 output unidirectional ports to neighboring sites, $\simeq$ 50, 1 Gbit/sec optical lines per switching site would replace the $\simeq$ 500, 100 Mbit/sec electrical lines while preserving the 100 Mhz communication rate per port.

The discussion above assumed equally active network links (i.e. a balanced network activity). If the routing algorithm assigns a single, unique path to each source/destination pair, conflicts for link access can induce considerable waiting time within the network. For balanced activity, the network must provide several alternative paths and its routing algorithm should dynamically and locally direct communications through lightly used portions of the network. A distributed routing algorithm implemented in the individual switching sites is assumed here. The routing algorithm locally evaluates its own directional activity and interacts with neighboring routers to generate a measure of the network activity. A physical analogy for this activity measure is an "activity potential" Ψ whose gradient $\nabla \Psi \cdot \mathbf{n}$ in the direction $\mathbf{n}$ measures the network activity in that direction. Algorithms to locally generate the spatial and time dependence of this activity potential Ψ over the entire network may be an interesting problem in communication network design. For example, Ψ might be an analog signal, generated and propagated by a "neural network" circuit with nodes at the switching sites.

The simple example of a routing algorithm sketched here suggests a VLSI-complexity function. Variations in path length among alternative end-to-end paths cause delay fluctuations which we wish to avoid. Alternative paths should therefore be restricted to lengths near the minimum path length. Figure 5.8a shows the several minimum-length paths between diagonally separated nodes but only a single minimum length path for horizontally or vertically separated nodes. Addition of diagonal connections to the basic mesh network (roughly corresponding to adding a mesh rotated by $\pi/4$ to the original network, as seen in Figure 5.8b) provides a density of alternative, minimum paths roughly independent of source-destination angle. Given the high wire density per port, direct implementation of the diagonal connections would be costly. However, a "virtual diagonal" connection can be obtained by implementing only diagonal access control connections. Data can then be propagated directly through a neighbor (e.g. North) to the diagonal node (e.g. through North then East for a Northeast move).

A relatively simple mathematical algorithm to choose allowed output ports for mesh networks with the added diagonals is obtained by (1) taking the distance measure from the present node (r_x, r_y) to the destination node (d_x, d_y) to be

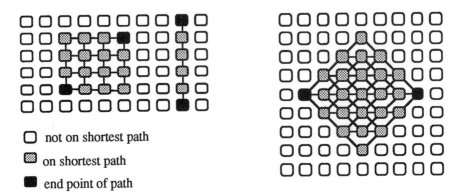

□ not on shortest path

▨ on shortest path

■ end point of path

Figure 5.8: Alternative minimum paths in mesh-connected arrays. (a) Orthogonal connections only. (b) Added diagonal connections.

$D \equiv \sqrt{\Delta x^2 + \Delta y^2}$ where $\Delta x = d_x - r_x$ and $\Delta y = d_y - r_y$, (2) requiring that each step decrease D, and (3) assuming the added virtual diagonal links. Here, the integer node indices (p,q) are positive with Δp and Δq having values ± 1 to neighboring vertical and horizontal nodes, respectively. The routing algorithm, not detailed here, obtains the set of allowed output directions from the signs of some differences between the local node's and the destination node's indices. Allowed output ports are weighted by the network activity function to determine the best output port. The regions within which routing is constrained is regenerated at each network node as the path is traversed and is large for large source-destination distances, decreasing as the destination is approached. A large area VLSI circuit would be needed for this routing algorithm, not only due to the parallel data pipelined computations but also due to connecting each input port's wide ($\simeq$ 60 bits) data path to all output ports. Placing this routing function on the top layer of the stacked wafer module, we consider next implementing the high density of lines extending between wafer-level modules. Three distinct approaches are briefly considered.

Figure 5.6 illustrated a a "micro-cable"-based interconnection scheme. The microcables are high density TAB-like cables (i.e. thin film interconnections patterned on polyimid or similar compliant plastic films). Precision silicon microconnectors (with etched V-groove alignment features) terminate the cable ends and are "plugged" into a similar silicon connector (also with etched V-grooves) monolitically fabricated on the top wafer of the module. Such connections allow replacement of faulty modules while also providing compliant but high density (expected to exceed 100 lines/inch) interconnections. Figure 5.7a shows a related scheme, in this case using a large area "monolithic" plane of TAB-like interconnection material on which the computing modules are directly mounted.

Electrical connections among modules as in Figures 5.6 and 5.7a have two serious

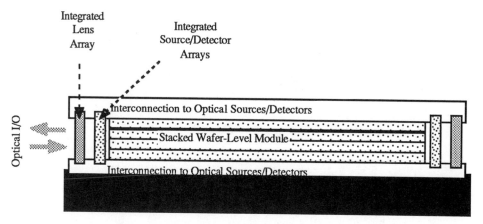

Figure 5.9: Example of a future optical interface.

limitations - (1) the mechanical connections are susceptible to failure and (2) arrays are generally restricted to planes by mechanical limits. Free space optical interconnections, briefly considered next, may relax these and other limitations. Figure 5.7b illustrated schematically a mesh network with free-space optical communications between the faces of the module (and possibly extendable to 3-D lattice arrays by adding optical links to the top and bottom faces) Figure 5.9 suggests an internal structure of a module, using arrays of optical sources and detectors and microintegrated lens arrays at source (and perhaps detector) ends of the free-space links. Assuming mechanical alignment of sources and detectors is unnecessary, the modules can be simply dropped into place. (With a higher density of detectors than of incident light beams, "electronic alignment" might eliminate precise mechanical alignment of sources and detectors.)

Such optical free-space links and the high density electrical connections above both suggest major improvements in concurrent computing machines by focusing technology on the communication rather than computation function. The mesh-organized array of time shared links and switches favors L-class communications. The very high speed bus considered next would support R-class communications in a particularly simple manner.

5.4.3 High- T_c Superconductor Bus Network

Starting with the discovery of superconductivity at critical temperatures $T_c \approx 30K$ last year, an increased effort seeking high transition temperature superconducting materials culminated in the discovery of superconductivity in a large number of XBaCuO composites at transition temperatures $T_c > 90K$, above the boiling point of nitrogen (see, for example [31],[32]). Subsequently, workers reported evidence for

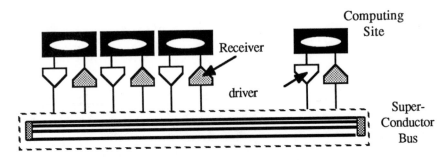

Figure 5.10: Superconductor bus and computer site interfaces.

superconductivity at $\simeq$ 240K [33] with speculation concerning superconductivity above room temperature [34]. Although considerable work will be required before their technology and understanding allow practical applications in electronics, they provide a particularly exciting new route to explore for higher performance digital systems. Josephson junction devices [35] (not yet demonstrated in the new materials) would play an important role as line drivers/receivers for very high bandwidth (10-100 GHz), low dispersion and low attenuation (at $T/T_c < 0.6$) superconducting interconnections [36].

Figure 5.10 illustrates a bus interconnection scheme for a massively parallel processing system, providing a guaranteed time slot for each attached processor at the access rate per processor of R_a. The time slot duration provided per attached processor is then $\tau_{slot} = 1/(2N_{cs}R_a)$, giving a bus communication rate

$$R_{bus} = 2N_{cs}R_a = \frac{2N_{cs}\beta}{N_{grn}}f_{clk},$$

using the parameters of Section 5.4.2. Assuming a fully parallel data bus and taking $f_{clk} = 20$ Mhz, $\beta = 0.1$ and $N_{grn} = 10$, a bus communication rate $R_{bus} = 10$ GHz would support 10,000 processing nodes while a 100 GHz rate would support 100,000 processing nodes. Fine-grained communication among such a large number of processing nodes is achieved with the following important attributes: (1) no complex routing algorithms through the "network", (2) absence of network deadlock, (3) absence of network blocking and (4) constant (single instruction time) source-to-destination network delay from any source to any destination.

One might expect such bus interconnections to be seriously limited by speed-of-light delays, i.e. a processor would have to retain control of the bus until its output data had propagated the full length of the bus. However, speed-of-light limits on data rates can be physically avoided by confining the spatial extent of the propagating data pulses to distances smaller than the distance between adjacent access points. A data pulse of width τ_p occupies a spatial extent $\lambda_p \approx c\tau_p/\sqrt{\epsilon_r}$, propagating down the line (in both directions from the source point) at a velocity $c/\sqrt{\epsilon_r}$. Once that data pulse has propagated past the adjacent processor's access point (and reached the next following access point), that adjacent processor can place its data pulse on the line. The bus data rate is

$$R_{bus} = \frac{1}{(1+\gamma)\tau_p},$$

(i.e. assuming spacing $\gamma\tau_p$ between successive, rightward propagating data pulses of width τ_p) and the physical separation d_{node} between adjacent processor access points on the bus is $d_{node} = c\tau_p/\sqrt{\epsilon_r}$. After the last processor has accessed the bus, the first processor can not access the bus until the data from the last processor has propagated the full length of the bus, i.e. back to the first processor. This is the speed-of-light limit, setting the minimum cycle time of local accesses. For local access rate R_a, the maximum bus line length $L_{\max}$ is about

$$L_{\max} \approx \frac{c}{2\sqrt{\epsilon_r}R_a}.$$

For $R_a = 1$ MHz, $L_{\max} \approx 50$ meters.

"Self-timed" clocking schemes propagating a clock pulse at half the data velocity to successive access points could provide the needed, high speed sequencing of the access of successive processors to the bus. Assuming proper clock timing, the source pulse generators must provide short pulsewidth, return-to-zero pulses while not distorting propagation of previously inserted pulses propagating on the bus. This might be achieved with a Josephson junction driver device (supporting low voltage pulses) or a very high speed semiconductor driver. Indirect pulse insertion, perhaps using inductive coupling of signals onto uninterrupted buss wires, would preserve the nearly ideal pulse propagation characteristics of the superconducting bus wires (though such coupling schemes have not been demonstrated).

The receivers must acquire information without distorting pulses propagating down the bus. Magnetic coupling between bus lines and receiver lines might again help in this regard, avoiding direct connection of receiver taps to the bus. The circuitry directly interfacing the bus would probably favor Josephson junction devices if similar devices are used for pulse insertion. Following the receiver's pulse detectors, each computing site interface must evaluate the address information at the full line data rate (using either very high speed semiconductor or superconductor logic or using slower logic following a high-speed demultiplexer) to determine whether the local processor is the data destination.

Overall, the ultra-high-speed bus application considered here provides motivation for specific studies and experimental evaluations of the role of superconductor-based interconnections for massively parallel, general purpose computer architectures.

5.5 Summary

The conditions under which the communication environment can significantly degrade the intrinsic computational concurrency of an algorithm was qualitatively considered, providing insight on the performance issues desired for communications within a massively concurrent computing system. By focussing advanced technologies on the communication function, innovative communication networks and systems providing the desired coordination of massive numbers of computing elements will emerge over the next decade. This focus is particularly important since the traditional approach to higher system performance (conventional sequential computing architecture with higher performance integrated devices) is rapidly nearing fundamental limits, requiring new approaches to continue the historic rate at which increased computing power at low cost has evolved.

Acknowledgements

We gratefully acknowledge numerous converstions with and suggestions by our colleagues, particularly A. Ligtenberg (who clarified several mesh network characteristics using a different routing algorithm), M. Hatamian, Sailesh Rao, Paul Franzon (who helped design a WSI testbed for fault-tolerance in arrays with large-grain cells) and B. Sugla.

References

[1] J. Zeman and G.S. Moschytz, *Systematic design and programming of signal processors using project management techniques*, IEEE Trans. Acoust., Speech, Sig. Proc., vol. ASSP-31, pp. 1536-1549 (1983).

[2] S.K. Tewksbury, *Hierarchically localized mapping of signal processing algorithms onto programmable DSP's*, to be published.

[3] H.V. Jagadish, T. Kailath, J.A. Newkirk and R.G. Mathews, *On hardware description from block diagrams*, Proc. 1984 Int. Conf. Acoust., Speech, Sig. Proc., pp. 8.4.1-8.4.4 (1984).

[4] S-Y. Kung, K.S. Arun, R.J. Gal-Ezer and D.V. Bhaskar Rao, *Wavefront array processor: language, architectures and applications*, IEEE Trans. Comp., vol. C-31, pp. 1054-1066 (1982).

[5] M. Broy, *A theory for nondeterminism, parallelism, communication and concurrency*, Theoretical Computer Science, vol. 45, pp. 1-61 (1987).

[6] P.H. Enslow and T.G. Saponas, *Distributed and decentralized control in fully distributed processing systems: a survey of applicable models*, Techn. Report GIT-ICS-81/02, Georgia Inst. of Technol. (Feb 1981).

[7] P.C. Trelevan, D.R. Brownbridge and R.P. Hopkins, *Data-driven and demand-driven computer architectures*, Comp. Surveys, vol. 14, pp. 93-143 (1982).

[8] D.B. Gannon and J. van Rosendale, *On the impact of communication complexity on the design of parallel numerical algorithms*, IEEE Trans. Comp., vol. C-33, pp. 1180-1194 (1984).

[9] J.D. Meindl, *Ultra-large scale integration*, IEEE Trans. Elect. Dev., vol. ED-31, pp. 1555-1561 (1984).

[10] R.W. Keyes, *Fundamental limits in digital information processing*, Proc. IEEE, vol. 69, pp. 267-278 (1981).

[11] R.C. Aubusson and I. Catt, *Wafer-scale integration - a fault tolerant procedure*, IEEE J. Solid-State Circuits, vol. SC-13, pp. 339-344 (1973).

[12] T. Mangir, *Sources of failure and yield improvement for RVLSI and WSI: Part I*, Proc. IEEE, vol. 72, pp. 690-708 (1984).

[13] J.W. Greene and A. El Gamal, *Configuuration of VLSI arrays in the presence of defects*, J. ACM, vol. 31, pp. 694-717 (1984).

[14] J.I. Raffel, et al., *A wafer-scale integrator using restructurable VLSI*, IEEE Trans. Elect. Dev., vol. ED-32, pp. 479-486 (1986).

[15] W.R. Moore and M.J. Day, *Yield-enhancement of a large systolic array chip*, Microelectronic Reliability, vol. 24, pp. 511-526 (1984).

[16] T. Leighton and C.E. Leiserson, *Wafer-scale integration of systolic arrays*, IEEE Trans. Comp., vol. C-34, pp. 448-461 (1981).

[17] C.D. Chesley, *Main memory wafer-scale integration*, VLSI Design, vol. 6(3), pp. 54-58 (1985).

[18] R.K. Spielberger, et al., *Silicon-on-silicon packaging*, IEEE Trans Compon., Hybr. and Manuf. Techn., vol. CHMT-7, pp. 193-196 (1984).

[19] S.K. Tewksbury, et al., *Chip alignment templates for multichip module assembly*, IEEE Trans. Compon., Hybr. and Manuf. Techn., vol. CHMT-10, pp. 111-121 (1987).

[20] M. Hatamian, S.K. Tewksbury, P. Franzon, L.A. Hornak, C.A. Siller and V.B. Lawrence, *FIR filters for high sample rate applications*, IEEE Communications, July 1987.

[21] P. Franzon, M. Hatamian, L.A. Hornak, T. Little and S.K. Tewksbury, *Fundamental Interconnection Issues*, AT&T Techn. J., Aug. 1987.

[22] P. Franzon and S.K. Tewksbury, *'Chip frame' scheme for reconfigurable mesh connected arrays*, Proc. IFIP Workshop on Wafer Scale Integration, Brunel Univ., (Sept. 1987).

[23] J. Grinberg, R.G.R. Mudd and R.D. Etchells, *A cellular VLSI architecture*, IEEE Computer, pp. 69-81 (Dec. 1984).

[24] A.V. Brown, *An overview of Josephson packaging*, IBM J. Res. Dev., vol. 24, pp. 167-171 (1980).

[25] J.W. Goodman, *Optical interconnects in microelectronics*, Proc. SPIE, vol. 456, pp. 72-85 (1984).

[26] J.W. Goodman, F.J. Leonberger, S-Y. Kung and R.A. Athale, *Optical interconnections for VLSI systems*, Proc. IEEE, vol. 72, pp. 850-866 (1984).

[27] L.A. Hornak and S.K. Tewksbury, *On the feasibility of through-wafer optical interconnects for hybrid wafer-scale integrated architectures*, IEEE Trans. Elect. Dev., vol. ED-34, pp. 1557-1563 (1987).

[28] S.R. Forrest, *Monolithic optoelectronic integration: a new component technology for lightwave communications*, J. Lightwave Technol, vol. LT-3, pp. 1248-1263 (1985).

[29] A. Rosenberg, *Three-dimensional VLSI: a case study*, J. ACM, vol. 30, pp. 397-416 (1983).

[30] F.P. Preparata, *Optimum three dimensional VLSI layouts*, Math Systems Theory, vol. 16, pp. 1-8 (1983).

[31] P.H.Hor, et al., *Superconductivity above 90K in the square-planar compound system $ABa_2Cu_3O_{6+x}$ with $X = Y$, La, Sm, Eu, Gd, Ho, Er and Lu*, Phys. Rev. Lett., vol. 58, pp. 1891-1894 (1987).

[32] A. Khurana, *Superconductivity seen above the boiling point of nitrogen*, Physics Today, vol. 40(4), pp. 17-23 (1987).

[33] J.T. Chen et al., *Observation of the reverse ac Josephson effect in Y-Ba-CU-O at 240K*, Phys. Rev. Lett., vol. 58, pp. 1972-1975 (1987).

[34] Eds, *Superconductivity at room temperature*, Nature, vol. 327, pg. 357 (June 4, 1987).

[35] T.R. Gheewala, *Josephson-logic devices and circuits*, IEEE Trans. Elect. Dev., vol. ED-27, pp. 1857-1869 (1980).

[36] R.L. Kautz, *Picosecond pulses on superconducting striplines*, J. Appl. Phys, vol. 49, pp. 308-314 (1978).

Chapter 6

Understanding Clock Skew in Synchronous Systems

Mehdi Hatamian [1]

6.1 Introduction

Clock distribution and synchronization in synchronous systems are important issues especially as the size of the system and/or the clock rate increase. Minimization of clock skew has always been a major concern for the designers. Many factors contribute to clock synchronization and skew in a synchronous system. Among the major factors are: the clock distribution network, choice of clocking scheme, the underlying technology, the size of the system and level of integration, the type of material used in distributing the clock, clock buffers, and the clock rate. To be able to get around the problems related to clock skew and synchronization , one has to understand the effect that clock skew can have on the operation of a given system. In this paper we derive simple and practical formulations of these effects in terms of a few time-parameters that can be considered as properties of the individual modules and the clock network in a synchronous system. Knowing these time-parameters, one can determine the maximum throughput of a given system as well as its reaction to a change in clock skew. Three different clocking schemes, namely, edge-triggered, single-phase level sensitive, and two-phase clocking are considered. However, using the approaches discussed in this paper, the effect of clock skew for any other clocking scheme can be analyzed and formulated.

1. t_{pl} –propagation delay of the individual modules (e.g., processing elements) in the synchronous system.

2. t_{sl} – settling time or the computation delay of the modules.

---

[1] AT&T Bell Laboratories, Holmdel, NJ

3. t_{pr} – propagation delay of registers involved in data transfer between modules.

4. t_{sr} – settling time of registers.

5. t_{ck} – the time of arrival of the clock signal at a given register involved in a synchronous data transfer.

6. t_{pi} – propagation delay of the interconnection between communicating modules.

Depending on the the direction of the data transfer between modules, clock skew can take on both positive and negative values. A negative clock skew occurs when the time of arrival of the clock signal at the source of a data transfer is greater than the time of arrival at the destination. Otherwise, clock skew is positive. We show that for the positive case, clock skew must be bounded by an upper limit determined by the propagation delay parameters defined above. If this is violated, the system will fail no matter how low the clock rate is. On the other hand, if clock skew is always negative, then the possibility of system failure does not exist but the maximum throughput is reduced. In a system where both positive and negative clock skews occur, the possibility of a failure mode always exists and the minimum clock period is increased by the maximum of the absolute value of the negative skews. These results are discussed in detail for various clocking schemes. Aside from studying the effect of clock skew, the formulations discussed in this paper are also useful in studying the role of the propagation delay of the interconnection in the operation of a system; a problem that is receiving wide spread attention especially as IC technologies are scaled down.

6.2   Clock Skew Formulation

Clock skew has often been termed the chief difficulty in designing high-speed synchronous systems [1]-[3]; a problem that grows with the speed and size of the system. However, we believe that clock skew is often blamed for impairing the high speed operation of a synchronous circuit while it can be handled by first understanding how it affects the operation of the system, and then by using proper buffering and distribution networks to overcome the difficulties.

Factors determining the clock skew in a synchronous system are a) resistance, capacitance and inductance of the interconnection material used in distributing the clock, b) the shape of clock distribution network, c) the number of modules (size of the system) and the load presented by each module, d) clock buffers and the buffering scheme used (global vs. distributed), e) fabrication process variations, and f) the rise and fall time and the frequency of the clock signal. While having zero clock skew between different modules of a synchronous system is desirable, it is not always a necessity. The way clock skew affects the operation of a synchronous system depends on the speed of the modules, propagation delay of the interconnection between modules involved in data transfer, the direction of the data transfer with respect to clock, and the clocking scheme used. To illustrate these effects we use the example of Fig. 1 where a data transfer takes place between two modules of a synchronous system. Each module is composed of a computation logic section CL, and a register R clocked by the global clock signal. The computation module can be a collection of gates on a chip, a chip on a printed circuit board (or wafer in case of WSI), a board in a system, etc.

The following time parameters are defined for the purpose of discussing the clock skew. t_{pl} and t_{sl} are the propagation delay and the settling time (computation delay) of the computation logic, t_{pr} and t_{sr} are the propagation delay and the settling time of the register, t_{ck} is the clock skew of the module with respect to a global reference, and t_{pi} is the propagation delay of the interconnection between the two modules. We would like to emphasize that the cell propagation delay t_{pl} is defined as the time it takes before at least one of the outputs of the cell starts a transition in response to a change at any one of its inputs. This time is generally much shorter than the settling time (computation delay) t_{sl} of the cell and plays

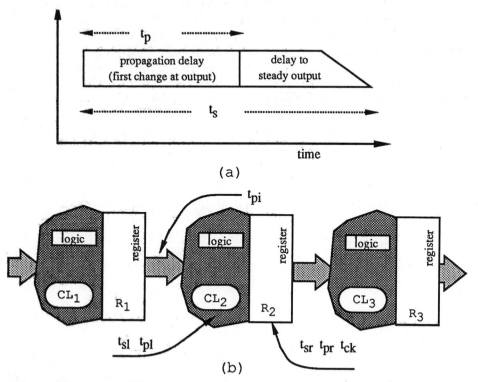

Figure 6.1: A linear pipelined array used as an example of a synchronous system for the purpose of clock skew discussion.

an important role in studying the effect of clock skew. The propagation delay and settling times of the register (t_{pr} and t_{sr}) are also defined in the same way.

6.2.1   Edge-Triggered Clocking

Consider the transfer of data between CL_1 and CL_2 in Fig. 1. At time $t = t_0$ the output of CL_1 begins to be loaded into R_1, and at time $t = t_0 + \delta$ the output of CL_2 begins to be loaded into R_2, where δ is the clock skew between the two modules defined as :

$$\delta = t_{ck_2} - t_{ck_1} \tag{1}$$

Also, at time $t = t_0 + t_{pr} + t_{pi} + t_{pl}$ the response to the change in R_1 has propagated all the way to the input of R_2. For the data transfer to take place properly, this event at $t = t_0 + t_{pr} + t_{pi} + t_{pl}$ must occur after the loading time of R_2 at $t_0 + \delta$. So the following condition must be satisfied:

$$\delta < t_{pr} + t_{pi} + t_{pl} \tag{2}$$

In other words, the clock skew must be less than the sum of the propagation delay times of the register, interconnection, and logic. Notice that according to the definition of δ in (1), the clock skew can be either positive or negative depending on the direction of the clock with respect to data transfer. In the case of negative δ, since the propagation delays are always positive, condition (2) is always satisfied. This might sound very surprising because no matter how large the skew is, the system still operates. However, the discussion will not be complete without considering the effects on the throughput of the system. The clock period T must be large enough to allow for the computation to take place. The total computation (or settling) time is $t_c = t_{sr} + t_{pi} + t_{sl}$. If the skew is negative ($t_{ck_2} < t_{ck_1}$) then we must have

$$T > t_c + |\delta| = t_{sr} + t_{pi} + t_{sl} + |\delta| \tag{3}$$

and if the skew is positive ($t_{ck_2} > t_{ck_1}$) the relation becomes:

$$T > t_c - \delta = t_{sr} + t_{pi} + t_{sl} - \delta \tag{4}$$

Combining (2) - (4) the conditions to be satisfied are as follows.

For $\delta \le 0$:

$$T > t_{sr} + t_{pi} + t_{sl} + |\delta|. \tag{5a}$$

For $\delta > 0$:

$$\delta < t_{pr} + t_{pi} + t_{pl}, \tag{5b}$$

$$T > t_{sr} + t_{pi} + t_{sl} - \delta. \tag{5c}$$

These relations can simply be interpreted as follows. If clock skew is negative, then condition (2) is always satisfied and the propagation delays of register, logic, and interconnection cannot cause disasters; the system will always operate properly provided the period is large enough. The price paid is loss of throughput because the clock period must be increased by $|\delta|$. On the other hand if clock skew is positive, then condition (2) must be satisfied to prevent system failure. However, throughput is improved because the period can be shortened by δ (a case where skew helps the throughput!). Obviously the improvement in throughput as δ is increased is very limited because soon δ violates condition (2). Also notice that in the case of positive skew, if condition (2) is violated, no matter how large T is, the system will fail and cannot be made operational by just reducing the clock frequency of the synchronous system unless the dependency of skew on the clock frequency is strong enough such that lowering the frequency can reduce the skew to where it satisfies the condition.

If the goal is maximizing the throughput, it seems that the case of positive clock skew should be preferred over the negative skew case (at the cost of increasing the possibility of system failure). In a general purpose synchronous system (especially those with feedback loops), the clock skew parameter δ could take on both positive and negative values throughout the system, depending on the direction of data transfers between modules. in that case, the worst possible situation must be considered. However, if there are no feedbacks involved, the clock distribution network can be designed with the data transfer directions in mind such that the clock skew can be made positive for all cases of data transfer. If there are feedbacks involved, usually the clock skew cannot be controlled to maintain the same sign throughout the system. However, knowing the settling time and propagation delay parameters defined above, the clock distribution network can be designed for maximum throughput using relations (5).

In their discussion of clock skew in synchronous systems, Franklin and Wann in [4], derive a single condition for the clock period similar to relation (3) (without using the absolute value function on δ) and do not consider the case of negative clock skew or derive any conditions similar to (2). As a result, the effect of clock skew is always that of reducing the throughput without any failure modes, while such is not always the case as we discussed above.

6.2.2 Single-Phase Clocking

Following the same approach as above for a single phase level-sensitive clocking scheme we arrive at the following condition :

$$\delta + t_\phi < t_{pr} + t_{pi} + t_{pl} \tag{6}$$

or

$$\delta < t_{pr} + t_{pi} + t_{pl} - t_\phi$$

where t_ϕ is the width of the clock pulse. Again δ can be either negative or positive, and whether the system fails or not is determined by the above relation. For example, if δ is positive and greater than $t_{pr} + t_{pi} + t_{pl}$ then we must have $t_\phi < 0$ which is impossible, resulting in system failure. Even if δ is smaller than $t_{pr} + t_{pi} + t_{pl}$, for a high speed system, t_ϕ might become so small that it makes the implementation impractical. On the other hand, if δ is negative the upper bound on t_ϕ becomes $t_{pr} + t_{pi} + t_{pl} + |\delta|$ and makes the system more practical, but with a loss of throughput in the same manner as we discussed in the edge-triggered clocking section.

Notice that parameter t_ϕ can also be entered into the discussion of edge-triggered clocking as the hold time of the registers, in which case, it should be treated as a module property parameter rather than one that is globally controlled, like the clock pulse-width.

6.2.3   Two-Phase Clocking

In this case, register "R" in Fig. 1 is considered as a two-phase-clock register. We use a dynamic two-phase (level sensitive) register widely used in MOS designs. t_{ϕ_1}, t_{ϕ_2} and $t_{\phi_{12}}$ are defined as the phase-1 pulse-width, phase-2 pulse-width and the delay time between phase-1 and phase-2, respectively. The clock period is T as before. The rest of the parameters remain the same except that we use t_{pr_1} and t_{sr_1} for the phase-1 register, and t_{pr_2} and t_{sr_2} for the phase-2 register. We can also assume that clock skew has the same pattern for both phases of the clock, which is not an impractical assumption. At time $t = t_0$ the output of CL_1 begins a transfer into the phase-1 register of R_1 and its effect appears at the phase-1 input of R_2 at time $t = t_0 + t_{\phi_1} + t_{\phi_{12}} + t_{pr_2} + t_{pi} + t_{pl}$. For the system to operate properly, this time should be greater than $t_0 + \delta + t_{\phi_1}$. So, the condition to be satisfied is :

$$\delta < t_{pr_2} + t_{pi} + t_{pl} + t_{\phi_{12}} \tag{7}$$

$$T > t_{\phi_1} + t_{\phi_{12}} + t_{sr_2} + t_{pi} + t_{sl} + |\delta|, \quad \delta \leq 0,$$

$$\tag{8}$$

$$T > t_{\phi_1} + t_{\phi_{12}} + t_{sr_2} + t_{pi} + t_{sl} - \delta, \quad \delta > 0.$$

We also must have $t_{\phi_1} > t_{sr_1}$ and $t_{\phi_2} > t_{sr_2}$. The discussion for positive and negative clock skew is very similar to the edge-triggered clocking scheme except that in the case of positive δ, one can always increase $t_{\phi_{12}}$ to satisfy condition (7) at the cost of reducing the throughput, as can be seen from condition (8). In other words, unlike the edge-triggered clocking scheme, in a two-phase clock approach, it is always possible to prevent failure mode by slowing down the clock frequency (increasing $t_{\phi_{12}}$). $t_{\phi_{12}}$ acts as a global parameter that can be used to control the clock skew.

The approaches taken in the above sections can be followed for any other clocking scheme or register arrangement to study and understand the effect of clock skew on the operation of a synchronous system. The important point to be made here is that, contrary to what most designers believe, clock skew does not always result in loss of throughput; at the same time it can also create failure modes that are independent of the clock frequency no matter how low it is. Another point to remember is that of differentiating between the propagation delay and the settling time (computation delay). Throughout this section we have made the assumption that each time-parameter of interest takes on the same value for all modules in Fig. 1. In a general purpose synchronous system this may not be true, in which case we have to consider the worst possible situation. The relations can be generalized by using the appropriate $min(\cdot)$ and $max(\cdot)$ functions on the parameters.

6.3   Clock Distribution

As mentioned before, one of the parameters that determines the clock skew picture in a given synchronous system is the shape of the clock distribution network. Probably the most widely studied distribution network that minimizes the clock skew is the H-tree network [2]-[6]. Fig. 2 shows an example of such a network for an 8 × 8 array of cells.

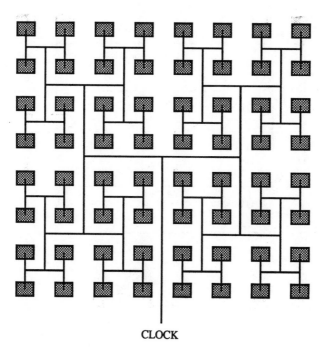

CLOCK

Figure 6.2: Example of an H-tree network for an 8 by 8 array of cells

It is obvious from this figure that if the cells are all identical in size and present
the same load to the network, then the paths from the clock source to all cells are
similar and clock skew is minimized. In [2], Kung and Gal-Ezer derive an equivalent
transfer function for the clock path from the clock source to each individual cell for
an $N \times N$ array of processing elements. They show that the equivalent time constant
of the distribution network is $O(N^3)$ and conclude that the clock pulse rise-time
and the clock skew associated with it are $O(N^3)$ as well. Therefore, as N (i.e., the
size of the system) increases, the clock skew should rapidly become a stumbling
block. We find this conclusion about clock skew rather surprising because it is the
skew between different modules that is the determining factor, not that between the
clock source and each processing element. As for the problem with the clock pulse
rise-time, it can be alleviated by using a distributed buffering scheme by introducing
intermediate buffers (repeaters) as is also suggested by Bakoglu and Meindl [5] [7].
The disadvantage of introducing the intermediate buffers in the H-tree network,
other than possible area overhead, is increasing the sensitivity of the design to
process variations. Despite their attractiveness in providing minimum clock skew,
H-tree networks are most useful when all processing cells in a synchronous system

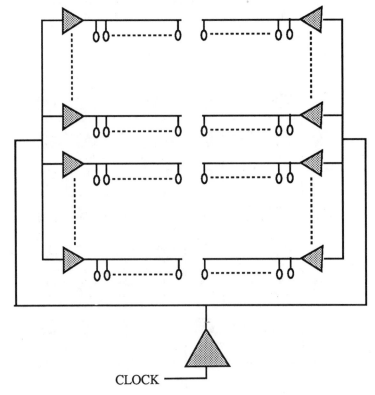

Figure 6.3: A clock distribution network using two-level distributed buffering. Circles represent computation cells

are identical and the clock can truly be distributed in a binary tree fashion. Cases where H-tree networks lose their attractiveness for hierarchical designs are discussed in [6]. An alternative to H-tree network, which is suitable for on-chip and on-wafer clock distribution, and is simple to design is the network shown in Fig. 3 [8]. This figure shows a two-level distributed buffering scheme. The number of intermediate buffering levels depends on the size of the system, the load presented by the modules to be clocked, the desired speed and the interconnection material used. Detailed discussion of this network and examples of fabricated VLSI designs in CMOS technologies can be found in [8] and [9]. This network, in theory, does not offer the minimum clock skew feature offered by an ideal H-tree network. However, it can prove to be equally useful because the extent of skew minimization depends on the amount of clock skew tolerated by the system, which is often not as strict as it is thought to be.

An important issue that requires special attention in designing clock distribution networks for very high-speed synchronous circuits and systems is the problem of timing simulation. In order to make sure that the clock skew criteria are satisfied at all cells, the clock network must be accurately modeled and a timing simulation must be performed. This can prove to be extremely time consuming or even impossible if a SPICE-type simulation of the whole circuit is undertaken. However, it should be noted that, as far as clock timing and skew are concerned, only the load presented by each cell to the clock network is of importance; details of the internal timings of the cell need not be known for this purpose. Therefore, the cell loads can be modeled by only considering the input circuit of their clocked section in the simulation. This leads to a reduction in the simulation time by a considerable factor depending on the size of the system. Examples of such simulations are given in [9]. Obviously, the results obtained from such timing simulations are only as close to reality as is the model of the network. The choices of circuit elements used to model the clock distribution network, or any other interconnection for that matter, can range from lumped capacitance to lumped RC to distributed RC and finally to a transmission line model. The choice directly depends on the length of the interconnection and the rise-time, fall-time and the frequency of the signal being propagated. For instance, if the rise-time of the clock pulse times the speed of light is less than the length of a given branch of the clock network, then the clock propagation on that branch should be studied with a transmission line model of the branch. However, if this figure is considerably greater than the length of the clock path, a distributed RC model of the path is sufficient. For example, for a clock signal having a rise-time of 500 ps, the latter will be the case for distributing clocks at the on-chip level, while at the wafer or printed circuit board level a transmission line model is more appropriate. Lumped capacitance models are useful for short metal interconnections (mostly at the on-chip level), and lumped RC models are generally useful for short high-resistance interconnects (e.g., polysilicon or diffusion wires). Work on automated extraction, modeling and simulation of interconnection networks (especially clock distribution networks) from VLSI layouts is currently being pursued by the author and will be reported in the near future.

References

[1] J. W. Goodman, F. I. Leonberger, S-Y. Kung, and R. A. Athale, *Optical interconnections for VLSI systems*, Proc. IEEE, vol. 72, no. 7, July 1984, pp. 850-866.

[2] S.Y. Kung and R.J. Gal-Ezer, *Synchronous Versus Asynchronous Computation in Very Large Scale Integrated Array Processors*, Proceedings of SPIE Symposium , Vol. 341 Real Time Signal Processing V, 1982, pp. 53-65.

[3] Allan L. Fisher and H.T. Kung, *Synchronizing Large VLSI Processor Arrays*, IEEE Transactions on Computers, Vol. C-34, No. 8, Aug. 1985, pp. 734-740.

[4] Donald F. Wann and Mark A. Franklin, *Asynchronous and Clocked Control Structures for VLSI Based Interconnection Networks*, IEEE Transactions on Computers, Vol. C-32, No. 3, March 1983, pp. 284-293.

[5] H.B. Bakoglu, J.T. Walker and J.D. Meindl, *A Symmetric Clock Distribution Tree and Optimized High-Speed Interconnections for Reduced Clock Skew in ULSI and WSI Circuits*, IEEE International Conference on Computer Design, Rye Brook, NY, Oct. 1986, pp. 118-122.

[6] E. G. Friedman and S. Powell, *Design and Analysis of a Hierarchical Clock Distribution System for Synchronous Standard Cell/Macrocell VLSI*, IEEE Journal of Solid-State Circuits, Vol. SC-21, No. 2, April 1986, pp. 240-246.

[7] H.B. Bakoglu and J.D. Meindl, *Optimal Interconnection Circuits for VLSI*, IEEE Transactions on Electron Devices, Vol. ED-32, No. 5, May 1985, pp. 903-909.

[8] M. Hatamian and G. L. Cash, *A 70-MHz 8-bit × 8-bit parallel pipelined multiplier in 2.5 micron CMOS*, J. Solid State Circuits, vol. SC-21, no. 4, Aug. 1986, pp. 505-513.

[9] M. Hatamian and G. L. Cash, *Parallel Bit-Level Pipelined VLSI designs for High Speed Signal Processing*, Proc. IEEE, vol. 75, No.9, pp. 1192-1202 (1987).

PART II
Theoretical Issues

Introduction

Technology issues lie at one extreme of the range of issues impacting development of highly parallel concurrent computing systems. These technology issues provide a basis for the physical design of concurrent computing systems. At the other extreme are the underlying theoretical issues which provide a consistent basis for the functional design and the programming of such systems.

Programmability for highly parallel, multi-purpose architectures remains a central issue limiting their development. Many distinctive architectures have been proposed and many benchmarks have been suggested. Comparisons among such a diverse set of architectures and benchmarks are difficult. **Rosenberg** addresses the issue of validating that a proposed architecture can, in fact, efficiently perform a variety of multi-purpose tasks. The important objective here is a theoretical basis, rather than simulation, for such validation.

Achieving a formal model of parallel computation, separate from implementation details, is essential in obtaining general programming and analysis techniques that can be used from system-to-system and from one generation machine to the next. **Ramachandran** reviews the PRAM model of parallel computation and its extensions, expressing restrictions on access to global memory. She then addresses the problem of parallel algorithms for reducible flow graphs (important, for example, in control mechanisms), drawing on concepts from network flow. The next two papers bear witness to the emergence of algorithmic paradigms for parallel computation. **Gazit et al.** address the problem of parallel tree contraction, using the EREW PRAM model. Techniques to manipulate the graph and achieve computational efficiencies (using time and number of processors as efficiency measures) are described. **Mayr** considers the dynamic tree expansion problem, illustrating its use with the familiar algebraic straightline programs. These papers provide a good overview of both general theoretical models and recent advances in the development and applications of new theoretical results in concurrent computation.

The computational efficiency of parallel algorithms often depends on the specific instance being executed. Worst case performance, for example, can depend critically on a specific local behavior arising only under special circumstances within some part of the parallel algorithm. Worst case performance can then be much worse than the average performance. **Rajasekaren and Reif** review the historical development of algorithm randomization, with particular steps randomized rather than proceeding deterministically. This technique, which reduces the sensitivity of the algorithm performance to local special case behavior, is illustrated using randomization of several well-known algorithms.

Many models of parallel computation, while considering the impact of global memory access, provide limited representations of communication delays. An exception is found in VLSI models of computation where various interconnection delay models have been discussed extensively. However, for more general systems, communication delays are usually modeled poorly. **Vitányi** discusses communication complexity (space and time) from some general principles, drawing on examples of popular networks (e.g. cube-connected cycles) for parallel computing. Very general results on the temporal behavior of networks are obtained, allowing communication time to be added to complexity models of large scale systems without requiring detailed representations of the specific network.

Drawing on experience in designing and modeling VLSI systems, **Subrahmanyam** considers operational semantics, based on linguistic primitives, for specifying concurrent system realizations. This leads to a formal model which can represent not only events with nonzero duration times (as opposed to instantaneously completed events in simpler models) but also concurrent execution of a multiplicity of events.

Jones and Sheeran also address algorithms from the perspective of their implementation. The objective is again to obtain a suitable language for abstracting (reasoning about) circuits and algorithms in a hierarchical manner, capable of capturing the low level details while representing the higher level behavior. Using such "system description languages," coherent strategies can be developed for implementing entire classes of algorithms. A design methodology which represents an array circuit's function in terms of manipulation of data objects is proposed, constraining the design according to a framework for synchronization and communications.

Chapter 7

On Validating Parallel Architectures via Graph Embeddings [1]

Arnold L. Rosenberg [2]

Abstract

A viable multi-purpose parallel architecture must exhibit many qualities, including: technological feasibility, low communication overhead, robustness in the face of faults, and programmability. The maturity and reliability of techniques for detecting and studying these desiderata decrease rapidly in the order of their listing. The issues of technological feasibility and communication overhead are relatively well understood; progress is being made on the problem of detecting and enhancing the robustness of proposed architectures; but the issue of programmability has not yet even been formulated in a generally accepted way. We extrapolate here from joint work with S. N. Bhatt, F. R. K. Chung, L. S. Heath, and F. T. Leighton, to propose an avenue for assessing the programmability of a proposed parallel architecture, using the formal tool of graph embeddings. As is common, we represent the communication structure of a parallel architecture as an undirected graph, and we seek efficient embeddings of a variety of specific graph families in the architecture graph. We choose the graph families to abstract the intertask dependency structures of popular algorithmic strategies, such as divide-and-conquer and convolution. We illustrate the approach by describing recent studies of optimal embeddings in the Hypercube architecture.

---

[1] This research was supported in part by NSF Grant DCI-87-96236.
[2] University of Massachusetts, Amherst, MA

99

7.1   Introduction

There has been considerable interest for several years in an architectural style that
is especially well suited for integrated circuit technology, namely arrays of identical processing elements (PEs). The literature is full of proposed instances of this
architectural style. In most cases, the proposed array structure is defined in terms
of an undirected graph whose vertices represent the PEs of the array and whose
edges represent the inter-PE communication links. Many of these proposals seek to
establish the feasibility of their proposed array structure by arguing that it enjoys:

1. *small communication overhead*

The most simple-minded measure of communication overhead is the diameter of the
underlying graph; more sophisticated studies consider also the ease of finding short
inter-PE routes, and the probability that routing paths will "block" one another.

2. *technological feasibility*

The most simple-minded measure of technological feasibility is the presence of small
vertex-degrees in the underlying graph; more sophisticated studies consider other
issues: for instance, in a VLSI implementation, one must be able to lay the proposed
array structure out on a chip or wafer in small area, with short runs of wire. A
significant subset of these proposals address also the issue of the

3. *robustness* of the array, i.e., its capacity for "working around" faults.

The issue of robustness is of particular importance when one contemplates implementing the array using large-scale VLSI or WSI technology since, at least for the
foreseeable future, any aggressive use of these technologies exposes one to having
a positive fraction of one's PEs not survive the fabrication process. One cannot
predict where these faulty PEs will lie in the array, so the capacity to avoid faulty
PEs is a major issue in the evaluation of a proposed architecture.

One final issue is addressed in only a small minority of these proposals, despite its
importance for any proposed multi-purpose architecture, namely,

4. *programmability*, i.e., the demonstration that the array can efficiently perform a
   variety of tasks.

Despite the complications mentioned above, there would seem to be rather general
agreement on how to study the first two of our four issues. The issue of robustness encounters less universal agreement, there being one school of thought that

advocates *a priori* avoidance of faults [7, 18, 19], by incorporating reconfigurability (via a network of switches) in the design, and one school that advocates *a posteriori* avoidance of faults [8, 9, 14], by salvaging a working sub-architecture. But even with these divergent approaches and philosophies, one is likely to obtain some level of agreement about the ability of an architectural design to survive faults; the arguments would more likely center on how efficiently the robustness is achieved.

The final issue in our list, programmability, is the most problematical since, despite its importance, no standard approaches to establishing the programmability of a proposed architectures have evolved. It is our intention here to propose and illustrate one approach to validating the programmability of a proposed parallel architecture, by means of graph embeddings.

A word about the competitors of our proposal: One finds in the literature two main approaches to establishing the programmability of an array structure, exemplified by the following. In [16] the viability of the Cube-Connected Cycles (CCC) architecture is argued by presenting a programming scheme that can be implemented efficiently on the CCC and demonstrating that a number of important computations can be specified efficiently using that scheme. This approach does indeed establish the programmability of an architecture for a variety of tasks; it does not, however, suggest how one can ascertain whether or not one's desired task can be programmed efficiently using the indicated scheme; moreover, no one has come up with an analogous scheme for most other architectures, suggesting that this approach may not be generally applicable. In [13] the viability of the Mesh-of-Trees (MT) architecture is argued by demonstrating how a variety of important computational problems can be solved efficiently on an MT. Johnsson use the same technique to show that grid-based algorithms can be implemented efficiently on the Hypercube architecture. This technique is the ultimate test for the specific algorithms studied, but it often gives no hint about how to assess the efficiency of the architecture for even closely related algorithms.

The proposal we espouse here is to try to argue the programmability of an architecture by looking at its performance on *classes* of algorithms, rather than on individual ones. The informal vehicle we propose to develop here builds on the *logical mapping problem* for parallel architectures, namely the problem of realizing the intertask communication structure of one's parallel algorithm on the idealized parallel architecture one has access to [2, 6]. The word "idealized" here indicates that the logical mapper, who plays the role of the programmer in a sequential environment, does not worry about a variety of physical issues such as possible faults in, or efficiency-enhancing compromises in the fabricated architecture.

We represent both algorithms and arrays as undirected graphs, with the following notations and interpretations: For parallel algorithms, the vertices V_G of the algorithm-graph G represent the tasks in the algorithm; the edges E_G of G represent the interdependencies among the tasks. These interdependencies might reflect data dependencies, as with grid- or convolution-based algorithms, or control dependencies, as with divide-and-conquer algorithms. As noted earlier, for processor arrays, the vertices V_H of the array-graph H represent the PEs of the array; the edges E_H of H represent the inter-PE communication links.

The goal of a logical mapping is to place interdependent tasks close to one another in the array, all the while utilizing array resources efficiently. Formally, we formulate a logical mapping as an *embedding* of the algorithm-graph G in the array-graph H:

- We associate tasks with PEs via a one-to-one mapping

$$\iota : V_G \rightarrow V_H$$

- We measure proximity via the *DILATION* of the embedding, namely, the farthest apart that interdependent tasks get separated in the array:

$$DILATION(\iota) = \max_{(u,v) \in E_G} \mathrm{Dist}_H(\iota(u), \iota(v))$$

- We measure efficiency of resource utilization via

  - the *LOAD-FACTOR* of the embedding, namely, the largest number of intertask paths that traverse a single array edge – a measure of congestion
  - the *EXPANSION* of the embedding, namely, the fraction of PEs that *do not* get used in the embedding; more conveniently:

$$EXPANSION(\iota) = |V_H|/|V_G|$$

The now-standard measures of DILATION and EXPANSION originated in [17].

An Aside. One might expect it to suffice to consider only one of DILATION and EXPANSION, since it is intuitive that minimizing one measure would minimize both. Such is not the case [4, 11]. For instance, one finds the following in [11]:

Theorem 1. *There is an embedding of the N-node complete ternary tree in the complete binary tree with*

$$\mathrm{DILATION} = 2$$

and

$$\mathrm{EXPANSION} = N^{.26}$$

Every embedding of the N-node complete ternary tree in the complete binary tree that has

$$\mathrm{EXPANSION} < 2$$

also has

$$\mathrm{DILATION} > (constant) \log \log N$$

Thus, attempting to optimize either measure causes the other to grow without bound.

Validation Proposal. We propose to establish the programmability of a proposed array by seeking the best embeddings in the proposed array of a variety of families of algorithm-graphs whose structures abstract the intertask dependency structures of a variety of important algorithmic paradigms. If, on the one hand, we establish the existence of embeddings with $O(1)$ DILATION, $O(1)$ LOAD-FACTOR, and $O(1)$ EXPANSION, then we shall conclude that the proposed architecture can be used to implement efficiently any algorithm in the class represented by the graph family: one simply uses that embedding as the logical mapping. If, on the other hand, we fail to find efficient embeddings, or even if we can prove that none exist, we shall not be able to draw any definitive conclusion therefrom, for our notion of embedding is static – tasks are preassigned to PEs and, once assigned, are never moved; even if no static assignment strategy exists, there might be some dynamic assignment strategy that leads to an efficient algorithm implementation. Thus the strengths of our proposed approach result from

- its forcing the array designer to focus on algorithms instead of programs

- its precision and rigor

The weaknesses of the approach are embodied in the problem of "false negatives," i.e., the fact that our proposal yields only a *semi*-decision procedure for the existence of efficient implementations.

To develop our proposal further, we begin a list of important graph families that would be of especial interest in any attempt to verify the programmability of a proposed multipurpose parallel architecture.

- *arbitrary binary trees:* (cf. Fig. 7.1)

    - For our purposes, a binary tree is a connected acyclic graph one of whose vertices is bivalent (the *root*) and all of whose other vertices are either

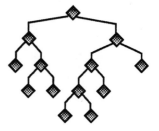

Figure 7.1: Sample binary trees; the lefthand one is complete

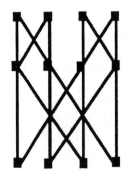
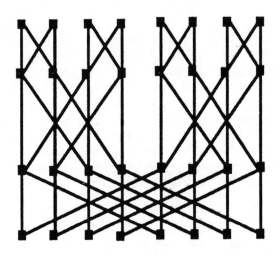

Figure 7.2: The FFT graphs F(2) and F(3)

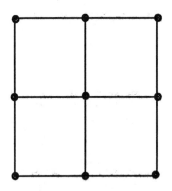
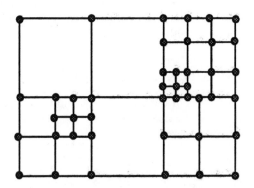

Figure 7.3: A sample rectangular grid (left) and refinable grid (right)

univalent (the *leaves*) or trivalent. The tree is *complete* if all root-to-leaf paths have the same length (which is called the *height* of the tree). We denote by $T(h)$ the height-h complete binary tree.

– Binary trees are paradigmatic for divide-and-conquer algorithms in the sense that each task of such an algorithm spawns either zero or two subtasks which never interact until they return control to their common parent.

• *FFT graphs:* (cf. Fig. 7.2)

– Let m be a positive integer. The 2^m-*input FFT graph* denoted $F(m)$, is defined as follows. $F(m)$ has vertex-set

$$V_m = \{0, 1, \cdots, m\} \times \{0, 1, \cdots, 2^m - 1\}$$

The subset $V_{m,\ell} = \{\ell\} \times \{0, 1, \cdots, 2^m - 1\}$ of V_m $(0 \le \ell \le m)$ is called the ℓth *level* of $F(m)$; vertices in $V_{m,0}$ are called *inputs*, and vertices in $V_{m,m}$ are called *outputs* (in deference to the algorithmic origin of the graph). The edges of $F(m)$ form *butterflies* (or, copies of the complete bipartite graph $K_{2,2}$) between consecutive levels of vertices. Each butterfly connects vertices

$$\langle \ell, \alpha 2^{\ell+1} + \beta \rangle \text{ and } \langle \ell, \alpha 2^{\ell+1} + \beta + 2^\ell \rangle$$

on level ℓ of $F(m)$ $(0 \le \ell < m;\ 0 \le \alpha < 2^{m-\ell-1};\ 0 \le \beta < 2^\ell)$ with vertices

$$\langle \ell+1, \alpha 2^{\ell+1} + \beta \rangle \text{ and } \langle \ell+1, \alpha 2^{\ell+1} + \beta + 2^\ell \rangle$$

on level $\ell + 1$. One can view $F(m)$, $m \ge 2$ ($F(1) = K_{2,2}$ being given), as being constructed inductively, by taking two copies of $F(m-1)$ and 2^m new output vertices, and constructing butterflies connecting the kth outputs of each copy of $F(m-1)$, on the one side, to the kth and $(k + 2^{m-1})^{\text{th}}$ new outputs, on the other side. Thus, $F(m)$ has $(m+1)2^m$ vertices and $m2^{m+1}$ edges.

– FFT graphs are paradigmatic for convolution-based algorithms since they reflect the data-dependency structure of the 2^m-input FFT algorithm – cf. [1, Ch. 7] – and related convolution-based algorithms.

• *rectangular grids:* (cf. Fig. 7.3)

– The $d_1 \times d_2 \times \cdots \times d_k$ *rectangular grid* has vertices

$$\{1, \cdots, d_1\} \times \{1, \cdots, d_2\} \times \cdots \times \{1, \cdots, d_k\}$$

and edges connecting each vertex of the form $\langle x_1, \ldots, x_i, \ldots, x_k \rangle$ with vertices $\langle x_1, \ldots, x_i \pm 1, \ldots, x_k \rangle$. Each d_i is called a *dimension* of the grid.

— A large family of numerical algorithms naturally have data dependencies that are based on the rectangular grid [12].

Other families of graphs can be identified by perusing the literature. Just one more example are the so-called *refinable rectangular grids* [15], which approximate the structure of finite-element algorithms.

We now present a case study to illustrate the application and implementation of our evaluation proposal. We focus here on one particular popular architecture, namely, the (Boolean) Hypercube. This architecture has been implemented commercially by BBN, Intel, N-cube, and Thinking Machines.

- Let d be a nonnegative integer. The *d-dimensional (Boolean) Hypercube $C(d)$* is the graph whose vertices are all binary strings of length d and whose edges connect each string-vertex x with the d strings that differ from x in precisely one bit (position); cf. Fig. 7.4. Thus, $C(d)$ has 2^d vertices and $d2^{d-1}$ edges.

7.2 Optimal Embeddings in the Hypercube

7.2.1 A Useful Sample Embedding

In order to prepare the reader for the more complicated embeddings in subsequent sections, we present here a simple embedding of the complete binary tree $T(d)$ in the Hypercube $C(d)$. This embedding, which comes from [3], has DILATION 2, LOAD-FACTOR 2, and EXPANSION 1. The embedding is "computed" in three steps.

1. Perform an *inorder* traversal of $T(d)$, numbering the nodes from 0 to $2^d - 2$ as one goes; cf. Fig. 7.5a. Recall that inorder traversal:

   - visits the left subtree in inorder;
   - visits the root;
   - visits the right subtree in inorder.

2. Convert the inorder numbering from Step 1 to a string-labelling by converting each number to its length-d binary representation (padding with leading 0's where necessary); cf. Fig. 7.5b.

3. Assign each vertex of $T(d)$ to the vertex of $C(d)$ indicated by the tree-vertex's string-label.

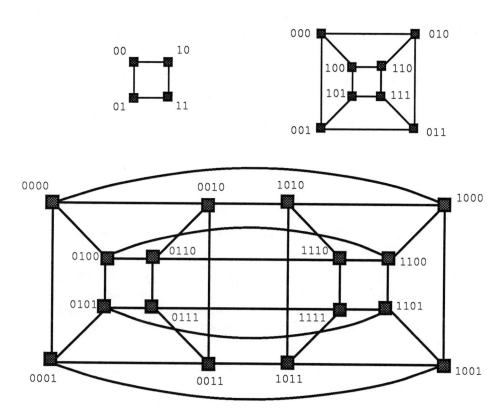

Figure 7.4: The Boolean Hypercubes C(2), C(3), and C(4)

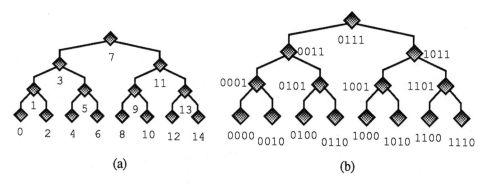

Figure 7.5: The (a) inorder numbering and (b) inorder labelling of the complete binary tree

It is obvious that the proposed mapping is one-to-one, hence an embedding of $T(d)$ in $C(d)$, and that the embedding has unit EXPANSION. To show that the embedding has DILATION 2 and LOAD-FACTOR 2, one can verify easily that the left and right children of the level-ℓ vertex

$$\beta = \beta_0 \beta_1 \cdots \beta_\ell \beta_{\ell+1} \cdots \beta_{d-1}$$

of $T(d)$ (the left child being visited before β in the inorder traversal and the right child being visited after β) are, respectively,

$$\beta_0 \beta_1 \cdots \beta_\ell \overline{\beta}_{\ell+1} \cdots \beta_{d-1}$$

and

$$\beta_0 \beta_1 \cdots \overline{\beta}_\ell \overline{\beta}_{\ell+1} \cdots \beta_{d-1}$$

Since adjacencies in the Hypercube flip single bits, our claims about DILATION and LOAD-FACTOR follow.

For this embedding problem, optimal (i.e., unit) DILATION and EXPANSION are not simultaneously accessible. To wit, $T(d)$ and $C(d)$ are both bipartite (i.e., composed of red and blue vertices, with all edges connecting a red vertex with a blue one), but the ratios between the numbers of red and blue vertices are very different for the two graphs.

The illustrated embedding itself will be useful later. But there are two other lessons to learn from this example:

1. Embeddings in Hypercubes can be specified via labellings with binary strings.

2. One must often settle for nearly optimal embeddings if one wants to consider more than one cost measure.

7.2.2   Divide-and-Conquer Algorithms

This section is devoted to a result by Bhatt, Chung, Leighton, and Rosenberg [3] that shows that the divide-and-conquer control structure can be implemented on the Hypercube architecture efficiently. This assertion translates in our framework to the following theorem.

Theorem 2. *Every binary tree can be embedded in a Boolean hypercube with constant DILATION, constant LOAD-FACTOR, and constant EXPANSION.*

Proof Strategy: The theorem is proved in three stages, using two auxiliary graphs, the bucket-tree and the thistle-tree.

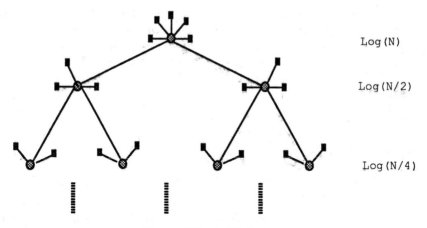

Log (N)

Log (N/2)

Log (N/4)

Figure 7.6: A thistle tree

An *N-vertex bucket-tree* is an N-vertex complete binary tree whose vertices are
super-nodes: there is a constant c such that each vertex at level ℓ of the bucket-tree
can, in an embedding, hold as many as

$$6 \log \frac{N}{2^\ell} + 18$$

vertices of the guest graph G.

A *height-h thistle-tree* is obtained from a height-h complete binary tree by appending
(via edges) to each level-ℓ vertex of the complete tree $h - \ell$ new leaf vertices called
thistles; cf. Fig. 7.6. The thistle-tree vertices that are inherited from the underlying
complete binary tree are called its *primary* vertices.

The three stages of the embedding of a given binary tree T in a Hypercube are:

1. Embed the given binary tree T in a bucket-tree.

2. Embed the resulting bucket-tree in a thistle-tree.

3. Embed the resulting thistle-tree in a Hypercube.

A. Embed a Binary Tree T in a Bucket-Tree

Lemma 1. *An N-node binary tree can be embedded with DILATION 3 in an N-
node bucket-tree.*

Proof Sketch.

1. Recursively node-bisect the binary tree T, thereby implicitly creating a decomposition tree for T: the root of the decomposition tree is T; the children of the root are the subgraphs T_1 and T_2 of T created by the first bisection; the children of the vertex representing T_i are the subgraphs T_{i1} and T_{i2} created by bisecting T_i; and so on.

2. Make the decomposition tree into a bucket-tree by converting each vertex into a super-node with the appropriate capacity.

3. Proceeding down the bucket-tree:

   - Place the bisector nodes from each bisection in the corresponding bucket (i.e., at the corresponding super-node).

   - Also place at that super-node any nodes of T that must be placed there in order to maintain the desired bound on DILATION.

In [3] it is shown how to accomplish Step 3 with super-nodes of the indicated sizes, using the *multi-color separator theorem for trees* [5]:

Theorem 3. *Let T be an N-node binary tree, with N_i nodes of color i*

$$1 \le i \le K; \quad \sum N_i = N.$$

By removing $\le K \log N$ nodes, one can bisect T so that, for each $1 \le i \le K$, the two halves have equally many i-colored nodes.

B. Embed a Bucket-Tree in a Thistle-Tree

Lemma 2. *Every bucket-tree can be "realized" with DILATION $O(1)$ by a thistle-tree.*

Proof Sketch: Distribute the contents of a given super-node of the bucket-tree among the thistles at the corresponding vertex of the thistle-tree. This places multiple tree-vertices at the higher-level vertices of the thistle-tree, with a corresponding deficit at the lower-level vertices. In [3] it is shown how to "push" the excess vertices down a few levels without increasing DILATION more than a bounded amount.

As an immediate corollary we have:

Lemma 3. *Any N-vertex binary tree can be embedded with DILATION $O(1)$ in an N-vertex thistle-tree.*

C. Embed a Given Thistle-Tree in a Hypercube

We embed a thistle-tree with $N = 2^n - 1$ primary vertices in the Hypercube $C(n+1)$ in two stages.

1. First, we take the n-vertex complete binary tree underlying the thistle-tree, and inorder-label it with binary strings, just as we did in the previous section.

Note that for each vertex v of the complete binary tree, the string labelling v differs in exactly one bit-position from each of the strings labelling the vertices along the right spine of the subtree rooted at v's left child.

2. Map the primary vertices of the thistle-tree on the corresponding vertices of the complete binary tree. Map the thistles attached to each vertex v in any manner on the vertices along the right spine of the subtree rooted at v's left child.

The described mapping clearly has DILATION 2, but it is not an embedding, since it places two vertices of the thistle-tree at each image vertex of the Hypercube. Our final step fixes this problem, at the cost of increasing EXPANSION.

3. Take a second copy of $C(n)$ (which is connected to $C(n)$ by a matching). Use each vertex of this shadow-Hypercube to hold the thistle-vertex that resides at the corresponding vertex of the original copy of $C(n)$.

By using this shadow Hypercube, we have thus embedded a thistle-tree in a Hypercube with DILATION 2 and EXPANSION 2.

Composing the mappings in subsections A, B, and C completes the proof concerning DILATION and EXPANSION. A more detailed analysis of the steps of the proof establishes the claim about LOAD-FACTOR also.

7.3 Convolution-Based Algorithms

The Hypercube is a very congenial host for convolution-based algorithms also, as the following result of Heath and Rosenberg [10] indicates.

Theorem 4. *Every FFT graph is a subgraph of the smallest Boolean hypercube that will hold it; this gives an embedding with simultaneous optimal DILATION, optimal LOAD-FACTOR, and optimal EXPANSION.*

Proof Strategy. Let us focus on embedding the FFT graph $F(m) = (V_m, E_m)$ in the Hypercube $C(m + \lceil \log_2(m + 1) \rceil)$. We specify the desired embedding by describing two labelling schemes.

- We assign each vertex $v \in V_m$ a unique d-bit label $L(v)$ (so that the labelling specifies an embedding).

- We assign each level ℓ of $F(m)$ a *bit-position pair* (a_ℓ, b_ℓ) so as to satisfy the following: For each edge (u, v), where u is at level ℓ of $F(m)$ and v is at level $\ell + 1$, the labels $L(u)$ and $L(v)$ differ either in bit-position a_ℓ or in bit-position b_ℓ.

For each $i \in \{1, 2, \cdots, m\}$, there is a *bit-pair* (a_i, b_i) of bit-positions that are used for assignments to edges between levels $i - 1$ and i of $F(m)$; i.e., all such butterfly edges "flip" the same pair of bits.

One verifies by induction that when we assign the d-bit label $L(v)$ to any single vertex v of $F(m)$, the labels of all remaining vertices can be specified uniquely by specifying the *levelled bit-pair sequence (LBPS)* $S = (a_1, b_1), (a_2, b_2), \ldots, (a_m, b_m)$. We can, therefore, specify the desired embedding by labelling some input of $F(m)$ with the string $00 \cdots 0$ and using an appropriate LBPS to specify the labelling of the remaining vertices.

It remains to select an LBPS so that the labelling L is injective. To this end, call the LBPS $S = (a_1, b_1), (a_2, b_2), \ldots, (a_m, b_m)$ *proper* if, for each i, at least one of a_i or b_i does not occur in $\{a_j, b_j \; : \; j < i\}$, i.e., at an earlier level; call an a_i or b_i satisfying this condition *new*. We shall find a proper LBPS to effect the desired labelling by visiting $F(m)$ level by level.

The advantage to using a new bit-position in our labelling is, of course, that labels obtained by flipping that bit-position are guaranteed never to have appeared at earlier levels. Since $F(m)$ has $(m + 1)2^m$ vertices, it is clear that we can always use one new bit-position at every level. Indeed, when the level we are visiting is a power of 2, then we can use two new bit-positions. Our problem reduces, therefore, to deciding how to choose the not-new bit-position at each level of $F(m)$ that is not a power of 2. We accomplish this by partitioning the levels into blocks between adjacent powers of 2. Assuming with no loss of generality that the a-position at each level is always new, we choose the b-position as follows.

- For $1 \leq i \leq 2^{k-1} - 1$, we set $b_{2^k + i} = a_{2^{k-1} + i}$.

- For $i = 2^{k-1}$, we set $b_{2^k + i} = a_{2^k - 1}$.

- For $1 \leq i \leq 2^{k-1} - 1$, we set $b_{2^k + 2^{k-1} + i} = b_{2^k + i}$.

The proof that the specified labelling is injective proceeds in two steps. First one shows that if the labelling is not injective, then it must fail to be so on the height-m complete binary tree rooted at vertex $00\cdots0$ (whose leaves are the level-m vertices of $F(m)$). Then one shows that if there are any repeated labels in this tree, then there must be an instance of repeated labels the first few levels of the tree. Finally, one verifies directly that such repetitions do not occur early in the tree.

7.4 Grid-Based Algorithms

Our treatment of rectangular grids is restricted to the important subcase wherein each dimension is a power of 2. This subcase occurs commonly since it often entails simplified programming.

For this special case, the following result is easily established.

Theorem 5. *For arbitrary integers $d_1, d_2, \ldots, d_k$, the $2^{d_1} \times 2^{d_2} \times \cdots \times 2^{d_k}$ grid is a subgraph of the Boolean Hypercube $C(d_1 + d_2 + \cdots + d_k)$; this gives an embedding with simultaneous optimal DILATION, optimal LOAD-FACTOR, and optimal EXPANSION.*

One proves the theorem easily by noting that if one folds the grid in half along any dimension, the edges connecting the two halves form a matching; but the same is true if one focusses on any one dimension of the Hypercube. Thus, successively folding the grid in half, and dedicating one dimension of the Hypercube to each fold, yields the sought embedding immediately; cf. Fig. 7.7.

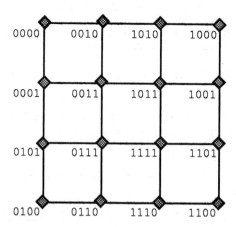

Figure 7.7: Ilustrating the folding of a rectangular grid and its embedding in the hypercube

7.5 Concluding Remarks

Despite all of its shortcomings, the proposed validation scheme has certain advantages, such as precision, rigor, and generality, not shared by the competitors that have thus far found their way into print. It should, therefore, be considered one useful implement in the toolbox of the architectural designer.

Acknowledgements

It goes without saying that this paper owes its existence to my friends and collaborators Sandeep Bhatt, Fan Chung, Lenny Heath, and Tom Leighton.

References

[1] A.V. Aho, J.E. Hopcroft, J.D. Ullman (1974): *The Design and Analysis of Computer Algorithms.* Addison-Wesley, Reading, MA.

[2] F. Berman and L. Snyder (1984): On mapping parallel algorithms into parallel architectures. *Intl. Conf. on Parallel Processing.*

[3] S.N. Bhatt, F.R.K. Chung, F.T. Leighton, A.L. Rosenberg (1986): Optimal simulations of tree machines. *27th IEEE Symp. on Foundations of Computer Science*, 274-282.

[4] S.N. Bhatt and F.T. Leighton (1984): A framework for solving VLSI graph layout problems. *J. Comp. Syst. Sci. 28*, 300-343.

[5] N. Blum (1983): An area-maximum edge length tradeoff for VLSI layout. *16th ACM Symp. on Theory of Computing*, 92-97.

[6] S.H. Bokhari (1981): On the mapping problem. *IEEE Trans. Comp., C-30*, 207-214.

[7] F.R.K. Chung, F.T. Leighton, A.L. Rosenberg (1983): DIOGENES – A methodology for designing fault-tolerant processor arrays. *13th Intl. Conf. on Fault-Tolerant Computing*, 26-32.

[8] J.W. Greene and A. El Gamal (1984): Configuration of VLSI arrays in the presence of defects. *J. ACM 31*, 694-717.

[9] J. Hastad, F.T. Leighton, M. Newman (1986): Reconfiguring a hypercube in the presence of faults. Typescript, MIT.

[10] L.S. Heath and A.L. Rosenberg (1987): An optimal mapping of the FFT algorithm onto the Hypercube architecture. Tech. Rpt., Univ. of Massachusetts; submitted for publication.

[11] J.-W. Hong, K. Mehlhorn, A.L. Rosenberg (1983): Cost tradeoffs in graph embeddings. *J. ACM 30*, 709-728.

[12] L. Johnsson (1985): Basic linear algebra computations on hypercube architectures. Tech. Rpt., Yale Univ.

[13] F.T. Leighton (1984): Parallel computation using meshes of trees. *1983 Workshop on Graph-Theoretic Concepts in Computer Science*, Trauner Verlag, Linz, pp. 200-218.

[14] F.T. Leighton and C.E. Leiserson (1985): Wafer-scale integration of systolic arrays. *IEEE Trans. Comp., C-34*, 448-461.

[15] R.J. Lipton, A.L. Rosenberg, A.C. Yao (1980): External hashing schemes for collections of data structures. *J. ACM 27*, 81-95.

[16] F.P. Preparata and J.E. Vuillemin (1981): The cube-connected cycles: a versatile network for parallel computation. *C. ACM 24*, 300-309.

[17] A.L. Rosenberg (1981): Issues in the study of graph embeddings. In *Graph-Theoretic Concepts in Computer Science: Proceedings of the International Workshop WG80*, Bad Honnef, Germany (H. Noltemeier, ed.) *Lecture Notes in Computer Science 100*, Springer-Verlag, New York 150-176.

[18] A.L. Rosenberg (1983): The Diogenes approach to testable fault-tolerant arrays of processors. *IEEE Trans. Comp., C-32*, 902-910.

[19] L. Snyder (1981): Overview of the CHiP computer. In *VLSI 81: Very Large Scale Integration* (J.P. Gray, ed.) Academic Press, London, pp. 237-246.

Chapter 8

Fast Parallel Algorithms for Reducible Flow Graphs [1]

Vijaya Ramachandran [2]

Abstract

We give parallel NC algorithms for recognizing reducible flow graphs (rfg's), and for finding dominators, minimum feedback vertex sets, and a depth first search numbering in an rfg. We show that finding a minimum feedback vertex set in vertex-weighted rfg's or finding a minimum feedback arc set in arc-weighted rfg's is P-complete. We present RNC algorithms for finding a minimum feedback arc set in an unweighted rfg, and for finding a minimum weight feedback set when arc or vertex weights are in unary; and we show that these problems are in NC if and only if the maximum matching problem is in NC.

8.1   Introduction

Reducible flow graphs (rfg's) are graphs that model the control structure of computer programs. They are used extensively in problems on code optimization and global data flow analysis. Several linear time sequential algorithms for these graphs are known, including algorithms for recognizing rfg's [Ta1, GaTa], for finding *dominators* [Ha] and for finding a *minimum feedback vertex set* (FVS) [Sh]. The basis

---

[1] Funded by National Foundation Grant ECS 8404866, the Semiconductor Research Corporation Grant 86-12-109 and the Joint Services Electronics Program Grant N00014-84-C-0149.

[2] University of Illinois, Urbana, IL

for all of these fast sequential algorithms is a depth first search on the input graph. Recently, polynomial-time algorithms for finding a minimum *weight* FVS in vertex-weighted rfg's and a minimum *feedback arc set (FAS)* in arc-weighted or unweighted rfg's have been developed [Ra1]. These algorithms make extensive use of algorithms for network flow [FoFu, La, PaSt, Ta2, GoTa]. It is also known that the sequential complexity of these latter problems is at least that of finding a minimum cut in a flow network [Ra1, Ra2].

In this paper we give parallel NC algorithms for recognizing rfg's, for finding dominators, and for finding a minimum FVS in an unweighted rfg. We also give an NC algorithm for finding a depth first search (DFS) numbering for an rfg; however, none of our other parallel algorithms make use of this DFS numbering.

We show that if arbitrary weights are allowed, the weighted FAS and FVS problems on rfg's are both P-complete. Hence fast parallel algorithms for these problems appear unlikely to exist. For the case when the weights are in unary, we present an RNC algorithm for the FAS problem on rfg's. We also give NC reductions between the weighted FAS problem, the unweighted FAS problem, the weighted FVS problem and the problem of finding a minimum cut in a flow network (when weights and capacities are in unary). Thus if any one of these problems is in NC, then all of them would be in NC. In particular, an NC algorithm for the maximum matching problem would give NC algorithms for these three problems on rfg's, and an NC algorithm for any one of these three problems would, in turn, give an NC algorithm for maximum matching.

This paper is organized as follows. In section 2 we provide the necessary background on parallel computation. In section 3, we give graph-theoretic definitions and basic properties of rfg's that we will use in our algorithms. Finally in section 4 we present our parallel algorithms.

8.2   Parallel Model of Computation

The parallel model of computation that we will use is the *PRAM (Parallel Random Access Machine)*. This consists of a collection of sequential RAM's [AhHoUl] computing synchronously through a shared memory (figure 8.1). At each step of the computation, each RAM (or *processor*) can read a location in shared memory, perform an unit-time local operation, and write into a location in memory. On an input of size n, the parallel computation time $t(n)$ is the number of steps needed to compute the result. The number of processors participating in the computation is $p(n)$.

For sequential computation, sequential time is the undisputed primary complexity measure, and polynomial time (or the class P) is universally acknowledged as characterizing feasible sequential computation. Analogously for parallel computation

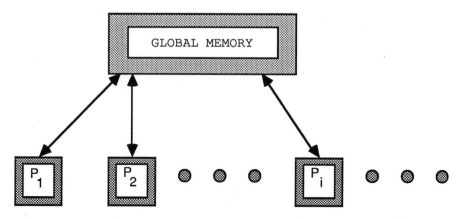

Figure 8.1: The PRAM model.

parallel time is clearly a key complexity measure. However, the amount of hardware used is also of importance since achieving small parallel time at the expense of very large amount of hardware does not seem desirable. Thus a simultaneous bound on parallel time and hardware appears to be the appropriate primary complexity measure for parallel computation [Co], and the class of feasible parallel computation, denoted by NC, is characterized by simultaneous bounds on parallel time and hardware (as defined below).

There are several different models of parallel computation, such as uniform families of boolean circuits, alternating turing machines, vector machines and the PRAM [Co, KarRa]. Fortunately, the class NC remains the same for all of these classes and represents the class of problems computable in *polylog* time using a polynomial amount of hardware in the size of the input; here, polylog(n) is defined to be $O(\log^k(n))$ for some fixed k. Thus with reference to PRAM computation NC is the class of problems computable in polylog(n) parallel time using poly(n) processors, where poly(n) is a fixed polynomial in n, and n is the size of the input. We use the PRAM model of parallel computation because it is the one most suitable for specifying complex algorithms.

PRAMs can be classified according to restrictions on global memory access. An EREW PRAM is a PRAM for which simultaneous access to any memory location by different processors is forbidden either for reading or for writing. In a CREW PRAM simultaneous reads are allowed but no simultaneous writes. A CRCW PRAM allows simultaneous reads and writes. In this case we have to specify how to resolve write conflicts. Some commonly used methods of resolving write conflicts are:

- All processors writing into the same location write the same value (COMMON model);

- Any one processor participating in a common write succeeds, and the algorithm should work correctly regardless of which one succeeds (the ARBITRARY model);

- There is a linear ordering on the processors, and the minimum numbered processor writes its value in a concurrent write (MINIMUM model).

Even though there is a variety of PRAM models, they do not differ very widely in their computational power. First, it should be clear that any algorithm in the COMMON model is also an algorithm in the ARBITRARY model, and similarly, any algorithm in the ARBITRARY model is an algorithm in the MINIMUM model. Further, it is not difficult to establish that any algorithm for the MINIMUM model can be simulated in the COMMON model with no loss in parallel time, and with only a squaring in the number of processors [Ku], and also by an EREW PRAM with the same number of processors and with the parallel time increased by only a factor of $O(\log p(n))$, where $p(n)$ is the number of processors ([Vi], in conjunction with [AjKoSz, Co]). The algorithms we develop in this paper are either for the COMMON PRAM or for the CREW PRAM. Using the relations described above, they can be transformed into algorithms for any of the other PRAM models.

From the definition of NC it should be clear that NC $\subseteq$ P. It is not known if NC=P. However, we can identify certain problems in P that are the hardest problems to parallelize in the sense that, if any one of these problems is in NC, then P=NC. These are the *P-complete* problems, and a problem is P-complete if every problem in P is log space reducible to it [HoUl, KarRa]. Examples of P-complete problems are the circuit value problem [La], linear programming [DoLiRe] and the maximum flow value problem with arc capacities in binary [GoShSt].

Once we have placed a problem in the parallel class NC, we would then like to obtain as efficient an algorithm as possible for it, i.e., an algorithm that runs in polylog time with as few processors as possible, and preferably with $s(n)$ processors, where $s(n)$ is the running time of the best sequential algorithm for the problem. For undirected graphs there are algorithms known for several problems that run in polylog time using a linear number of processors on a PRAM; these problems include connectivity [HiChSa, ShVi], biconnectivity [ChLaCh, TaVi], triconnectivity [MiRa], four-connectivity (assuming adjacency matrix representation) [KanRa], $s-t$ numbering [MaScVi], planarity [KlRe], etc. For directed graphs unfortunately, such processor efficient algorithms are not known mainly due to the *transitive closure bottleneck*. The best parallel method known at present to test reachability from one vertex to another in a directed graph is to find the transitive closure of the adjacency matrix of the graph. To compute this in polylog time requires n^α processors (to within a polylog factor), where α is the *matrix multiplication exponent*, which is currently 2.375 [St, CoWi] (but for practical computations should be taken as 3). Thus, since rfg's are directed graphs, all of the algorithms we present in this paper are affected by the transitive closure bottleneck, and use more than a linear number of processors.

Some problems that are not known to be in NC and are not known to be P-complete are however known to be in the class *RNC* or *Random NC*. RNC is the class of problems that have algorithms that run in polylog time with a polynomial number of processors with probability $1 - \epsilon$, for any fixed $\epsilon > 0$. Problems known to be in RNC (but not known to be in NC) include finding a maximum weight matching in a graph with edge weights in unary [KaUpWi, MuVaVa], finding a maximum flow in a network with arc capacities in unary and finding a depth first search tree in an undirected graph [AgAn].

The algorithms we develop in this paper make use of some well-known basic parallel algorithms as subroutines. We conclude this section with a brief review of these algorithms.

1. *Boolean matrix multiplication and transitive closure:* The standard matrix multiplication algorithm can be parallelized to give a constant time algorithm using n^3 processors on a COMMON PRAM to find the product of two $n \times n$ boolean matrices. Since $(I + B)^m$, for $m \geq n$, gives the transitive closure B^* of an $n \times n$ boolean matrix B, B^* can be computed using $\log n$ stages of boolean matrix multiplication by repeated squaring. The more sophisticated matrix multiplication algorithms lend themselves to parallelization in a similar manner. Thus B^* can be obtained in $O(\log n)$ time using $M(n) = O(n^\alpha)$ processors on a COMMON PRAM.

2. *Prefix sums:* Let $+$ be an associative operation over a domain D. Given an ordered list $< x_1, \ldots, x_n >$ of n elements from D, the prefix problem is to compute the $n - 1$ prefix sums $S_i = \sum_{j=1}^{i} x_i, i = 1, \ldots, n$. This problem has several applications. For example, consider the problem of compacting a sparse array, i.e., we are given an array of n elements, many of which are zero, and we wish to generate a new array containing the nonzero elements in their original order. We can compute the position of each nonzero element in the new array by assigning value 1 to the nonzero elements, and computing prefix sums with $+$ operating as regular addition.

The n element prefix sums problem can be computed in $O(\log n)$ time using n processors on a CREW PRAM (see e.g., [LaFi]), assuming unit time for a single $+$ operation.

3. *List Ranking:* This is a generalization of prefix sums, in which the ordered list is given in the form of a linked list rather than an array. The term list ranking is usually applied to the special case of this problem in which all elements have value 1, and $+$ stands for regular addition (and hence the result of the prefix sums computation is to obtain, for each element, the number of elements ahead of it in the list, i.e., its rank in the list); however the technique easily adapts to our generalization. List ranking on n elements can be computed in $O(\log n)$ time using n processors on a CREW PRAM (see e.g., [Vi2]).

4. *Tree contraction:* There are several applications that require computation on a rooted tree (see next section for a definition of a rooted tree); one such application

is the evaluation of an arithmetic expression. A standard sequential method of performing this evaluation is to compute from the leaves upward. However, if the tree is highly imbalanced, i.e., its height is large in relation to its size, then this method performs poorly in parallel.

Tree contraction is a method of evaluating tree functions efficiently in parallel. The method transforms the input tree using two operations Rake and Compress. The operation Rake removes leaves from the tree. The operation Compress halves the lengths of chains in the tree (a chain is a sequence of vertices with exactly one incoming and outgoing arc in the tree) by removing alternate nodes in the chain and linking each remaining node to the parent of its parent in the original tree. The Contract operation is one application of Rake followed by one application of Compress. It can be shown [MiRe] that $O(\log n)$ applications of the Contract operation to an n node tree are sufficient to transform the tree into a single vertex. This contraction can be done in $O(\log n)$ time with n processors on a CREW PRAM.

By associating appropriate computation with the Rake and Compress operations, we can evaluate the tree function during tree contraction. For instance, for expression evaluation, the Rake operation will compute the value at the parents of leaves, and the Compress operation will do partial computation corresponding to the computation associated with the deleted nodes.

Some of the algorithms we will present in this paper will use a modified tree contraction method. We should mention that for certain simple tree functions, such as computing preorder or postorder numbering, computing the number of descendants of each vertex, etc., a simpler technique, based on list ranking, called the *Euler tour technique on trees* [TaVi] can be used.

8.3   Graph-Theoretic Definitions and Properties of Rfg's

A *directed graph* $G=(V,A)$ consists of a finite set of vertices (or nodes) V and a set of arcs A which is a subset of $V \times V$. An arc $a = (v_1, v_2)$ is an *incoming* arc to v_2 and an *outgoing* arc from v_1. Vertex v_1 is a *predecessor* of v_2 and v_2 is a *successor* of v_1; v_1 is the *tail* of a and v_2 is its *head*. Given a directed graph $G = (V, A)$ and a set of arcs C, we will sometimes use the notation $C \cap G$ to denote the set $C \cap A$. An *arc-weighted (vertex-weighted) directed graph* is a directed graph with a real value on each arc (vertex).

A *directed path* p in G from vertex u to vertex v is a sequence of arcs $a_1, \ldots, a_r$ in A such that $a_i = (w_i, w_i + 1), i = 1, \ldots, r$ with $w_1 = u$ and $w_{r+1} = v$. The path p *passes through* each $w_i, i = 1, \ldots, r + 1$. A directed path p from u to v is a *cycle* if $u = v$. A directed graph G is *acyclic* if there is no cycle in G. A *rooted directed graph* or a *flow graph* $G = (V, A, r)$ is a directed graph with a distinguished vertex r such that there is a directed path in G from r to every vertex v in $V - r$.

A *rooted tree* is a flow graph $T = (V, A, r)$ in which every vertex in $V - \{r\}$ has exactly one incoming arc. If (u, v) is the unique incoming arc to v then u is the *parent* of v, and v is a *child* of u. A *leaf* is a vertex in a tree with no outgoing arc. The *height* of a vertex v in a tree is the length of a longest path from v to a leaf. The *height of a tree* is the height of its root. A *forest* is a collection of trees.

Let $G = (V, A, r)$ be a rooted DAG. A vertex u is a descendant of vertex v if either $u = v$ or there is a directed path from v to u in G. The vertex u is a *proper* descendant of v if $u \neq v$ and u is a descendant of v.

Let $G = (V, A)$ be an arc-weighted directed graph. A set $F \subseteq A$ is a *feedback arc set (FAS)* for G if $G' = (V, A - F)$ is acyclic. The set F is a *minimum FAS* if the sum of the weights of arcs in F is minimum. Analogous definitions hold for a *feedback vertex set*.

Let $G = (V, A)$ be a directed graph, and let $V' \subseteq V$. The *subgraph of G induced by V'* is the graph $G_s(V') = (V', A')$, where $A' = A \cap V' \times V'$. The graph $G - V'$ is the subgraph of G induced on $V - V'$.

A *reducible (flow) graph* (or *rfg*) is a rooted directed graph for which the rooted depth first search DAG [Ta2] is unique. Thus, the arcs in a reducible graph can be partitioned in a unique way into two sets as the *DAG* or *forward* arcs and the *back* arcs.

An alternate definition of a reducible graph (due to [HeUl]) is stated below.

Definition 3.1 [HeUl] Let $G = (V, A, r)$ be a flow graph. We define two transformations on G :

> Transformation T_1 : Given an arc $a = (v, v)$ in A remove a from A.

> Transformation T_2 : Let v_2 be a vertex in $V - \{r\}$ and let it have a single incoming arc $a = (v_1, v_2)$. T_2 replaces v_1, v_2 and a by a single vertex v. Predecessors of v_1 become predecessors of v. Successors of v_1 and v_2 become successors of v. There is an arc (v, v) if and only if there was formerly an arc (v_2, v_1) or (v_1, v_1).

G is a *reducible flow graph (rfg)* if repeated applications of T_1 and T_2 (in any order) reduce G to a single vertex.

Another characterization of rfg's based on a forbidden subgraph is given in Lemma 3.1.

Lemma 3.1: [HeUl] Let (*) denote any of the flow graphs represented in figure 8.2, where the wiggly lines denote node disjoint (except for endpoints) paths; nodes a, b, c and r are distinct, except that a and r may be the same.

A flow graph is not an rfg if and only if it contains a (*)-subgraph.

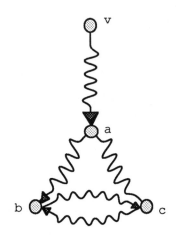

Figure 8.2: The forbidden (*)-subgraph.

In the following we assume that the vertices of G are numbered in postorder (with respect to a depth first search rooted at r). Let $G = (V, A, r)$ be a reducible graph and let $b = (u, v)$ be a back arc in G. Then b *spans* vertex w (or w is in the *span* of b) if there exists a path from v to u in the DAG of G that passes through w. Given two vertices $u, v \in V$, vertex u *dominates* vertex v if every path from r to v passes through u (note that u dominates itself).

Lemma 3.2 [HeUl] Let $b_1 = (u_1, v_1)$ and $b_2 = (u_2, v_2)$ be two back arcs in a reducible graph G that span a common vertex w. Then either v_1 dominates v_2 or v_2 dominates v_1 : if $v_1 \geq v_2$ in postorder numbering, then v_1 dominates v_2, otherwise v_2 dominates v_1.

Lemma 3.3 [HeUl] Let $G = (V, A, r)$ be a reducible graph and let $b = (u, v)$ be a back arc in G. Then v dominates every vertex w in the span of b.

It is well-known [AhUl] that the dominator relation can be represented in the form of a tree rooted at r, the root of the flow graph G. This tree is called the *dominator tree* T of G. The descendents of a vertex v in T are the vertices dominated by v in G. A vertex v' is *immediately dominated* by v if it is a child of v in T.

Given a set $V' \subseteq V$ the *dominator forest* $F_{V'}$ for V' represents the dominator relation restricted to the set V'. Let $V_h = \{v \in V \mid v \text{ is the head of a back arc in } G\}$. Since r is the head of a back arc in G it is easy to see that F_{V_h} is a tree; we call it the *head dominator tree of* G and denote it by T_h. This tree can be constructed from T by applying transformation T_2 of Definition 3.1 to each vertex v in T that is not a head of a back arc in G.

8.4 Parallel Algorithms for RFG's

This section is divided into three subsections. In section 4.1 we present NC algorithms to test if a rooted directed graph is an rfg, to construct the head dominator tree for an rfg, and to find a DFS tree in an rfg; we also introduce a modified tree contraction method in this section. In section 4.2 we develop an NC algorithm to find a minimum FVS in an unweighted rfg. Finally in section 4.3 we consider finding a minimum FAS in an unweighted rfg and finding minimum weight feedback sets in weighted rfg's, and we present some RNC algorithms as well as P-completeness results.

8.4.1 NC Algorithms for Preprocessing a Flow Graph

Algorithm 1.1: Testing flow graph reducibility

Input: $G = (V, A, r)$ with adjacency matrix B.

1. Test if G is a flow graph, i.e., test if every vertex in $V - \{r\}$ is reachable from r.

   Form B^*, the transitive closure of B and check if every nondiagonal element in row r has a 1.

2. Partition the arcs of G into forward (f) and back (b) arcs, assuming G is an rfg.

   In parallel, for each vertex u, remove u from G and test, for each arc (v, u) in G, if v is reachable from r. If *not*, then mark (v, u) as b and v as a head, otherwise mark (v, u) as f.

3. Delete all arcs marked b and check if resulting graph G' is acyclic. G is an rfg if and only if G' is acyclic.

   Form transitive closure of the adjacency matrix B' of G' and check that for every $i < j \leq n$, one of the two entries in position (i, j) and position (j, i) in B'^* is zero.

Lemma 4.1 If G is an rfg, then the arcs marked f in step 2 are the forward arcs, and the arcs marked b in step 2 are the back arcs.

Proof If (v, u) is a forward arc, then v is reachable from r by a path that does not include u and hence v is reachable from r in $G - \{u\}$.

If (v, u) is a back arc, then u dominates v (by Lemma 3.3) and hence every path from r to v passes through u in G. Hence v is not reachable from r in $G - \{u\}$. □

Lemma 4.2 The graph G' in step 3 is acyclic if and only if G is an rfg.

Proof If G is an rfg, then by Lemma 3.1, the set of arcs marked b is precisely the set of back arcs in G and hence $G - \{a \in A \mid a$ is marked $b\}$ is acyclic.

Suppose G' is acyclic but G is not an rfg. Hence G has a (*)-subgraph (Lemma 3.1 and figure 8.2). We observe that for every arc (u, v) in this subgraph, u is reachable from r when v is deleted. Hence all of these arcs will be marked f in step 2. But this implies that G' contains a cycle. Hence G must be an rfg if G' is acyclic. □

Steps 1 and 3 take $O(\log n)$ time using $M(n)$ processors on a COMMON PRAM (recall that $M(n)$ is the number of processors required to multiply two $n \times n$ matrices in constant time on a COMMON PRAM). Step 2 requires $O(\log n)$ time using $n \cdot M(n)$ processors on a COMMON PRAM. Hence the complexity of this algorithm is $O(\log n)$ parallel time using $n \cdot M(n)$ processors.

Algorithm 1.2: Forming T_h, the head dominator tree for G

1. Use algorithm 1.1 to construct DAG G'.

2. Use the algorithm in [PaGoRa] to compute the dominator tree T for DAG G'.

3. Use tree contraction [MiRe] to extract the head dominator tree T_h from T.

On a COMMON PRAM, step 1 takes $O(\log n)$ time with $n \cdot M(n)$ processors, step 2 takes $O(\log n)$ time with $M(n)$ processors, and step 3 takes $O(\log n)$ time with $n + m$ processors. Hence step 1 dominates the complexity of this algorithm.

Algorithm 1.3: NC algorithm for finding a DFS numbering for an rfg $G = (V, A, r)$

1. Use algorithm 1.1 to construct the DAG G'.

2. Find a DFS tree in DAG G' as follows.

   i) Identify a vertex v with more than $n/2$ descendents for which every child has at most $n/2$ descendents:

   Find the transitive closure B'^* of B'. Determine the number of descendents of each vertex as the sum of the nondiagonal entries in its row in B'^*. Each arc in G' compares the number of descendents of its head with the number of descendents of its tail, and *marks* its tail if it is not the case that the head has at most $n/2$ descendents and the tail has more than $n/2$ descendents. The (unique) unmarked vertex is v.

ii) Find a path P from root r to v :

Find a directed spanning tree for G' by making each vertex with an incoming arc choose one such arc as its tree arc. Form P as the path in this tree from r to v.

iii) Associate each descendent v' of v with the largest numbered child of v (numbering according to some fixed order) from which it is reachable; associate each vertex v' not reachable from v with the lowest vertex in path P from which it is reachable:

Use list ranking to number the vertices on P in increasing order from the root, followed by the children of v in some fixed order. Replace all nonzero entries in the columns of B'^* corresponding to these nodes by their new number. For each row, find the maximum numbered entry in that row, and identify it as the vertex with which the row vertex is to be associated.

iv) Recursively solve problem in subdags rooted at the newly numbered vertices, together with their descendants as computed in step iii.

Lemma 4.3 Let $G = (V, A, r)$ be a DAG, with $|V| = n$. There exists a unique vertex $u \in V$ with more than $n/2$ descendants for which every child has at most $n/2$ descendants.

Proof Straightforward, and is omitted.   ☐

Lemma 4.4 Algorithm 1.3 correctly finds a DFS tree in an rfg.

Proof We observe that the algorithm constructs a DFS tree consisting of the initial path P to v, followed by a DFS on the vertices reachable from the children of the largest numbered child of v, followed by vertices reachable from the second largest numbered child of v (but not reachable from the largest child of v),..., followed by vertices reachable from the smallest numbered child of v (but not reachable from larger numbered children of v), followed by vertices reachable from nodes on $P - \{v\}$ in reverse order of their occurrence on P. It is not difficult to see that this is a valid depth first search.   ☐

Step 2i takes $O(\log n)$ time using $M(n)$ processors on a COMMON PRAM. Step 2ii is very efficient: it takes constant time using a linear number of processors on a CREW PRAM. Step 2iii takes $O(\log n)$ time using n^2 processors on a CREW PRAM. Finally the recursive steps take $\log n$ stages since each new subproblem is at most half the size of the previous problem; further the sum of the sizes of the new problems is less than the size of the previous problem and hence the processor count is dominated by the first stage. Thus the algorithm takes $O(\log^2 n)$ using $M(n)$ processors on a COMMON PRAM.

Other NC algorithms for finding a DFST in a DAG are known [At, GhBh].

In the next two sections we present parallel algorithms to find minimum feedback sets in rfg's. Our algorithms require computation on the head dominator tree T_h of the input rfg $G = (V, A, r)$. For this we will use a variant of tree contraction [MiRe]. We conclude this section with a description of this modified tree contraction method.

Recall the a chain in a directed graph G is a path $< v_1, \ldots, v_k >$ such that each v_i has exactly one incoming arc and one outgoing arc in G. A *maximal chain* is one that cannot be extended. A *leaf chain* $< v_1, \ldots, v_{l-1}, v_l >$ in a rooted tree $T = (V, A, r)$ consists of a maximal chain $< v_1, \cdots, v_{l-1} >$, with v_l the unique child of v_{l-1}, and with v_l, a leaf.

The two tree operations we use in our modified tree contraction method are Rake and Shrink. As before, the Rake operation removes leaves from the tree. The Shrink operation shrinks each maximal leaf chain in the current tree into a single vertex.

Lemma 4.5 In the modified tree contraction method, $O(\log n)$ applications of Rake followed by Shrink, suffice to transform any n node tree into a single vertex.

Proof Consider another modified tree contraction algorithm in which the Shrink operation shrinks all maximal chains, including leaf chains, into a single vertex (one for each chain). This modification certainly requires no more steps than regular tree contraction, and hence by the result in [MiRe], transforms any n node tree into a single vertex in $O(\log n)$ time. But the number of applications of Rake followed by Shrink in the above modified tree contraction method is exactly the same as that in our modified tree contraction method, since the only difference is that a chain gets shrunk in several stages, rather than all at once.   ☐

In our algorithms for minimum feedback sets, we will associate appropriate computation with the Rake and Shrink operations in order to obtain the desired result.

8.4.2   NC Algorithm for Finding a Minimum FVS in an Unweighted Rfg

We first review the basic ideas in Shamir's polynomial time sequential algorithm [Sh]. Given an rfg $G = (V, A, r)$ together with a partial FVS S for G, a head v in G is *active* if there is a DAG path from v to a corresponding tail, which is not cut by vertices in S. A *maximal active head* v is an active head such that none of its proper DAG descendants in G is an active head.

The following theorem is established in [Sh].

Theorem 4.1 [Sh] Let $G = (V, A, r)$ be an rfg, and let S be a subset of a minimum FVS in G. If v is a maximal active head in G with respect to S, then $S \cup v$ is also a subset of a minimum FVS in G.

Theorem 4.1 gives us the following algorithm to construct a minimum FVS for an rfg. This algorithm can be viewed as an implementation of Algorithm A in [Sh] using the head dominator tree.

Minimum FVS Algorithm

Input: An rfg $G = (V, A, r)$ together with its head dominator tree T_h.

Output: A set $S \subseteq V$ which is a minimum FVS for G.

1. Initialize $S \leftarrow \phi$.

2. *Repeat*

   a) $S \leftarrow S \cup L$, where L is the set of leaves in T_h.

   b) $G \leftarrow G - L$, $T_h \leftarrow T_h - L$.

   c) Find U, the set of heads in current G that are not reachable from themselves.

   d) Remove all vertices in U from T_h, and for each vertex v not in U, find its closest proper ancestor w that is not in U, and make w the parent of v.

   until $T_h = \phi$.

We implement the above tree computations using our modified tree contraction method. The computation associated with a Rake step is exactly one application of step 2 of the above algorithm: we add the leaves of the current tree to S, and delete them from both G and T_h. We then test reachability from themselves for heads in the current G using transitive closure to obtain U. We then implement step 2d using regular tree contraction. The complexity of this computation is dominated by step 2c, and hence the implementation of the Rake step for finding minimum FVS takes $O(\log n)$ time using $M(n)$ processors on a COMMON PRAM.

Now consider the operation Shrink, which shrinks each maximal leaf chain into a single vertex. During the Shrink operation we will identify the vertices in these leaf chains that belong to S based on the following observation. Let $C = \langle v_1, \ldots, v_l \rangle$

be a leaf chain, where v_l is the leaf. Suppose v_i is in the minimum FVS S. Then the largest $j < i$ such that v_j is in S (if such a j exists) is immediately determined as the largest $j < i$ for which v_j is reachable from itself in $G - \{v_i\}$. This is because v_i dominates all $v_k, k > i$, hence any cycle extending above v_i which is cut by a vertex below v_i is certainly cut by v_i; thus v_j will become a maximal active head if v_i is added to S.

Our computation for the Shrink operation determines for each v_i in C, the largest $j < i$ (if it exists) such that v_j is in S if v_i is in S. Then for each i for which j exists, we place a pointer from v_j to v_i. This defines a forest F on $v_1, \cdots, v_l$. Since v_l is a leaf in T_h it belongs to S. Hence the vertices in C that belong to S are precisely those from which v_l is reachable in F. We identify these vertices using regular tree contraction and add them to S.

The most costly step in this computation is determining v_j for each v_i. It requires a transitive closure computation on each $G - \{v_i\}, i = 2, \ldots, l$. Hence the Shrink operation can be implemented in $O(\log n)$ time using $n \cdot M(n)$ processors on a COMMON PRAM.

By Lemma 4.3, $O(\log n)$ applications of Rake and Shrink operations suffice to contract T_h to a single vertex, and at this point we will have constructed a minimum FVS S. Thus we have presented a parallel algorithm to find a minimum FVS in an rfg in $O(\log^2 n)$ time using $n \cdot M(n)$ processors on a COMMON PRAM.

8.5   Finding a Minimum FAS in an Unweighted Rfg and Related Problems

We first state some definitions and results from [Ra1], which gives a polynomial time sequential algorithm for finding a minimum FAS in an rfg. We then give an RNC algorithm for this problem and related results.

A *flow network* $G = (V, A, s, t, C)$ is an arc-weighted directed graph with vertex set V and arc set A, where s and t are vertices in V called the *source* and *sink* respectively, and C is the *capacity function* on the arcs which specifies the arc weights, which are always nonnegative. The *maximum flow* problem asks for a *flow* of maximum value from s to t (see [Ev, FoFu, PaSt, Ta2] for definition of a flow). A *cut* C separating s and t is a set of arcs that breaks all paths from s to t. The *capacity* of C is the sum of the capacities of arcs in C. A *minimum cut* separating s and t is a cut of minimum capacity. It is well-known that the value of a maximum flow is equal to the value of a minimum cut [FoFu].

Let $G = (V, A, r)$ be an arc-weighted reducible graph and let v be the head of a back arc in G. Let $b_1 = (u_1, v_1), \ldots, b_r = (u_r, v_r)$ be the back arcs in G whose heads are dominated by v. The *dominated back arc vertex set of* v is the set

$V_v = \{v' \in V \mid v'$ lies on a DAG path from v to some $u_i, i = 1, \ldots, r\}$. It is easy to see that v dominates all vertices in V_v.

Definition 4.1 Let $G = (V, A, r)$ be an arc-weighted reducible graph with nonnegative arc weights, and let v be the head of a back arc in G; hence $G_s(V_v) = (V_v, A_v)$ is the subgraph of G induced by the dominated back arc vertex set of v. The *maximum flow network of G* with respect to head v is a flow network $G_m(v)$ formed by splitting each head h in $G_s(V_v)$ into h and h' (see figure 8.3). All DAG arcs entering or leaving the original head h will enter or leave the newly formed h; all back arcs entering the original h will enter h'. There will be an arc of infinite capacity from h' to a new vertex t. All other arcs will inherit their capacities from their weights in G. We will interpret v as the source and t as the sink of $G_m(v)$.

Note that the set of arcs in $G_s(V_v)$ is exactly the set of arcs of finite capacity in $G_m(v)$.

Definition 4.2 Let $G = (V, A, r)$ be an arc-weighted reducible graph. We define $G_{mm}(v)$, the mincost maximum flow network with respect to head v inductively as follows:

a. If v dominates no other head in G then $G_{mm}(v) = G_m(v)$.

b. Let $v_1, \ldots, v_r$ be the heads immediately dominated by v in G and let the capac-

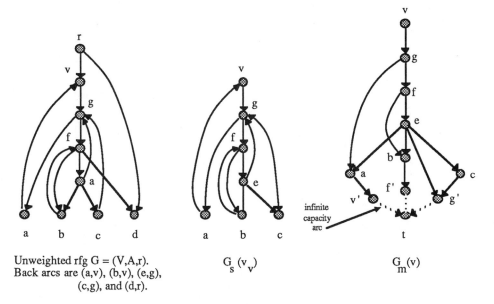

Unweighted rfg $G = (V,A,r)$.
Back arcs are (a,v), (b,v), (e,g),
(c,g), and (d,r).

$G_s(v_v)$

$G_m(v)$

Figure 8.3: Constructing $G_m(\nu)$ from $G = (V, A, r)$.

ity of a minimum cut in $G_{mm}(v_i)$ be $c_i, i = 1, \ldots, r$. Then $G_{mm}(v) = (V, A)$ where V is the same as the vertex set for $G_m(v)$ and $A = \{\text{arcs in } G_m(v)\}$ $\cup\{\text{arcs in } G_{mm}(v_i), i = 1, \ldots, r\} \cup F_v$, where $F_v = \{f_{vv_i} = (v, v_i) \mid i = 1, \ldots, r, \text{with capacity of } f_{vv_i} = c_i\}$.

We call F_v the *mincost-arc set for head v*; if j is a head immediately dominated by head i then f_{ij} is the *mincost arc from head i to head j*.

Figure 8.4 gives an example of $G_{mm}(v)$.

It is established in [Ra1] that following algorithm determines the cost of a minimum FAS in an arc-weighted reducible graph G.

Minimum FAS Algorithm for Reducible Flow Graphs

Input: A reducible graph $G = (V, A, r)$ with nonnegative weights on arcs.

Output: The cost of a minimum FAS for G. *begin*

1. Preprocess G : Label the heads of back arcs in G in postorder. Derive the head dominator tree T_h for G. Introduce a pointer from each vertex i in T_h (except r) to its parent h_i. Let the number of heads be h.

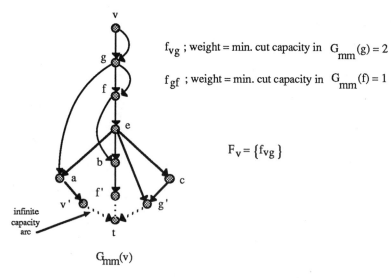

Figure 8.4: The mincost maximum flow network with respect to head ν for the graph $G = (V, A, R)$ of Figure 8.3.

2. For $i = 1, \ldots, h$ process head i :

    a. Find the capacity of minimum cut, c_i, in $G_m(i)$.

    b. If $i \neq h$ then introduce an arc of weight c_i from h_i to i in G. (Note that G changes during the execution of the algorithm so that $G_m(i)$ is the same as $G_{mm}(i)$ if G were unchanged.)

3. Output c_h as cost of minimum FAS for G.

end.

We implement the above tree computations once again using our modified tree contraction method. For a Rake step, we form $G_m(l)$, for each leaf l in T_h and compute c_l, the capacity of a minimum cut in $G_m(l)$. If $l \neq h$ then we place an arc of capacity c_l from h_l to l. Finally we delete all leaves from the current T_h.

The Shrink operation is a little more involved. We assume some familiarity with [Ra1]. Let $C = < v_1, \ldots, v_l >$ be a leaf chain, with v_l the leaf. During the Shrink operation, we will determine the capacity of a minimum cut in $G_{mm}(v_i), i = 1, \ldots, l$. We can then use this to construct to $G_{mm}(v_1)$. We only describe how to compute c_{v_1}; the other computations are entirely analogous. For simplicity assume $l = 2^r + 1$, for some r.

The algorithm proceeds in $r + 1$ phases. In the first phase we divide C into $l - 1$ subchains, $C_{1,1}, \ldots, C_{1,l-1}$, where $C_{1,j} = < v_j, v_{j+1} >$. In general in the i th phase we will have $(l - 1)/2^{i-1}$ chains with $C_{i,j} = < v_{i,(j-1)2^{i-1}+1}, \ldots, v_{i,j2^{i-1}+1} >$.

With each $C_{i,j}$ we maintain the following information. Let $x = (j-1)2^{i-1} + 1$ and $y = j2^{i-1} + 1$, hence $C_{i,j} = < v_x, \cdots, v_y >$. We call v_x the *head* and v_y the *tail* of chain $C_{i,j}$. The arcs in chain $C_{i,j}$ are those arcs whose tails are dominated by v_x and not dominated by v_y, if $y > l$, and are all arcs whose tails are dominated by v_x, if $y = l$.

Let $B(x, y)$ be the set of back arcs in $C_{i,j}$. For $p = y - 1, y - 2, \ldots, x$, for $q = x, x - 1, \ldots, 1$, let $m_i(p, q)$ be the minimum cost of cutting cycles due to back arcs in $B(x, y)$ with heads to v_s, for $s \geq q$, using only arcs in $C_{i,j}$ and $B(x, y)$, assuming that v_p is separated from v_{p+1} by the remaining separating arcs, and p is the largest such index in $C_{i,j}$ for which this holds. Let $m_i(y, q)$ be analogous costs for the case when no v_p in $C_{i,j}$ is separated from v_{p+1} in the minimum cost FAS.

Our algorithm for the Shrink operation computes these $x \cdot (y - x + 1) = O(l^2)$ values for $v_i(p, q)$ for each $C_{1,j}$, and combines these values in successive stages until at the final stage we can obtain the capacity of minimum cut in $G_{mm}(v_1)$ as $\min_p(m_{r+1}(p, 1))$ for $C_{r+1,1}$.

The computation is performed as follows. For the base case $i = 1$, $C_{1,j} = \langle v_j, v_{j+1} \rangle$, and we need to compute $m_1(j+1, q)$ and $m_1(j, q)$ for $q = j, j-1, \ldots, 1$. The value $m_1(j, q)$, when $q > l$ is computed by deleting the arcs dominated by v_{j+1} in G, by forming $G_m(v_q)$ in the resulting graph, making v_j the source of this flow network, and computing the value of minimum cut in this network, call it $H_{1,j}(j, q)$. The value $m_1(j + 1, q)$ is computed by augmenting $H_{1,j}(j, q)$ to include an arc of infinite capacity from v_{j+1} to sink t, and computing the value of minimum cut in this network $H'_{1,j}(j, q)$. For $j = l$ we delete no arcs in G when forming $H_{1,j}(j, q)$ and $H'_{1,j}(j, q)$ and we use $G_{mm}(v_l)$. Thus the base case can be computed with $O(n^2)$ computations of minimum cut in flow networks.

Observe that a minimum cut in $H_{1,j}(j, q)$ gives the minimum number of arcs, all in $C_{i,j}$, that need to be deleted in order to separate tails of back arcs to v_s, for $q \leq s$, from their heads. For $m_1(j + 1, q)$ we put in an arc of infinite capacity from v_{j+1} to t and this further forces v_j to be separated from v_{j+1} by the minimum cut as required. Hence this computation correctly gives the $m_1(p, q)$.

Assume inductively that we have the values $m_i(p, q)$ for some i. In order to compute them for $i + 1$ we will combine the $m_i(p, q)$ values for the two chains $C_{i,j}$ and $C_{i,j+1}$, for j odd into $m_{i+1}(p, q)$ values for $C_{i+1,(j+1)/2}$ by the relation $m_{i+1}(p, q) = \min_{q'}(m_i(p, q') + m_i(q', q))$, where $j \cdot 2^{i-1} + 1 \leq q' \leq q$.

Let M be a minimum cardinality set of arcs that cuts back arcs in $C_{i+1,(j+1)/2}$, using only arcs in $C_{i+1,(j+1)/2}$ assuming v_p is separated from v_{p+1} in the minimum cut; thus $m_{i+1}(p, q) = |M|$.

Let v_{q_0} be the vertex with largest index in $C_{i,j}$ which is separated by M from the vertex it immediately dominates in $C_{i+1,(j+1)/2}$. Then observe that the back arcs with tails in $C_{i,j+1}$ and heads dominated by v_{q_0+1} must have their cycles cut by arcs in $C_{i,j+1}$. But the minimum cost for this is $m_i(p, q_0)$. The cost of cutting the remaining cycles that contribute to $m_{i+1}(p, q)$ is $m_i(q_0, q)$ since the back arcs of these cycles are in $C_{i,j}$. Thus the total cost is $m_i(p, q_0) + m_i(q_0, q)$, assuming that v_{q_0} is the vertex with largest index in $C_{i,j}$ which is separated from v_{q_0+1} by M. Clearly q_0 will contribute the minimum value to the sum $m_i(p, q') + m_i(q', q)$ when q' ranges over indices of vertices in $C_{i,j+1}$. This establishes the correctness of the Shrink algorithm.

The FAS algorithm, using the above Rake and Shrink operations, uses a polynomial number of applications of a minimum cut algorithm on a unit capacity network, applied in $O(\log n)$ stages. Hence it is an RNC algorithm.

Finally we present some results on the parallel complexity of finding feedback sets in weighted rfg's.

Lemma 4.6 The following problems are reducible to one another through NC reductions.

1) Finding a minimum FAS in an unweighted rfg.

2) Finding a minimum weight FAS in an rfg with unary weights on arcs.

3) Finding a minimum weight FVS in an rfg with unary weights on vertices.

4) Finding a minimum cut in a flow network with capacities in unary.

Proof: NC reductions between 1), 2), and 3) follow from results in [Ra1]. We show that 4) reduces to 2): We use the NC reduction in [Ra2] from the problem of finding a minimum cut in a general flow network G to the problem of finding a minimum cut in an acyclic flow network N. Minimum cut for N can be obtained by finding a minimum weight FAS in graph G' derived from N by coalescing source and sink [Ra1].

We also have the result that 2) reduces to 4) since the RNC algorithm given above uses $O(\log n)$ applications of an algorithm for 4) together with some additional NC computation.   □

Lemma 4.7 The following two problems are P-complete:

1) Finding a minimum weight FAS in an rfg with arbitrary weights on arcs.

2) Finding a minimum weight FVS in an rfg with arbitrary weights on vertices.

Proof: It is established in [Ra2] that finding minimum cut in acyclic networks is P-complete. The theorem then follows using the results in Theorem 3.1 in [Ra1]. □

References

[AgAn] A. Aggarwal, R. Anderson, "A random NC algorithm for depth first search," *Proc. 19th Ann. ACM Symp. on Theory of Computing*, New York, NY, May 1987, pp. 325-334.

[AhHoUl] A. V. Aho, J. E. Hopcroft, J. D. Ullman, *The Design and Analysis of Computer Algorithms*, Addison Wesley, 1974.

[AhUl] A. V. Aho, J. D. Ullman, *Principles of Compiler Design*, Addison Wesley, 1977.

[AjKoSz] M. Ajtai, J. Komlos, E. Szemeredi, "Sorting in $c \log n$ parallel steps," *Combinatorica,* vol. 3, 1983, pp. 1-19.

[Al] F. E. Allen, "Program optimization," *Annual Rev. Automatic Prog.,* vol. 5, Pergamon Press, 1969.

[At] M. Attalah, private communication, 1985.

[ChLaCh] F. Y. Chin, J. Lam. I. Chen, "Efficient parallel algorithms for some graph problems," *Comm. ACM,* vol. 25, 1982, p. 659-665.

[Col] R. Cole, "Parallel merge sort," *Proc. 27th Ann. IEEE Symp. on Foundations of Comp. Sci.,* Toronto, Canada, October 1986, pp. 511-516.

[Co] S. A. Cook, "Towards a complexity theory for synchronous parallel computation," *Enseign. Math., vol. 27, 1981, pp. 99-124.*

[Co2] S. A. Cook, "A taxonomy of problems with fast parallel algorithms," *Inform. Control,* vol. 64, 1985, pp. 2-22.

[CoWi] D. Coppersmith, S. Winograd, "Matrix multiplication via arithmetic progressions," *19th Ann. ACM Symp. on Theory of Computing,* New York, NY, May 1987, pp. 1-6.

[DoLiRe] D. Dobkin, R. Lipton, S. Reiss, "Linear Programming is log space hard for P," *Inform. Proc. Lett.,* vol. 8, 1979, pp. 96-97.

[Ev] S. Even, *Graph Algorithms,* Computer Science Press, Potomac, MD, 1979.

[Fl] R. W. Floyd, "Assigning meanings to programs," *Proc. Symp. Appl. Math.,* 19, pp. 19-32, 1967.

[FoFu] L. R. Ford, Jr., D. R. Fulkerson, *Flows in Networks,* Princeton Univ. Press, Princeton, NJ, 1962.

[GaTa] H. Gabow, R. E. Tarjan, "A linear-time algorithm for a special case of disjoint set union," *Proc. 15th Ann. ACM Symp. on Theory of Computing,* 1983, pp. 246-251.

[GhBh] R. K. Ghosh, G. P. Bhattacharya, "Parallel search algorithm for directed acyclic graphs," BIT, 1982.

[GoShSt] L. Goldschlager, R. Staples, J. Staples, "Maximum flow problem is log space complete for P," *TCS, 1982, pp. 105-111.*

[GoTa] A. Goldberg, R. E. Tarjan, "A new approach to the maximum flow problem," *Proc. 18th Ann. ACM Symp. on Theory of Computing,* Berkeley, CA, May 1986.

[Ha] D. Harel, "A linear algorithm for finding dominators in flow graphs and related problems," *17th Ann. Symp. on Theory of Computing*, Providence, RI, May 6-8, 1985, pp. 185-194.

[HeUl] M. S Hecht, J. D. Ullman, "Characterization of reducible flow graphs," *JACM*, 21.3, 1974, pp. 167-175.

[HiChSa] D. S. Hirschberg, A. K. Chandra, D. V. Sarwate, "Computing connected components on parallel computers," *Comm. ACM*, vol. 22, no. 8, 1979, pp. 461-464.

[HoUl] J. E. Hopcroft, J. D. Ullman, *Introduction to Automata Theory, Languages, and Computation*, Addison Wesley, 1979.

[KanRa] A. Kanevsky, V. Ramachandran, "Improved algorithms for graph four connectivity," *Proc. 28th Ann. IEEE Symp. on Foundations of Comp. Sci.*, Los Angeles, CA, October 1987.

[KarRa] R. M. Karp, V. Ramachandran, "Parallel Algorithms," in *Handbook of Theoretical Computer Science*, J. Van Leeuwan, ed., North Holland, 1988, to appear.

[KaUpWi] R. M. Karp, E. Upfal, A. Wigderson, "Constructing a perfect matching is in Random NC," *Proc. 17th ACM Ann. Symp. on Theory of Computing*, 1985, pp. 22-32.

[KlRe] P. Klein, J. H. Reif, "An efficient parallel algorithm for planarity," *Proc. 27th Ann. IEEE Symp. on Foundations of Computer Science*, Toronto, Canada, 1986, pp. 465-477.

[Ku] L. Kucera, "Parallel computation and conflicts in memory access," *Inform. Processing Lett.*, 1982, pp. 93-96.

[La] R. E. Ladner, "The circuit value problem is log space complete for P," *SIGACT News*, vol. 7, no. 1, pp. 18-20.

[MaScVi] Y. Maon, B. Schieber, U. Vishkin, "Parallel ear decomposition search (EDS) and *st* -numbering in graphs," *Proc. Aegean Workshop on Computing*, Loutraki, Greece, 1986, Springer Verlag.

[MiRa] G. L. Miller, V. Ramachandran, "A new graph triconnectivity algorithm and its parallelization," *Proc. 19th Ann. ACM Symp. on Theory of Computing*, New York, NY, May 1987.

[MiRe] G. L. Miller, J. H. Reif, "Parallel tree contraction and its application," *FOCS*, 1985.

[MuVaVa] K. Mulmuley, U. Vazirani, V. Vazirani, "Matching is as easy as matrix inversion," *Proc. 19th Ann. ACM Symp. on Theory of Computing,* May 1987.

[PaSt] C. H. Papadimitriou, K. Steiglitz, *Combinatorial Optimization: Algorithms and Complexity,* Prentice-Hall, Englewood Cliffs, NJ, 1982.

[PaGoRa] S. R. Pawagi, P. S. Gopalakrishnan, I. V. Ramakrishnan, "Computing dominators in parallel," *Inform. Proc. Lett.,* vol. 24, 1987, pp. 217-221.

[Ra1] V. Ramachandran, "Finding a minimum feedback arc set in reducible graphs," *Journal of Algorithms,* to appear.

[Ra2] V. Ramachandran, "The complexity of minimum cut and maximum flow problems in an acyclic network," *Networks,* 1987, pp. 387-392.

[Sh] A. Shamir, "A linear time algorithm for finding minimum cutsets in reducible graphs," *SIAM J. Computing,* 8:4, 1979, pp. 645-655.

[ShVi] Y. Shiloach, U. Vishkin, "An $O(\log n)$ parallel connectivity algorithm," *J. Algorithms,* vol. 3, 1982, pp. 57-63.

[St] V. Strassen, "The asymptotic spectrum of tensors and the exponent of matrix multiplication," *Proc. 27th Ann. Symp. on Foundations of Comp. Sci.,* 1986, pp. 49-54.

[Ta1] R. E. Tarjan, "Testing flow graph reducibility," *JCSS,* 9:3, 1974, pp. 355-365.

[Ta2] R. E. Tarjan, *Data Structures and Network Algorithms,* SIAM, Philadelphia, PA, 1983.

[TaVi] R. E. Tarjan, U. Vishkin, "Finding biconnected components and computing tree functions in logarithmic parallel time," *FOCS,* 1984, pp. 12-22.

[Vi] U. Vishkin, "Implementations of simultaneous memory address access in models that forbid it," *J. Algorithms,* vol. 4, 1983, pp. 45-50.

[Vi2] U. Vishkin, "Randomized speed-ups in parallel computation," *Proc. 16th ACM Ann. Symp. on Theory of Computing,* 1984, pp. 230-239.

Chapter 9

Optimal Tree Contraction in the EREW Model [1]

Hillel Gazit [2]
Gary L. Miller [2]
Shang-Hua Teng [2]

Abstract

A deterministic parallel algorithm for parallel tree contraction is presented in this paper. The algorithm takes $T = O(n/P)$ time and uses P ($P \leq n/\log n$) processors, where $n =$ the number of vertices in a tree using an Exclusive Read and Exclusive Write (EREW) Parallel Random Access Machine (PRAM). This algorithm improves the results of Miller and Reif [MR85,MR87], who use the CRCW randomized PRAM model to get the same complexity and processor count. The algorithm is *optimal* in the sense that the product $P \cdot T$ is equal to the input size and gives an $O(\log n)$ time algorithm when $P = n/\log n$. Since the algorithm requires $O(n)$ space, which is the input size, it is *optimal* in space as well. Techniques for prudent parallel tree contraction are also discussed, as well as implementation techniques for fixed-connection machines.

[1] This work was supported in part by National Science Foundation grant DCR-8514961.
[2] University of Southern California, Los Angeles, CA

139

9.1   Introduction

In this paper we exhibit an optimal deterministic Exclusive Read and Exclusive Write (EREW) Parallel Random Access Machine (PRAM) algorithm for parallel tree contraction for trees using $O(n/P)$ time and P ($P \leq n/\log n$) processors. For example, we can dynamically evaluate an arithmetic expression of size n over the operations of addition and multiplication using the above time and processor bounds. In particular, suppose that the arithmetic expression is given as a tree of pointers where each vertex is either a leaf with a particular value or an internal vertex whose value is either the sum or product of its children's values. These time and processor bounds also apply to arithmetic expressions given as a tree, as previously described. One can reduce an expression given as a string to one given in terms of a pointer by using the results of Bar-On and Vishkin [BV85]. There are many other applications of our parallel programming technique for problems that possess an underlying tree structure, such as the expression evaluation problem [MR]. The goal of this paper is to improve this paradigm so that the $O(\log n)$ time $n/\log n$ processor algorithm can easily and efficiently be constructed for a wide variety of problems.

Our algorithm has two stages. The first stage uses a new reduction technique called Bounded (Unbounded) Reduction. A bounded (unbounded) degree tree of size n is reduced to one of size P in $O(n/P)$ time, using P processors on an EREW PRAM. The second stage uses a technique called Isolation to contract a tree of size P to its root in $O(\log P)$ time, using P processors on an EREW PRAM.

The constants are small since techniques notorious for introducing large constants, such as expander graphs and the more general workload balancing techniques of Miller and Reif [MR85,MR87], are not used. Instead, only one simple load balance is needed to support our procedure, *UNBOUNDED-REDUCTION*. Wyllie's technique for list ranking, when used to carry out the Compress operation, performs many unnecessary function compositions [MR87]. In Section 9.6.4 we show that these techniques provide a solution for prudently compressing chains . In Section 9.3 we discuss how to implement these procedures on a fixed connection machine, which minimizes the size of constants.

Miller and Reif [MR85,MR87] give a deterministic Concurrent Read and Concurrent Write (CRCW) PRAM algorithm for tree contraction, using $O(\log n)$ time and n processors, and an 0-sided randomized version (CRCW) of the algorithm using $n/\log n$ processors. By attaching a complete binary tree to each path of the original tree to guide the application of Compress, Dekel et al. [DNP86] present a tree contraction algorithm on an EREW PRAM in $O(\log n)$ time using n processors. But their methods and proofs are complicated and difficult to follow. This paper, on the other hand, presents a parallel-tree contraction algorithm for an EREW PRAM, which reduces the processor count to $n/\log n$ and simplifies their algorithm.

This paper consists of six sections. Section 9.2 contains basic graph theoretic results and definitions that are needed in Section 9.4 to reduce the problem of size n to one of size n/P, where P is the number of processors. In Section 9.3 we discuss the relationship between the List-Ranking problem and the All-Prefix-Sums problem. In Section 9.4 we show how our reduction of a tree of size n to one of size $n/\log n$ can be performed with only one List-Ranking. Section 9.5 reviews the work of Miller and Reif [MR85] that is used in Section 9.6. Section 9.5 also includes definitions of Rake and Compress, the two basic tree contraction operations, presented by Miller and Reif [MR85,MR87]. In Section 9.6, the isolation technique is used to implement parallel tree contraction on a deterministic EREW PRAM in $O(\log n)$ time by using $n/\log n$ processors. In that section, we demonstrate that the isolation technique can be use for prudent parallel tree contraction.

9.2 Basic Graph Theoretic Results

The main graph theoretic notions needed in this paper are defined in this section. We also present a simple, yet important, structural theorem for trees which is used to reduce a tree of size n to one of size $O(n/P)$.

We begin with some basic definitions. Throughout this paper a **tree**, $T = (V, E)$, is defined as a directed graph, in which every vertex except the root points to its unique parent. The **weight** of a vertex v in T is the number of vertices in the subtree rooted at v, denoted by $W(v)$. If n equals the number of vertices in T, the weight of the root r is n.

In this section we also consider the decomposition of a tree T into subtrees by partitioning the tree at its vertices, which partitions the edges of T in a natural way. Subtrees (subgraphs) are then formed out of each set of edges by reintroducing the end-point vertices. These subgraphs are known as **bridges**. The standard formal definition of a bridge will be presented next.

Let C be a subset of V in a graph, $G = (V, E)$. Two edges, e and e' of G, are C-**equivalent** if there exists a path from e to e' which avoids the vertices C. The induced graphs, formed from the equivalent classes of the C-equivalent edges, are called the **bridges** of C. A bridge is **trivial** if it consists of a single edge. The **attachments** of a bridge B are the vertices of B in C.

Let m be any integer such that $1 < m \leq n$. One could think of $m = n/P$, where P is the number of processors; but the fact that $m = n/p$ is not used in this section. A vertex is m-**critical** if it belongs to the following set of vertices of T:

$$C = \{v \in V \mid \lceil \frac{W(v)}{m} \rceil \neq \lceil \frac{W(v')}{m} \rceil \text{ for all children } v' \text{ of } v\}.$$

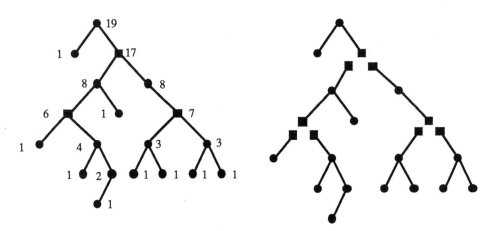

Figure 9.1: The Decomposition of a Tree into its 5-Bridges.

The m-**bridges** are those bridges of C in T where C is the set of m-critical vertices of T. Note that an attachment of an m-bridge B is either the root of B or one of its leaves. In Figure 9.1 we give a tree and its decomposition into its 5-bridges. The vertices represented by boxes are the 5-critical vertices, and the numbers next to these vertices are their weights. Next, we show that B can have at most one attachment which is not its root.

Lemma 9.2.1 *If B is an m-bridge of a tree T, then B can have at most one attachment, i.e., a leaf of B.*

Proof: The proof is by contradiction. We assume that B is an m-bridge of a tree T, and v and v' are two leaves of B that are also m-critical. We prove that this is impossible. Let w be the lowest common ancestor of v and v' in T. Since B is connected, w must be a vertex of B and cannot be m-critical, because if that were true, then both v and v' would not belong to B. On the other hand, w must be m-critical since w has at least two children of weight m. $\square$

From Lemma 2.1 one can see that there are three types of m-bridges: (1) a leaf bridge which is attached by its root; (2) an edge bridge which is attached by its root and one leaf; and (3) a top bridge, containing the root of T, which exists only when the root is not m-critical. Except for the top bridge, the root of each m-bridge has a unique child. The edge from this child to its root is called the **leading edge** of the bridge.

Lemma 9.2.2 *The number of vertices of an m-bridge is at most $m + 1$.*

Proof: Consider the three types of m-bridges: leaf, edge, and top. If B is a leaf bridge then its root r' is the first and only vertex in B with weight $\geq m$. Since

$m > 1$, there must exist a unique vertex w in B which is the only child of r'. Thus, B consists of the subtree rooted at w plus r'. Since the weight of w is less than m, the number of vertices of B is at most m. If B is a leaf m-bridge with m-critical root r' and m-critical leaf u, then r' will have a unique child w in R. Therefore, the number of vertices in B is $W(w) - W(u) + 2$. $W(w) - W(u) < m$ since w is not m-critical. Thus, the number of vertices in $B \leq m + 1$. The case for a top bridge follows by similar arguments. $\square$

Although it is desirable to have few m-bridges [i.e., $O(n/m)$], that is not the case here. This fact can be seen in the example of an unbounded degree tree of height 1 where $m < n$ and every edge is an m-bridge. However, the number of m-critical vertices is not large.

Lemma 9.2.3 *The number of m-critical nodes in a tree of size n is at most $2n/m - 1$ for $n \geq m$.*

Proof: Let n_k be the number of nodes in a minimum size tree with k m-critical nodes. The lemma is equivalent to the statement:

$$n_k \geq (\frac{k+1}{2})m \quad \text{for} \quad k \geq 1. \tag{9.2.1}$$

Inequality 9.2.1 is proven by induction on k. If v is m-critical, then its weight must be at least m. This proves 9.2.1 for $k = 1$. Suppose that 9.2.1 is true for $k \geq 1$ and all smaller values of k. We prove 9.2.1 for $k + 1$. Suppose that T is a minimum size tree with $k + 1$ m-critical nodes. The root r of T must be m-critical for it to be of minimal size, because we can discard all of the tree above the first m-critical node (the root bridge) without affecting the number of critical nodes. Assuming r is m-critical, there are two possible cases for the children of root r: (1) r has two or more children, $u_1, \ldots, u_t$, and each of their subtrees contains an m-critical node; or (2) r has exactly one child u whose subtree contains an m-critical node.

We first consider Case 1. Let n_i be the number of vertices, and k_i the number of m-critical nodes in the subtree of u_i for $1 \leq i \leq t$. Since T is of minimum size, $u_1, \ldots, u_t$ must be the only children of r. Thus, $k + 1 = \sum_{i=1}^{t} k_i$ and $\sum_{i=1}^{t} n_i \leq n_{k+1}$. Using these two inequalities and the inductive hypothesis we get the following chain of inequalities:

$$n_{k+1} \geq \sum_{i=1}^{t} n_i \geq \sum_{i=1}^{t} \left(\frac{k_i+1}{2}\right) m \geq \left(\frac{(\sum_{i=1}^{t} k_i)+t}{2}\right) m,$$
$$\geq \left(\frac{k+t}{2}\right) m \geq \left(\frac{k+2}{2}\right) m \geq \left(\frac{(k+1)+1}{2}\right) m.$$

This proves Case 1.

In Case 2 the subtree rooted at u contains a unique maximal node w which is m-critical, and the subtree of w contains k m-critical nodes. Thus, the induction hypothesis shows that $W(u) \geq \lceil \frac{k+1}{2} \rceil m$. Since $W(r)$ is an integral multiple of m greater than $W(u)$,

$$W(r) \geq (\lceil \frac{k+1}{2} \rceil + 1)m \geq \lceil \frac{k+2}{2} \rceil m.$$

▯

The m-**contraction** of a tree, T with root r is a tree, $T_m = (V', E')$, such that the vertices V' are the m-critical vertices of T union r. Two vertices, v and v' in V', are connected by an edge in T_m, if there is an m-bridge in T which contains both v and v'. Note that every edge in T_m corresponds to a unique m-bridge in T, which is either an edge bridge or the top bridge. Thus by Lemma 9.2.3, T' is a tree with at most $2n/m$ vertices. In the next section we show how to reduce a tree to its m-contraction, where $m = n/P$ in $O(m + \log n)$ time.

9.3   List-Ranking Versus All-Prefix-Sums

There are two problems which are very similar, but their complexity is quite different. The first is the List-Ranking problem, where one is given a linked list of length n packed into consecutive memory locations. The goal is to compute for each pointer its distance from the beginning of the list. The second problem is **All-Prefix-Sums**, where we are given a semi-group $(S, *)$ and a string of elements, $s_1 \cdots s_n$. We may request that the n elements be loaded into memory in some convenient order. The solution is to replace each element s_i with $t_i = s_1 * \cdots * s_i$. It is easy to see how to generalize the List-Ranking problem to include the All-Prefix-Sums problem by storing s_i in the ith pointer and requiring all prefixes to be computed. All known algorithms for List-Ranking solve this generalized problem for semi-groups at no extra cost. Thus, we could view any All-Prefix-Sums problem as a List-Ranking problem; but this may increase the running time.

It can be shown that an All-Prefix-Sums problem can be computed in $6 \log n$ time on a binary N-cube parallel computer where $N = n/\log n$. On the other hand, the results for List-Ranking use the PRAM model. The problem was first introduced by Wyllie [Wyl79]. He gave an $O(\log n)$ time n processor algorithm for an EREW PRAM. This result has been improved upon by many authors. Miller and Reif [MR85] give the first $O(\log n)$ time, optimal number of processors algorithm for this problem. Their algorithm uses randomization and requires the CRCW model. Cole and Vishkin [CV86b] give the first $O(\log n)$ time deterministic EREW algorithm, using an optimal number of processors. Both of these algorithms involve very large constants. Miller and Anderson [AM87] give a simple deterministic EREW algorithm and an even simpler, randomized EREW $O(\log n)$ time, optimal algorithm for List-Ranking.

While the List-Ranking problem can be performed in $O(\log n)$ time on an EREW

PRAM using $n/\log n$ processors, it translates into an $O(\log^2 n)$ algorithm on a binary N-cube, i.e., $N = n/\log n$. Therefore, if the ultimate purpose of a parallel algorithm is to run it on a fixed connection machine, then we should minimize the number of List-Rankings we perform; and, whenever possible, replace the List-Ranking procedure with the All-Prefix-Sums procedure. Karlin and Upfal [KU86] show that once the numbering is known, then the values can be loaded into consecutive memory locations in $O(\log n)$ time by using a randomized algorithm. Readers who are interested in the subject should see the improved results of Ranade [Ran87].

We shall present our code so that, whenever possible, we can perform the All-Prefix-Sums problem on strings stored in consecutive memory locations. Thus, if the user implements these sums on a fixed connection machine, he can implement them only in $O(\log n)$ time. We shall refer to All-Prefix-Sums as all prefix sums over consecutive memory locations.

Tarjan and Vishkin [TV85] define the notion of a Euler tree tour of an ordered tree. Recall that we have defined a tree as a directed graph consisting of directed edges from child to parent. Let T' be the tree T where we have added in the reverse edges. A **Euler tree tour** is the path in T' from root to root which traverses the edges around the outside of T' in a clockwise fashion, when T' is drawn in the plane in an order preserving way.

Figure 9.2 shows how to compute the weights of all nodes in a tree, in parallel, using the Euler tree tour. This algorithm for computing weights has been derived from Tarjan and Vishkin [TV85].

It is an interesting open question whether parallel tree contraction can be performed with only one List-Ranking or not. In the next section we show that the m-contraction of a tree of size n can be constructed with only one List-Ranking of the Euler tree tour.

Procedure $WEIGHTS(T)$

1. Number every tree edge 1, and its reverse 0.
2. Compute the All-Prefix-Sums of the Euler tree tour.
3. Compute the weight of each vertex v as the difference between the prefix sums when we first visited v and when we last visited v.

Figure 9.2: Computing the Weights.

9.4   Reduction From Size n to Size n/m

In this section we show how to contract a tree of size n to one of size n/m in $O(m)$ time using n/m processors, for $m \geq \log n$. If we set $m = \lceil n/P \rceil$, then this gives us a reduction of a problem of size n to one of size P. In Section 9.6 we show how to contract a tree of size P to a point. In that section we consider the special case of when tree T is of bounded degree. In Subsection 9.4.2 the general case of when the tree may be of unbounded degree is discussed.

Let us assume that a tree is given as a set of pointers from each child to its parent and that the tree is ordered, so that the children of a vertex are ordered from left to right. Further, assume that each parent has a consecutive block of memory cells, one for each child, so that each child can write its value, when known, into its location. This last assumption permits us to use the All-Prefix-Sums procedure to compute the associative functions of each set of siblings. This assumption is used to determine the m-critical sets.

9.4.1   The Bounded Degree Case

From Section 9.2, we learned that there are at most $2n/m$, m-critical vertices. Since we assume in this section that the tree is of bounded degree d, there can be at most d of the m-bridges common to and below an m-critical vertex of T. We also know that each m-bridge has a size of at most $m + 1$. To perform the reduction, we need only find the m-bridges and efficiently assign them to processors in order to evaluate them. A processor, assigned to each m-critical vertex, computes the value (function) of the m-bridges below it. A processor is also assigned to each existing root bridge, and computes the function for each root bridge. This algorithm is given in procedural form in Figure 9.3.

We discuss in more detail how to implement the steps in Procedure *BOUNDED-REDUCTION*. We start with Step 2 in which the Euler tree tour of T is loaded into consecutive memory locations. (This step is described in Section 9.3.) This representation is used in all steps except Step 4. Step 3 is also described in Section 9.3. To compute the m-critical vertices (Step 4), we copy the weight of each vertex back to the original representation. There, we compute the maximum value of each set of siblings by using the All-Prefix-Sums procedure on this representation in the natural way. The maximum value is then returned to the right-most sibling. Note that the maximum value could have been returned to all siblings, which would have allowed all vertices to determine if their parents were m-critical with no extra message passing. The right-most sibling could then determine whether its parent is an m-critical vertex or not. To enumerate the m-critical vertices, we can use either representation of the tree.

Procedure *BOUNDED-REDUCTION(T)*

1. set $m \leftarrow \lceil n/2P \rceil$
2. Compute a List-Ranking of the Euler tree tour of T. Use these values to map the ith edge into memory location i.
3. Using the All-Prefix-Sums procedure, compute the weight of every vertex in T.
4. Using the All-Prefix-Sums procedure over the original representation, determine the m-critical vertices in T.
5. Using the All-Prefix-Sums procedure, assign a processor to each m-critical vertex and one to the root.
6. Require each processor assigned in Step 5 to compute the value of the leaf bridges below it and the unary function for the edge or top bridges below it.
7. Return the m-contraction of T and store the Euler tree tour in consecutive memory locations.

Figure 9.3: A Procedure that Contracts a Bounded Degree Tree of Size n to One of Size n/P.

In Step 6 we note that each leaf of the m-bridge is stored in at most $2m$ consecutive memory locations: i.e., there is one memory location for each edge or its reverse. On the other hand, the edge and root m-bridges consist of two consecutive runs of memory locations with a total size of $2m$. If we are implementing this algorithm on a fixed connection machine, then the memory cells of each m-bridge are contained in the memory of a constant number of processors. In Step 7 we note that the Euler tree tour of the m-contraction of T can be constructed without using List Ranking.

9.4.2 The Unbounded Degree Case

In this subsection we show how to compute the m-contraction of an unbounded degree tree. In the unbounded case the number of leaf m-bridges may be much larger than the number of processors. On the other hand, the number of bridges that are either a top bridge or an edge bridge is bounded by the number of m-critical vertices, which, in turn, is bounded by $2n/m$. In the unbounded degree case, a processor may be required to evaluate many small leaf bridges, since there may be a large number of them. The procedure, *UNBOUNDED-REDUCTION*, is given in Figure 9.4.

Steps 1–4 are identical to those used in the *BOUNDED-REDUCTION* procedure. Steps 5 and 6 assign a processor to a collection of leaf bridges. Step 5 is a straightforward, All-Prefix-Sums calculation. Note that all leaf bridges in an interval, $[(i -$

Procedure *UNBOUNDED-REDUCTION(T)*

1. set $m \leftarrow \lceil n/2P \rceil$
2. Compute a List-Ranking of the Euler tree tour of T and map the ith edge into memory location i.
3. Using the All-Prefix-Sums procedure, compute the weight of every vertex in T.
4. Using the All-Prefix-Sums procedure over the original representation, determine the m-critical vertices in T.
5. Assign to each leading edge of a leaf bridge a value equal to the weight of its bridge. To all other edges, assign a value of zero. Compute All-Prefix-Sums of value; let $S(e)$ be the sum up to e.
6. Assign processor i to all leaf bridges with leading edge e so that $(i - 1)m \leq S(e) < im$.
7. Using the All-Prefix-Sums procedure, assign a new processor to each edge or root m-bridge.
8. Require each processor assigned in Step 7 to compute the value (unary function) of the leaf bridge (edge or root) for its assigned bridge.
9. Return the m-contraction of T and store the Euler tree tour in consecutive memory locations.

Figure 9.4: A Procedure that Contracts an Arbitrary Tree of Size n to One of Size n/P.

$1)m, im)$, are stored in consecutive memory locations. Therefore, we need assign a processor only to the first leaf bridge in each interval. After Step 5, the first leading edge of each interval knows that it is the first leading edge. Therefore, we can do one more All-Prefix-Sums calculation to enumerate the leading edges that are the first ones in their interval. Using this information, we can then assign the processors per Step 6. Note, in Step 8, that each processor must evaluate at most $2m$ vertices.

Another way of assigning processors to bridges is to compute the weight of all m-bridges and use Step 6 to assign processors to all bridges–not just leaf bridges. The weight of a leaf bridge is the difference between its bottom attachment and its top attachment; and in the Euler tree tour, there are no attachments between the bottom attachment and the top attachment for a leaf bridge. Therefore, we can compute the weights of all leaf bridges by using one All-Prefix-Sums procedure. Similarly, we can compute the top bridge weight. This approach may give us a better implementation in practice.

9.5 Rake and Compress Operations

In this section, we review the Rake and Compress operations that Miller and Reif [MR85,MR87] use for their parallel tree contraction algorithm. Let *Rake* be the operation which removes all leaves from a tree T. Let a *chain* be a maximal sequence of vertices, $v_1, ..., v_k$ in T, such that v_{i+1} is the only child of v_i for $1 \leq i < k$, and v_k has exactly one child. Let *Compress* be the operation that replaces each chain of length k by one of length $\lceil k/2 \rceil$. One possible approach to replacing a chain of length k by one of length $\lceil k/2 \rceil$ is to identify v_i with v_{i+1} for i odd and $1 \leq i < k$.

Let Contract be the simultaneous application of Rake and Compress to the entire tree. Miller and Reif [MR85,MR87] show that the Contract operation need be executed only $O(log\ n)$ times to reduce T to its *root*. They prove the following theorem.

Theorem 9.5.1 *After $\lceil log_{5/4} n \rceil$ executions of Contract are performed, a tree of n vertices is reduced to its root.*

In this section, we present the important definitions, used in the proof of Theorem 9.5.1, that are needed later on in this paper.

Let V_0 be the leaves of T, V_1 be the vertices of T with only one child, and V_2 be those vertices of T with two or more children. Next, partition the set V_1 into C_0, C_1, and C_2 according to whether its child is in V_0, V_1, or V_2, respectively. Similarly, partition the vertices C_1 into GC_0, GC_1, and GC_2, according to whether the grandchild is in V_0, V_1, or V_2, respectively. Let $Ra = V_0 \cup V_2 \cup C_0 \cup C_2 \cup GC_0$ and $Com = V - Ra$.

Miller and Reif [MR85] show that Rake reduces the size of Ra by at least a factor of $1/5$ in each application, while Compress reduces the size of Com by a factor of $1/2$ in each application. We use similar techniques in this paper.

9.6 Isolation and Deterministic EREW Tree Contraction

In Section 9.4 we showed that if we find an EREW parallel tree contraction algorithm which takes $O(\log n)$ time and uses n processors, then we get an $O(\log n)$ time, $n/\log n$ processor, EREW PRAM algorithm for parallel tree contraction. Thus, we may restrict our attention to $O(n)$ processor algorithms. In this section, a technique called **isolation** is presented and used to implement parallel tree contraction on an EREW PRAM without increasing time and processor count. We also present a method for prudent tree contraction which is important in more general tree problems (for example, see Miller and Teng [MT87]).

begin
    while $V \neq \{r\}$ **do**
        In Parallel, for all $v \in V - \{r\}$, **do**
            if v is a leaf, **mark** its parent and remove it;   **(Rake)**
            isolate all the chains of the tree;   **(Isolation)**
            Compress each chain in isolation;   **(Local Compress)**
            If a chain is a single vertex **then** unisolate it.   **(Integration)**
end

Figure 9.5: Isolate and Compress for Deterministic Parallel Tree Contraction when $P = n$.

It is important to understand why the deterministic parallel tree contraction, presented by Miller and Reif [MR85,MR87], does not work on the EREW model. The problem arises at the parent v of a chain (a node in V_2 with a child in V_1 is called the **parent** of a chain). Using the pointer-jumping algorithm of Wyllie [Wyl79], we encounter the problem that, over time, many nodes may eventually point to this parent. Now, if v becomes a node in V_1, then all these nodes must determine the parent of v and point to it, which seems to require a concurrent read. We circumvent this problem by isolating the chain until it is Compressed to a point. At that point, we then let it participate in another chain. (See Isolate and Compress presented in Subsection 9.6.1.)

Theorem 9.6.1 (Main Theorem) *Tree contraction can be performed, deterministically, in $O(n/P)$ time using P processors on an EREW PRAM for all $P \leq n/\log n$.*

9.6.1 Isolation and Local Compress

Figure 9.5 displays a high level description of our deterministic algorithm for parallel tree contraction on an EREW PRAM which uses $O(\log n)$ time and n processors which we call **Isolate and Compress**.

The difference between the contraction phase used in this algorithm and the dynamic contraction phase presented by Miller and Reif [MR85,MR87] is that the Compress operation here is replaced by two operations: Isolation and Local Compress. Each Local Compress operation applies one conventional Compress operation to an isolated chain during each contraction phase.

Lemma 9.6.1 *After each Isolate and Compress $|Com|$ decreases by a factor of at least $1/4$.*

Proof: By the way the steps Isolation and Integration are implemented, each isolated chain has a length of at least 2. Moreover, no two consecutive nodes are singletons. Thus, a chain consists of an alternating sequence of an isolated chain and a singleton. We view a chain as a set of pairs where each pair consists of an isolated chain and a singleton. Each pair is decreased in size by at least a factor of 1/4. The worst case is a pair containing an isolated chain of size 3. □

Since step Rake removes 1/5 of Ra and steps Isolation through Intergration removes 1/4 of Com, together they must remove 1/5 of the vertices. This gives the following theorem.

Theorem 9.6.2 *A tree of n vertices is reduced to its root after one applies Isolate and Compress $\lceil \log_{5/4} n \rceil$ times.*

The next two sections present a more detailed implementation of Isolate and Compress.

9.6.2 Implementation Techniques

In this subsection, we present one method of implementing the generic contraction phase on an EREW model in $O(\log n)$ time, using n processors. Another implementation method is presented in the next subsection. The complexity of these two implementation methods differs only by a constant factor. However, they have different scopes of application.

We view each vertex, which is not a leaf, as a function to be computed where the children supply the arguments. For each vertex v with children, $v_1 \ldots v_k$, we will set aside k locations, $l_1 \ldots l_k$, in common memory. Initially, each l_i is empty or **unmarked**. When the value of v_i is known, we assign it to l_i and denote it by **mark** l_i. Let $Arg(v)$ denote the number of unmarked l_i. Then, initially, $Arg(v) = k$, the number of children of v. We need one further notation: Let **vertex $(\mathbf{P(v)})$** be the vertex associated with the sole parent of v with storage location $P(v)$. All vertices shall be **tagged** with one of four possibilities: G, M, R, or $\emptyset$. Vertices with a nonempty tag belong to an isolated chain. When a chain is first isolated, the root is tagged R, the tail is tagged G, and the vertices between the root and the tail are tagged M. A vertex v is **free** if $Arg(v) = 1$ and $Tag(v) = \emptyset$; otherwise, v is **not free**.

To determine whether or not a child is free, we assign each vertex a new variable that is read only by the parent. To determine if a parent is free, we require that each vertex keep a copy of this variable for each child. Initially, all copies indicate that the vertex is not free. When the parent vertex becomes free, it need only change the copy of the variable associated with the remaining child, since there are no other children that can read the other variables.

Procedure *ISOLATE-COMPRESS*

    In parallel, for all $v \in V - \{r\}$, **do**

        case $Arg(v)$ **equals**

        0)   Mark $P(v)$ and delete v ;;

        1)   **case** $Tag(\mathrm{v})$ **equals**

            $\emptyset$) **if** child is not free and parent is free **then** $Tag(v) \leftarrow G$

                   if parent is not free and child is free **then** $Tag(v) \leftarrow R$

                   if parent is free and child is free **then** $Tag(v) \leftarrow M$;;

            G) **if** $Tag(P(v)) = R$ **then** $Tag(v) \leftarrow \emptyset$

                   $P(v) \leftarrow P(P(v))$;;

            M) **if** $Tag(P(v)) = R$ **then** $Tag(v) \leftarrow R$

                   $P(v) \leftarrow P(P(v))$;;

            esac

        esac

    od

Figure 9.6: The First Implementation of the *ISOLATE-COMPRESS* Procedure.

9.6.3   The Expansion Phase

If we use procedure *ISOLATE-COMPRESS* to, say, evaluate an arithmetic expression it will not return the value of all subexpression. There is one a one-to-one correspondence between subexpression and subtrees of the expression tree. For many application it is necessary to compute the value of all subexpressions. This is usually done by running the contraction phase "backwards" which is called parallel tree expansion, see [MR87]. To insure that the expansion phase only uses exclusive reads we must be a little careful, since many nodes may need the value of the same node in order to compute its value. Thus one solution requires each node to maintain a queue of $O(\log n)$ pointers. We store one pointer at node v for each node w which needs the value of v to determine its value. A solution using only a constant amount of space per node can be achieved by several methods.

One easy to describe method can be obtained by reversing the pointers in each isolated chain and compressing these chain based one the reverse pointers. In this case, the original root is now the tail and is tagged G while the original tail is now the root and is tagged R. Otherwise, we run Procedure *ISOLATE-COMPRESS* as in Figure 9.6. Each time we apply the procedure we will increment a counter which we think of as the time. When a node v obtains a R tag (except the new root node of the chain) it records the time (the value of the counter) using variable t_v and the node n_v with tag R that it is now pointing. In the expansion phase node v determines the value of node n_v at time t_v with the clock running "backwards".

One can also perform the expansion phase without reversing the pointers in an isolated chain by simulating the above method directly on the original forward pointers. In this case, during the contraction phase if a node v is pointing to a node w and values have been assigned to the variables t_v and n_v or the tag of v is G then node v sets t_w equal to the value of the counter and n_w to v. This method is basically the same as the first method. The expansion is the same as in the first method.

9.6.4 Prudent Parallel Tree Contraction

One disadvantage of the Local Compress procedure is that, during each Compression stage, one useless chain is produced out of each chain. For generic tree contraction this disadvantage appears harmless; but for more general tree problems, such as those involved in evaluating min-max-plus-times-division trees where the cost to represent the function on each edge doubles with each functional composition, a factor of n is added to the number of processors used, and a factor of $\log n$ is added to the running time [MT87].

The Compress procedure, where no useless chain is produced, is called **prudent Compress**. Using prudent Compress, as much as a factor of n may be saved in the number of processors used in certain applications (see Miller and Teng [MT87] for further details).

Our idea is very simple: first, isolate each chain; second, use Wyllie's [Wyl79] List-Ranking algorithm to rank the vertices in the chain; and third, use this ranking of the vertices in the chain to determine the order in which pairs of vertices are identified. The procedure is written assuming that the parallel tree contraction is run asynchronously: i.e., a block of processors and memory are assigned the task of evaluating a subprocedure which is then performed independently of the rest of the processors and memory. It is unnecessary to write the procedure this way, but it is easier to follow. The Compress part of the procedure is written; the Rake part remains the same.

Procedure *PRUDENT-COMPRESS*

> 1) **Form** isolated chains from free vertices.
> 2) **In parallel**, for all isolated chains, C in 1), **do**
>   *COLLAPSE(C)*.

COLLAPSE(C) is a procedure which computes the ranks of all nodes in an isolated chain C and uses this information to Compress the chain in such a way that no useless subchain is produced. $COLLAPSE(C)$ can be specified as follows:

> **Procedure** $COLLAPSE(C)$
>
> 1) **Run** List-Ranking on C.
> 2) **While** C is not a singleton **do**
>   **Compress** C by combining the vertex of rank $2i - 1$ with
>   the vertex of rank $2i$, forming a new vertex of rank i.
> 3) **Set** C free.

Theorem 9.6.3 *When the PRUDENT-COMPRESS operation is incorporated, the resulting prudent parallel tree contraction algorithm contracts a tree of n nodes in $O(\log n)$ time, using $O(n/\log n)$ processors on an EREW PRAM.*

Proof: The $PRUDENT\text{-}COMPRESS$ operation collapses a chain of length l to a single vertex in $O(\log l)$ time using l processors. It takes constant time with l processors to form an isolated chain of length l from the free vertices. After the $PRUDENT\text{-}COMPRESS$ operation is performed, the $COLLAPSE$ procedure can then Compress a chain of l nodes in $O(\log l)$ time, using $O(l)$ processors.

It is worth noting the technical subtlety in proving the theorem. The $PRUDENT\text{-}COMPRESS$ procedure, in fact, slows down the Compressing of each isolated chain. Thus, the tree may contract in a dramatically different way from other algorithms we have given. Nevertheless, the theorem can be proven by amortizing the cost of collapsing an isolated chain over time. Amortizing arguments similar to those used in the analysis of procedure, *asynchronous parallel tree contraction* by Miller and Reif [MR87], can be used.

More specifically, we can assign a weight of 1 to each free vertex in the tree and a weight, equal to the proportion of work done in collapsing it to a singleton, to each isolated chain. Thus, a chain whose original length was l has a weight that geometrically decreases from l to 1. The weight of the tree decreases by a constant factor after each step. For example, if the weight of the tree is 1, the tree has been reduced to its root. $\square$

9.6.5   Unbounded Contraction

Parallel tree contraction for trees of unbounded degree has been discussed in two previous papers. In the first paper, Miller and Reif [MR87] show that if one uses a Rake operation which (instead of working in unit time) Rakes k siblings in $O(\log k)$

time, then a natural modification of parallel tree contraction, called Asynchronous Parallel Tree Contraction, contracts the tree in $O(\log n)$ time. These techniques work with the Isolation techniques for Compress.

The second way of performing the Rake operation for trees of unbounded degree is to identify consecutive pairs of leaf siblings as we did when we used Compress for parent-child pairs [MR]. A **run** of leaves is a maximal sequence of leaves, $l_1, \ldots, l_k$, which are consecutive siblings. We assume that the siblings are cyclically ordered (i.e., the left-most child follows the right-most child). The operation, **Rake Restricted to Runs**, replaces each run of length k by one of length $\lceil k/2 \rceil$ for $k \geq 2$; and, any run of length 1 is removed completely. We also remove both leaves when they are the only siblings.

Theorem 9.6.4 *Parallel Tree Contraction, where Rake is restricted to runs and Compress uses Isolation, reduces a tree of size n to its root in $\lceil \log_{7/6} n \rceil$ applications.*

Proof: The proof is a straight forward calculation based upon techniques presented by Miller and Reif [MR87,MR].

References

[AM87]   Richard Anderson and Gary L. Miller. *Optimal Parallel Algorithms for List Ranking.* Technical Report , USC, Los Angeles, 1987.

[BV85]   I. Bar-On and U. Vishkin. Optimal parallel generation of a computation tree form. *ACM Transactions on Programming Languages and Systems*, 7(2):348–357, April 1985.

[CV86a] R. Cole and U. Vishkin. Approximate and exact parallel scheduling with applications to list, tree, and graph problems. In *27th Annual Symposium on Foundations of Computer Science*, pages 478–491, IEEE, Toronto, Oct 1986.

[CV86b] Richard Cole and Uzi Vishkin. Deterministic coin tossing with applications to optimal list ranking. *Information and Control*, 70(1):32–53, 1986.

[DNP86] Eliezer Dekel, Simeon Ntafos, and Shie-Tung Peng. *Parallel Tree Techniques and Code Opimization*, pages 205–216. Volume 227 of *Lecture Notes in Computer Science*, Springer-Verlag, 1986.

[KU86]   Anna Karlin and Eli Upfal. Parallel hashing–an efficient implementation
         of shared memory. In *Proceedings of the 18th Annual ACM Symposium
         on Theory of Computing*, pages 160–168, ACM, Berkeley, May 1986.

[MR]   Gary L. Miller and John H. Reif. Parallel tree contraction part 2: further
         applications. *SIAM J. Comput.* submitted.

[MR85]   Gary L. Miller and John H. Reif. Parallel tree contraction and its ap-
         plications. In *26th Symposium on Foundations of Computer Science*,
         pages 478–489, IEEE, Portland, Oregon, 1985.

[MR87]   Gary L. Miller and John H. Reif. *Parallel Tree Contraction Part 1: Fun-
         damentals.* Volume 5, JAI Press, 1987. to appear.

[MT87]   Gary L. Miller and Shang-Hua Teng. Systematic methods for tree based
         parallel algorithm development. In *Second International Conference on
         Supercomputing*, pages 392–403, Santa Clara, May 1987.

[Ran87] A. Ranade. How to emulate shared memory. In *28th Annual Symposium
         on Foundations of Computer Science*, pages 185–194, IEEE, Los Angeles,
         Oct 1987.

[TV85]   R. E. Tarjan and U. Vishkin. An efficient parallel biconnectivity algo-
         rithm. *SIAM J. Comput.*, 14(4):862–874, November 1985.

[Wyl79] J. C. Wyllie. *The Complexity of Parallel Computation.* Technical Re-
         port TR 79-387, Department of Computer Science, Cornell University,
         Ithaca, New York, 1979.

Chapter 10

The Dynamic Tree Expression Problem [1]

Ernst W. Mayr [2]

Abstract

We present a uniform method for obtaining efficient parallel algorithms for a rather large class of problems. The method is based on a logic programming model, and it derives its efficiency from fast parallel routines for the evaluation of expression trees.

10.1 Introduction

Computational complexity theory tries to classify problems according to the amount of computational resources needed for their solution. In this sense, the goal of complexity theory is a refinement of the basic distinction between *decidable* (solvable, recursive) and *undecidable* (unsolvable, non-recursive) problems. There are basically three types of results: (i) algorithms or upper bounds showing that a certain problem can indeed be solved using a certain amount of resources; (ii) lower bounds establishing that some specified number of resources is insufficient for the problem to be solved; and (iii) reductions of one problem to another, showing that the first problem is essentially no harder than the second. Complexity classes are collections

---

[1] This work was supported in part by a grant from the AT&T Foundation, ONR contract N00014-85-C-0731, and NSF grant DCR-8351757.
[2] Stanford University, Stanford, CA

157

of all those problems solvable within a specific bound, and complete problems in a complexity class are distinguished problems such that every other problem in the class is efficiently reducible to them. In this sense, they are hardest problems in the class.

A very important goal of computational complexity theory is to characterize which problems are *feasible*, i.e. solvable efficiently in an intuitive sense. For (standard) sequential computation, the class $\mathcal{P}$ of those problems solvable in polynomial time has become accepted as the "right" class to represent feasibility. For parallel computation, on a wide number of machine models (idealized and realistic), the class $\mathcal{NC}$ is considered to play the same role. It is the class of problems solvable (in the PRAM model, discussed below) in time polylogarithmic in the size of the input while using a number of processing elements polynomial in the input size.

In this paper, we show an interesting relationship between efficient parallel computation and the existence of reasonably sized *proof trees*. While the reader may feel reminded of the well-known characterization of the class $\mathcal{NP}$ as those problems permitting short (polynomial length) proofs of solutions, proof trees are considerably more restricted in the sense that if some intermediate fact is reused the size of its (sub) proof tree is counted with the appropriate multiplicity. Drawing on, and generalizing, work reported in [14], [18], and [21], we give a very general scheme to obtain efficient parallel algorithms for many different problems in a way that abstracts from details of the underlying parallel architecture and the necessary synchronization and scheduling.

The remainder of the paper is organized as follows. In the next section, we discuss a few techniques for parallel algorithm design appearing in the literature. Then we introduce our new technique, the *dynamic tree expression problem* or *DTEP*, and present some of its applications and extensions. In particular, we show an adaptation of the basic DTEP paradigm to the evaluation of straightline programs computing polynomials. The final section summarizes our results and mentions some open problems.

10.2   Techniques for Parallel Algorithm Design

General techniques for the design of efficient parallel algorithms are hard to come by. To a large degree, this is due to the fact that there are so many vastly different parallel machine architectures around, and no compilers between the programmers and the machines that could provide a uniform interface to the former. Obviously, this is a situation quite different from sequential computation. We may even ask ourselves why we should be looking for such a compiler, or whether parallel algorithms shouldn't be developed for specific architectures in the first place. Even more so, we might ask whether parallelism is any good for solving problems in general, or whether there are just a few selected, and highly tuned, applications that could benefit from heavy parallelism. We believe that there is a large number of problems and applications out there which should be parallelized individually, designing architectures permitting the highest speedup for these applications. There is also a large class of applications whose data structures and flow of control are extremely regular and can easily be mapped onto appropriate parallel machine architectures, like systolic arrays or n-dimensional hypercubes. It is our impression, however, that many other problems, particularly in Artificial Intelligence, exhibit the same specifications (with respect to parallelization) as we would expect from general purpose parallel computation. It is the interest in these, and other, applications that makes us convinced that general purpose parallelism is relevant, and that in fact it carries immense potential for areas of general importance.

The design of parallel algorithms may be guided by several, sometimes conflicting, goals. The first, and maybe most obvious, goal in using parallel machines may be to obtain fast, or even the fastest possible, response times. Such an objective will be particularly important in security sensitive applications, like warning or monitoring systems; weather forecasters also seem to be interested in this particular aspect of parallel computation. Another option seems to worry about the marginal cost factor: how much real computational power do we gain by adding one (or k) additional processors to the system? While optimal speedup can be achieved over a certain range of execution times (depending on the problem), fastest execution times may be obtained, for certain problems, only by using a very large number of parallel processors, thus decreasing, in effect, the actual speed-up.

Another consideration is what could be termed the *Law of Small Numbers*. Many of the very fast parallel algorithms have running times which are some power of the logarithm of the input size n, like $O(\log^2 n)$. If we compare this running time with one given by, say, $O(\sqrt{n})$, and assume that the constant factor is 1 in both cases for simplicity, we get that the *polylogarithmic* algorithm is actually slower than the *polynomial* algorithm for n greater than 2 and less than $65,536$.

10.2.1 Parallel Machine Models

We shall base most of our discussions onto a theoretical machine model for parallel computation called the *Parallel Random Access Machine*, or *PRAM* [4]. This machine model has the following features. There is an unbounded number of identical *Random Access Machines* (or *RAM*'s), and an unbounded number of global memory cells, each capable of holding an arbitrarily big integer. The processors work synchronously, controlled by a global clock. Different processors can execute different instructions. Each processor can access (read or write) any memory cell in one step. Our model also allows that more than one processor read the same memory cell in one step (*concurrent read*), but it disallows *concurrent writes* to the same memory cell. For definiteness, we assume that all reads of global memory cells in a step take place before all writes (resolving *read/write conflicts*).

Most of our algorithms actually do not use concurrent reads, and can be run on a somewhat weaker machine model in which concurrent reads are forbidden. In any case, algorithms in our model can always be simulated on this weaker model with a very small slowdown.

The shared memory feature of the PRAM model, allowing unbounded access to memory in a single step, is somewhat idealistic. A more realistic machine model consists of a *network* of (identical) processors with *memory modules* attached to them. The processors are connected via point-to-point communication channels. Each processor can directly access only cells in its own memory module, and it has to send messages to other processors in order to access data in their modules. To respect technological constraints, the number of channels per processor is usually bounded or a very slowly growing function of the number of processors. Examples for such networks of processors are the Hypercube [20], the Cube-Connected-Cycles network [17], or the Ultracomputer (RP3) [15], [19].

10.2.2 Parallel Complexity Measures

The two most important parameters determining the computational complexity of algorithms in the PRAM model are the time requirements and the number of processors used by the algorithm. Both are usually expressed as functions of the input size n. Note that another factor determining the practical efficiency of a parallel algorithm — its communication overhead — is of no importance in the PRAM model. As in sequential computation, and its associated complexity theory, complexity classes are defined based on bounds of the available resources, like time or number of processors. One such class has become particularly important in the theoretical study of parallel algorithms and the inherent limits of parallelism. This class is called $\mathcal{NC}$ and is defined as the class of problems solvable on a PRAM in time *polylogarithmic* (polylog for short) in the size of the input while using a number of processors *polynomial* in the input size. Thus, a problem is in $\mathcal{NC}$ if there is a PRAM algorithm and some constant $c > 0$ such that the algorithm runs in time $O(\log^c n)$ and uses $O(n^c)$ processors on instances of size n.

The resource constraints for $\mathcal{NC}$ reflect two goals to be achieved by parallelism: a (poly)logarithmic running time represents an "exponential" speedup over sequential computation (requiring at least linear time); and the restriction on the number of processors limits the required hardware to be "reasonable". Of course, both these considerations have to be taken in the asymptotic sense, and are thus of a rather theoretical nature (see the Law of Small Numbers above).

However, the class $\mathcal{NC}$ is *robust* in the sense that it does not change with small modifications in the underlying machine model. In fact, it is the same for the PRAM model(s) outlined above as well as the Hypercube or Ultracomputer architecture. It was originally defined for boolean circuits in [16]. Membership in $\mathcal{NC}$ has become (almost) synonymous with "efficiently parallelizable".

Real parallel machines have, of course, a fixed number of processors. Even though some parallel algorithm may have been designed for a (theoretical) parallel machine with a number of processors depending on the input size, the following theorem states that it is always possible to reduce, without undue penalty, the number of processors:

Simulation Theorem: An algorithm running in time $T(n)$ on a p-processor PRAM, can be simulated on a p'-processor PRAM, $p' \leq p$, in time $O(\lceil p/p' \rceil T(n))$. □

If an algorithm runs in time $T(n)$ on a sequential architecture, and we port it to a parallel architecture with p processors we cannot expect a parallel running time better than $\Omega(T(n)/p)$ — unless our machine models exhibit some strange quirks which are of no concern here. Conversely, we say that a problem of sequential time complexity $T(n)$ exhibits *optimal speedup* if its parallel time complexity on a PRAM using p processors is in fact $O(T(n)/p)$, for p in an appropriate range.

10.2.3 Parallel Programming Techniques

Only relatively few parallel programming schemes and/or techniques can be found in the literature. This is due in part to the fact that there is universal parallel architecture, and no commonly accepted way of hiding architectural details from parallel programs. Nonetheless, the PRAM model can, to a certain degree, play this role. Certain operations have been identified to be fundamental or important for a large number of parallel applications, and efficient algorithms have been developed for them. Examples are the parallel prefix computation [10], census functions [11], (funnelled) pipelines [7], parallel tree evaluation [13], [14], and sorting [2]. As an illustration, we state a procedural implementation of an algorithm for the parallel prefix problem, in a PASCAL like notation. Given an array $s_1, \ldots, s_n$ of elements from some arbitrary domain S, and a binary operator ∘ over that domain, this procedure computes all partial "sums" $s_1 \circ \cdots s_i$, for $i = 1, \ldots, n$.

procedure *parallel_prefix*(n, *start*, *result*, ∘)

int n; gmemptr *start, result*; binop ∘;

co n is the number of elements in the input array starting at position *start* in global memory; *result* is the starting position for the result vector **oc**

begin

    local type_of_S: *save, save2*; int: *span, myindex*;

if PID $< n$ **then**

    $span := 1;$

    $myindex := start + \text{PID};$

    $save := M_{myindex};$

    while $span \leq \text{PID}$ **do**

        $M_{myindex} := M_{myindex-span} \circ M_{myindex};$

        $span := 2 * span$

    od;

    $save2 := M_{myindex};$

    $M_{myindex} := save;$

    $myindex := result + \text{PID};$

    $M_{myindex} := save2$

  fi

end *parallel_prefix*.

This algorithm takes, on a PRAM, $O(\log n)$ steps and uses n processors. As such, its speedup is suboptimal since the straightforward sequential algorithm runs in linear time. However, by grouping the input elements into contiguous groups of size $\lceil \log n \rceil$ (the last group may be smaller), we can solve the parallel prefix problem using just $\lceil n/\log n \rceil$ processors. Using one processor for every group, we first compute the prefix "sums" within every group. Then we run the above algorithm with the group prefix "sums" as input, and finally we extend the partial sums over the groups inside the groups, again using one processor per group. The time for this modified algorithm is still $O(\log n)$ (with a somewhat larger constant).

The parallel prefix problem has many applications. It occurs as a subproblem in algorithms for *ranking, linearization of lists, graph traversals, processor scheduling*, and others.

In the next section, we discuss another fundamental algorithm for parallel computation.

10.3   The Dynamic Tree Expression Problem

In [18], Alternating Turing machines using logarithmic space and a polynomial size computation tree are studied. These machines can be thought of as solving a recognition problem by guessing a proof tree and recursively verifying it. Each internal node in the proof tree is replaced, in a universal step, by its children in the tree while leave nodes correspond to axioms which can be verified directly. In every step, the machine is allowed to use only logarithmic storage (at every node) to record the intermediate derivation step.

We in effect turn the top-down computation of ATM's as discussed in [18] around into computations proceeding basically bottom-up, similar to the approach taken in [13], [14], and [21]. We present a uniform method containing and extending the latter results.

10.3.1   The Generic Problem

For the general discussion of the Dynamic Tree Expression Problem, we assume that we are given

i. a set P of N boolean *variables*, $p_1, \ldots, p_N$;

ii. a set I of *inference rules* of the form

$$p_i \; :- \, p_j p_k \quad \text{or} \quad p_i \; :- p_j.$$

Here, juxtaposition of boolean variables is to be interpreted as logical AND, and "$:-$" is to be read as "if". In fact, the two types above are Horn clauses with one or two hypotheses, written in a PROLOG style notation: $(p_j \wedge p_k) \Rightarrow p_i$ and $p_j \Rightarrow p_i$, respectively. We note that the total length of I is polynomial in N.

iii. a distinguished subset $Z \subseteq P$ of *axioms*.

Definition 10.3.1 *Let* (P, I, Z) *be a system as above. The minimal model for* (P, I, Z) *is the minimal subset* $M \subseteq P$ *satisfying the following properties:*

(i) *the set of axioms, Z, is contained in M;*

(ii) *whenever the righthand side of an inference rule is satisfied by variables in M, then the variable on the lefthand side is an element of M:*

$$
\begin{aligned}
p_j, p_k \in M, \; p_i \; :- p_j p_k \in I \\
p_j \in M, \; p_i \; :- p_j \in I
\end{aligned}
\quad \Rightarrow \quad p_i \in M.
$$

We say that (P, I, Z) *implies* some fact p if the boolean variable p is in the minimal model M for (P, I, Z). For each such p, there is a *derivation* or *proof tree*: This is a (rooted) tree whose internal vertices have one or two children, and whose vertices are labelled with elements in P such that these four properties are satisfied:

1. the labels of the leaves are in Z;

2. if a vertex labelled p_i has one child, with label p_j, then $p_i :- p_j$ is a rule in I;

3. if a vertex labelled p_i has two children, with labels p_j and p_k, then $p_i :- p_j p_k$ is a rule in I;

4. the label of the root vertex is p.

The *Dynamic Tree Expression Problem* consists in computing the minimal model for a given system (P, I, Z), or, formulated as a decision problem, in deciding whether, given (P, I, Z) and some $p \in P$, whether p is in the minimal model for (P, I, Z). The algorithm below can be used to solve DTEP on a PRAM.

algorithm $DTEP(N, P, I, Z)$;

int N; set of boolean P, Z; set of inference rules I;

co N is the number of boolean variables in P; Z is the subset of P distinguished as axioms; I is a set of Horn clauses with at most two variables on the righthand side; P is implemented as an array of boolean variables; the algorithm sets to **true** exactly those elements of the array P corresponding to elements in the minimal model **oc**

begin

    array $DI[1..N, 1..N]$;

    co initialization **oc**

    $P[i] :=$ **true** for all $p_i \in Z$, else **false**;

    $DI[j, i] :=$ **true** for $i = j$ and all $p_i :- p_j \in I$, else **false**;

    do l times   **co** l will be specified below **oc**

        for $i \in \{1, \ldots, N\}$ with $P[i] =$ **false do in parallel**

            if $((P[j] = P[k] =$ **true**$) \wedge (p_i :- p_j p_k \in I)) \vee ((P[j] =$ **true**$) \wedge (DI[j, i] =$ **true**$))$

    then $P[i] :=$ **true**

    fi;

    if $(P[k] =$ **true**$) \wedge (p_i \; :- p_j p_k \in I)$ **then** $DI[j, i] :=$ **true fi**

  od;

  $DI := DI \cdot DI$

od

end *DTEP*.

We first remark about the intuition behind the array DI used in the DTEP algorithm. If entry (i, j) of this array is **true** then it is known that p_i implies p_j. This is certainly true at the beginning of the algorithm, by way of the initialization of DI. Whenever for an inference rule $p_i \; :- p_j p_k$ with two antecedents, one of them, say p_k, is known to be **true**, the rule can be simplified to $p_i \; :- p_j$. But this means that p_j implies p_i, as recorded in the assignment to $DI[j, i]$. Also, the net effect of squaring the matrix DI in the last step of the outer loop is that chains of implications are shortened. Note that entry (i, j) of the matrix becomes **true** through the squaring of the matrix only if there was a chain of implications from p_i to p_j before the squaring.

Lemma 10.3.1 *If p_i is in the minimal model M, and if there is a derivation tree for p_i in (P, I, Z) of size m then $P[i] =$ **true** after at most $l = 2.41 \log m$ iterations of the outer loop of the DTEP algorithm. $P[i]$ never becomes **true** for $p_i \notin M$.*

Proof: The proof is by induction on the number of iterations of the main loop. We use the following induction hypothesis:

> At the end of each iteration of the main loop, there is a derivation tree for p_i of size at most $3/4$ of the size at the end of the previous iteration, allowing as inference rules the rules in I and the "direct implications" as given by the current values of the elements of DI, and as axioms the p_j with $P[j]$ currently **true**.

This induction hypothesis is trivially satisfied before the first iteration of the loop. For the r-th iteration, let $m = |T|$ be the size of a derivation tree T for p_i, using the current axioms and derivation rules. We distinguish two cases:

1. T has at least $m/4$ leaves: since the parallel loop in the DTEP algorithm set $P[k] = \text{true}$ if the children of a vertex labelled p_k are all leaves, this case is trivial.

2. Let r_k be the number of maximal chains in T with k internal vertices, and let b be the number of leaves of T, both at the start of the loop. A chain in T is part of a branch with all internal vertices having degree exactly 2. Then we have:

$$
\begin{aligned}
m &= 2b - 1 + \sum_{k \geq 1} k \cdot r_k; \\
m' &\leq b - 1 + \sum_{k \geq 1} \lceil k/2 \rceil r_k \\
&\leq b - 1 + \frac{1}{2}\left(\sum_{k \geq 1} k r_k + \sum_{k \geq 1} r_k \right) \\
&\leq b - 1 + \frac{1}{2}(m + 1 - 2b) + \frac{1}{2} \cdot 2b \\
&< \frac{3}{4} m \ .
\end{aligned}
$$

Here, m' is the size of the derivation tree after execution of the loop, using the current set of axioms and inferences. The first inequality follows since squaring of the matrix DI halves the length of all chains. The third inequality stems from the observation that in a tree with b leaves, the number of (maximal) chains can be at most $2b - 1$ as can be seen by assigning each such chain to its lower endpoint.

There is a trivial derivation tree for p_i after at most $\log_{4/3} m \leq 2.41 \log m$ (note that all logarithms whose base is not mentioned explicitly, are base 2). $\square$

Theorem 1 *Let (P, I, Z) be a derivation system with the property that each p in the minimal model M has a derivation tree of size*

$$
\leq 2^{\log^c N}.
$$

Then there is an $\mathcal{N}C$-algorithm to compute M.

Proof: It follows immediately from the above Lemma that under the stated conditions the DTEP algorithm runs in polylogarithmic time. More precisely, for the bound on the size of derivation trees given in the Theorem, the running time of the algorithm is

$$
O(\log^{c+1} N),
$$

and it uses N^3 processors. $\square$

10.3.2 Applications of DTEP

Longest Common Substrings

As a very simple example of a DTEP application we consider the *longest common substring* problem: given two strings $a_1 \cdots a_n$ and $b_1 \cdots b_m$ over some alphabet Σ, we are supposed to find a longest common substring $a_i \cdots a_{i+r} = b_j \cdots b_{j+r}$.

To obtain a DTEP formulation of this problem, we introduce variables $p_{i,j,l}$ whose intended meaning is:

$$p_{i,j,l} \text{ is true if } a_i \cdots a_{i+l-1} = b_j \cdots b_{j+l-1}.$$

Thus, we have the following axioms:

$$p_{i,j,1} \text{ iff } a_i = b_j;$$

and these inference rules:

$$p_{i,j,l} \;:-\; p_{i,j,\lceil l/2 \rceil} \, p_{i+\lceil l/2 \rceil, j+\lceil l/2 \rceil, \lfloor l/2 \rfloor}.$$

An easy induction on the length l of a common substring shows that $p_{i,j,l}$ is true in the minimal model for the above derivation system if and only if the substring of length l starting at position i in the first string is equal to the substring of the same length at position j in the second string. A similar induction can be used to show that for every common substring of length l, there is a derivation tree of size linear in l in the above derivation system. These two observations together establish the longest common substring problem as an instance of DTEP.

Other Applications

It is possible to rephrase a number of other problems as instances of the *Dynamic Tree Expression Problem*. A few examples are contained in the following list:

- transitive closure in graphs and digraphs

- the non-uniform word problem for context-free languages

- the circuit value problem for planar monotone circuits

- DATALOG programs with the *polynomial fringe property*

- ALT($\log n$, expolylog n); these are the problems recognizable by log-space bounded Alternating Turing machines with a bound of $2^{\log^c n}$ for the size of their computation tree, for some constant $c > 0$. This class is the same as $\mathcal{NC}$.

For the DATALOG example, we refer the interested reader to [21]. We briefly discuss the third and the last example in the above list.

The Planar Monotone Circuit Value Problem

A *circuit* is a directed acyclic graph (dag). Its vertices are called *gates*, and the arcs correspond to *wires* going from outputs of gates to inputs. The gates are either input/output gates connecting the circuit to its environment, or they are combinational mapping the values on the input wires of the gate to values on the output wires, according to the function represented by the gate. In a *monotone* circuit, all combinational gates are AND or OR gates. Therefore, the output of the circuit, as a function of the inputs, is a monotone function. A *planar* circuit is a circuit whose dag can be laid out in the plane without any wires crossing one another, and with the inputs and output(s) of the circuit on the outer face.

We assume that circuits are given as follows. The description of a circuit with n gates is a sequence $\beta_0, \ldots, \beta_{n-1}$. Each gate β_i is

1. 0-INPUT or 1-INPUT (also abbreviated as 0 and 1, respectively); or

2. AND(β_j, β_k), OR(β_j, β_k), or NOT(β_j), where $j, k < i$; in this case, the gate computes the logical function indicated by its name, with the values on the input wires to the gate as arguments, and puts this function value on all of the gate's output wires.

3. The output wire of β_{n-1} carries the output of the circuit.

The *circuit value problem* (CVP) consists in computing the output of a circuit, given its description. It is well known that CVP is complete for $\mathcal{P}$ under log-space reductions, and hence is a "hardest" problem for the class of polynomial time algorithms [9]. Even if we restrict ourselves to monotone circuits (allowing only AND and OR gates), or to planar circuits, the corresponding restricted CVP remains $\mathcal{P}$-complete [5]. However, the circuit value problem can be solved efficiently in parallel for circuits which are monotone *and* planar [3], [6]. While the original solutions were indirect and technically quite involved, the DTEP paradigm gives us a relatively simple approach.

For ease of presentation, we assume that the description $\beta_0, \ldots, \beta_{(n-1)}$ of a planar monotone circuit satisfies the following additional properties:

- The circuit is arranged in *layers*; the first layer is constituted by the inputs to the circuit; the last layer consists of the output of the circuit; for every wire in the circuit, there is a (directed) path from an input to the output of the circuit, containing that wire.

- All wires run between adjacent layers, without crossing one another.

- All gates on even layers are AND gates with at least one input.

- All gates on odd layers (except the first) are OR gates.

These restrictions are not essential. Given a general type description of a circuit, there are well-known $\mathcal{NC}$-algorithms to test whether it represents a monotone and planar circuit, and to transform it into a description of a functionally equivalent circuit satisfying the restrictions listed above.

To obtain a DTEP formulation, it is helpful to look at *intervals* of gates. Formally, let the triple (l, i, j) denote the interval from the ith through the jth gate (in the order in which they appear in the circuit description) on the lth layer of the circuit. We shall introduce boolean variables $p(l, i, j)$ with the intended meaning

$$p_{l,i,j} \text{ is \textbf{true} iff the output of every gate in } (l, i, j) \text{ is \textbf{true}.}$$

We simply observe that an interval of AND-gates has outputs all **true** if and only if all inputs to this interval of gates are **true**, and that these inputs come from an interval of gates on the next lower layer. For intervals of OR-gates, the situation is a bit more complicated since in order to obtain a **true** output from an OR-gate, only one of its inputs needs to be **true**. We must therefore be able to break intervals of **true** OR-gates down into smaller intervals which are **true** because a contiguous interval of gates on the next lower layer is **true**. Because wires don't cross in a planar layout, such a partition is always possible.

More formally, the inference rules for a given instance of the planar monotone circuit value problem are:

1. $p_{l,i,j} \; :- \; p_{l-1,i',j'}$ for every (nonempty) interval of AND-gates; here, i' denotes the first input of the first gate, and j' the last input of the last gate in the interval;

2. $p_{l,i,j} \; :- \; p_{l,i,k}\, p_{l,k+1,j}$, for all $i \leq k < j$; and

3. $p_{l,i,j} \; :- \; p_{l-1,i',j'}$, for all layers l of OR-gates and all pairs (i', j') where $i' \leq j'$ and $\beta_{i'}$ is an input of β_i, and $\beta_{j'}$ is an input of β_j.

The axioms are given by all intervals of **true** inputs to the circuit.

It is easy to see that the total length of the representation for the axioms and for all inference rules is polynomial in n. A straightforward induction, using the planarity of the circuit as outlined above, also shows that $p_{l,i,i}$ is **true** in the minimal model if and only if the ith gate on layer l has output **true** in the circuit.

Suppose $p_{l,i,j}$ is in the minimal model. To see that there is a linear size derivation tree for $p_{l,i,j}$ we consider the sub-circuit consisting of all gates from which a gate in (l,i,j) is reachable. Whenever we use an inference rule of type 1 or of type 3 in a derivation tree for $p_{l,i,j}$ we charge the vertex in the derivation tree (corresponding to the left-hand side of the rule) to the interval (l,i,j). We observe that we need to use an inference rule of type 2 for an OR-gate on some layer l' only if the interval of gates on layer $l' - 1$ feeding into this interval contains more than one maximal sub-interval of **true** gates. In this case, we can split the interval (l',i,j) into sub-intervals (l',i,k) and $(l',k+1,j)$ in such a way that one of the inputs to gate k on layer l' is **false**. We charge the left-hand side of the type 2 inference rule to the interval of **false** gates on layer $l' - 1$ containing this input. Because of the planarity of the circuit, rules used to derive $p_{l',i,k}$ and $p_{l',k+1,j}$ are charged to disjoint sets of intervals. Therefore, the number of maximal intervals of gates with the same output value is an upper bound on the size of a smallest derivation tree for any element in the minimal model. We conclude

Theorem 2 *There is an $\mathcal{NC}$ algorithm for the planar monotone circuit value problem.*

⬜

Tree Size Bounded Alternating Turing Machines

Alternating Turing Machines are a generalization of standard (nondeterministic) Turing Machines [8], [1]. They are interesting because their time complexity classes closely correspond to (standard) space complexity classes, and because their computations directly correspond to first order formulas in prenex form with alternating quantifiers.

For an informal description of Alternating Turing Machines (ATM's), we assume that the reader is familiar with the concept of standard (nondeterministic) Turing machines [8]. An ATM consists of a (read-only) input tape and some number of work or scratch tapes, as in the standard model, as well as a finite control. The states of the finite control, however, are partitioned into *normal, existential,* and *universal* states. Existential and universal states each have two successor states. Normal states can be *accepting, rejecting,* or *non-final*. An *instantaneous description* or *id* of an ATM consists of a string giving the state of the finite control of the ATM, the position of the read head on its input tape, and head positions on and contents of the work tapes. It should be clear that, given the input to the machine, an id completely determines the state of the machine.

A computation of an ATM can be viewed as follows. In any state, the machine performs whatever action is prescribed by the state it is in. If the state is a normal non-final state, the finite control then simply goes to its unique successor state. Accepting or rejecting states have no successors. If the state is existential or universal, the machine clones itself into as many copies as there are successor states, with each clone starting with an id identical to that of the cloning id, except that the state of the cloning id is replaced by that of the successor state. The cloning machine disappears. It is best to view the computation of an ATM as a tree, whose nodes correspond to id's of the computation, in such a way that children of a node in the tree represent successor id's.

Given this interpretation, each computation of an ATM can be mapped into a (binary) tree. This tree need not necessarily be finite, since there can be non-terminating branches of the computation. To define whether an ATM *accepts*, or, dually, *rejects* some input, we look at the corresponding computation tree as defined above. A *leaf* in this tree is called *accepting* iff the state in its id is accepting. An internal node whose id contains a normal state is accepting iff its (unique) descendent in the tree is accepting. An internal node with an existential state is accepting iff at least one of its immediate descendents is accepting, and a node with a universal state iff all of its immediate descendents are accepting. An id of the ATM is called *accepting* if it is attached to an accepting node in the computation tree. An ATM is said to accept its input if and only if the root of the corresponding computation tree is accepting by the above definition, or, equivalently, if the initial configuration of the machine is accepting.

To cast ATM computations into the DTEP framework we note that if the space used by the ATM on its worktapes is bounded by $O(\log n)$ then the number of distinct id's for the machine is bounded by some polynomial in n. Assume that $id_0, \ldots, id_r$ is an (efficient) enumeration of these id's. Our instance of DTEP will have boolean variables p_i, for $i = 0, \ldots, m$, with the intended meaning

$$p_i \text{ is \textbf{true} iff } id_i \text{ is accepting.}$$

The axioms are given by those id's containing accepting states, and we will have inference rules

- $p_i :- p_j p_k$ iff p_i contains a universal state with two successor states, and id_j and id_k are the two immediate successor id's of id_i; and

- $p_i :- p_j$ in all other cases where id_j is a (the) immediate successor id of id_i.

A straightforward induction shows that there is an exact correspondence between derivation trees for the p_i corresponding to the initial id of the ATM, and certain subtrees of the ATM's computation tree. These subtrees are obtained by removing, for each node whose id contains an existential state, all but one of the children of that node, together with their subtrees. Also, the leftover child has to be an accepting node.

Since the computation tree of the ATM is of expolylog size, so is any derivation tree obtained by the above construction. This concludes the (informal) proof for

Theorem 3 *Problems in the complexity class ALT(*$\log n$, *expolylog* n*) are instances of the Dynamic Tree Expression Problem. Since this class is the same as* $\mathcal{NC}$*, every problem in* $\mathcal{NC}$ *can be cast as an instance of DTEP.*

☐

10.4   Application to Algebraic Straightline Programs

Straightline programs consist of a sequence of assignment statements executed in order. Each variable referenced in a statement is either an input variable, or it has been assigned a value in an earlier assignment. In *algebraic straightline programs*, the expressions on the righthand side of the assignment are basic unary or binary expressions involving the arithmetic operators $+$, $-$, $\times$, and $/$, as well as constants and input or computed variables. Each non-input variable is assigned a value in exactly one statement in the straightline program. In what follows, we are only concerned with straightline programs containing the operators $+$, $-$, and $\times$. Any such straightline program computes some (multivariate) polynomial of the input variables. Conversely, every multivariate polynomial can be expressed as an algebraic straightline program.

A straightline program corresponds, in a natural way, to a *directed acyclic graph* or *dag*. In this dag, the vertices correspond to the assignment statements, the *sources* of the dag (*i.e.*, vertices of indegree zero) correspond to the input variables of the program, and arcs signify the use of variables defined in one statement by some other statement. We define the *size* of a straightline program to be its number of assignment statements. Figure 10.1 shows a straightline program for the polynomial $((x_1 + x_2)^3 + x_1^2 x_3)(x_1 + x_2)$ with indeterminates x_1, x_2, and x_3, and Figure 10.2 gives the corresponding computation dag. In Figure 10.2, all edges represent arcs directed upward.

We assume without loss of generality that the operation in the last assignment in a straightline program is a multiplication.

To make straightline programs more uniform and easier to deal with, we first wish to transform them into an (algebraically equivalent) sequence of *bilinear forms* of

$$
\begin{aligned}
x_4 &= x_1 + x_2 & x_5 &= x_4 \times x_4 \\
x_6 &= x_5 \times x_4 & x_7 &= x_1 \times x_1 \\
x_8 &= x_3 \times x_7 & x_9 &= x_6 \times x_8 \\
x_{10} &= x_4 \times x_9
\end{aligned}
$$

Figure 10.1: Straightline program for polynomial $((x_1 + x_2)^3 + x_1^2 x_3)(x_1 + x_2)$.

the form

$$
y_i = \Big(\sum_{j<i} c_{ij}^{(l)} y_j\Big) \cdot \Big(\sum_{j<i} c_{ij}^{(r)} y_j\Big).
$$

Again, the y_j are either input variables or assigned a value in an earlier assignment. In fact, each non-input variable y_i is in exact correspondence with some x_{k_i} in the original straightline program which is the result of a multiplication operation. Assuming that $i = k_i$ for the input variables, the coefficients $c_{ij}^{(l)}$ (resp., $c_{ij}^{(r)}$) give the number of distinct paths from x_{k_j} to the left (resp., right) multiplicand of x_{k_i} in the dag belonging to the original straightline program. These coefficients can be obtained in a straightforward way by repeatedly squaring a matrix A obtained as follows. We first construct, from the computation dag for the straightline program, an auxiliary dag by splitting every multiplication node v into three nodes, v^1, v^2, and v^3: v^1 receives the arc from the left multiplicand, v^2 the arc from the right multiplicand. All arcs originally leaving v are reattached to v^3. In the resulting digraph, the original input nodes and the type 3 nodes are sources, the type 1 and type 2 nodes are the sinks. The matrix A mentioned above is the adjacency matrix of this digraph, after we attach a loop (directed cycle of length one) to every sink. Further details are left to the reader. The coefficients in the bilinear forms can be computed on a PRAM in time $O(\log^2 n)$, using $M(n)$ processors. Here, n is the size of the original straightline program, and $M(n) \le n^3$ is the number of processors required to perform matrix multiplication in time $O(\log n)$.

Figure 10.3 shows a graphical representation of the transformed "bilinear straight-line program" for the straightline program resp. computation dag given in Figures 10.1 and 10.2. Each bilinear form is represented by a flat 'V', with the terms in the first (second) factor given by the arcs ending in the left (right) arm of the 'V'. The coefficient belonging to a term is attached to the arc unless it is one or zero. In the latter case, no arc is drawn at all.

For the digraph B of a bilinear straightline program, we can define operations analogous to the DTEP operations. The boolean variables are replaced by pairs (c_i, v_i) for every node i in B. The first element, c_i, is a boolean indicating whether the value of the polynomial corresponding to vertex i has been computed; if c_i is **true** then v_i holds this value. Initially, the values of all the input variables are known. We call a node of B a *1-node* iff it is not a sink (*i.e.* a node of outdegree zero) and the value of exactly one of its factors is known (*i.e.* the values of all nodes

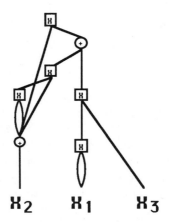

Figure 10.2: Computation dag for straightline program in Figure 10.1

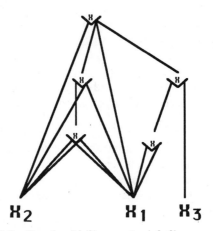

Figure 10.3: Graph of bilinear straightline program

with arcs to the corresponding "leg" have been computed). All other nodes whose value is still unknown are called *2-nodes*.

We also have several matrices describing the arcs in the graph, and the coefficients attached to them. The set of arcs and coefficients, and consequently the entries in these matrices, change during the course of the algorithm.

- the (i, j)th entry of $C^{(l)}$ contains the coefficient of the arc from the ith node of B to the *left leg* of the jth node, for all nodes i and j such that j is a 2-node; the other entries are zero;

- $C^{(r)}$ is defined correspondingly for the righthand factors;

- the (i, j) entry of the matrix D, for j a 1-node, is the coefficient of the arc from node i to node j;

- the matrix $\bar{C}$ is a projection matrix; all its entries are zero except for those on the diagonal positions corresponding to 2-nodes.

Each phase of the adapted DTEP algorithm consists of the following

    co one phase of adapted DTEP **oc**

    for all i with $c_i = $ **false** and both factors known **do in parallel**

        compute v_i and set $c_i = $ **true**;

    for all nodes i which just became new 1-nodes **do in parallel**

        multiply the coefficients of all arcs entering the unknown "leg" by the other, known factor;

    update all matrices so as to reflect the changes in the sets of 1-nodes and 2-nodes;

    $C^{(l)} := (\bar{C} + D) \cdot C^{(l)}; \quad C^{(r)} := (\bar{C} + D) \cdot C^{(r)};$

    $D := (\bar{C} + D) \cdot D;$

    update B to correspond to the newly computed matrices;

    co end of phase **oc**

It is straightforward to verify that the bilinear program given by the modified graph after execution of one phase computes the same values as the original bilinear program. Also, one phase can be executed on a PRAM in time $O(\log n)$ employing $M(n)$ processors, where $M(n) \leq n^3$ is the number of processors required to multiply two $n \times n$ matrices in $O(\log n)$ parallel steps.

To obtain a bound on the required number of phases, it is not sufficient to unfold the dag B into a tree and then use our general results about DTEP since there are easy examples where the unfolded tree is of exponential size. However, the following *worst-term* argument works.

Define, for each node in the graph of a bilinear straightline program, its *formal degree* inductively as follows: The original input variables have formal degree 1, and each other variable has a formal degree equal to the sum of the formal degrees of its two factors. Here, the formal degree of a factor is defined to be the maximum of the formal degrees of all its terms.

For every factor of a non-input variable y_i in a bilinear straightline program, we define a *worst term*, derived from the execution of the above algorithm: determine, for the corresponding "leg" of the node i in the sequence of graphs generated by the algorithm, which other node j with an arc to that leg is the latest to have its value computed. If j has an arc to i in the first graph of the sequence, y_j becomes the worst term for the factor of y_i under consideration, and we distinguish the arc from j to i. Otherwise, distinguish any path from j to the appropriate leg of i in the first graph of the sequence. Every node on this path corresponds to the worst term of the appropriate factor of the next node on the path.

Consider the subgraph B' induced by the distinguished arcs. Assume without loss of generality that B' has only one sink. Clearly, B' computes some monomial. If we unfold B' into a tree T then T's size is exactly twice the formal degree of this monomial, minus one. The reason is that all non-leaf nodes in T are multiplication nodes, with exactly one term per factor.

A simple induction shows that the adapted DTEP algorithm computes the values of nodes in T in exactly the same phases as it computes them when run on B, the full graph. The analysis of the general DTEP algorithm applies to the adapted algorithm, executed on T. We obtain that the adapted DTEP algorithm computes the value of each y_i in a number of phases proportional to the logarithm of the formal degree of y_i. Since the initial transformation to a bilinear straightline program is basically a transitive closure computation, we obtain the

Theorem 4 *Suppose S is a straightline program containing n variables and computing a polynomial of formal degree d. The adapted DTEP algorithms computes the value of the polynomial given by S, at a specified point, in time $O(\log n(\log n + \log d))$ using $M(n)$ processors on a unit cost PRAM.*

□

Thus, polynomials given by straightline programs can be evaluated by an $\mathcal{NC}$ algorithm as long as their formal degree is expolylog in n. This result was originally obtained in [13]. Our presentation is intended to give evidence for the wide applicability of the DTEP paradigm, and maybe to supply a somewhat simpler, less technical proof.

10.5 Conclusion

The DTEP paradigm establishes an interesting connection between fast parallel computation and small proof or derivation trees. In more technical terms, the instances of DTEP and $\mathcal{NC}$ are characterized as ALT($\log n$, expolylog n) or the set of problems recognizable by log space bounded alternating Turing machines with "expolylogarithmically" bounded computation trees. DTEP is not necessarily intended to provide optimal parallel algorithms for a given problem. Instead, its primary function is to provide a convenient, high level way for the specification of an efficient and highly portable parallel algorithm. The input to DTEP is in a format very similar to logic programs, rather declarative than procedural, and thus allows to abstract away many peculiarities of any underlying parallel architecture.

The DTEP algorithm can be efficiently implemented on various "real world" parallel architectures, like binary hypercubes, butterfly networks, shuffle-exchange based networks, and multi-dimensional meshes of trees [12]. The algorithm can also be tailored and made more efficient (both in terms of time and number of processors) for more restricted instances of the dynamic tree expression problem. Some of these optimizations are also discussed in [12].

We have also shown how to extend the basic DTEP paradigm to problems over algebraic domains, in particular the parallel evaluation of algebraic computation dags, or straightline programs. Further interesting extensions include the possibility of handling non Horn clauses in the basic formulation of DTEP, along the lines discussed in [21].

Another interesting line of research is the characterization of more classes of problems for DTEP in independent terms, in particular the derivation of conditions for the existence of small proof trees.

References

[1] A. Chandra, D. Kozen, and L. Stockmeyer. Alternation. *J.ACM*, 28(1):114–133, 1981.

[2] R. Cole. Parallel merge sort. In *Proceedings of the 27th Ann. IEEE Symposium on Foundations of Computer Science (Toronto, Canada)*, pages 511–516, 1986.

[3] P. Dymond. *Simultaneous resource bounds and parallel computation*. Ph.D. Thesis, Department of Computer Science, University of Toronto, 1980.

[4] S. Fortune and J. Wyllie. Parallelism in random access machines. In *Proceedings of the 10th Ann. ACM Symposium on Theory of Computing (San Diego, CA)*, pages 114–118, 1978.

[5] L. Goldschlager. The monotone and planar circuit value problems are log-space complete for $\mathcal{P}$. *SIGACT News*, 9(2):25–29, 1977.

[6] L. Goldschlager. A space efficient algorithm for the monotone planar circuit value problem. *Information Processing Letters*, 10(1):25–27, 1980.

[7] P. Hochschild, E. Mayr, and A. Siegel. Techniques for solving graph problems in parallel environments. In *Proceedings of the 24th Ann. IEEE Symposium on Foundations of Computer Science (Tucson, AZ)*, pages 351–359, 1983.

[8] J. Hopcroft and J. Ullman. *Introduction to Automata Theory, Languages, and Computation*. Addison-Wesley, 1979.

[9] R. Ladner. The circuit value problem is log-space complete for $\mathcal{P}$. *SIGACT News*, 7(1):583–590, 1975.

[10] R. Ladner and M. Fischer. Parallel prefix computation. *J.ACM*, 27(4):831–838, 1980.

[11] R. Lipton and J. Valdes. Census functions: An approach to VLSI upper bounds (preliminary version). In *Proceedings of the 22nd Ann. IEEE Symposium on Foundations of Computer Science (Nashville, TN)*, pages 13–22, 1981.

[12] E. Mayr and G. Plaxton. *Network Implementations of the DTEP Algorithm*. Technical Report STAN-CS-87-1157, Department of Computer Science, Stanford University, 1987.

[13] G. Miller, V. Ramachandran, and E. Kaltofen. *Efficient parallel evaluation of straight-line code and arithmetic circuits*. Technical Report TR-86-211, Computer Science Dept., USC, 1986.

[14] G. Miller and J. Reif. Parallel tree contraction and its application. In *Proceedings of the 26th Ann. IEEE Symposium on Foundations of Computer Science (Portland, OR)*, pages 478–489, 1985.

[15] G. Pfister. *The architecture of the IBM research parallel processor prototype (RP3)*. Technical Report RC 11210 Computer Science, IBM Yorktown Heights, 1985.

[16] N. Pippenger. On simultaneous resource bounds. In *Proceedings of the 20th Ann. IEEE Symposium on Foundations of Computer Science (San Juan, PR)*, pages 307–311, 1979.

[17] F. Preparata and J. Vuillemin. The cube-connected-cycles: A versatile network for parallel computation. In *Proceedings of the 20th Ann. IEEE Symposium on Foundations of Computer Science (San Juan, PR)*, pages 140–147, 1979.

[18] W. Ruzzo. Tree-size bounded alternation. In *Proceedings of the 11th Ann. ACM Symposium on Theory of Computing (Atlanta, GA)*, pages 352–359, 1979.

[19] J. Schwartz. Ultracomputers. *ACM Transactions on Programming Languages and Systems*, 2(4):484–521, 1980.

[20] C. Seitz. The cosmic cube. *CACM*, 28(1):22–33, 1985.

[21] J. Ullman and A. van Gelder. Parallel complexity of logical query programs. In *Proceedings of the 27th Ann. IEEE Symposium on Foundations of Computer Science (Toronto, Canada)*, pages 438–454, 1986.

Chapter 11

Randomized Parallel Computation

Sanguthevar Rajasekaran [1] [2]
John H. Reif [1] [3]

Abstract

This paper surveys randomized parallel algorithms found in the literature for various problems in computer science. In particular we will demonstrate the power of randomization as a tool for parallelizing sequential algorithms and introduce the reader so some of the techniques employed in designing randomized parallel algorithms. We consider representative problems from the following areas of computer science and describe how randomized parallel algorithms for these problems have been obtained: 1) routing and sorting, 2) processor load balancing, 3) algebra, and 4) graph theory. Finally we discuss methods of derandomizing randomized parallel algorithms.

11.1 Introduction

11.1.1 History

The technique of randomizing an algorithm to improve its efficiency was first introduced in 1976 independently by [Ra76] and [SS76]. Since then, this idea has been used to solve myriads of computational problems successfully. And today randomization has become a powerful tool in the design of both sequential and parallel algorithms.

---

[1] Authors supported by ONR ContractN00014-80-C-0647 and NSF grant DCR-85-03251.
[2] Harvard University, Cambridge, MA
[3] Duke University, Durham, NC

Even though the idea of randomization is at least as old as Hoare's quicksort algorithm [Ho62], these previous approaches presuppose a distribution on the space of all possible inputs, which is not a valid assumption at all. For example, Hoare's quicksort algorithm may run for an indefinitely long period of time on certain inputs. But all such bad input permutations are only a small fraction. If we assume (which indeed Hoare does) that each input permutation is equally likely to occur, then quicksort algorithm is very well practical because with very high probability the given input permutation will not be a bad one and hence the algorithm will terminate quickly. But, Hoare's assumption of a uniform distribution on the input space is questionable, since the input distribution may vary quite unpredictably. [Ra76] and [SS76] rectify this problem by introducing randomization into the algorithm itself.

Informally, a randomized algorithm (in the sense of [Ra76] and [SS76]) is one which bases some of its decisions on the outcomes of coin flips. We can think of the algorithm with one possible sequence of outcomes for the coin flips to be different from the same algorithm with a difference sequence of outcomes for the coin flips. Therefore, a randomized algorithm is really a family of algorithms. For a given input, some of the algorithms in this family might run for an indefinitely long time. The objective in the design of a randomized algorithm is to ensure that the number of such bad algorithms in the family is only a small fraction of the total number of algorithms. If for *any* input we can find at least $(1 - \epsilon)$ (ϵ being very close to 0) portion of algorithms in the family that will run quickly on that input, then clearly, a random algorithm in the family will run quickly on any input with probability $\geq (1 - \epsilon)$. In this case we say that this family of algorithms (or this randomized algorithm) runs quickly with probability at least $(1 - \epsilon)$. ϵ is called the error probability. Observe that this probability is independent of the input and the input distribution.

To give a flavor for the above notions, we now give an example of a randomized algorithm. We are given a polynomial of n variables $f(x_1, \ldots, x_n)$ over a field F. It is required to check if f is identically zero. We generate a random n -vector $(r_1, \ldots, r_n)$ $(r_i \in F, i = 1, \ldots, n)$ and check if $f(r_1, \ldots, r_n) = 0$. We repeat this for k independent random vectors. If there was at least one vector on which f evaluated to a nonzero value, of course f is nonzero. If f evaluated to zero on all the k vectors tried, we conclude f is zero. It can be shown (see section 11.5.1) that the probability of error in our conclusion will be very small if we choose a sufficiently large k. In comparison, the best known deterministic algorithm for this problem is much more complicated and has a much higher time bound.

11.1.2   Advantages of Randomization

Advantage of randomized algorithms are manyfold. Two extremely important advantages are their simplicity and efficiency. A major portion of randomized algorithms found in the literature are extremely simpler and easier to understand than the best deterministic algorithms for the same problems. The reader would have already got a feel for this from the above given example of testing a polynomial for identity. Randomized algorithms have also been shown to yield better complexity bounds. Numerous examples can be given to illustrate this fact. But we won't enlist all of them here since the algorithms described in the rest of the paper will convince the reader.

A skeptical reader at this point might ask: How dependable are randomized algorithms in practice, after all there is a nonzero probability that they might fail? This skeptic reader must realize that there is a probability (however small it might be) that the hardware itself might fail. [AM77] remark that if we can find a fast algorithm for a problem with an error probability $< 2^{-k}$ for some integer k independent of the problem size, we can reduce the error probability far below the hardware error probability by making k large enough.

11.1.3 Contents of this Paper

The tremendously low cost of hardware nowadays has prompted computer scientists to design parallel machines and algorithms to solve problems very efficiently. In an early paper [Re84] proposed using randomization in parallel computation. In this paper he also solved many algebraic and graph theoretic problems in parallel using randomization. Since then a new area of CS research has evolved that tries to exploit the special features offered by both randomization and parallelization. The scope of this paper is to survey randomized parallel algorithms available for various computational problems. We do not claim to give a comprehensive list of all the randomized parallel algorithms that have been discovered so far. We only discuss representative examples which demonstrate the special features of both randomization and parallelization. The areas of CS we consider are: 1) routing and sorting, 2) processor load balancing, 3) algebra, and 4) graph theory. We also discuss in this paper various ways of derandomizing randomized parallel algorithms. The reader is assumed to be familiar with some basic concepts of probability theory and graph theory. For a beautiful treatise on probability theory read [Fe68]. Some of the standard text books on graph theory are [Ha69], [De74], and [RND77].

The rest of this paper is organized as follows. Section 11.2 contains descriptions of standard parallel machine models and some relevant definitions. In sections 11.3, 11.4, 11.5, and 11.6, we discuss problems in the areas of sorting and routing, processor load balancing, algebra, and graph theory, respectively. In section 11.7 we explain some derandomization techniques. Finally, in section 11.8 we provide some concluding remarks.

11.2 Preliminaries

11.2.1   Parallel Machine Models

A large number of parallel machine models have been proposed. Some of the widely accepted models are: 1) fixed connection machines, 2) shared memory models, 3) the boolean circuit model, and 4) the parallel comparison trees. In the randomized version of these models, each processor is capable of making independent coin flips in addition to the computations allowed by the corresponding deterministic version. The *time complexity* of a parallel machine is a function of its input size. Precisely, time complexity is a function $g(n)$ that is the maximum over all inputs of size n of the time elapsed when the first processor begins execution until the time the last processor stops execution.

Just like the big- O function serves to represent the complexity bounds of deterministic algorithms, $\tilde{O}$ serves to represent the complexity bounds of randomized algorithms. We say a randomized algorithm has resource (time, space, etc.) bound $\tilde{O}(g(n))$ if there exists a constant c such that the amount of resource used by the algorithm (on any input of size n) is no more than $c\alpha g(n)$ with probability $\geq 1-1/n^{\alpha}$. We say a parallel algorithm is *optimal* if its processor bound P_n and time bound T_n are such that $P_n T_n = \tilde{O}(S)$ where S is the time bound of the best known sequential algorithm for that problem. Next we describe the machine models.

Fixed connection networks

A fixed connection network is a directed graph $G(V, E)$ whose nodes represent processors and whose edges represent communication links between processors. Usually we assume that the degree of each node is either a constant or a slowly increasing function of the number of nodes in the graph.

Shared memory models

In shared memory models, a number of processors (call it P) work synchronously communicating with each other with the half of a common block of memory accessible by all. Each processor is a random access machine [AHU74]. Each step of the algorithm is an arithmetic operation, a comparison, or a memory access. Several conventions are possible to resolve read or write conflicts that might arise while accessing the shared memory. EREW PRAM is the shared memory model where no simultaneous read or write is allowed on any cell of the shared memory. CREW PRAM is a variation which permits concurrent read but not concurrent write. And finally, CRCW PRAM model allows both concurrent read and concurrent write. Read or write conflicts in the above models are taken care of with a priority scheme. For a discussion of further variations in PRAM models see [FW78].

Boolean circuits

A Boolean circuit on n input variables is directed acyclic graph (dag) whose nodes of indegree 0 are labeled & with variables & their negations, and the other nodes (also called gates) are labeled with a boolean operation (like $\land$, $\lor$, etc.). The size of a circuit is defined to be the number of gates in the circuit. Fan-in and depth of a circuit C are defined to be the maximum indegree of any node, and the depth of the dag, respectively. Let Σ be an alphabet set, and consider any language over this alphabet set, $W \subseteq \Sigma^n$. We say that the Boolean circuit C accepts W if on any input $\omega \in W$, C evaluates to 1. The circuit size complexity of W is defined to be the size of the smallest circuit accepting W.

Let $A \subseteq \Sigma^*$. The circuit size complexity of A is a function $g: n \to n$ such that $g(n)$ is the circuit size complexity of A^n where $A^n = A \cap \Sigma^n$. The above definition implies that a recognizer for a language is a family of Boolean circuits $< \alpha_n >$, each circuit of which accepts strings of a particular length. We say A has polynomial size circuits if its circuit size complexity is bounded above by a polynomial in the input size. The Boolean circuits defined above are language recognizers and have a single output. The notion of a Boolean circuit can easily be extended to Boolean circuits that compute an arbitrary function $f: \{0,1\}^n \to \{0,1\}^m$. The definitions of depth and size remain unchanged.

Parallel comparison trees

A parallel comparison tree model (proposed by Valiant [Va75]) is the same as a sequential comparison tree model of {AHU74}, except that in the parallel model at each node P comparisons are made. The computation proceeds to a child of the node, the child being determined by the outcomes of the comparisons made. This model is a much more powerful model than the ones seen before.

11.2.2   Some Definitions

We say a Boolean circuit family $< \alpha_n >$ is uniform if the description of α_n (for any integer n) can be generated by a deterministic Turing machine (TM) in space $O(\log n)$ with n as the input.

With the definition of uniformity at hand, we can define a new complexity class, viz., the set of all functions f computable by circuits of "small" depth. We define NC^k to be the set of all functions f computable by a uniform circuit family $< \alpha_n >$ with size $n^{0(1)}$ and depth $O(\log^k n)$. Also define $NC = \cup_k NC^k$.

A randomized Boolean circuit is the same as a Boolean circuit except that each node of the former, in addition to performing a Boolean operation can also make a coin flip. A randomized Boolean circuit on n inputs is said to compute a function f

if its outputs the correct value of f with probability $> 1/2$. Define RNC^k to be the set of all fuinctions computable by a uniform family of randomized Boolean circuits $< \alpha_n >$ with size $n^{O(1)}$ and depth $O(log^k n)$. Also define $RNC = \cup_k RNC^k$.

It should be mentioned that NC and RNC contain the same set of functions even when any of the standard shared memory computer models are used to define them. All the problems studied in this paper belong to RNC.

Different types of randomized algorithms

Two types of randomized algorithms can be found in the literature: 1) those that always output the correct answer but whose run-time is a random variable with a specified mean. These are called Las Vegas algorithms; and 2) those that run for a specified amount of time and whose output will be correct with a specified probability. These are called Monte Carlo algorithms. Primality testing algorithm of Rabin [Ra76] is of the first type.

The error of a randomized algorithm can either be 1-sided or 2-sided. Consider a randomized algorithm for recognizing a language. The output of the circuit is either *yes* or *no* . There are algorithms which when outputting *yes* will always be correct, but when outputting *no* they will be correct with high probability. These algorithms are said to have 1-sided error. Algorithms that have nonzero error probability on both possible outputs are said to have 2-sided error.

The above definitions and preliminaries will help the reader to better understand the algorithms that follow.

11.3   Randomized Parallel Algorithms for Routing and Sorting

How fast a parallel computer can run is determined by two factors: 1) how fast the individual processors can compute, and 2) how fast the processors can communicate among themselves. The rapid advance in VLSI technology makes it possible to increase the computational powers of individual processors arbitrarily and hence the time efficiency of a parallel machine is essentially determined by the interprocessor communication speed. one of the main problems in the study of parallel computers has been to design an N -processor realistic parallel computer whose communications time is as low as permitted by the topology of its interconnection. A realistic parallel machine is understood to be a fixed connection network. In-degree and out-degree in this network must be low due to physical limitations.

Since a realistic parallel computer is not only a parallel machine on its own right, but also it can be used to simulate any of the ideal computers (like PRAMs) very efficiently. One step of communication in such a machine is: Each processor has a distinct packet of information that it wants to send to some other processor. The task (also called routing) is to route the packets to their destinations such that at the most one packet passes through any wire at any time, and all the packets arrive at their destinations quickly. A communication step ends when the last packet has reached its destination.

The problem of designing a realistic computer is then to come out with a topology for a fixed connection network and to design an efficient routing algorithm for that topology. A fixed connection network topology together with its routing function is called a *communication scheme* .

Valiant [Va82] was the first to present an efficient randomized parallel algorithm for a fixed connection network called the binary n -cube. If N is the number of nodes in the network, the degree of the binary n -cube is $n = O(\log N)$ and his routing algorithm runs in the $\tilde{O}(\log N)$. This work was followed by Upfal's [Up84] who gave a routing algorithm with the same time bound for a constant degree network. Both these algorithms had a queue size (i.e., the maximum number of packets that any node will have at any time during execution) of $\tilde{O}(log N)$. Pippenger's [Pi84] routing algorithm for a constant degree fixed connection network not only had the same time complexity but also it had a queue size of $O(1)$. Details of these algorithms are given in section 11.3.1.

In section 11.3.2 we will describe some of the important randomized sorting algorithms found in the literature. Sorting is the process of rearranging a sequence of values in ascending or descending order. Many application programs like compilers, operating systems, etc., use sorting extensively to handle efficiently tables and lists. Both due to its practical value and theoretical interest, sorting has been an attractive area of research in CS.

A large number of deterministic parallel algorithms have been proposed for special purpose networks called *sorting networks* (see [Ba68] and [Vo71]). Of these the most efficient algorithm is due to Batcher [Ba68]. His algorithm sorted N numbers in time $O(\log^2 N)$ time using $N \log^2 N$ processors. Designing an algorithm to sort N numbers in $O(\log N)$ parallel time using a linear number of processors was a long open problem. Reischuk [Reis81] discovered a randomized algorithm that employed N PRAM processors and ran in time $\tilde{O}(\log N)$. This algorithm is impractical owing to its large memory requirements. It was Reif and Valiant [RV82] who first gave a practical sorting algorithm (FLASHSORT) that was optimal and ran on a network called cube connected cycles (CCC). This was a randomized algorithm that used N processors to sort N numbers in time $\tilde{O}(\log N)$. Since $\Omega(N \log N)$ is a well known lower bound [AHU74] for sequential comparison sorting of N keys, their algorithm is indeed optimal. A summary of this algorithm appears in section 11.3.2.

When the keys to be sorted are from a finite set, sorting becomes simple. Bucketsort algorithm [AHU74] can be used to sort N integer keys (keys in the range $[1,N]$)

in N sequential steps. Does there exist an optimal parallel algorithm for sorting integer keys (INTEGER SORT)? In other words, is there a parallel algorithm that can sort N integer keys in $O(\log N)$ time using only $N/\log N$ processors? This question is answered in the affirmative by the randomized PRAM algorithm of Rajasekaran and Reif [RR85]. Details of this algorithm will also appear in section 11.3.2.

It should be mentioned here that deterministic algorithms have been found to sort N general keys on a constant degree network of linear size in time $O(\log N)$ by [AKS83] and Leighton [Le84]. Unfortunately these algorithms have large constants in their time bounds making them impractical. This is a good instance where randomization seems to help in obtaining practical parallel algorithms (remember [RV82]'s FLASHSORT has better constants). Recently [Co86] has given a deterministic optimal sorting algorithm for the CRCW PRAM model that has a small constant in its time bound. But obtaining an optimal sorting algorithm (either deterministic or randomized) on any fixed connection network with a small constant in the time bound is still an open problem.

11.3.1   Routing Algorithms

Valiant's algorithm

Valiant's routing algorithm [Va82] runs on an n-cube. Each node in the cube is named by an n-bit binary vector $(x_1, \ldots, x_n)$. There are 2^n nodes in the network and the degree of each node is n. For each node $e = (x_1, \ldots, x_n)$, let $\bar{e}_i$ stand for $(x_1, \ldots, x_{i-1}, \bar{x}_i, x_{i+1}, \ldots, x_n)$, where $\bar{x}_i$ is the complement of x_i. Every node e is connected to n neighbors one in each dimension, i.e., e is connected to $\bar{e}_i$, $i = 1, \ldots, n$.

The algorithm consists of two phases. In the first phase, each packet is sent to a random destination and in the second phase packets in the random destinations are sent to their correct destinations. Each phase runs in time $\tilde{O}(n)$ and hence the whole algorithm runs in the same time bound.

In phase I, at each time unit, every packet p chooses one of the dimensions at random. It decides to traverse along this direction with probability $1/2$. Once a packet chooses a dimension, it will not choose it again. If a packet decides to traverse along the chosen dimension, it will enter a queue in its current location. In phase II, each packet does the same thing, except that now the set of dimensions to be traversed by a packet is determined by the shortest path between its destination and the node that it starts its second phase from. Also, once a packet chooses a dimension in phase II, it will traverse along it.

Each node contains n queues, one for packets to be transmitted along each dimension. Each packet p is associated with a set $U_p \subseteq \{1, 2, \ldots, n\}$. In phase I, it consists of the set of dimensions along which possible transmissions have not yet been considered, and in phase II it consists of dimensions along which transmission still has to take place. Each of the phases is said to be dimensions along which transmission still has to take place. Each of the phases is said to be finished when U_p is empty for every p. One of the special features of Valiant's algorithm is that it is *oblivious*, i.e., the route taken by any packet is independent of the routes taken by the other packets. This property is important for an algorithm to be applied in a distributed environment.

Upfal's algorithm

Upfal's [Up84] routing algorithm is applicable to a class of communication schemes he calls *Balanced Communication Schemes* (BCS). Here, we will consider only an example of a BCS. The topology of this communication scheme is a 2-way Digit Exchange Graph (2DEG). The number of nodes in the graph (N) is $m2^m$. Here again each processor named by a binary vector of length $\log m + m$. The m rightmost bits of a node's name is called its address and the other bits constitute its prefix. The degree of each node is 2. The two edges leaving the processor with an address $b_0, \ldots, b_a, \ldots, b_{m-1}$ and a prefix α are connected to processors with the addresses $b_0, \ldots, b_\alpha, \ldots, b_{m-1}$ and $b_0, \ldots, \overline{b_\alpha}, \ldots, b_{m-1}$ and prefixes $(\alpha + 1) \bmod m$. Each group of processors with the same prefix form one *stage* of the network.

Upfal's algorithm is also a two phase algorithm, where packets are sent to random destinations in the first phase and from random destinations to their actual destinations in the second phase. Packets are given priority numbers, which are integers in the range $[1, 3m - 1]$. The priority of a packet strictly increases on transitions. At any node packet with the least priority numbers is given precedence over the others when there is a contention for an outgoing edge. In other words, packets that have traveled less distance are given priority over those that have traveled more.

Consider a packet initially located in a processor with prefix α and which is destined for a processor with address $b_0, \ldots, b_{m-1}$ and prefix β. The first phase of the algorithm is performed in two steps. First step of phase A takes the packet to a random destination with the same prefix α. The packet undergoes m transitions. At each time, the packet chooses randomly one of the edges that leave the node the packet is currently in. Since there are m stages in the network, the packet will end up in a node with prefix α. The the second step of phase A, the packet traverses to a random address in stage (prefix) β. This is done by $(\beta - \alpha) \bmod m$ transitions. Here again, the packet chooses a random edge leaving its current location.

Phase B takes the packet to its final destination. At a transition leaving processor with prefix i, the packet can enter either a processor with the same address or a processor with an address different in the i th bit. The packet traverses according to the i th bit in its destination address. Thus, m transitions are sufficient in phase B. The analysis of this algorithm is done using a new technique called *critical delay sequences* . The algorithm is shown to run in time $\tilde{O}(m)$.

Pippenger's [Pi84] communication scheme has a topology of d -way Digit Exchange Graph. It is very similar to a 2DEG. It has kd^k nodes and degree d. His algorithm successfully routes N packets in time $\tilde{O}(\log N)$. This algorithm also has a maximum queue size of $O(1)$. Both Upfal's and Pippenger's algorithm are oblivious.

11.3.2   Sorting Algorithms

In this section we will discuss [RV82]'s FLASHSORT and [RR85]'s INTEGERSORT algorithms. The sorting problem is: Given a set K of N keys $\{k_1,\ldots,k_N\}$ and a total ordering $<$ on them. To find a permutation $\sigma = (\sigma(1),\ldots,\sigma(N))$ such that $k_{\sigma(1)} < \cdots < k_{\sigma(N)}$. Define the rank of a key k to be $\text{rank}(k) = |\{k \in K/k' < k\}|$.

FLASHSORT

FLASHSORT runs on a fixed connection network called the cube connected cycles (CCC) have N nodes with labels from $\{0, 1, \ldots, N-1\}$ and constant degree. A CCC is nothing but an n -cube with each one of its nodes being replaced by a cycle of n nodes. The n edges leaving each n -cube node now leave from distinct nodes of the corresponding cycle. The CCC thus has $n2^n$ nodes.

Each of the nodes of the CCC initially contains a key. FLASHSORT sorts these keys by routing each packet $k \in K$ to a node $j = \text{rank}(k)$. The algorithm consists of 4 steps. Step 1 finds a set of $2^n/n^\epsilon, \epsilon < 1$ elements called the *splitters* that divide K when regarded as an ordered set into roughly equal intervals. In step 2, keys in each interval (determined by the splitters) are routed to the sub-cube they belong to. This routing task is achieved using the two phase algorithm of [Va82] described in section 11.3.1. After the second step, they keys will be approximately sorted. In step 3, the rank of each key is determined and finally in step 4, each packet is routed to the node corresponding to its rank.

In the above algorithm it is assumed that each node has a local memory of $O(\log N)$. The algorithm has a time bound of $\tilde{O}(\log N)$ and is optimal.

INTEGERSORT

The problem of INTEGERSORT is to sort N integer keys in the range $[1, N]$. [RR85] present an optimal PRAM algorithm for this problem that uses $N/\log N$ processors and runs in time $\tilde{O}(\log N)$. The main idea behind INTEGERSORT is radix sorting. The problem of sorting N integer keys is tackled in two phases. In the first phase, the keys are sorted with respect to their least significant $\log N - 2 \log \log N$ binary bits and in the second phase the output from phase I is *stable* sorted with respect to the most significant $2 \log \log N$ bits of the keys. Thus, the first phase sorts N keys in the range $[1, N/ \log^2 N]$ and the second phase sorts N keys in the range $[1, \log^2 N]$. (A sorting algorithm is said to be *stable* if given the keys $k_1, \ldots, k_N$, the algorithm outputs a sorting permutation σ of $(1, \ldots, N)$ where for all $i, j \in [1, N]$ if $k_i = k_j$ and $i < j$, then $\sigma(i) < \sigma(j)$). Each phase and the whole algorithm runs in time $\tilde{O}(\log N)$ on a PRAM with $N/ \log N$ processors.

In [RR84], a survey of randomized sorting algorithms can be found. They also explain how to derive and analyze randomized sorting algorithms using some interesting results from sampling theory.

11.4 Randomized Algorithms for Load Balancing

If the number of arithmetic (or comparison) operations to be performed by an algorithm is S and if we employ P processors to run the algorithm in parallel, then the best run time possible is T/P. This best run time occurs when the work load is shared equally by all the processors. In this case we obtain an optimal algorithm. Optimal parallel algorithms with run time $O(\log N)$ are of special importance and these are called linear parallel algorithms. In this section we will discuss some of the linear parallel algorithms found in the literature.

11.4.1 Parallel Tree Contraction

Tree problems arise frequently in CS. A simple example is evaluating an arithmetic expression, which can be viewed as tree evaluation. Miller and Reif [MR85] give an optimal algorithm for parallel tree contraction. Some of the many applications of this algorithm are: dynamic expression evaluation, testing for isomorphism of trees, list ranking, computing canonical forms for planar graphs, etc. For the list ranking problem [Vi86] has presented an optimal deterministic algorithm. For evaluating arithmetic expressions of length N Brent [Br74] has given a parallel algorithm that employs N PRAM processors and runs in time $O(\log N)$. His algorithm needs a lot of preprocessing. If no preprocessing is for free, then his algorithm seems to require $O(\log^2 N)$ time. The problem of evaluating an arithmetic expression without any preprocessing is called *dynamic expression evaluation* .

An arithmetic expression is defined as follows. 1) Constants are arithmetic expressions; 2) If E_1 and E_2 are arithmetic expressions, then so are $E_1 \pm E_2$, $E_1 \times E_2$, and E_1/E_2. As one can see, an arithmetic expression can be represented as a binary tree whose internal nodes are operators and whose leaves are constants. This tree is called an expression tree.

[Br74]'s algorithm first finds (by preprocessing) a *separator set* for the expression tree. A *separator set* is a set of edges which when removed from the tree will divide the tree into almost equal sized disjoint subtrees. The algorithm evaluates the subexpressions defined by the separator set in parallel and combines the results. Since the size of the subtrees decreases by a constant factor at each step, only $O(\log N)$ steps are needed. This algorithm, clearly, is a top down algorithm. [MR85] give a bottom up algorithm for dynamic expression evaluation.

They define two operators viz., RAKE and COMPRESS on rooted trees such that a most $O(\log N)$ of these operations are needed to reduce any tree with N nodes to a single node. They also give an efficient implementation of these operations on a CRCW PARM.

Let $T = (V, E)$ be a rooted tree with N nodes and root r. Let RAKE be the operation of removing all the leaves from T (this operation intuitively corresponds to evaluating a node if all of its children have been evaluated or partially evaluating a node if some of its children have been evaluated). We say a sequence of nodes $v_1, \ldots, v_k$ is a chain if v_{i+1} is the only child of v_i for $1 \le i \le k$ and v_k has exactly one child that is not a leaf. Let COMPRESS be the operation on T that contracts al the maximal chains of T in one step. If a node has been partially evaluated except for one child, then the value of the node is a linear function of the child, say $aX + b$, where X is a variable. Thus a chain is a sequence of nodes each of which is a linear function of its child.

Let CONTRACT be the simultaneous application of RAKE and COMPRESS. [MR85] prove only $O(\log N)$ CONTRACT operations are necessary to reduce T to its root. Their CRCW PRAM randomized algorithm runs in time $\tilde{O}(\log N)$ employing $O(N/\log N)$ processors.

11.4.2   Parallel Hashing

PRAM models of parallel computation are powerful, elegant, and algorithms written for PRAM machines are simpler. But these models are not practical. One of the problems in the theory of parallel computation is to simulate shared memory on a more realistic computer, viz., a fixed connection network. [KU86]'s randomized scheme for implementing shared memory on a butterfly network enables N processors to store and retrieve an arbitrary set of N data items in $\tilde{O}(\log N)$ parallel steps. Since a butterfly network has a constant degree, this time bound is optimal. A number of authors have worked on this problem before. The best previous result was due to [AHMP85] who showed that T arbitrary PRAM steps can be simulated in a bounded degree network in $\tilde{O}(T \log^2 N)$ time. [KU86]'s algorithm improves this bound to an optimal $\tilde{O}(T \log n)$.

[KU86]'s PRAM simulation is based on a scheme they call *parallel hashing* . The scheme combines the use of universal hash functions [CW79] for distributing the variables among the processors and a randomized algorithm (very much similar to the one given in the section on Upfal's algorithm in section 11.3.1) for executing memory requests. Simulation of a single step of PRAM step involves communication between each processor that wishes to access a variable and the processor storing that variable. A single PRAM instruction is executed by sending messages to the processor storing the variable that each processor wishes to access and back in the base of read instruction.

Other examples of optimal algorithms are: INTEGERSORT of [RR85], connectivity algorithm of [Ga86].

11.5 Parallel Randomized Algorithms in Algebra

Examples considered in this section are: (1) testing a polynomial for identity; (2) testing if $AB = C$ for integer $N \times N$ matrices A, B, and C; and (3) testing for the existence of a perfect matching in a graph.

11.5.1 Testing for Polynomial Identity

Given a polynomial $Q = Q(x_1, ..., x_N)$ in N variables over the field F. It is desired to check if Q is identically 0. If I is any set of elements in the field F such that $|I| \geq c$ degree(Q), then the number of elements of $I \times \cdots \times I$ which are zeros of Q is at most $|I|^N/c$. This fact suggests the following randomized algorithm for solving the given problem: (1) choose an I such that $|I| >$ degree(Q) and $c > 2$; (2) choose at random m elements of $I \times ... \times I$ and on each vector of values check if the polynomial is nonzero. If on all these m vectors Q evaluates to zero, conclude Q is identically zero. If on at least one vector Q evaluates to a nonzero value, then Q is nonzero.

If $c > 2$ and Q is nonzero, probability that a random vector of $I \times ... \times I$ is a zero of Q is at most $1/2$. Therefore, probability that only zeros of Q are picked in m independent steps is at most 2^{-m}. If $m = O(\log N)$, this probability is at most $N^{-\alpha}$, $\alpha > 1$. This randomized algorithm can easily be parallelized. We have $N \log N$ collections of processors; each collection chooses a random N-vector from $I \times ... \times I$ and evaluates Q on it. The results are combined later. Since we can evaluate a multivariate polynomial in $O(\log N)$ time using a polynomial number of processors, the entire algorithm runs in time $\tilde{O}(\log N)$ using $N^{O(1)}$ processors.

11.5.2 Is $AB = C$?

Given $N \times N$ integer matrices A, B, and C. To test if $AB = C$. Reif [Re84] shows that a randomized PRAM with time bound $\tilde{O}(\log N)$ and processor bound

$N^2/\log N$ can solve this problem. The idea is to choose randomly and independently m column vectors $x \in \{-1, 1\}^N$ and test if $A(Bx) = Cx$. This test is done by a randomized PRAM within time $\tilde{O}(\log N)$ and $(N^2/\log N)$ processors by forming $N/\log N$ binary trees of processors, each of size $2N$ and depth $O(\log N)$ and pipelining the required dot products.

Freivalds [Fr79] shows that if $AB \neq C$, for a random x, probability $[A(Bx) = Cx] < 1/2$. And hence, the probability of error in the above algorithm can be made arbitrarily small, by choosing a sufficiently large m.

11.5.3 Testing for the Existence of a Perfect Matching

Given an undirected graph $G = (V, E)$ with vertices $V = \{1, ..., N\}$. To test if G has a perfect matching. Tutte [Tu47] showed that a graph has a perfect matching iff a certain matrix of indeterminates called the *Tutte matrix* is non-singular. The Tutte matrix M is defined as: (1) $M_{ij} = x_{ij}$ if $(i, j) \in E$ and $i < j$; (2) $M_{ij} = -x_{ij}$ if $(i, j) \in E$ and $i > j$; and (3) $M_{ij} = 0$ otherwise. Here x_{ij}, $j = 1, ..., N$ and $i = 1, ..., j - 1$ are indeterminates. Since the determinant of M is a multivariate polynomial, we can test if it is identically zero in parallel quickly using the algorithm of section 11.5.1.

11.6 Randomized Parallel Graph Algorithms

Graph theory finds application in every walk of life, more so in the field of computer science. Efficient sequential algorithms have been found for numerous graph problems. But unfortunately, not many efficient parallelization of these algorithms have been made. Randomized seems to play an important role in parallelizing graph algorithms as evidenced from the literature. In this section we will demonstrate this with representative examples. In particular, we will look at: (1) Symmetric complementation games of Reif [Re184]; (2) the random mating lemma of [Re85]; (3) the depth first search algorithm of Aggarwal and Anderson [AA86]; and (4) the maximal independent set algorithm of Luby [Lu85].

11.6.1 Symmetric Complementation

Reif [Re184] has discovered a class of games he calls *symmetric complementation games* . These games are interesting since their related complexity classes include many well known graph problems. Some of these problems are: (1) finding minimum spanning forests; (2) k -connectivity; (3) k -blocks; (4) recognition of chordal graphs, comparability graphs, interval graphs, split graphs, permutation graphs, and constant valence planar graphs. For all these problems [Re184] gives sequential algorithms requiring simultaneously logarithmic space and polynomial time. Furthermore, he also gives randomized parallel algorithms requiring simultaneously logarithmic time and a polynomial number of processor thus showing that all these problems are in RNC.

11.6.2 The Random Mating Lemma and Optimal Connectivity

Let $G = (V, E)$ be any graph. Suppose we assign for each vertex $v \in V$, independently and randomly SEX$(v) \in \{male, female\}$. Let vertex v be active if there exists at least one departing edge $(v, u) \in E$ ($v \neq u$). We say the vertex v is mated if SEX$(v) = male$ and SEX$(u) = female$ for at least one edge $(v, u) \in E$. Then, the Random Mating Lemma of [Re85] states that with probability $1/2$ the number of mated vertices is at least $1/8$ of the total vertices.

The random mating lemma naturally leads to an elegant randomized algorithm [Re85] for the connectivity problem, i.e., the problem of computing the connected components of a graph, that has a run time of $\tilde{O}(\log |V|)$ on the PRAM model using $|V| + |E|$ processors. In an earlier paper [SV83] have presented a deterministic algorithm for connectivity that has the same resource bounds. But, the random mating lemma has been used by [Ga86] to obtain an optimal randomized algorithm for connectivity that runs in time $\tilde{O}(\log |V|)$ and uses $(|V|+|E|)/\log(|V|)$ processors.

11.6.3 Depth First Search

The problem of performing Depth First Search (DFS) in parallel was studied by many authors ([Re83],[EA77], etc.) and they conjectured that DFS was inherently sequential. However, NC algorithms were given for DFS of some restricted class of graphs by Smith [Sm84] for planar graphs, and Ghosh and Bhattacharjee [GB84] for directed acyclic graphs. It was Aggarwal and Anderson [AA86] who first gave an RNC algorithm for DFS.

The DFS problem is: Given a graph $G = (V, E)$ and vertex r, construct a tree T that corresponds to a depth first search of the graph starting from the vertex r. The DFS algorithm of [AA86] is a divide and conquer algorithm. They first find an initial portion of the DFS tree which allows them to reduce the problem to DFS in trees of less than half the original size. This algorithm has $O(\log N)$ levels of recursion.

An initial segment is a rooted subtree T' that can be extended to some DFS tree T. They give a RNC algorithm to construct an initial segment with the property that the largest component of $V - T'$ (i.e., G after removal of T' from it) has size at most $N/2$ (N being the size of the original graph). This initial segment T' is then extended to a DFS as follows. Let C be a connected component of $V - T'$. There is a unique vertex $x \in T'$ of greatest depth that is adjacent to some vertex of C. Let $y \in C$ be adjacent to x. Construct a DFS tree for C rooted at y. Connect this tree to T' using the edge from x to y. This is performed independently for each component. This algorithm is applied recursively to finding DFS of components of $V - T'$. Since finding an initial segment is in RNC, the whole algorithm is in RNC.

11.6.4 Maximal Independent Set

Given an undirected graph $G(V, E)$. The problem is to find a maximal collection of vertices such that no two vertices in the collection are adjacent. There is a trivial linear time sequential algorithm for this problem [AHU74]. An efficient algorithm did not exist for this problem until Karp and Wigderson [KW84] presented a randomized algorithm that utilized $O(N^2)$ EREW PRAM processors and ran in time $\tilde{O}(\log^4 N)$. Luby [Lu85] later gave a simplified randomized algorithm with a time bound of $\tilde{O}(\log^2 N)$ and processor bound $O(M)$ (where M is the number of edges and N is the number of vertices in G). This algorithm also has the property that it can be made deterministic. Some details of this property will be discussed in Section 11.7.2.

A summary of [Lu85]'s algorithm for the maximal independent set (MIS) problem: This is an iterative algorithm. At each step of the algorithm, certain number of nodes will be decided to be in the MIS and certain number of edges will be deleted from G. The algorithm iterates on the resultant graph $G'(V', E')$. At each step of the algorithm at least a constant fraction of edges will be removed. Therefore, the algorithm will terminate after $\tilde{O}(\log N)$ iterations.

At any iteration of the algorithm that starts with $G'(V', E')$, every node $v \in V'$ decides (independently) to add itself to the MIS with probability $1/2d(v)$, $d(v)$ being its degree. Let I' be the set of all nodes that decide to add themselves to the MIS in this step. Check if there is an edge between any pair of vertices in I'. For every such edge, the node with the smaller degree is deleted from I'. A tie is broken arbitrarily. Add the resultant I' to the MIS. And finally delete all the nodes in I' and its neighbors from G'. The resultant graph is what the next iteration of the algorithm starts with.

Since each step of the algorithm gets rid of a constant fraction of edges, the algorithm terminates in $\tilde{O}(\log N)$ iterations. Also notice that at any step of the algorithm no more than $O(M)$ processors are needed. Each step of the algorithm takes $\tilde{O}(\log N)$ time and hence the whole algorithm runs in time $\tilde{O}(\log^2 N)$.

11.7   Derandomization of Parallel Randomized Algorithms

Computer Scientists have made some effort to make randomized algorithms deterministic owing mainly to the fact that our computers are deterministic and are unable to generate truly random numbers. Four techniques have been discovered so far: (1) the usage of pseudo-random numbers in the place of truly random numbers; (2) reducing the size of the probability space so an exhaustive search in the space can be done; (3) deterministic coin tossing; and (4) combinatorial construction. No work has been done to parallelize the fourth technique. This section is devoted to introducing the reader to the first three methodologies as applicable to parallel algorithms.

11.7.1 Pseudo-Random Number Generation

Informally, a pseudo-random sequence is a sequence of bits generated by a deterministic program from a random *seed* such that a computer with "limited" resources won't be able to distinguish it from a truly random sequence. Examples of pseudo-random number generators include linear congruential generators, additive number generators, etc. [Kn81]. All of the randomized algorithms that are currently being used in practice use only pseudo-random generators of one type or the other in the place of true random generators.

In [RT85] a parallel NC algorithm is given which can generate N^c (forany $c > 1$) pseudo-random bits from a seed of N^ϵ, ($\epsilon < 1$) truly random bits. This takes polylog time using $N^{\epsilon'}$ processors where $\epsilon' = k\epsilon$ for some fixed constant $k > 1$. The pseudo-random bits output by this algorithm cannot be distinguished from truly random bits in polylog time using a polynomial number of processors with probability $\geq \frac{1}{2} + \frac{1}{N^{O(1)}}$ if the multiplicative inverse problem cannot be solved in RNC.

As a corollary to their algorithm, they also show that given any parallel algorithm (over a wide class of machine models) with time bound $T(N)$ and processor bound $P(N)$, it can be simulated by a parallel algorithm with time bound $T(N) + O(\log N \log \log N)$, processor bound $P(N)N^{\epsilon'}$, and only using N^ϵ truly random bits.

11.7.2 Probability Space Reduction

Any randomized algorithm generates certain number (say R) of mutually independent random numbers (say in the range $[1, L]$) during its execution. Any randomized algorithm can be made deterministic by running it on every possible sequence of R numbers (in the range $[1, L]$). The set of all these sequences (that constitute a probability space) usually is of exponential size. There are algorithms [Lu85] which will run even if the random numbers generated are only pairwise independent. When the random numbers are only pairwise independent, the size of the probability space might reduce tremendously enabling us to do an exhaustive search quickly.

In [Lu85]'s randomized algorithm for MIS (section 11.6.3), remember, if the program starts an iteration with the graph $G'(V', E')$, each node $v \in V'$ decides to add itself

to the MIS with a probability $1/2d(v)$, $d(v)$ being the degree of v. It was assumed
that the decision of the nodes were mutually independent. It turns out that the
algorithm runs correctly with the same time and processor bounds even if they are
only pairwise independent.

[Lu85] also constructs a probability space of size $O(N^2)$ (N being the number of
nodes in the input graph) and defines N pairwise independent events $E_1, ..., E_N$
in this space such that Prob. $[E_v] = 1/2d(v)$ for each $v \in V$. This probability
space then makes his algorithm deterministic with the same time bound and a
processor bound of $O(MN^2)$. This idea has been extended to k -wise ($k \geq 2$)
independent events by [KUW85]. They also show that using $k \log N$ random bits
we can construct a sequence of N random bits such that every k bits are mutually
independent.

11.7.3   Deterministic Coin Tossing

Consider the following problem: Given a connected directed graph $G(V, E)$. The
indegree and outdegree of each vertex is 1. Such a graph forms a directed circuit
and is called a *ring* . We define a subset U of V to be an *r-ruling set* of G if: (1)
no two vertices of U are adjacent; and (2) for each vertex $v \in V$ there is a directed
path from v to some vertex in U whose edge length is at the most r. The problem
is to find an r -ruling set of G for a given r.

We can use the following simple randomized algorithm: (1) in the first step, each
node in the graph chooses to be in the ruling set with a probability of $1/r$; and (2)
in the second step each group of adjacent (in G) nodes chosen in step 1, randomly
choose one of them to be in the ruling set.

[CV86] give a deterministic algorithm for obtaining an r -ruling set. We will consider
here only the case $r = \log n$. Their algorithm finds a $\log n$ -ruling set in $O(1)$ time
on the EREW PRAM model.

input representation: The vertices are given in an array of length n. The entries
of the array are numbered from 0 to $n - 1$ (each being a $\log n$ bit binary number).
Each vertex has a pointer to the next vertex in the ring.

algorithm: Processor i is assigned to entry i of input array (for simplicity entry i
is called the vertex i). Let $SERIAL_0(i) = i$ for $i = 0, ..., n - 1$. Let i_2 be the vertex
following i in the ring and j be the index of the right most bit in which i and i_2
differ. Processor i sets $SERIAL_1(i) = j$, $i = 0, ..., n - 1$.

Note that for all i, $SERIAL_1(i)$ is a number between 0 and $\log n - 1$. We say
$SERIAL_1(i)$ is a local minimum if $SERIAL_1(i) \leq SERIAL_1(i_1)$ and $SERIAL_1(i) \leq$
$SERIAL_1(i_2)$ (where i_1 and i_2 are vertices preceding and following i respectively).

Define a local maximum similarly. It can be seen that the number of vertices in the shortest path from any vertex in G to the next local extremum is at the most $\log n$. Thus the extrema satisfy condition 2 for the $\log n$ ruling set. From among local extrema, there might be *chains* of successive vertices in G. [CV86] pick a unique node deterministically from each such chain. Thus a subset of all the extrema forms a $\log n$ -ruling set.

[CV86] call the above technique *deterministic coin tossing*. They use this technique to solve many graph problems like list ranking, selecting the nth smallest out of n elements, finding minimum spanning forest in a graph, etc.

11.8   Conclusion

Randomization is a powerful tool in developing parallel algorithms which have small processor and time bounds. It also appears to significantly simplify the structure of parallel algorithms. This fact has been demonstrated in this paper. Techniques for making randomized algorithms deterministic have also been explored.

References

[AA86]   Aggarwal, A., and Anderson, R., "A Random NC Algorithm for Depth First Search," to appear in Proc. of the 19th annual ACM Symposium on Theory of Computing, 1987.

[AHMP85] Alt, H., Hagerup, T., Mehlhorn, K., and Preparata, F. P., "Simulation of Idealized Parallel Computers on more Realistic ones," Preliminary Report.

[AHU74]   Aho, A. U., Hopcroft, J. E., Ullman, J. D., *The Design and Analysis of Computer Algorithms* , Addison-Wesley Publishing Company, Massachusetts, 1977.

[AKS83]   Ajtai, M., Komlós, J., and Szemerédi, E., "An $O(n \log n)$ Sorting Network," Combinatorica, 3, 1983, pp. 1-19.

[AM77]   Adleman, L. and Manders, K., "Reducibility, Randomness and Untractability," Proc. 9th ACM Symposium on Theory of Computing, 1977, pp. 151-163.

[Ba68]   Batcher, K. E., "Sorting Networks and Their Applications," Proc. 1968 Spring Joint Computer Conference, Vol. 32, AFIPS Press, 1968, pp. 307-314.

[Br74]   Brent, R. P., "The Parallel Evaluation of Generalized Arithmetic Expressions," Journal of ACM, Vol. 21, No. 2, 1974, pp. 201-208.

[Co86]   Cole, R., "Parallel Merge Sort," Proceedings of the IEEE Foundations of Computer Science, 1986, pp. 511-516.

[CV86]   Cole, R., and Vishkin, U., "Approximate and Exact Parallel Scheduling with Applications to List, Tree, and Graph Problems," Proc. of the IEEE Foundations of Computer Science, 1986, pp. 478-491.

[CW79]   Carter, L., and Wegman, M., "Universal Class of Hash Functions," Journal of CSS, Vol. 18, No. 2, 1979, pp. 143-154.

[De74]   Deo, N., *Graph Theory With Applications to Engineering and Computer Science* , Prentice Hall Publishing Company, New York, 1974.

[EA77]   Eckstein, D., and Alton, D., "Parallel Graph Processing Using Depth First Search," Proc. Conference on Theoretical Computer Science at Univ. of Waterloo, 1977, pp. 21-29.

[Fe68]   Feller, W., *An Introduction to Probability Theory and Its Applications* , John Wiley & Sons Publishing Company, New York, 1968.

[Fr79]   Freivalds, "Fast Probabilistic Algorithms," 8th MFCS, 1979.

[FW78]   Fotune, S., and Wyllie, J., "Parallelism in Random Access Machines," Proc. 10th Annual ACM Symposium on Theory of Computing, 1978, pp. 114-118.

[Ga86]   Gazit, H., "An Optimal Randomized Parallel Algorithm for Finding Connected Components in a Graph," Proc. Foundations of Computer Science Conference, 1986, pp. 492-501.

[GB84]   Ghosh, R. K., and Bhattacharjee, G. P., "A Parallel Search Algorithm for Directed Acyclic Graphs," BIT, Vol. 24, 1984, pp. 134-150.

[Ha69]   Harary, F., *Graph Theory* , Addison-Wesley Publications, Massachusetts, 1969.

[Ho62]   Hoare, C. A. R., "Quicksort," Computer Journal, Vol. 5, No. 1, 1962, pp. 10-15.

[Kn81]   Knuth, D. E., *The Art of Computer Programming: Vol. 2, Seminumerical Algorithms*, Addison-Wesley Publications, Massachusetts, 1981.

[KU86]   Karlin, A. R., and Upfal, E., "Parallel Hashing - An Efficient Implementation of Shared Memory," Proc. 18th Annual ACM Symposium on Theory of Computing, 1986, pp. 160-168.

[KUW85] Karp, R. M., Upfal, E., and Wigderson, A., "The Complexity of Parallel Computation on Matroids," IEEE Symposium on Foundations of Computer Science, 1985, pp. 541-550.

[KW84] Karp, R. M., and Wigderson, A., "A Fast Parallel Algorithm for the Maximal Independent Set Problem," Proc. 16th Annual ACM Symposium on Theory of Computing, 1984, pp. 266-272.

[Le84] Leighton, T., "Tight Bounds on the Complexity of Parallel Sorting," Proc. 16th Annual ACM Symposium on Theory of Computing, 1984, pp. 71-80.

[Lu85] Luby, M., "A Simple Parallel Algorithm for the Maximal Independent Set Problem," Proc. 17th Annual ACM Symposium on Theory of Computing, 1985, pp. 1-10.

[MR85] Miller, G. L., Reif, J. H., "Parallel Tree Contraction and Its Applications," Proc. IEEE conference on Foundations of Computer Science, 1985, pp. 478-489.

[Pi84] Pippenger, N., "Parallel Communication With Limited Buffers," Proc. IEEE Symposium on Foundations of Computer Science, 1984, pp. 127-136.

[Ra76] Rabin, M. O., "Probabilistic Algorithms," in: Traub, J. F., ed., *Algorithms and Complexity* , Academic Press, New York, 1976,, pp. 21-36.

[Re83] Reif, J. H., "Depth First Search is Inherently Sequential," Information Processing Letters, Vol. 20, No. 5, June 1985, pp. 229-234.

[Re84] Reif, J. H., "On Synchronous Parallel Computations With Independent Probabilistic Choice," SIAM Journal on Computing, Vol. 13, No. 1, 1984, pp. 46-56.

[Re184] Reif, J. H., "Symmetric Complementation," Journal of the ACM, Vol. 31, No. 2, 1984, pp. 401-421.

[Re85] Reif, J. H., "Optimal Parallel Algorithms for Integer Sorting and Graph Connectivity," Technical Report TR-08-85, Aiken Computing Lab., Harvard University, Cambridge, Massachusetts 02138, 1985.

[RND77] Reingold E., Nievergelt, J., and Deo, N., *Combinatorial Algorithms: Theory and Practice* , Prentice Hall Publishing Company, New York, 1977.

[Reis81] Reischuk, R., "A Fast Probabilistic Parallel Sorting Algorithm," Proc. IEEE Symposium on Foundations of Computer Science, 1981, pp. 212-219.

[RR84]   Rajasekaran, S., and Reif, J. H., "Derivation of Randomized Algorithms,"
          Aiken Computing Lab. Technical Report TR-16-84, 1984.

[RR85]   Rajasekaran, S., and Reif, J. H., "Optimal and Sub-Logarithmic Time
          Sorting Algorithms," Aiken Computing Lab. Technical Report, 1985.
          Also appeared as "An Optimal Parallel Algorithm for Integer Sorting,"
          in the Proc. of the IEEE Symposium on FOCS, 1985, pp. 496-503.

[RT85]   Reif, J. H., and Tygar, J. D., "Efficient Parallel Pseudo-Random Number
          Generation," Aiken Computing Lab. Technical Report, 1984.

[RV82]   Reif, J. H., and Valiant, L. G., "A Logarithmic Time Sort for Linear Size
          Networks," Proc. of the 15th Annual Symposium on Theory of Comput-
          ing, 1983, pp. 10-16, also to appear in JACM 1987.

[Sm84]   Smith, J. R., "Parallel Algorithms for Depth First Searches: I. Planar
          Graphs," International Conference on Parallel Processing, 1984.

[SS76]   Solovay, R., and Strassen, V., "A Fast Monte-Carlo Test for Primality,"
          SIAM Journal of Computing, Vol. 6, 1977, pp. 84-85.

[SV83]   Shiloach, Y., and Vishkin, U., "An $O(\log n)$ Parallel Connectivity Algo-
          rithm," Journal of Algorithms, Vol. 3, 1983, pp. 57-67.

[Tu47]   Tutte, W. T., "The Factorization of Linear Graphs," Journal of the Lon-
          don Mathematical Society, Vol. 22, 1947, pp. 107-111.

[Up84]   Upfal, E., "Efficient Schemes for Parallel Communication," Journal of
          the ACM, Vol. 31, No. 3, 1984, pp. 507-517.

[Va75]   Valiant, L. G., "Parallelism in Comparison Problems," SIAM Journal of
          Computing, Vol. 14, 1985, pp. 348-355.

[Va82]   Valiant, L. G., "A Scheme for Fast Parallel Communication," SIAM Jour-
          nal of Computing, Vol. 11, No. 2, 1982, pp. 350-361.

[Vo71]   Voorhis, V., "On Sorting Networks," Ph.D. Thesis, Stanford University
          CS Department, 1971.

[We83]   Welsh, D. J. A., "Randomized Algorithms," Discrete Applied Mathemat-
          ics, Vol. 5, 1983, pp. 133-245.

Chapter 12

A Modest Proposal for Communication Costs in Multicomputers [1]

Paul M.B. Vitányi [2]

Abstract

Getting rid of the 'von Neumann' bottleneck in the shift from sequential to non-sequential computation, a new communication bottleneck arises because of the interplay between locality of computation, communication, and the number of dimensions of physical space. As a consequence, realistic models for non-sequential computation should charge extra for communication, in terms of time and space. We give a proposal for this that is more subtle, but *mutatis mutandis* similar to the transition from unit cost to logarithmic cost in the sequential Random Access Machine model. The space cost of communication is related with the topology of the communication graph. We use a new lower bound on the average interconnect (edge) length in terms of diameter and symmetry of the topology. This lower bound is technology independent, and shows that many interconnection topologies of today's multicomputers do not scale well in the physical world with 3 dimensions.

---

[1] The initial part of this work was performed at the MIT Laboratory for Computer Science, supported in part by the Office of Naval Research under Contract N00014-85-K-0168, by the Office of Army Research under Contract DAAG29-84-K-0058, by the National Science Foundation under Grant DCR-83-02391, and by the Defense Advanced Research Projects Agency (DARPA) under Contract N00014-83-K-0125.

[2] Centrum voor Wiskunde en Informatica, Kruislaan 413, 1098 SJ Amsterdam, The Netherlands.

12.1   Multicomputers and Physical Space

As multicomputers with large numbers of processors start to be conceived, the need for a formal framework to analyze their computations becomes more pressing. Such a computational model should preferably be technology independent. In many areas of the theory of parallel computation we meet graph structured computational models. These models suggest the design of parallel algorithms where the cost of communication is largely ignored. Yet it is well known that the cost of computation - in both time and space - vanishes with respect to the cost of communication in parallel or distributed computing. Here we shall argue that, while getting rid of the so called 'von Neumann' bottleneck [3], in the shift from serial to non-serial computing, we run into a new *communication* bottleneck due to the three dimensionality of physical space. We will analyze the time cost and space cost due to communication, and present a proposal to measure those costs. The present bind we find us in with respect to multicomputers is not unlike the introduction of the random access machine (RAM) as model for sequential computation. One of the features which distinguish the RAM from the ubiquitous Turing machine are the unlimited length registers in memory. In a single step, such a register can be accessed, and its contents operated on. This corresponds neatly to the von Neumann concept.[3] The theoretical registers are unlimited because usually computer programs use the physical registers that way. Yet charging unit cost for a single operation on unlimited length binary strings is obviously unrealistic and leads to misleading theoretical results. Thus, a *two register* RAM is already computation universal, and a RAM *with multiplication operation* can solve NP-complete problems in polynomial time. This is not exactly what we can do with the computers for which the RAM is supposed to be a model. To curb the theoretical power of the RAM to intuitively acceptable proportions, therefore, we hit it with the logarithmic cost measure. This cost measure charges the binary length of the operands per operation, and accounts for the hierarchy of fast to slow memories (the slower they are, the larger they get) by charging the bit length of the address containing the operand. Now the actual costs by this measure are way too low to account for the practical overhead. However, the theoretical model and the general results are brought in line. Importantly, the cost measure is technology independent and based on information theory, and is easy to comprehend and apply. We aim for a solution for communication cost in multicomputers that has similar desirable properties. First we analyze geometric consequences of the fact that the physical space we put our computers in has but 3 dimensions. See also [6,7]. We shall use these results as firm foundation on which to base the communication cost measures for multicomputers.

---

[3] When the operations of a computation are executed serially in a single Central Processing Unit, each one entails a 'fetch data from memory to CPU; execute operation in CPU; store data in memory' cycle. The cost of this cycle, and therefore of the total computation, is dominated by the cost of the memory accesses which are essentially operation independent. This is called the 'von Neumann' bottleneck, after the brilliant Hungarian mathematician John von Neumann.

12.2 A Folklore Item about Non-Sequential Computation Time

Models of parallel computation that allow processors to randomly access a large shared memory, such as P-RAMs, or rapidly access a large number of processors, such as NC computations, can provide new insights in the inherent parallellizability of algorithms to solve certain problems. They can *not* give a reduction from an asymptotic exponential time best algorithm in the sequential case to an asymptotic polynomial time algorithm in *any* parallel case. At least, if by 'time' we mean time. This is a folklore fact dictated by the Laws of Nature. Namely, if the parallel algorithm uses 2^n processing elements, regardless of whether the computational model assumes bounded fan-in and fan-out or not, it cannot run in time polynomial in n, because *physical space* has us in its tyranny. Viz., if we use 2^n processing elements of, say, unit size each, then the tightest they can be packed is in a 3-dimensional sphere of volume 2^n. Assuming that the units have no "funny" shapes, e.g., are spherical themselves, no unit in the enveloping sphere can be closer to all other units than a distance of radius R,

$$R = \left(\frac{3 \cdot 2^n}{4\pi} \right)^{1/3} \tag{12.1}$$

Unless there is a major advance in physics, it is impossible to transport signals over $2^{\alpha n}$ ($\alpha > 0$) distance in polynomial $p(n)$ time. In fact, the assumption of the bounded speed of light says that the lower time bound on *any* computation using 2^n processing elements is $\Omega(2^{n/3})$ outright. Or, for the case of NC computations which use n^α processors, $\alpha > 0$, the lower bound on the computation time is $\Omega(n^{\alpha/3})$ [4]. Science fiction buffs may want to keep open the option of embedding circuits in hyper dimensions. Counter to intuition, this does not help - at least, not all the way, see [7].

12.3 Non-Sequential Computation and Space

At present, many popular multicomputer architectures are based on highly symmetric communication networks with small diameter. Like all networks with small diameter, such networks will suffer from the communication bottleneck above, i.e., necessarily contain *some* long interconnects (embedded edges). However, we can demonstrate that the desirable fast permutation properties of symmetric networks don't come free, since they require that the average of *all* interconnects is long. (Note that 'embedded edge,' 'wire,' and 'interconnect' are used synonymously.) To

---

[4] It is sometimes argued that this effect is significant for large values of n only, and therefore can safely be ignored. However, in computational complexity, virtually were all results are of asymptotic nature, i.e., hold only for large values of n, so the effect is especially relevant there.

preclude objections that results like below hold only asymptotically (and therefore can be safely ignored for practical numbers of processors), or that processors are huge and wires thin (idem), I calculate precisely without hidden constants [5]. We recapitulate some new results [6,7]. and assume that wires have length but no volume and can pass through everything. It is consistent with the results that wires have *zero* volume, and that *infinitely* many wires pass through a unit area. Such assumptions invalidate the previous VLSI related arguments, see e.g [5]. Theorem 2 below expresses a lower bound on the *average* edge length for *any* graph, in terms of certain symmetries and diameter. The new argument is based on graph automorphism, graph topology, and Euclidean metric. For each graph topology I have examined, the resulting lower bound turned out to be sharp. Concretely, the problem is posed as follows. Let $G = (V, E)$ be a finite undirected graph, without loops or multiple edges, *embedded* in 3-dimensional Euclidean space. Let each embedded node have unit *volume* . For convenience of the argument, each node is embedded as a sphere, and is *represented* by the single point in the center. The *distance* between a pair of nodes is the Euclidean distance between the points representing them. The *length* of the embedding of an edge between two nodes is the distance between the nodes. How large does the *average* edge length need to be?

We illustrate the novel approach with a popular architecture, the *binary n-cube*. Recall, that this is the network with $N = 2^n$ nodes, each of which is identified by an n -bit name. There is a two-way communication link between two nodes if their identifiers differ by a single bit. The network is represented by an undirected graph $C = (V, E)$, with V the set of nodes and $E \subseteq V \times V$ the set of edges, each edge corresponding with a communication link. There are $n2^{n-1}$ edges in C. Let C be embedded in 3-dimensional Euclidean space, each node as a sphere with unit volume. The distance between two nodes is the Euclidean distance between their centers. Let x be any node of C. There are at most $2^n/8$ nodes within Euclidean distance $R/2$ of x, with R as in (1.1). Then, there are $\geq 7 \cdot 2^n/8$ nodes at Euclidean distance $\geq R/2$ from x. Construct a spanning tree T_x in C of depth n with node x as the root. Since the binary n -cube has diameter n, such a shallow tree exists. There are N nodes in T_x, and $N - 1$ paths from root x to another node in T_x. Let P be such a path, and let $|P|$ be the *number of edges* in P. Then $|P| \leq n$. Let $length(P)$ denote the Euclidean length of the embedding of P. Since 7/8 th of all nodes are at Euclidean distance at least $R/2$ of root x, the average of $length(P)$ satisfies

$$(N - 1)^{-1} \sum_{P \in T_x} length(P) \geq \frac{7R}{16}$$

The average Euclidean length of an embedded edge *in a path* P is bounded below as follows:

$$(N - 1)^{-1} \sum_{P \in T_x} \left(|P|^{-1} \sum_{e \in P} length(e) \right) \geq \frac{7R}{16n} \qquad (12.2)$$

---

[5] Ω is used sometimes to simplify notation. The constant of proportionality can be reconstructed easily in all cases, and is never very small.

This does *not* give a lower bound on the average Euclidean length of an edge, the average taken *over all edges* in T_x. To see this, note that if the edges incident with x have Euclidean length $7R/16$, then the average edge length *in each path* from the root x to a node in T_x is $\geq 7R/16n$, even if all edges not incident with x have length 0. However, the average edge length *in the tree* is dominated by the many short edges near the leaves, rather than the few long edges near the root. In contrast, in the case of the binary n-cube, because of its symmetry, if we squeeze a subset of nodes together to decrease local edge length, then other nodes are pushed farther apart increasing edge length again. We can make this intuition precise.

Lemma 1. *The average Euclidean length of the edges in the 3-space embedding of C is at least $7R/(16n)$.*

Proof.   Denote a node a in C by an n-bit string $a_1 a_2 \cdots a_n$, and an edge (a, b) between nodes a and b differing in the k th bit by:

$$(a_1...a_{k-1}a_k a_{k+1}...a_n, a_1...a_{k-1}(a_k \oplus 1)a_{k+1}...a_n)$$

where $\oplus$ denotes modulo 2 addition. Since C is an undirected graph, an edge $e = (a, b)$ has two representations, namely (a, b) and (b, a). Consider the set A of automorphisms $\alpha_{v,j}$ of C consisting of

1. modulo 2 addition of a binary n-vector v to the node representation, followed by

2. a cyclic rotation over distance j.

Formally, let $v = v_1 v_2 \cdots v_n$, with $v_i = 0, 1$ ($1 \leq i \leq n$), and let j be an integer $1 \leq j \leq n$. Then $\alpha_{v,j} : V \to V$ is defined by

$$\alpha_{v,j}(a) = b_{j+1} \cdots b_n b_1 \cdots b_j$$

with $b_i = a_i \oplus v_i$ for all i, $1 \leq i \leq n$.

Consider the spanning trees $\alpha(T_x)$ isomorphic to T_x, $\alpha \in A$. The argument used to obtain (1.2) implies that for *each* α in A separately, in each path $\alpha(P)$ from root $\alpha(x)$ to a node in $\alpha(T_x)$, the average of $length(\alpha(e))$ over all edges $\alpha(e)$ in $\alpha(P)$ is at least $7R/16n$. Averaging (1.2) additionally over all α in A, the same lower bound applies:

$$(N \log N)^{-1} \sum_{\alpha \in A} \left[(N-1)^{-1} \sum_{P \in T_x} \left(|P|^{-1} \sum_{e \in P} length(\alpha(e)) \right) \right] \geq \frac{7R}{16n} \qquad (12.3)$$

Now fix a particular edge e in T_x. We average $length(\alpha(e))$ over all α in A, and show that this average equals the average edge length. Together with (1.3) this will yield the desired result. For each edge f in C there are $\alpha_1, \alpha_2 \in A$, $\alpha_1 \neq \alpha_2$, such that $\alpha_1(e) = \alpha_2(e) = f$, and for all $\alpha \in A - \{\alpha_1, \alpha_2\}$, $\alpha(e) \neq f$. (For $e = (a, b)$ and $f = (c, d)$ we have $\alpha_1(a) = c$, $\alpha_1(b) = d$, and $\alpha_2(a) = d$, $\alpha_2(b) = c$.) Therefore, for each $e \in E$,

$$\sum_{\alpha \in A} length(\alpha(e)) = 2 \sum_{f \in E} length(f)$$

Then, for any path P in C,

$$\sum_{e \in P} \sum_{\alpha \in A} length(\alpha(e)) = 2|P| \sum_{f \in E} length(f) \qquad (12.4)$$

Rearranging the summation order of (1.3), and substituting (1.4), yields the lemma.
□

The symmetry property yielding such huge edge length is 'edge-symmetry.' To formulate the generalization of Lemma 1 for arbitrary graphs, we need some mathematical machinery. We recall the definitions from [1]. Let $G = (V, E)$ be a simple undirected graph, and let Γ be the automorphism group of G. Two edges $e_1 = (u_1, v_1)$ and $e_2 = (u_2, v_2)$ of G are similar if there is an automorphism γ of G such that $\gamma(\{u_1, v_1\}) = \{u_2, v_2\}$. We consider only connected graphs. The relation 'similar' is an equivalence relation, and partitions E into nonempty equivalence classes, called orbits , $E_1, \ldots, E_m$. We say that Γ acts transitively on each E_i, $i = 1, \ldots, m$. A graph is edge-symmetric if every pair of edges are similar ($m = 1$).

Additionally, we need the following notions. Let $D < \infty$ be the diameter of G. If x and y are nodes, then $d(x, y)$ denotes the number of edges in a shortest path between them. For $i = 1, \ldots, m$, define $d_i(x, y)$ as follows. If (x, y) is an edge in E_i then $d_i(x, y) = 1$, and if (x, y) is an edge not in E_i then $d_i(x, y) = 0$ Let Π be the set of shortest paths between x and y. If x and y are not incident with the same edge, then $d_i(x, y) = |\Pi|^{-1} \sum_{P \in \Pi} \sum_{e \in P} d_i(e)$. Clearly,

$$d_1(x, y) + \cdots + d_m(x, y) = d(x, y) \leq D$$

Denote $|V|$ by N. The i th orbit frequency is

$$\delta_i = N^{-2} \sum_{x, y \in V} \frac{d_i(x, y)}{d(x, y)},$$

$i = 1, \ldots, m$. Finally, define the orbit skew coefficient of G as

$$M = \min\{|E_i|/|E|: 1 \leq i \leq m\}.$$

Consider a d -space embedding of G, with embedded nodes, distance between nodes, and edge length as above. Let R be the radius of a d -space sphere with volume N, e.g. (1.1) for $d = 3$. We are now ready to state the main result. Just in case the reader doesn't notice, (i) is the most general form.

Theorem 2. *Let graph G be embedded in d -space with the parameters above, and let $C = (2^d - 1)/2^{d+1}$* [6].

(i) *Let $l_i = |E_i|^{-1} \sum_{e \in E_i} l(e)$ be the average length of the edges in orbit E_i, $i = 1, \ldots, m$. Then, $\sum_{1 \le i \le m} l_i \ge \sum_{1 \le i \le m} \delta_i l_i \ge CRD^{-1}$.*

(ii) *Let $l = |E|^{-1} \sum_{e \in E} l(e)$ be the average length of an edge in E. Then, $l \ge CRMD^{-1}$.*

The proof is a generalization of the argument for the binary n -cube, and is given in [7]. Let us apply the theorem to a few examples.

Binary n-Cube

Let Γ be an automorphism group of the binary n -cube, e.g., A in the proof of Lemma 1. Let $N = 2^n$. The orbit of each edge under Γ is E. Substituting R, D, $m = 1$, and $d = 3$ in Theorem 2 (i) proves Lemma 1. Denote by L the *total* edge length $\sum_{f \in E} l(f)$ in the 3-space embedding of C. Then

$$L \ge \frac{7RN}{32} \qquad (12.5)$$

Recapitulating, the sum *total* of the lengths of the edges is $\Omega(N^{4/3})$, and the *average* length of an edge is $\Omega(N^{1/3} \log^{-1} N)$. (In 2 dimensions we obtain in a similar way $\Omega(N^{3/2})$ and $\Omega(N^{1/2} \log^{-1} N)$, respectively.)

Cube-Connected Cycles

The binary n -cube has the drawback of unbounded node degree. Therefore, in the fixed degree version of it, each node is replaced by a *cycle* of n trivalent nodes [5], whence the name *cube-connected cycles* or CCC. If $N = n2^n$, then the CCC version, say $CCC = (V, E)$, of the binary n -cube has N nodes, $3N/2$ edges, and diameter $D < 2.5n$.

Corollary 3.   *The average Euclidean length of edges in a 3-space embedding of CCC is at least $7R/(120n)$.*

Proof.   Denote a node a by an n -bit string with one marked bit,

$$a = a_1 \cdots a_{i-1} \bar{a}_i a_{i+1} \cdots a_n.$$

---

[6] This constant C can be improved, for d = 3, from 7/16 to 3/4. Similarly, in 2 dimensions we can improve C from 3/8 to 2/3.

There is an edge (a, b) between nodes

$$a = a_1 \cdots a_{i-1} \bar{a}_i a_{i+1} \cdots a_n$$

and

$$b = a_1 \cdots a_{j-1} \bar{b}_j a_{j+1} \cdots a_n,$$

if either $i \equiv j \pm 1 \bmod n$, $a_i = b_i$ and $a_j = b_j$ (edges in cycles), or $i = j$ and $a_i \neq b_i$ (edges between cycles). There are two orbits: the set of *cycle* edges, and the set of *non-cycle* edges. Since there are $N/2$ non-cycle edges, N cycle edges, and $3N/2$ edges altogether, the coefficient M is $1/3$. Substitution of R, D, M, and orbit skew $d = 3$ in Theorem 2 (ii) yields the Corollary.   □

I.e., the *total* edge length is $\Omega(N^{4/3} \log^{-1} N)$ and the *average* edge length is $\Omega(N^{1/3} \log^{-1} N)$. (In 2 dimensions $\Omega(N^{3/2} \log^{-1} N)$ and $\Omega(N^{1/2} \log^{-1} N)$, respectively.) I expect that similar lower bounds hold for other fast permutation networks like the *butterfly-* , *shuffle-exchange-* and *de Bruijn* graphs.

Edge-Symmetric Graphs

Recall that a graph $G = (V, E)$ is *edge-symmetric* if each edge is mapped to every other edge by an automorphism in Γ. We set off this case especially, since it covers an important class of graphs. (It includes the binary n -cube but excludes CCC.) Let $|V| = N$ and $D < \infty$ be the diameter of G. Substituting R as in (1.1), $m = 1$, and $d = 3$ in Theorem 2 (i) we obtain:

Corollary 4. *The average Euclidean length of edges in a 3-space embedding of an edge-symmetric graph is at least $7R/(16D)$.*

For the complete graph K_N, this results in an average wire length of $\geq 7R/16$. I.e., the average wire length is $\Omega(N^{1/3})$, and the total wire length is $\Omega(N^{7/3})$.

For the complete bigraph $K_{1,N-1}$ (the star graph on N nodes) we obtain an average wire length of $\geq 7R/32$. I.e., the average wire length is $\Omega(N^{1/3})$, and the total wire length is $\Omega(N^{4/3})$.

For a N -node δ -dimensional mesh with wrap-around (e.g., a ring for $\delta = 1$, and a torus for $\delta = 2$), this results in an average wire length of $\geq 7R/(8N^{1/\delta})$. I.e., the average wire length is $\Omega(N^{(\delta-3)/3\delta})$, and the total wire length is $\Omega(\delta N^{(4\delta-3)/3\delta})$.

Complete Binary Tree

The complete binary tree T_n on $N-1$ nodes ($N = 2^n$) has $n-1$ orbits $E_1, \ldots, E_{n-1}$. Here E_i is the set of edges at level i of the tree, with E_1 is the set of edges incident with the leaves, and E_{n-1} is the set of edges incident with the root. Let l_i and l be as in Theorem 2 with $m = n - 1$. Then $|E_i| = 2^{n-i}$, $i = 1, \ldots, n-1$, the orbit skew coefficient $M = 2/(2^n - 2)$, and we conclude from Theorem 2 (ii) that l is $\Omega(N^{-2/3} \log^{-1} N)$ for $d = 3$. This is consistent with the known fact l is $O(1)$. However, we obtain significantly stronger bounds using the more general part (i) of Theorem 2. In fact, we can show that 1-space embeddings of complete binary trees with $o(\log N)$ average edge length are impossible.

Corollary 5. *The average Euclidean length of edges in a d-space embedding of a complete binary tree is $\Omega(1)$ for $d = 2, 3$, and $\Omega(\log N)$ for $d = 1$.*

Proof. Consider d-space embeddings of T_n, $d \in \{1, 2, 3\}$ and $n > 1$. By Theorem 2,

$$\sum_{i=1}^{n-1} \delta_i l_i \geq CRD^{-1} \tag{12.6}$$

It turns out that $\delta_i \leq (n-1)^{-1}$, for $i = 1, \ldots, n-1$, see [7]. Substitution of $|E_i|$, $|E|$, m in the expression for l in Theorem 2 (i) gives

$$l = (2^n - 2)^{-1} \sum_{i=1}^{n-1} 2^{n-i} l_i \tag{12.7}$$

Substitute in (1.6) the values of C, R (depending on d) and $D = 2(n-1)$. Next, substitute $(n-1)^{-1}$ for δ_i, and multiply both sides with $n-1$. Use the resulting expression to substitute in (1.7), after rearranging the summation as follows:

$$l = (2^n - 2)^{-1} \left(\sum_{j=1}^{n-1} 2^{j-1} \sum_{i=1}^{n-j} l_i + \sum_{i=1}^{n-1} l_i \right)$$

$$\in \Omega \left(2^{-(1-1/d)n} \sum_{j=1}^{n-1} 2^{(1-1/d)j} \right)$$

Therefore, for $d = 2, 3$, we obtain l is $\Omega(1)$. However, for $d = 1$, l is $\Omega(\log N)$. $\square$

12.3.1 Optimality Conjecture

There is evidence that the lower bound of Theorem 2 is optimal. Viz., it is within a constant multiplicative factor of an upper bound for several example graphs of various diameters. Consider only 3-dimensional Euclidean embeddings, and recall the assumption that wires have length but no volume, and can pass through nodes. For the *complete graph* K_N with diameter 1, the lower bound on the average wire length is $7R/16$, while $2R$ is a trivial upper bound. For the *star* graph on N nodes the bounds are $7R/32$ and $2R$, respectively. The upper bound on the total wire length to embed the *binary* n -*cube* requires more work. Let $N = 2^n$.

The construction is straightforward. For convenience we assume now that each node is embedded as a 3-space cube of volume 1. Recursively, embed the binary n -cube in a cube of 3-dimensional Euclidean space with *sides* of length S_n. Use 8 copies of binary $(n-3)$ -cubes embedded in Euclidean $S_{n-3} \times S_{n-3} \times S_{n-3}$ cubes, with $S_{n-3} = S_n/2$. Place the 8 small cubes into the large cube by fitting each small cube into an octant of the large cube. First connect the copies pairwise along the first coordinate to form four binary $(n-2)$ cubes. Connect these four pairwise along the second coordinate to form two binary $(n-1)$ cubes, which in turn are connected along the third coordinate into one binary n -cube. The total length of wire required turns out to be $L(n)(= L) < 4N^{4/3}$, cf [7].

Similarly, for *cube-connected cycles* with $N = n2^n$ nodes, the total interconnect length is $\leq 4N^{4/3} \log N$ [7]. For δ -dimensional meshes with wrap-around, with $\delta = 1, 2, 3$ and diameter $N^{1/\delta}$, a lower bound of $\Omega(1)$ follows from Theorem 2, and the upper bound is $O(1)$ by the obvious embedding. Note that $\delta = 1$ is the ring, and $\delta = 2$ the torus. To prove that Theorem 2 is optimal up to a multiplicative constant in the general case, one may try to generalize the upper bound construction for the binary n -cube.

12.3.2   Robustness

Theorem 2 is robust in the sense that if $G' = (V', E')$ is a subgraph of $G = (V, E)$, and the theorem holds for either one of them, then a related lower bound holds for the other. Essentially, this results from the relation between the orbit frequencies of G, G'. Let us look at some examples, with $d = 3$.

Let a graph G have the binary n -cube C as a subgraph, and $N = 2^n$. Let G have $N' \leq 8N$ nodes, and at most $N' \log N'$ edges. The lower bound on the total wire length $L(G)$ of a 3-space embedding of G follows trivially from $L(G) \geq L(C)$, with $L(C) \geq 7RN/32$ the total wire length of the binary n -cube. Therefore, expressing the lower bounds in N' and radius R' of a sphere with volume N', yields $L(G) \geq 7R'N'/512$, and the average edge length of G is at least $7R'/(512 \log N')$.

Let the binary n -cube C have a subgraph G with $n2^{n-1} - 2^{n-5}$ edges. The lower bound on the total wire length $L(G)$ of a 3-space embedding of G follows from the observation that each deleted edge of C has length at most twice the diameter R of (1.1). I.e., $L(G) \geq L(C) - 2^{n-4}R$ with $L(C)$ as above. Note that G has $N' \geq 2^n - (2^{n-6}/n)$ nodes. Therefore, expressing the lower bounds in N' and radius R' of a sphere with volume N', yields $L(G) \geq 5RN/32 \geq 5R'N'/32$, and the average edge length of G is at least $5R/16n \sim 5R'/(16 \log N')$.

12.4 Interconnect Length and Volume

An effect that becomes increasingly important at the present time is that most space in the device executing the computation is taken up by the wires. Under very conservative estimates that the unit length of a wire has a volume which is a constant fraction of that of a component it connects, we can see above that in 3-dimensional layouts for binary n-cubes, the volume of the $N = 2^n$ components performing the actual computation operations is an asymptotic fastly vanishing fraction of the volume of the wires needed for communication:

$$\frac{\text{volume computing components}}{\text{volume communication wires}} \in o(N^{-1/3})$$

If we charge a constant fraction of the unit volume for a unit wire length, and add the volume of the wires to the volume of the nodes, then the volume necessary to embed the binary n-cube is $\Omega(N^{4/3})$. However, this lower bound ignores the fact that the added volume of the wires pushes the nodes further apart, thus necessitating longer wires again. How far does this go? A rigorous analysis is complicated, and not important here. The following intuitive argument indicates what we can expect well enough. Denote the volume taken by the nodes as V_n, and the volume taken by the wires as V_w. The total volume taken by the embedding of the cube is $V_t = V_n + V_w$. The total wire length required to lay out a binary n-cube as a function of the volume taken by the embedding is, substituting $V_t = 4\pi R^3/3$ in (1.5),

$$L(V_t) \geq \frac{7N}{32} \left(\frac{3V_t}{4\pi}\right)^{1/3}$$

Since $\lim_{n\to\infty} V_n/V_w \to 0$, assuming unit wire length of unit volume, we set $L(V_t) \sim V_t$. This results in a better estimate of $\Omega(N^{3/2})$ for the volume needed to embed the binary n-cube. When we want to investigate an upper bound to embed the binary n-cube under the current assumption, we have a problem with the unbounded degree of unit volume nodes. There is no room for the wires to come together at a node. For comparison, therefore, consider the fixed degree version of the binary n-cube, the CCC (see above), with $N = n2^n$ trivalent nodes and $3N/2$ edges. The same argument yields $\Omega(N^{3/2} \log^{-3/2} N)$ for the volume required to embed CCC with unit volume per unit length wire. It is known, that every small degree N-vertex graph, e.g., CCC, can be laid out in a 3-dimensional grid with volume $O(N^{3/2})$ using a unit volume per unit wire length assumption [2]. This neatly matches the lower bound.

Because of current limitations to layered VLSI technology, previous investigations have focussed on embeddings of graphs in 2-space (with unit length wires of unit volume). We observe that the above analysis for 2 dimensions leads to $\Omega(N^2)$ and $\Omega(N^2 \log^{-2} N)$ volumes for the binary n-cube and the cube-connected cycles, respectively. These lower bounds have been obtained before using *bisection width* arguments, and are known to be optimal [5]. Recall, in [4,8] it is shown that we cannot always assume that a unit length of wire has $O(1)$ volume. (For instance, if we want to drive the signals to very high speed on chip.)

12.5   A Modest Proposal

The lower bounds on the wire length above are *independent* of the ratio between the volume of a unit length wire and the volume of a processing element. This ratio changes with different technologies and granularity of computing components. Previous results may not hold for optical communication networks, intraconnected by optical wave guides such as glass fibre or guideless by photonic transmission in free space by lasers, while ours do. The arguments we have developed are purely geometrical, apply to any graph, and give optimal lower bounds in all cases we have examined. Our observations are mathematical consequences from the noncontroversial assumption of 3 dimensional space. They form a firm underpinning for the cost measures we propose.

Currently, *wiring problems* start to plague computer designers and chip designers alike. Formerly, a wire had magical properties of transmitting data 'instantly' from one place to another (or better, to many other places). A wire did not take room, did not dissipate heat, and did not cost anything - at least, not enough to worry about. This was the situation when the number of wires was low, somewhere in the hundreds. Current designs use many millions of wires (on chip), or possibly billions of wires (on wafers). In a computation of parallel nature, most of the time seems to be spent on communication - transporting signals over wires. While the von Neumann bottleneck has been conquered by non-sequential computation, a Non-von Neumann communication bottleneck looms large.

It is clear that these communication mishaps will influence the architecture and the algorithms to be designed for the massive multiprocessors of the future. The present analysis allows us to see that any reasonable model for multicomputer computation must charge for communication. The communication cost will impact on both physical time and physical space costs. In contrast with sequential cost measures where space is more stable than time, here we see that space is more susceptible to variations in topology of the communication graph than time. Let us make a modest initial proposal. So as to make the proposal technology independent, I use the lower bounds on edge length of dimensionless wires, rather than lower bounds on volume. Below we apply the results of the previous discussion for 3 dimensions. However, in some applications 2 or 1 dimensions may be more appropriate.

12.5.1   Communication and Time

For the communication cost in terms of time, the topology of the communication network is irrelevant for the measure, only the number of nodes count. Every computation which utilizes n nodes, costs $\Omega(n^{1/3})$ time extra. The change in time complexity, from not charging for communication to charging for it, will depend on the particular algorithm involved. At the lower end of the scale there are algorithms which do not use the processors in parallel at all, but use them actually in sequence. Then, we assume that the communication cost is already reflected in the sequential cost. At the other end of the scale, there are subcomputations which use many processors each, and do not overlap in time. We define the communication time cost as follows.

A set S of nodes is *used* in a subcomputation if there is a single node $x \in S$ such that each node y in S could have transmitted information to x during the subcomputation. Let an algorithm A on an instance of problem X execute k subcomputations, the i th subcomputation using u_i processors, such that these subcomputations do not overlap in time. The *communication time cost* CT is the maximum of $\sum_{1 \le i < k} u_i^{1/3}$ over all possible such divisions of A. This is the "physical" time overhead which is incurred by communication requirements. (In two dimensions, like in layered VLSI, we replace 1/3 by 1/2.) The *communication time complexity* $CT(n)$ of A for problem X is the maximum CT over all problem instances of X of size n.

Let us give a simple example. It is well known, that we can solve NP-complete problems like graph coloring in polynomial 'time', i.e., number of causally related steps. In particular, using a $N = 2n^n - 1$ node tree, there is an algorithm A that takes $T(n) = n^2$ time to solve the 'minimum cost coloring' problem for n -node graphs [3], exclusive communication cost. The computation fans out from root to the leaves and fans in again to the root. Therefore, A uses $CT(n) \in \Omega(n^{n/3})$, which is superexponential in n !

12.5.2   Communication and Space

The space cost of embedding a multicomputer depends on the number of nodes, but ultimately on the interconnect topology. We want the cost measure to be technology independent. We associate the communication space cost to the algorithm concerned. To assure the technology independence, we do not aim at the physical *volume* taken by a computing aggregate as its communication space cost, but rather at the *total interconnect length* . Moreover, we count only the part of the computing aggregate required for the communications of the algorithm. We define that algorithm A *uses* a subgraph $G = (V, E)$, of the interconnect graph of the multicomputer it runs on, for a instance of problem X, if A uses the set of nodes

V and the set of edges E in that computation. The *communication space cost* CS of A for that problem instance is defined as the minimal total edge length of G as defined in the previous sections. The *communication space complexity* $CS(n)$ of A for problem X is the maximum CS over all problem instances of X of size n.

Let us look at another simple example: an algorithm for sorting on a cube-connected cycles network with n nodes of a list of size n. Straightforward application of Theorem 2 yields $CS(n)$ is $\Omega(n^{4/3} \log^{-1} n)$

Acknowledgement

Remarks by Andries Brouwer, Evangelos Kranakis, F. Tom Leighton, Lambert Meertens, Yoram Moses were helpful.

References

[1] Harary, F., *Graph Theory* , Addison-Wesley, Reading, Mass., 1969.

[2] Leighton, F.T. and A. Rosenberg, "Three-dimensional circuit layouts," Tech. Rept. TM-262, MIT, Laboratory for Computer Science, Cambridge, Mass., June, 1984.

[3] Mead, C. and L. Conway, *Introduction to VLSI Systems* , Addison-Wesley, Reading, Mass., 1980.

[4] Mead, C. and M. Rem, "Minimum propagation delays in VLSI," *IEEE J. on Solid State Circuits* , vol. SC-17, 1982, 773 - 775; Correction: *ibid* , vol. SC-19, 1984, 162.

[5] Ullman, J.D., *Computational Aspects of VLSI* , Computer Science Press, Rockville, Maryland, 1984.

[6] Vitányi, P.M.B., "Non-sequential computation and Laws of Nature," in *VLSI Algorithms and Architectures* , Lecture Notes in Computer Science, vol. 227, Springer Verlag, Berlin, 1986, 108-120.

[7] Vitányi, P.M.B., "Locality, Communication, and Interconnect Length in Multicomputers," *SIAM J. on Computing* , to appear.

[8] Vitányi, P.M.B., "Area penalty for sublinear signal propagation delay on chip," *Proceedings 26th Annual IEEE Symposium on Foundations of Computer Science* , 1985, 197-207.

Chapter 13

Processes, Objects and Finite Events: On a formal model of concurrent (hardware) systems

P. A. Subrahmanyam [1]

Abstract

This paper discusses aspects of the operational semantics of a set of linguistic primitives for specifying hardware systems. The primitives allow such systems to be viewed as a combination of objects and processes. Two important features of the development are the ability to deal with *finite events* (*i.e.*, events having extended durations, as opposed to abstract point events), and accommodate *true concurrency*. These features allow the development of a notion of observational equivalence that allows systems to be viewed during appropriate "windows" of time, and that is closer to the notion used in dealing with real hardware systems. While the major applications motivating this work arose in modelling and designing hardware (VLSI) systems, many of the issues discussed are relevant in the context of general concurrent and distributed systems, and calculii for such systems.

[1] AT&T Bell Laboratories, Holmdel, NJ

13.1   Introduction

The information involved in the design of a non-trivial VLSI system is both vast and conceptually complex. Our experience (and that of several others) indicates that an object-oriented paradigm is quite useful in developing automated design tools for assisting the synthesis and analysis of VLSI circuits[17], [3]. Additionally, an object-oriented perspective is useful in modelling the hardware systems themselves (as contrasted with developing tools for designing them). However, many of the developments in these contexts have hitherto had an informal flavor. Thus, they do not satisfactorily address issues that arise in the context of formally reasoning about relevant properties of hardware systems. The intent of this paper is to explore aspects of a theoretical framework for integrating related issues in concurrency models, object-oriented systems, type-theory, and equational-logic programming. The availability of such frameworks in turn enables progress toward a rigorous basis for building robust computer-aided design (CAD) systems, by providing a basis for developing tools that assist in reasoning about the behavioral, temporal, and structural characteristics of hardware systems.

More specifically, this paper discusses aspects of (1) some linguistic primitives for specifying hardware systems viewed as a combination of objects and processes, (2) an underlying formal model (based on event structures) for computations in such systems, and (3) an operational semantics (based on transition systems). Distinguishing features of the description primitives and operational model discussed here include the ability to deal with *finite events* (*i.e.*, events having extended durations, as opposed to abstract point events) and *true concurrency* (using partial orderings amongst events as opposed to viewing concurrent operations as nondeterministic interleavings of the constituent operations). In particular, the approach accommodates both synchronous and asynchronous (*e.g.*, self-timed systems[15]), and supports hierarchical timing specifications[19].

As remarked above, the major applications motivating this work arose in the context of modelling VLSI systems, and building object-oriented computer-aided design systems for VLSI. However, several issues discussed in this paper have relevance beyond the VLSI context. In particular, they relate to developments in the theory of concurrent systems and distributed systems, *e.g.*, Milner's calculii for communicating systems (CCS[11] and SCCS[10]) and Hoare's communicating sequential processes (CSP[9]). As a consequence, we use the terms "hardware module", "hardware system", "process", "distributed system", and "concurrent system" almost interchangeably in this paper[2]. Before delving into any details, however, we first recall some of the relevant aspects of hardware modules, object-based systems, processes, and finite events.

---

[2]One of the pertinent differences is that the processes in a hardware system are typically finite state automata (Moore or Mealy), whereas this is not necessarily true in software. On occasions where this difference is important, we will appropriately qualify our statements.

13.1.1 Hardware Modules as Objects

A hardware module may be viewed as having an external *interface* consisting of its input/output ports, and (optionally) an internal state. It communicates with its environment (or other hardware modules) by means of values that are input and/or output at its ports. These values may be viewed as being instances of object classes such as signals, booleans, integers, or other more "abstract" object classes, potentially parameterized, such as stacks, lists, streams, and trees. The internal state of a hardware module may similarly be thought of as an instance of an appropriate object class e.g., a registerfile or a stack.

The (dynamic) *behavior* of such a hardware module refers to how it responds to various external stimuli. For example, given a set of input objects, a module might provide appropriate output values at one or more of its output ports, and optionally update its internal state.

The internal *structure* of such a module may typically be described hierarchically in terms of the submodules and their interconnections. The "leaf" level modules in such a hierarchical description have a specified behavior, but are ascribed no further structure.

Given the structure of a module, it is often desirable to be able to infer its behavior. This is possible using the semantics of composition operators discussed in this paper. In addition to behavior and structure, a module may have several other attributes of interest, such as timing parameters and geometry. At another level, a hardware module may itself be viewed as an instance of a generic class of all hardware modules. Such a class may have common attributes such as input/output ports, be "composable" with other modules, etc.

Inheritance hierarchies in object-based paradigms

One of the main reasons that an object-oriented paradigm is useful in the development of CAD systems is that objects and object classes may be organized into hierarchies, and may inherit attributes from other objects in any one of several ways (depending upon the specifics of the object inheritance mechanisms provided and the desired behavior). From a software engineering viewpoint, this yields a powerful facility for organizing large systems, by enabling the reuse of common attributes and functions. From a theoretical viewpoint, an object-oriented framework seems quite suited for an amalgamation of the concepts of types, data abstractions, and inheritance/polymorphism [2].

13.1.2 Concurrent Aspects of a Hardware System

The "concurrent aspects" of a hardware system may be isolated into its synchronization skeleton (by viewing all of the sequential computations that do not involve any interaction with other modules as atomic actions without any internal structure). Such a synchronization skeleton may arise either from explicit synchronizer modules (that react to certain events) or may be abstracted from computations involving operations on data (the so-called "data-path" components of a system). However, an important aspect of dealing with real systems includes accounting for the various temporal attributes.

13.1.3   Computational Model

The computational framework developed here views the concurrent skeletons of hardware modules as *processes*; these processes are defined by a process algebra. The syntactic primitives for defining processes in this algebra are introduced in section 13.2. The operational semantics of the processes that can be described using these primitives is described as a labeled transition system, appropriately modified to deal with finite events. The denotation of such processes is indicated using a semantic domain of labelled event structures.

The states and data values are viewed as being objects; these objects may be described in a conventional manner, e.g., via algebraic axioms such as those used in specifying data abstractions[4]. The act of "sending messages to objects" is viewed as "applying processes to objects". Processes may share objects such as channels (enabling message based communication) or variables representing semaphores (enabling shared memory communication). Finite events are accounted for in the transition system by associating temporal assertions with each action.

13.1.4   Finite Events

Most concurrent algebras view events as being *point events* - events that have a time of occurrence but no duration. The only possible relations between two point events are those of *precedence* (when one event occurs before another) and *simultaneity* (when two events occur at the same instant).

However, almost all real physical events have a finite duration. In general, there can be a fairly complex interplay between sets of finite events. The abstraction of finite events as point events is acceptable only if the relationships of interest between the real events are restricted to be either precedence or simultaneity.

Unfortunately, in considering the operation of hardware modules, it becomes necessary to deal with finite events that overlap in ways that cannot be described by a finite event abstraction. For example, a circuit design may require that certain input data lines be held stable for the duration that the clock line is high; further, many other events of interest may occur during this period. If an event a "overlaps" events b and c, and event b precedes c, this cannot be described directly in an algebra using point event abstractions.

There are at least two ways to address this deficiency. One is to develop an algebra that can express all the possible relations amongst finite events, and then use some appropriate (sub)set of these relations to describe a system. The other is to explicitly state the duration of events, for example by indicating their start and end. Such durations can be stated, and reasoned about, using primitives such as those found in various temporal logics. We will illustrate the second approach here. There are two potential advantages of this approach. One is that existing models and results concerning point events can be used without major modification. The second is that a system specification using explicit durations *may* be simpler than one that explicitly enumerates the possibly complex inter-relations between various pairs of events: we do not have as yet enough empirical data to either support or refute this intuition.

13.2 Syntax and Process Interaction Semantics: Examples

In this section, we informally introduce some of the concepts relevant to this paper via simple examples. Specifically, we sketch the syntax for specifying process behavior and structure, comment on process-object interactions (including input/output and broadcast semantics), causal consistency, and finite events specifications.

13.2.1 Syntax

A module (synonymously, process) definition consists of a module header and a module body.

External Interface

A module header in turn consists of a module name, its input and output ports, its internal state, and (optionally) other relevant attributes, such as timing parameters.

Example. A stack-module with input ports {d, op}; output ports {q, full, empty}; and internal state S (Figure 13.1) is denoted

```
stack-module{d,op -> q, full, empty}(S)
```

The input/output ports and state may be explicitly typed, where the types themselves may be parameterized. For example,

```
stack-module{d:T, op:{reset,push,pop,hold} ->
              q:T, full,empty:Boolean}(S:Stack(T,N))
```

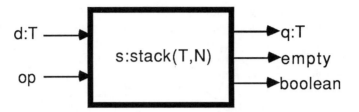

Figure 13.1: Stack Module: I/O structure

For instance, the internal state S is an instance of a type (synonymously, object class) Stack(T,N). This type characterizes stacks of maximum size N, containing elements of type T, and may be defined using standard type definition techniques (see, for instance, [2]).

Behavior

The rest of the module definition describes its behavior. This "module body" may consist of

- a *sequence* of actions, denoted A1 ; ... ; AN

- a *parallel composition* of actions, denoted A1 || ... || AN, and sometimes abbreviated [A1, ..., AN];

- a *choice* among alternative actions, denoted A1 + ... + AN. In describing hardware systems, this choice is usually *deterministic* [9], i.e., all of the choices are mutually exclusive.

Recursion

The expression syntax permits recursion in the module definition. In dealing with descriptions of hardware modules, our use of recursion is restricted to those forms that can be converted to bounded iteration, e.g., tail recursion. As indicated in section 13.4.2, syntactic constructs such as conditionals and case statements can be defined using this facility.

Input/Output Actions

Primitive actions (events) involve input or output of values at a port. We will adopt a CSP-like syntax for input/output in this paper. Thus, output of a value E at a port P is denoted P!E. Correspondingly, the action of input at a port P and its assignation (binding) to a variable X is denoted P?X.

```
stk-module{d:T,op:{reset, push, hold, pop}->q:T,full,empty:Boolean}(S:Stack(T,N)) =
   [op?OP,d?D];
   case OP of
     hold:  [q!top(S), full!isfull(S), empty !isempty(S), stk-module(S)]
     push:  if isfull(S)
            then [q!top(S), full!T, empty!F, stk-module(S)]
            else [q!top(push(S,D)), full!isfull(push(S,D)), empty!F,
               stk-module(push(S,D))]
     pop:   [q!top(pop(S)), full!F, empty!isempty(pop(S)), stk-module(pop(S))]
     reset: [q!top(EmptyStack), full!F, empty!T, stk-module(Emptystack)]
```

Figure 13.2: Stack Module: Input-Output Behavior

Example. The stack-module first inputs (reads) the values at the ports op,d. Depending upon the operation desired (the value of op), the module then updates its internal state denoted by, for example pop(S) (in the case of pop) and (in parallel) provides appropriate outputs at the ports {q, full, empty}.

Computations on objects

Sequential computations on objects and associated classes may be defined in a conventional fashion, and we will not elaborate upon details here. These objects and functions may also be characterized algebraically. For examples, see [2], [4].

Example - Stack Module Behavior   The "black box" behavior of a stack module may be described as shown in Figure 13.2.1.

Structure

Modules may be composed with ("connected to") other modules; this is denoted P1 | P2 ... | PN. Ports and signals (synonymously, wires) may be renamed; they may also be hidden (synonymously, abstracted) so as to preclude external accessibility.

Example. Consider the read-write-flip-flop (rwff) circuit in Figure 13.3. It consists of 4 submodules, namely a multiplexor (denoted mux, implementing *if-then-else*), a static latch (denoted static-latch, implementing a "storage" primitive), an inverter (denoted inv), and a complementary switch (denoted cs2, implementing *if-then*). This structure may be described as follows:

```
rwff{r,w,d -> q}(S) =
   Hide signals s, next-s in
      mux{w,d,s -> next-s}
      |static-latch{next-s -> s}(S)
      |inv{r -> rbar} | cs2{r,rbar,s -> q}
```

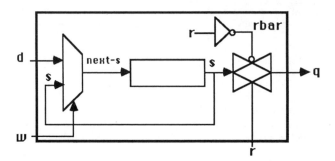

Figure 13.3: A read-write flip-flop

13.2.2   Associating processes and objects

Objects may be *named* once they have been created. "Messages" may be sent to (named) objects. We identify the act of *sending (a message) m to (the object named) n* with *applying (a process) m to (the object named) n*. The body of the message *i.e.,* the actual process description, may be arbitrarily complicated.

When sending a message to an object, it is often useful to ensure that the message is processed only if certain preconditions are met. It is possible to classify such preconditions into two broad categories.

- The first class of preconditions is concerned with the question "Does the object to which a message is being sent have the right set of attributes?", and serve to ensure that the message is applied only if the appropriate *constraints on the attributes of the object* are met.

- The second class of preconditions concerns the *temporal appropriateness* of the message application. In other words, they are concerned with the question "Is it appropriate to send the message to this object *at this time* (in the context of the overall computation)?"; we refer to such preconditions as ensuring the *causal consistency* of a message application.

We will next elaborate briefly on the intuition underlying these classes of preconditions; a formal discussion is contained in section 13.5.

Guards

The first category of preconditions are essentially "semantic guards" that constrain the attributes that an object must have. Such guards may be arbitrarily complex predicates defined on the state of an object; on the other hand, they may be relatively simple syntactic checks relating to the type of the object. For example

```
send-message m to s1 of type Stack
```

restricts process the message **m** to apply to an object named **s1** provided that **s1** alludes to an instance of the class of Stack objects i.e., is an instance of the type Stack.

Of course, in the extreme case, no preconditions might need to be checked in order for a message to be sent, *i.e.*, the guards may be empty.

Notation. `send-message m to (object) n` is denoted `m:n`. `n::s` denotes an object named (or labelled) **n** in state **s**. Thus, `send-message m to (object) n in state s` is denoted `m:<n::s>`. The letters l, n, o are typically used as names of objects.

Causal consistency

In order to informally introduce the notion of causal consistency, we need the notion of a computation. Intuitively, a *computation* corresponds to a finite segment of the execution of a process, wherein a specific set of choices amongst competing alternatives have been made along the way.

Since processes are allowed to operate on specific objects, it is useful to be able to separate the parts of a process p that apply to an object named n from those parts that do not. We will denote by $project(u, n)$ the parts of a computation u that apply to the object named n. We denote by $hide(u, n)$ the parts of a computation u that apply to objects other than n, that is, those parts of u that pertain to n are essentially "hidden".

If $project(u, n)$ is a prefix of u,[3] we say that it is *causally consistent* for u to be applied to n at this point, and denote this by $cc(u, n)$. For example, if a module is being synthesized from its behavior, it does not make sense for physical layout operations to be performed until a structure is available for the module. Similarly, in considering a semaphore that is used to control access to a resource shared between two processes, one may want to insist that a "release" (V) operation should *not* be performed on a semaphore unless an "acquire" (P) operation was performed earlier.

13.2.3 Interaction between processes

Input and Output on channels

The input/output actions of a module can be interpreted as actions that are performed on "objects" of type *channel* (or equivalently, a port, signal or node). In fact, it is possible to consider multiple subclasses of channels, such as ordinary wires and tristate busses. In such a case, the semantics of the send and receive actions may be viewed as being part of the definition of the type channel. The actions of

---

[3]While the notion of the prefix of a computation is formally defined later, an formal interpretation of the prefix as alluding to an initial segment of a computation will suffice here.

input/output at a channel can be viewed as affecting the environment through a set of $< variable, value >$ bindings.[4] The functionality of these operations may then be construed as being:

> Output: Channel [, OutputExpression] [, Env] → Env
> Input: Channel [, Variable] [, Env] → Env
> Synchronized-output: Channel [, Value] → Env

The presence of an optional output expression in the case of an output operation may be used to distinguish a *data assertion* from a *synchronization signal*. Stated differently, a channel *s* used only for synchronization may be thought of as encoding only Boolean values. The presence of a synchronization signal on such a channel, *i.e.*, the output *s*!1, is abbreviated *s*!, whereas the value 0 on the channel implies the absence of a synchronization signal. The variable associated with the corresponding input operation may similarly be dropped as long as the semantics embody the desired synchronzation effect. This is discussed further below.

Broadcast Semantics

Motivated by the fact that a signal on a wire can typically be sensed by all modules connected to it, we adopt a *broadcast semantics* for input/output actions. In the event that more than one value is output onto a channel, a "combining" function is used to determine the outcome. This mechanism can be used, for instance, to account for a tri-state bus, and several·other situations[16]. Thus, in the case of a tri-state bus, a high impedance output (a "floating" node) can be combined with other values that pull the node up or down; however, the output of two conflicting (high and low) values will result in an incorrect circuit.

Recall that we abbreviate Output(P) by P!, Output(P,E) by P!E, Input(P) by P?, and Input(P,X) by P?X. Further, we abbreviate synchronized-output(P) by P*. Note that some signals are hidden *i.e.*, "abstracted" and not made available to external modules. It is conceivable that hiding an unsynchronized action may lead to deadlock, whereas hiding a synchronized action may make it transparent to the operation of a system[5].

Assuming a broadcast semantics, the interactions between these operations may be expressed as follows:

1. p? ‖ p! ⇒ p*

2. p? ‖ p? ⇒ p?

3. p! ‖ p! ⇒ p!

---

[4]For the purpose of this discussion the environment may be viewed as a set of $< variable, value >$ bindings.

4. p! || p* ⇒ p*

5. p!E || p?X ⇒ p*E, X==E (*i.e.*, X is bound to E in the environment)

6. p!E1 || p!E2 ⇒ p!combine(E1,E2) (where combine (E1,E2) is a user defined operation on values)

13.2.4   Finite events

There may be a finite duration associated with each action. Moreover, the duration of an action may overlap two or more actions. Such durations can be explicitly indicated by a temporal specification (which may be viewed as annotating the behavioral specification). In many instances, the durations of concern may be inferred [19], given high level temporal models of the submodules used in a design. When dealing with finite events, the (most intuitive) temporal partial order of the associated "abstracted" point events is typically given by the order of the times of initiation of the corresponding finite events.

Another abstraction that is useful in some circumstances is that of a set of operations that can be executed in "zero" time (and all at the same "time instant"), but whose data dependencies need to preserved. An example is the operation of the combinational part of a sequential circuit. At some level of abstraction, such combinational circuits are assumed to provide their results "instantaneously". Clearly, however, the execution of each level of a multi-level combinational circuit takes finite time; further, the data dependencies between the computations are mirrored in the interconnections of such a circuit, and need to be preserved in order for the circuit to achieve the desired function. The abstraction that is relevant in such a case is to view all of the combinational operations as being executed during the same time step.

13.2.5   Reasoning about behavior and timing

Given a module or process description, several questions are relevant.

- What computations are denoted by the behavioral description? The concrete denotational domain we adopt for interpreting process descriptions is a variant of labeled event structures [20] that accommodates finite events, and is outlined in section 13.3.

- What is an execution model for the process? An execution model based on labeled transition systems is reviewed in section 13.5, and serves as the operational semantics for module/process descriptions.

- What is the temporal behavior of the system? The temporal assertions that are associated with actions in the transition system ("below the arrow") enable the overall temporal behavior of to be inferred from a behavioral description.

Given a structural description, it is of course important to be able to infer the overall behavior of the module. An "expansion theorem" assists in this process, and serves as an algorithm for inferring behavior from structure. Furthermore, given two process descriptions, an appropriate notion of their equivalance is needed? This issue is addressed in section 13.7.

An operational semantics provides a way of interpreting the behavioral description of a module. The operational semantics we discuss here views the specification as a labelled transition system (or rewriting system). A concurrent system is viewed as consisting of processes that act upon objects.[5] The operational semantics of both of these parts may therefore be considered individually.[6] We first discuss the operational semantics for process specifications, and then discuss their interaction with object specifications.

13.3 Denotations for process descriptions

A process description denotes a set of possible computations. We adopt a slight variant of *labelled event structures* [20] as our concrete model of hardware modules. These structures embody the distinct relations of causal dependence, concurrency (causal independence), and conflict (choice) amongst events. Corresponding to these relations, there are 3 operations sequential composition, parallel composition, and choice that provide a way of constructing event structures. This is done by juxtaposing two event structures and setting the corresponding relation between their events.

We next formally define labeled event structures (the semantic domain), and then discuss the correlation between the process descriptions (syntactically denoted by terms in an algebra), event structures and computations.

13.3.1 Labelled Event Structures

Definition 13.3.1 *Let A be a nonempty set (denoting the atomic actions performed in the system). An event structure labeled by A (synonymously, an A-labeled event structure) is a quintuple (A, E, $\leq$, #, L) where*

1. $E \subseteq \{0,1\}^*$ *is the set of events;*

2. $\leq \subseteq E \times E$ *is a partial order on E, denoting the causality relation;*

3. $\# \subseteq E \times E$ *($\leq \cup \geq$) is the symmetric "conflict" relation;*

4. *L:E$\rightarrow$A is the labelling function that names events.*

---

[5] Of course, processes may also be viewed as objects at some "higher" level. The point of view explored here attempts to preserve their distinction. In some sense, the processes constitute the active part, and the objects the passive part.

[6] This potentially provides for more flexibility by allowing for two different modes of presentation of the operational semantics.

Two events in E are *concurrent* (denoted a $\simeq$ b) if they are neither ordered nor in conflict. That is,

$$\simeq \ =_{def} \ \text{E} \times \text{E} - (\leq \cup \geq \cup \#)$$

This relation $\simeq$ is symmetric and irreflexive. By definition, the relations $\simeq, \leq, \geq,$ $\#$ induce a partition on E $\times$ E.

When it is necessary to account explicitly for finite events, we will refer to the start and end events of an action $a \in A$ by a_s and a_e respectively. In this paper, such start and end events are viewed as "abstract" (virtual) point events. We define A_s $= \{ a_s \mid a \in A \}$, $A_e = \{ a_e \mid a \in A \}$. The set of actions A is augmented with A_s and A_e whenever needed.

Examples. For all actions $a \in A$, it is always the case that $a_s \leq a_e$. If an event a overlaps an event $(b; c)$, then $a_s \leq b_s$ and $a_e \geq c_e$.

We denote by Time(a) the instant of occurrence of an event $a \in A$. If $\text{Time}(a_s) = t_1$, and $\text{Time}(a_e) = t_2$, then we say that the temporal *duration* of a finite event a is t_1 to t_2, and denote this by $\text{duration}(t_1, t_2)(a)$. For example, if a signal s is held stable at a voltage denoting the value 1 from t_1 to t_2, denoted $\text{stable}(t_1, t_2)(s)(1)$ then $\text{duration}(t_1, t_2)(s=1)$.

Event structures that are similar except for renaming of the actions are said to be *isomorphic*; we will denote this renaming isomorphism by $\sim_L$.

13.3.2 Composing Labelled Event Structures

We denote the finite A-labelled event structures by $L(A)$, and the infinite structures by $L(A)^\infty$. These sets have a natural algebraic structure arising from the 3 relations that can exist amongst primitive events, namely causal dependence, concurrency (causal independence) and conflict. The 3 corresponding ways of combining event structures correspond to 3 ways of combining processes, viz., via sequential composition (;), parallel composition ($\|$), and selection (+). Recall that it is intended that event structures represent all of the possible computations of a (concurrent) process. If S1 and S2 are A-labelled event structures, then the juxtaposition S = S1 *op* S2, *op*$\in \{\leq, \simeq, \#\}$ denotes the result combining S1 and S2 using *op*. If *op* = $\leq$, then S is called the sequential composition of S1 and S2, and denoted S1;S2. If *op* = $\simeq$ then S is called the parallel composition S1 $\|$ S2. If *op* = $\#$, then S is called the sum S1+S2.

13.3.3   Interpreting Finite Processes

Let **1** denote a terminated computation (or a "nil" process); we also used **1** to denote an empty event structure. Let $T(A)$ be the set of finite terms, denoting processes, generated by $\{1 , A\}$ that are closed under $\{;, \|, +\}$. The interpretation of a process $p \in T(A)$ is given by a map $ES:T(A) \to L(A)/ \sim_L$.

$$ES(1) = (A, \emptyset, \emptyset, \emptyset, \emptyset)$$
$$ES(a) = (A, \{\epsilon\}, = , \emptyset, \lambda), \text{ with } L(\epsilon) = a$$
$$ES(p;q) = (ES(p);ES(q))$$
$$ES(p\|q) = (ES(p)\|ES(q))$$
$$ES(p+q) = (ES(p)+ES(q))$$

In other words, a process p described by a term in $T(A)$ is viewed as denoting the (isomorphism class of) the event structure $ES(p)$, written $[ES(p)]$.

13.3.4   Interpreting recursive processes

Given a set of identifiers X, let $T^{rec}(A \cup Y)$ denote the set of terms formed by using $\{1 , A\}$, the set $Y \subset X$ of identifiers, and containing recursion (denoted by $\mu x.p$). A map $ES^\infty : T^{rec}(A \cup Y) \to L(A)^\infty/\sim_L$ is the interpretation of recursive terms that denote processes involving recursion.

A structure on $L(A)^\infty$

In the set $L(A)^\infty$, the operations $\{;, \|, +\}$ are associative, and have **1** as an identity. Further, the operations $\|$ and $+$ are commutative.

In particular, $L(A)^\infty$ obeys the following axioms characterizing a triple monoid structure. For all $p,q,r \in L(A)^\infty$,

1. $(L(A)^\infty, ;, 1)$ is a monoid

   A1: $(p;(q;r)) = ((p;q);r)$
   I1: $(p;1) = p = (1 ;p)$

2. $(L(A)^\infty, \|, 1)$ is a commutative monoid

   A2: $(p\|(q\|r)) = ((p\|q)\|r)$
   I2: $(p\|1) = p = (1 \|p)$
   C2: $(p\|q) = (q\|p)$

3. $(L(A)^\infty, +, 1)$ is a commutative monoid

   A3: $(p+(q+r)) = ((p+q)+r)$
   I3: $(p+1) = p = (1 +p)$
   C3: $(p+q) = (q+p)$

Figure 13.4: Excluded Event Structures ∇ and N

13.3.5   A subclass of A-labelled event structures

The set of A-labelled event structures is quite rich. It turns out that not all A-LES's correspond to processes that can be expressed using the operations ;, || and +. In particular, the fragment of event structures shown in Figure 13.4 below (labelled N and ∇ respectively because of their "geometric" appearance) cannot be expressed by means of the syntactic primitives defined earlier. Consequently, it is necessary to consider an appropriate subclass of event structures that can be used as a domain underlying processes that can be expressed using our syntactic primitives. This class of event structures is reasonably well understood in the context of net theory; for the sake of completeness, we include its definition next.

Let $R \subseteq E \times E$ be a relation on a set E.

1. $R^s = R \cup R^{-1}$, is the symmetric closure of R.

2. $R^c = R^s \cup R^0$, is the R-comparability relation.

3. $R^i = (E \times E)\text{-}R^c$ is the R-incomparability relation.

Lemma 1 *An A-LES is N- and ∇- free if and only if* $\forall$ *U, $V \in \{\leq, \#, \simeq\}$, $U \neq V$, and*

- *if e_0 U e_1 and e_0 U^i e_2, where $U^i = (E \times E)\text{-}(U \cup U^{-1} \cup U^0$)*
- *if e_2 U e_3 and e_1 U^i e_3*
- *then e_0 V $e_3 \Rightarrow \{ e_0, e_1 \} \times \{ e_2, e_3 \} \subseteq V$ $\square$*

This condition is depicted in Figure 13.5, where solid lines indicate assumptions, and dotted lines indicate the consequences implied by the assumptions.

The class of A-labeled event structures of interest is therefore those that are N-free and ∇-free. These are denoted NG(A) and $NG(A)^\infty$ respectively.

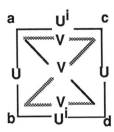

Figure 13.5: Condition obeyed by a subclass of labeled event structures

13.3.6 A Characterization Theorem

We will denote by $=_{ES}$ the semantic equality induced over T(A) and $T^{rec}(A \cup Y)$ by the equality of the corresponding event structures, and by $=_T$ the equality induced by the axioms of a triple monoid. The following theorem then formally characterizes the correspondence between processes described by terms in $T^{rec}(A \cup Y)$ and labeled event structures in $L(A)^\infty$.

Theorem 2 *The structure $(NG(A)/ \sim_L, \; ;, \; \|, \; +, \; 1 \;)$ is the free triple monoid generated by A, i.e., it obeys the axioms A1-A3, I1-I3, C1-C2.*

1. $s \in NG(A) \Leftrightarrow \exists \; p \in T(A), \; ES(p) \sim_L s$, and $T(A) = NG(A)/ \sim_L$

2. $p =_{ES} q \Leftrightarrow p =_T q$ □.

Further, $ES(T^{rec}(A)) \subseteq NG(A)^\infty/\simeq$.

13.4   More on Syntax

13.4.1   Interactions between Process and Objects

Objects and related abstractions may be defined using any of the standard techniques. For example, the functions that operate on objects may be defined via algebraic equations. We will not elaborate here on the details of such definitional mechanisms.

Objects may be named once they have been created, and subsequently be referred to explicitly by name in functions and processes. We extend the syntax of T^{rec} to allow for the act of "*sending a message m to an object n*"; this is synonymous with "*applying a function/process m to an object n*". As described earlier, such messages may be associated with preconditions that must be ensured prior to the actual application of a message.

13.4.2 Conditionals, Case Statements and Iteration

We next introduce some useful notational abbreviations that use guards.

Conditional Expressions

A conditional expression such as `cond(b,p,q)` = `if b then p else q` can be defined by using the notion of guards introduced above. Since the 2 states of a boolean variable are either true or false, we can test for these conditions (denoted c_0 and c_1 respectively). The conditional expression is then defined by `cond(b,p,q)` $\equiv_{def}$ `(b::` c_0 `);p +` `(b::` c_1 `);q`. This says that the process `cond(b,p,q)` offers two alternatives. If the state of the boolean variable is true, the process `p` is executed, otherwise process `q` is executed. The two choices are mutually exclusive.

Case Statement

The conditional expression defined above can easily be generalized to a case statement. Thus, a syntax similar to

```
case(s, c₁,p₁,...,cₙ,pₙ ) =
   case s of
        c₁:p₁
        ...
        cₙ:pₙ
   end
```

may be used to the express a process defined as follows:

`case(s, `$c_1,p_1,\ldots,c_n,p_n$` )` $\equiv_{def}$ `(s:: `$c_1;p_1$` ) +` $\ldots$ `+ (s:: `$c_n;p_n$` )`

Indefinite loops

The process `p` defined as `p = while b do q` may be interpreted as the recursive process $\mu.\mathbf{x}.$`((b:: `c_0` );p;x + (b:: `c_1` ))`. Note that this is meaningful only in a context where `b` is interpreted as being a boolean.

13.5 Computations and operational semantics

Recall that a process description denotes a set of computations, and that we have used labeled event structures as a concrete model for this set of computations. Externally observable behavior is influenced only by specific computations, rather than a combination of all possible computations. Thus, it is desirable to be able to treat as equivalent, for example, the processes $a;(b+c)$ and $a;b+a;c$.

A *specific* (finite) computation is a *conflict-free* prefix of an event structure, i.e., one in which specific choices have been made amongst the competing alternatives. For example, $a;b$ and $a;c$ are the two computations of the process $a;(b+c)$. We will here confine ourselves to the subset of A-labeled event structures that are N- and ∇-free. This is trivially true of conflict-free (and therefore deterministic) structures, denoted D(A).

Thus, given an event structure S=(A, E, $\leq$, #, L) and a finite set E1$\subseteq$E of events, a computation is defined as a conflict-free restriction of S to E1, and denoted $S_{|E1}$. The part of S that remains after computation C is denoted S/C, and called the *quotient* or the *residual* of S modulo C. Formally, $S/C \equiv_{def} S_{|(E-E1\cup E1\#)}$ where $E1^{\#} = \{e \mid \exists e' \in E1 \text{ and } e\#e'\}$.

Definition 13.5.1 *A labeled transition system Σ is a triple (S,C, Σ_S) where S is a set of states, C is a set of computations, and $\Sigma_S \subseteq S \times C \times S$ is a transition relation.*

The transition relation Σ_{ES} on A-LES's is given by p $\xrightarrow[\Sigma_{ES}]{u}$ p' iff u is a computation of p, and p' is the residual of the event structure after u has been executed.

Since we are primarily interested in the processes that are described by terms in $T^{rec}(A)$, it is of interest to seek a syntactic transition relation on the terms in $T^{rec}(A)$ that mirrors the semantic transition relation on A-LES's defined above.

Specifically, the syntactic actions include: identity (inaction); atomic actions; actions that are parts of sequential compositions, parallel compositions, or alternative actions; and actions that are parts of recursively defined processes.

In particular, the syntactic transition relation on terms, denoted Σ_T is defined by the least subset of $T^{rec}(A) \times D(A) \times T^{rec}(A)$ that satisfies A1, S1, S3, P2, C1, C2, and FP amongst the transitions listed in Figure 13.5.

The correspondence between the two transition systems is then expressed via the following theorem:

Theorem 3 *(Validity and Completeness) For all terms t_1, $t_2 \in T^{rec}(A)$:*

1. $t_1 \xrightarrow[\Sigma_T]{w} t_2 \Rightarrow \exists\ W \ni ES(w) \sim_L W,\ ES^{\infty}(t_1) \xrightarrow[\Sigma_{ES}]{W} (ES^{\infty}(t_1)/W)$ *and* $ES^{\infty}(t_2) \sim_L (ES^{\infty}(t_1)/W)$

2. $ES^{\infty}(t_1) \xrightarrow[\Sigma_{ES}]{W} S \Rightarrow \exists w \ni ES(s) \sim_L W,$ *and* $ES^{\infty}(t_2) \sim_L S$ *and* $t_1 \xrightarrow[\Sigma_T]{w} t_2$.
   $\square$

[AA1: Atomic Action] $a \in A \vdash a \xrightarrow{a} 1$

[S1: Sequential Composition 1] $P \xrightarrow{u} P' \vdash (P;Q) \xrightarrow{u} (P';Q)$

[S2: Sequential Composition 2] $P \equiv 1$, $Q \xrightarrow{v} Q' \vdash (P;Q) \xrightarrow{v} Q'$

[S3: Sequential Composition 3] $P \xrightarrow{u} P' \equiv 1$, $Q \xrightarrow{v} Q' \vdash (P;Q) \xrightarrow{u;v} Q'$

[P1: Parallel Composition 1] $P \xrightarrow{u} P' \vdash (P\|Q) \xrightarrow{u} (P'\|Q)$

[P2: Parallel Composition 2] $P \xrightarrow{u} P'$, $Q \xrightarrow{v} Q' \vdash (P\|Q) \xrightarrow{(u\|v)} (P'\|Q')$

[P3: Parallel Composition 3] $Q \xrightarrow{v} Q' \vdash (P\|Q) \xrightarrow{v} (P\|Q')$

[C1: Choice 1] $P \xrightarrow{u} P' \vdash (P+Q) \xrightarrow{u} P'$

[C2: Choice 2] $Q \xrightarrow{v} Q' \vdash (P+Q) \xrightarrow{v} Q'$

[FP: Fixpoint] $P[\mu X.P/X] \xrightarrow{u} P' \vdash \mu X.P \xrightarrow{u} P'$

[OA: Application of a process P to a named object N]

$P \xrightarrow{u} P' \vdash$ (send-message P to N) $\xrightarrow{send-message\ u\ to\ N}$ (send-message P' to N)

[QA: Qualified Application of a process P to a named object N]

$P \xrightarrow{u} P'$ & cc(u,N) and $s \mid \xrightarrow{project(u,N)} s' \vdash$
(send-message P to N in state s) $\xrightarrow{hide(u,N)}$ (send-message P' to N in state s')

Figure 13.6: Transition rules for point events

13.5.1   Examples of derived rules of inference

Using the definition of the conditional in terms of the guarded selection operator, we can infer rules for other syntactic abbreviations, e.g., a conditional operator.

[Conditional T] $P \xrightarrow{u} P'$, $Q \xrightarrow{v} Q' \vdash$ (if *true* then P else Q) $\xrightarrow{u} P'$

[Conditional F] $P \xrightarrow{u} P'$, $Q \xrightarrow{v} Q' \vdash$ (if *false* then P else Q) $\xrightarrow{v} Q'$

The rules for hiding, renaming, and handling data assertions may be given as usual, and we do not elaborate upon them here.

13.6   Temporal Assertions Associated with Finite Events

In order to accommodate finite events, a transition is modified to take the form

$$P \xrightarrow[temporal\ assertion]{action} P'$$

Thus, for each action, there is an associated temporal assertion. Such temporal assertions state properties about the durations of the actions, and any associated constraints.

Actions fall into different categories, such as atomic actions, concurrent actions, and actions leading to interprocess co-operation, e.g., synchronized input-output actions. The temporal assertions associated with each category of action may be different, and must be specified for all actions. We list some examples below.

For an input action, the temporal assertion may take the form:

$$(i?I;p) \xrightarrow[stable(t_1,t_2)(s_i)(I)\ \&\ t_2-t_1 \geq k]{i?I} p'$$

This implies that it is required that an input signal (at the port i) be stable for at least time k (a constant), and, in particular, that it is assumed that the signal s_i is stable at I between times t_1 and t_2. t_1 and t_2 are free variables in the above expression, such that $t_2 - t_1 \geq k$. Note however, that t_1 and t_2 might typically be bound (or constrained) in a larger context by events that precede this input action.

Consider the action of function evaluation f(X). Assuming the value of X is already available, and that the parameters d_1^f, d_2^f model the delays introduced by the evaluation of the function f[19], the temporal assertion for f(X) might take the form shown below. (We abusively use $stable(t_1,t_2)(X)$ to denote the fact that the signal that corresponds to the variable X is held stable between t_1 and t_2.)

$$f(X);p \xrightarrow[stable(t_1,t_2)(X)\ \&\ t_2-t_1 \geq k^f \Rightarrow stable(t_1+d_1^f,t_2+d_2^f)(s_f)(f(X))]{f(X)} p$$

If an evaluation action uses two or more values, then it may be required that all of these values be stable during a common interval of time. We abbreviate $stable(t_1, t_2)(X)$ & $stable(t_1, t_2)(Y)$ by $stable(t_1, t_2)(X, Y)$.

$$\text{f(X,Y);p} \xrightarrow[\;stable(t_1,t_2)(X,Y) \;\&\; t_2-t_1 \geq k^f \Rightarrow stable(t_1+d_1^f, t_2+d_2^f)(s_f)(f(X,Y))\;]{f(X,Y)} \text{p}$$

In other words, this implies that if prior inferences lead to unrelated time intervals over which X and Y were stable, a computation involving f(X,Y) would lead to a constraint that imposed a minimum overlap of the two durations.

13.7 Semantic Equivalence

We are interested here in developing a notion of equivalence on hardware (concurrent) system specifications with respect to externally observable behavior. There are two flavors of equivalences that are relevant when considering the behavior of hardware systems: one that takes into account the intervals of time during which inputs/outputs are made available (or observed), and another that abstracts away the details pertaining to actual durations of events and considers only the partial orders among them.

In practice, especially when considering synchronous hardware systems, the actual intervals during which signals are observed is quite important in developing an appropriate notion of equivalence. However, in this situation, it becomes necessary to establish a correlation between a higher level behavioral description that deals only with partial orders and a more detailed description of a circuit. In this context, a useful abstraction function from finite events to partial orders is one that reflects the order between the *starting* times of events that have a finite duration.

In this section, we will first summarize a notion of observable equivalence that does not explicitly consider temporal durations. To account for temporal durations, we use a transition system augmented with temporal assertions that are appropriate to the atomic actions involved. In addition, depending upon the complexity of such assertions, a suitable logic for reasoning about the temporal assertions themselves may be needed, e.g., [12],[19]. We will, however, not elaborate on the details of such temporal reasoning here.

Since our development so far has viewed a hardware system as consisting of concurrent processes that interact by performing computations on objects, our notion of observable equivalence uses both the notion of observable equivalence on the skeletal concurrent processes and that of observable equivalence on the computations themselves. Specifically, we recall here an equivalence of concurrent systems that is defined by bisimulation [14], [10] and observable equivalence defined on data abstractions. We also comment briefly on the associated proof techniques.

Given any transition system, there are several notions of equivalence that are feasible. Among these are trace semantics [9], failure semantics [1], testing equivalences [7], logical equivalences [8], and bisimulation [14], [10]. A bisimulation is a relation over that states of a transition system. The bisimulation relation "relates" states that have similar behaviors i.e., that are equivalent.

Definition 13.7.1 *Consider a labeled transition system* $\Sigma = (S, C, \Sigma_S)$. *Let* $R \subseteq S \times S$ *be a relation over states, and* $H \subseteq C \times C$ *be a relation over the computations. The pair (R,H) is said to be invariant with respect to* Σ_S *iff* $s_1 \ R \ s_2 \ \& \ s_1 \xrightarrow[\Sigma_S]{c} s_1' \Rightarrow \exists \ s_2', \ c'$ *such that* $c \ H \ c', \ s_2 \ R \ s_2'$ *and* $s_2 \xrightarrow[\Sigma_S]{c'} s_2'$. *An invariant pair (R,H) is called a bisimulation if both R and H are symmetric relations, and an equisimulation if both R and H are equivalence relations.*

Given an equivalence relation $\sim_C$, the relation $\sim_{\sim_C, \Sigma_S}$ is defined as follows:

$$\text{p} \sim_{\sim_C, \Sigma_S} \text{q} \equiv_{def} \exists \text{ a bisimulation (R,H) such that p R q \& H} \subseteq \sim_C.$$

It is a standard fact that $(\sim_{\sim_C, \Sigma_S}, \sim_C)$ is an equisimulation, and that it is the coarsest equisimulation (R,H) such that $H \subseteq \sim_C$.

If $\sim_C$ is the equivalence relation on D(A) (i.e., $\sim_L$), and Σ_S is defined by the syntactic rewriting system on terms, then $\sim_{\sim_C, \Sigma_S}$ becomes the semantic equality of terms. This is abbreviated $\sim_T$.

Under this notion of equivalence, the terms a||b, (a;b+b;a), and (a||b) + (a;b) + (b;a) are all pairwise distinct: the first does not allow either a;b or b;a, the second does not allow a||b, while the third allows both.

The equivalence in CCS is defined by a trivial equality for computations and the least transition relation satisfying AA1, S1, P1, P3, and {C1, C2, FP}. The equivalence of SCCS is defined by the least transition relation satisfying AA1, S1, P2, and {C2, C2, FP}.

13.7.1   Observable equivalence of data types

Observable equivalence on objects of a data type is typically defined with reference to a set of "extractor" or "observer" operations that return objects of a type other than that being defined. The proof method usually relies on some form of generator induction cf., [13], [6]. The induction hypothesis assumes that the two objects to be proved equivalent (say l_1 and l_2) are observably equivalent. In particular, it is assumed that l_1 and l_2 return equivalent results when an observer experiment is performed on them i.e., when an extractor operation is applied. The induction step consists of applying each of the constructor operations on both l_1 and l_2, and then ensuring that the resulting objects are observably equivalent.

13.7.2   Proof theory for finite terms

The recursion that is present in the description of many hardware modules typically arises from the fact that such module operate on (potentially infinite) streams of input objects. However, the overall behavior of a hardware system can sometimes be discerned by its behavior during one "cycle" of its operation. In such cases, the proof of equivalence of two hardware systems can be reduced to induction arguments combined with a proof of equivalence of finite terms that are derived from an unfolding of the corresponding recursive definitions. A proof theory for *finite terms* that essentially derives from the Hennessy-Milner theorem[8] is given by the equations (F1-F5) below. For the sake of brevity, we omit technical details of this technique here. However, note that the expansion equations still have the flavour of interleaving actions, something that we believe is undesirable.

[F1] $a;1 = a$ for all $a \in A$.

[F2] $(\sum_i \alpha_i; p_i); q = \sum_i ((\alpha_i; p_i); q)$

[F3] $(\alpha; 1); (\sum_j \beta_j; q_j) = \alpha; (\sum_j \beta_j; q_j) + \sum_j (\alpha; \beta_j); q_j$

[F4] $(\alpha; (\sum_i \alpha_i; p_i)); q = \alpha; (\sum_i (\alpha_i; p_i); q)$

[F5] $(\sum_i \alpha_i; p_i \| \sum_j \beta_j; q_j) = \sum_i \alpha_i; (p_i \| \sum_j \beta_j; q_j) + \sum_{i,j} ((\alpha_i \| \beta_j); (p_i \| q_j))$
     $+ \sum_j \beta_j; (\sum_i \alpha_j; p_i \| q_j)$

13.7.3   An alternative proof technique

An alternative way of approaching the proof of equivalance of two systems is to view them both as being instances of "process" object classes, and to characterize the operations that affect the input/output behavior at the ports as extractor operations [18]. The proof process can then be viewed entirely as one of proving equivalence of data abstractions, given the appropriate input stimuli. We do not as yet have enough empirical data to comment on the relative merits of these two techniques.

13.8   Inferring behavioral/temporal characteristics

13.8.1   An example

In this section, we use a simple example to indicate how the operational semantics described earlier can be used to infer the temporal and behavioral characteristics of a system, given its structural specification.

13.8.2   A level-sensitive dynamic latch

Consider a level-sensitive dynamic latch (Figure 13.7) that uses complementary switches (denoted cs) and inverters (denoted inv). Its structure may be described as follows:

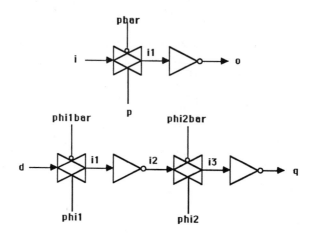

Figure 13.7: A level sensitive dynamic latch

dff{d, phi1, phi2 → q}(S) =
    Hide signals i1, i2, i3 in
        cs{phi1, d → i1}
        | inv{i1→i2}
        | cs{phi2, d → i2}
        | inv{i3→q}

We will next indicate how various characteristics of such a structure may be inferred using the operational semantics described earlier. We begin by outlining the behavior assumed of the two submodules, namely an inverter and a switch.

13.8.3   Inverter

The behavior of an inverter (denoted inv) may be described as follows:

$$\text{inv}\{i{\rightarrow}o\} = i?I; \ o!\text{not}(I); \ \text{inv}$$

Assume that the inverter is modelled with two delay parameters d_1^{inv}, d_2^{inv}. The execution of an inverter can then be inferred as follows:

$$\text{inv} \xrightarrow[stable(t_1,t_2)(s_i)(I)]{i?I} o!\text{not}(I) \ ; \ \text{inv} \xrightarrow[stable(t_1+d_1^{inv},t_2+d_2^{inv})(s_o)(not(I))]{o!I} \text{inv}$$

The temporal characteristics of an inverter are therefore summarized by

$$\text{inv}\{i{\rightarrow}o\} \equiv stable(t_1,t_2)(s_i)(I) \Rightarrow stable \ (t_1 + d_1^{inv}, t_2 + d_2^{inv})(s_o)(not(I))$$

13.8.4 A complementary switch

The behavior of a complementary switch may be described as follows:

$$cs\{p,i\to o\} = [p?P,i?I]; \text{ if } P \text{ then } o!I; \text{ cs}$$

If we assume that a switch is modelled with two delay parameters d_1^{cs}, d_2^{cs}, its "execution" may be automatically inferred using the operational semantics in the manner shown above.

Note. The behavior of a complementary switch (cs2) is more accurately expressed as

$$cs2\{p,pbar,i\to o\} = [p?P,pbar?PBAR,i?I]; \text{ if } (P \& not(PBAR)) \text{ then } o!I; \text{ cs2}$$

where it is arranged (or assumed) that PBAR=not(P). Given this assumption, the behavior of a module cs'=(Hide signal pbar in cs2) simplifies to that of cs as defined above, after use of the Boolean identity $(P \& not(not(P))) = P$. The analysis in this section can be carried through using the definition of cs2 instead of cs, if so desired. *End of Note.*

13.8.5   cs | inv

Consider now the circuit c1 whose structure is defined as follows:

$$c1\{d,phi1\to o\} = cs\{phi1,d\to i1\} \mid inv\{i1\to o\}$$

Given this structural description, its execution may be simulated as follows:

$$c1\{d,phi1\to o\} =$$
$$cs\{phi1,d\to i1\} \mid inv\{i1\to o\}$$
$$([phi1?PHI1,d?D]; \text{ if } PHI1 \text{ then } o!D; \text{ cs}) \mid (i1?I1; o!not(I1); inv)$$
$$\xrightarrow[stable(t_1,t_2)(s_d)(D) \ \& \ stable(t_1,t_2)(s_{\phi 1})(\Phi 1)]{[d?D,\phi 1?\Phi 1]}$$
$$(\text{if } PHI1 \text{ then } (i1!D; \text{ cs})) \mid (i1?I1; o!not(I1); inv)$$
$$\xrightarrow[stable(t_1,t_2)(s_{\phi 1})(true)]{\Phi 1=true}$$
$$(i1!D; \text{ cs}) \mid (i1?I1; o!not(I1); inv)$$
$$\xrightarrow[stable(t_1+d_1^{cs},t_2+d_2^{cs})(s_{i1})(D)]{((i1!D)\|(i1?I1))}$$
$$= \text{cs} \mid (o!not(D); inv)$$
$$\xrightarrow[stable(t_1+d_1^{cs}+d_1^{inv},t_2+d_2^{cs}+d_2^{inv})(s_o)(not(D))]{o!not(D)}$$
$$\text{cs} \mid inv$$
$$= c1$$

Extracting only the input/output actions and the conditional values on the top of the arrows, the behavior of module c1 is as follows:

c1{d,phi1→o} = [phi1?PHI1,d?D]; if PHI1 then o!not(D); c1

By accumulating the temporal details under the arrows that pertain to the externally visible ports, the temporal behavior of the module c1 is obtained as

$$stable(t_1, t_2)(s_d)(D) \ \& \ stable(t_1, t_2)(s_{\phi 1})(true) \Rightarrow$$
$$stable(t_1 + d_1^{cs} + d_1^{inv}, t_2 + d_2^{cs} + d_2^{inv})(s_o)(not(D))$$

13.8.6   dff behavior

Using the primitive c1 defined above, the structure of dff may be rewritten as

dff{d, phi1, phi2 → q}(S) = Hide signals i1 in c1{d, phi1 → i1} | c1{i1,phi2→q}

It is possible to think of the two instances of the modules c1 as two distinct objects in the same class, named, say, c1-1 and c1-2.

c1-1{d,phi1→o} = [phi1?PHI1,d?D]; if PHI1 then o!not(D); c1-1
c1-2{i1,phi2→q} = [phi2?PHI2,i1?I1]; if PHI2 then o!not(I1); c1-2

The behavior of dff may then be computed by using the operational semantics; we omit details for the sake of brevity.

13.9   Summary

We have outlined aspects of the operational semantics of a set of linguistic primitives for specifying concurrent (hardware) systems. The primitives allow such systems to be viewed as a combination of objects and processes. The operational semantics was based on a labeled transition system, and used an underlying formal model (based on a variant of event structures) for computations in such systems. Two important features of the development here are the ability to deal with *finite events* (*i.e.*, events having extended durations, as opposed to abstract point events), and accommodate "true concurrency". These features allow the development of a notion of observational equivalence that allows systems to be viewed during appropriate "windows" of time, and that is closer to the intuitive notion used in the context of dealing with hardware systems. While a detailed theory

References

[1] S. Brookes, C.A.R. Hoare, and A. Roscoe. A theory of communcating sequential processes. *JACM*, 31:560–599, 1984.

[2] L. Cardelli and P. Wegner. On understanding types, data abstraction, and polymorphism. *ACM Computing Surveys*, 17(4):471–522, December 1985.

[3] J. D. Gabbe and P. A. Subrahmanyam. An object-based representation for the evolution of VLSI designs. *International Journal of Artificial Intelligence in Engineering*, 2(4):204–223, 1987.

[4] J. A. Goguen. Parameterized programming. *IEEE Trans. on Software Engg.*, SE-10:528–552, September 1984.

[5] G. Gopalakrishnan. Personal communication. 1987.

[6] J. V. Guttag, E. Horowitz, and D. R. Musser. Abstract data types and software validation. *Communications of the ACM*, 21(2):1048–1064, December 1978.

[7] M. Hennessy and R. de Nicola. Testing equivalences for processes. *Theoretical Computer Science*, 34:83–133, 1984.

[8] M. Hennessy and R. Milner. Algebraic laws for nondeterminism and concurrency. *JACM*, 32:137–161, 1985.

[9] C. A. R. Hoare. *Communicating Sequential Processes*. Prentice-Hall International Series in Computer Science, 1985.

[10] R. Milner. *Calculii for Synchrony and Asynchrony*. Technical Report, University of Edinburgh, April 1982.

[11] R. Milner. *A Calculus of Communicating Systems*. Springer Verlag, LNCS 92, 1980.

[12] B. Moszkowski. A temporal logic for multilevel reasoning about hardware. *IEEE Computer*, 18(2):10–19, February 1985.

[13] D. R. Musser. Abstract data type verification in the AFFIRM system. *IEEE Transactions on Software Engineering*, SE-6(1), January 1980.

[14] D. Park. Concurrency and automata on infinite sequences. In *5th GI Conference, Lecture Notes in Comput. Sci. 104*, pages 167–183, 1981.

[15] C. L. Seitz. Self-timed VLSI Systems. In *Proc. Caltech Conference on VLSI*, 1979.

[16] P. A. Subrahmanyam. LCS – A leaf cell synthesizer employing formal deduction techniques. In *24th ACM/IEEE Design Automation Conference*, pages 459–465, July 1987.

[17] P. A. Subrahmanyam. Synapse: an expert system for VLSI design. *IEEE Computer*, 19(7):78–89, July 1986.

[18] P. A. Subrahmanyam. Synthesizing VLSI circuits from behavioral specifications: a very high level silicon compiler and its theoretical basis. In F. Anceau, editor, *VLSI 83: VLSI Design of Digital Systems*, pages 195–210, North Holland, August 1983.

[19] P. A. Subrahmanyam. Toward a framework for dealing with system timing in very high level silicon compilers. In G. Birtwistle and P. A. Subrahmanyam, editors, *VLSI Specification, Verification and Synthesis*, pages 159–215, Kluwer Academic Publishers, 1988.

[20] G. Winskel. Event structure semantics for CCS and related languages. In *Proc. of the 9th ICALP, LNCS 140*, pages 561–576, Springer Verlag, 1982.

Chapter 14

Timeless Truths about Sequential Circuits

Geraint Jones [1]
Mary Sheeran [2]

Abstract

We suggest the use of a declarative programming language to design and describe circuits, concentrating on the use of higher-order functions to structure and simplify designs. In order to describe sequential circuits, we use a language, μFP, which abstracts from temporal iteration. The practicalities of VLSI design make regularity attractive, and we describe the use of familiar higher order functions to capture spatial iteration.

By reasoning about circuits rather than signals (programs rather than data) one abstracts from the sequential nature of a circuit. By reasoning about forms of circuit (higher order functions) one can devise implementation strategies for whole classes of algorithms. Reasoning about μFP is formally quite similar to reasoning about FP.

In this paper we identify the semantic content of the formal similarity between FP and μFP. This makes it possible to carry over from conventional functional programming those intuitions we have about algorithm design. It also makes it possible to conduct parts of a design in the simpler world of static calculations, with confidence in the correctness of the corresponding sequential circuit.

---

[1] Oxford University Computing Laboratory, Oxford University
[2] Dept. of Computing Science, University of Glasgow

14.1   Concurrent systems

We begin from the premise that designing a logic circuit is essentially the same activity as writing a program, but that the constraints on the programmer are in this case rather unusual. The efficient use of VLSI resources requires the designer to write a highly concurrent program, consisting of a very large number of simple but simultaneously operating processes which cooperate to achieve the desired effect. Economic considerations dictate that many of these processes have to be identical, or of a very few types. The interaction between these processes has, moreover, to be achievable within a limited network of communication links that can be laid out with short wires in a few layers on a flat surface.

Although the behaviour of the whole circuit is certainly *caused* by the actions of the individual components, it is difficult to understand an explanation of an array of processes which begins from a description of the possible actions of the components. Bottom-up design of a circuit in a state-based language which gives the programmer complete freedom to organise synchronization is correspondingly difficult.

The natural description of the function of an array circuit is in terms of the data objects which it manipulates, although these may be spread out in space over the array, and in time over the duration of the calculation[9]. We advocate a design method which elaborates a high-level specification of the data manipulation to be achieved by the whole circuit, into an implementation which is constrained by a predetermined framework of synchronization and communication routes[7,6].

14.2   Combinational circuits

We represent a circuit in the first instance by its behaviour — that function which its output signal is of the signal at its input. The behaviour of a compound circuit can be induced from the behaviour of its components, and the way that they are connected. Thus if the input of an F circuit is driven by the output of a G circuit, they behave together as the composition $F \circ G$

$$z = Fy = F(Gx) = (F \circ G)x$$

There is a hierarchy: zeroth order objects, signals like x, y, z; first order functions, circuit behaviours like F and G; and higher order functions (functions of functions), like composition.

Since we will be interested in laying out our circuits, we note that there is an alternative interpretation of this functional programming langauge. The base objects are wires, first order functions are circuits, and higher-order functions are circuit structures. This interpretation gives a formula for constructing a circuit which (if the interpretation is *sound*) will implement the behavioural interpretation. If a function is applied to a named argument, then the circuit has its input connected to the corresponding wire, and so on.

If wires can be named, and those names mentioned at will in a circuit description, then there is no guarantee that the circuit has a simple layout. There is certainly no simple and systematic way of laying out such circuits. Notice however, that we can describe the composition $F \circ G$ without needing to name wires, and that we can, given layouts of F and G, combine them systematically to lay out the composition.

To preserve this systematic laying out, we use Backus' FP[1] as a circuit description language, and we do not name signals and wires. Our reasoning will all be about the application of structures (higher order functions) to circuits (first order functions). We will need a repertoire of higher order functions adequate to describe interesting circuit layouts, but we will then confine ourselves to just those circuits which can be *constructed* by these structures.

Buses and collections of signals are produced by a constructed circuit, in which each component circuit generates the corresponding component of the output signal.

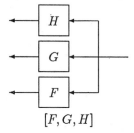

$$[F, G, H]$$

Since tuples may contain tuples, collections of signals can be given any required structure. A circuit with several input wires will be modelled by a function of a tuple, and a circuit with several output wires by a function returning a tuple.

Tuples can be analysed and synthesised by left- and right-handed constructors, app_L and app_R and their inverses:

$$
\begin{aligned}
\mathsf{app}_L \circ [p, [q, r, \ldots]] &= [p, q, r, \ldots] \\
\mathsf{app}_R \circ [[\ldots, p, q], r] &= [\ldots, p, q, r] \\
\mathsf{app}_L^{-1} \circ [p, q, r, \ldots] &= [p, [q, r, \ldots]] \\
\mathsf{app}_R^{-1} \circ [\ldots, p, q, r] &= [[\ldots, p, q], r]
\end{aligned}
$$

and because pairs are so important we define the following higher order abbreviations

$$
\begin{aligned}
(F \parallel G) \circ [a, b] &\stackrel{\text{def}}{=} [F \circ a, G \circ b] \\
\mathsf{fst}\ F &\stackrel{\text{def}}{=} (F \parallel \mathsf{id}) \\
\mathsf{snd}\ F &\stackrel{\text{def}}{=} (\mathsf{id} \parallel F)
\end{aligned}
$$

where id is the identity function, $\mathsf{id} \circ a = a$.

We will have loops in the language which allow us to duplicate a circuit over a tuple, either by *mapping*

$$
\begin{aligned}
\propto\! F \circ [\,] &\stackrel{\text{def}}{=} [\,] \\
\propto\! F \circ \mathsf{app}_L &\stackrel{\text{def}}{=} \mathsf{app}_L \circ (F \parallel \propto\! F)
\end{aligned}
$$

or by making a *right triangle*

$$\triangle_R F \circ [\,] \overset{\text{def}}{=} [\,]$$

$$\triangle_R F \circ \text{app}_L \overset{\text{def}}{=} \text{app}_L \circ \text{snd } (\propto F \circ \triangle_R F)$$

or, of course, the corresponding left handed construct.

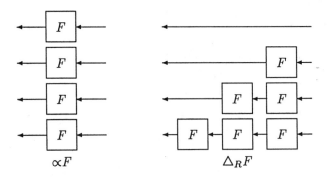

$$\propto F \qquad\qquad\qquad \triangle_R F$$

Getting those signals which are to interact to meet is a crucial part of the design of a circuit. We introduce left- and right-handed plumbing functions to *distribute* a signal across a bus

$$\text{dist}_L \circ [p, [q, r, \ldots]] \overset{\text{def}}{=} [[p, q], [p, r], \ldots]$$

$$\text{dist}_R \circ [[\ldots, p, q], r] \overset{\text{def}}{=} [\ldots, [p, r], [q, r]]$$

so that, for example, $\propto add \circ \text{dist}_L \circ [a, [b, c, d]]$ adds the output of a independently to the outputs of each of b, c, d. We will also find the need of a generic interleaving circuit, zip,

$$\text{zip} \circ [[a_0, b_0, c_0, \ldots], [a_1, b_1, c_1, \ldots], [a_2, b_2, c_2, \ldots], \ldots] \overset{\text{def}}{=}$$
$$[[a_0, a_1, a_2, \ldots], [b_0, b_1, b_2, \ldots], [c_0, c_1, c_2, \ldots], \ldots]$$

Interleaving arises naturally in the process of elaborating a word-level design to bit level. Since a large interleaving is prohibitively expensive, occupying an area quadratic in the number of signals through it, it will be in the designer's interests to eliminate large interleavings from his circuit.

Ideally, for minimum communciation cost, most of the interactions in a circuit will be between adjacent circuit elements. Horizontal and vertical forms of *reduction* implement continued sums, and products, and similar forms of loop. For example $/_H add \circ [[x_0, x_1, x_2, \ldots], 0] = \sum x_i$, where 0 is the function that produces a constant

zero output. Horizontal and vertical *arrays* are generalisations of reduce that yield
partial results as outputs.

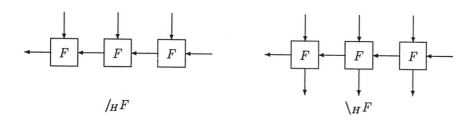

$$/_H F \qquad\qquad \backslash_H F$$

14.3   Equational reasoning

The equations defining various constructs, and indeed any other equations relating
our FP expressions, are statements about the behavioural semantics of the language.
That F is equal to G means that given the same input signal each will produce the
same output. In other ways, F and G may be different; they may, for example, have
radically different interpretations under the layout semantics. Consider zip o zip = id
which states that the effect of two consecutive interleaving circuits can be achieved
by straight wiring; the former has a greater area and complexity than the latter.
F may be a natural specification of a desired signal transformation, but its natural
layout may be prohibitively expensive; it may be harder to see immediately what
transformation is implemented by G, whereas it has a simpler and smaller layout.
In that case, their equality guarantees that G can safely be used efficiently to
implement the specification F.

Consider the development of a circuit which performs some arithmetic on integers. Its specification will, initially, be given in terms of integer signals. In the
course of implementation the integers will have to be represented by a number of
bits, and the circuits that process integers elaborated into circuits that process those
arrays of bits. Addition of integers, for example, can be implemented by an array
of full-adders, *fadd*, taking as input the interleaving of the representations of the
integers,

$$[sum, c_{out}] \quad = \quad \backslash_V fadd \text{ o snd zip o } [c_{in}, [x, y]]$$
$$\text{where } fadd \text{ o } [c_{in}, [x, y]] = [s, c_{out}]$$

Circuits in which there are several adders would naturally be described in terms of arrays of full-adders and expensive interleavings; the equality

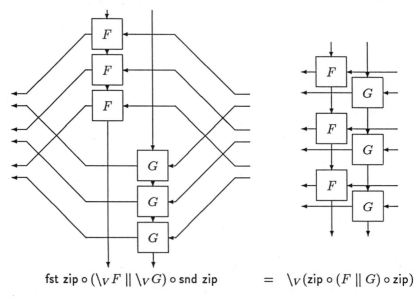

$$\text{fst zip} \circ (\backslash_V F \parallel \backslash_V G) \circ \text{snd zip} \quad = \quad \backslash_V (\text{zip} \circ (F \parallel G) \circ \text{zip})$$

permits the transformation of such a description into the familiar, if less obvious, interleaved data-path structure which occupies considerably less area, has shorter wires, and fewer wire-crossings. Notice that this equation is a statement about any circuits F and G, and can be used to improve the layout of any circuit of the given form involving large interleavings and arrays, irrespective of the function which it implements, and of the widths of the arrays.

There are also useful observations which can be made about circuit structures, conditional on properties of the components, for example

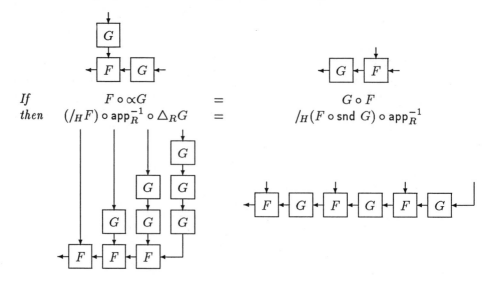

If $F \circ_\alpha G$ $=$ $G \circ F$

then $(/_H F) \circ \text{app}_R^{-1} \circ \triangle_R G$ $=$ $/_H (F \circ \text{snd } G) \circ \text{app}_R^{-1}$

Again, this theorem can be used to reduce the size and component count of a circuit by eliminating triangles, provided that the components satisfy its hypothesis. This theorem can be proved, like the first, by induction on the sizes of the loops. Given a sufficiently rich collection of such theorems, a circuit development can proceed by successive use of the equations as rewriting rules.

14.4 Sequential circuits

In what has gone before, we have been talking of combinational circuits and signals which have been constant in time. We will now develop the framework for discussing *synchronous* circuits. In such a circuit, each signal takes on a sequence of values, and there is some mechanism to ensure that component circuits which share a common signal agree about the times at which the signal is valid.

A synchronous circuit can be thought of as having a single global clock, which runs slowly enough to allow the circuit to settle on each clock cycle. The sequence of values representing a signal is the sequence of stable values of the signal at the ends of consecutive clock cycles. Notice, however, that we have abstracted from the mechanism that achieves synchronization, so the implementation could equally well be self-timed, or partly clocked and partly synchronized by handshaking.

We will use another functional programming language, μFP, superficially similar to FP, to describe synchronous circuits. (The name derives from a μ constructor introduced in [5] for constructing state-machines from state-transition functions.) In this language, the base objects are infinite sequences of signal values; the first order functions represent clocked circuits; and the higher order functions will again be circuit structures.

We will explain the meaning of μFP by giving a translation into FP. Where confusion might arise in what follows, μFP appears between double roman brackets, and the text outside the brackets is FP.

For a function F that represents a combinational circuit in FP, we will use the same name in μFP to stand for the effect of providing the (stateless) circuit with a sequence of values.

$$\llbracket F \rrbracket \stackrel{\text{def}}{=} \propto F$$

F and $\llbracket F \rrbracket$ are corresponding functions. We would draw the same circuit diagram for each of them. F maps a value to a value whereas $\llbracket F \rrbracket$ maps an infinite sequence of values to an infinite sequence of values. The definition says that $\llbracket F \rrbracket$ applies the function F to each element of the input sequence. If *not* is the combinational circuit which negates its input, then $\llbracket not \rrbracket$ is the sequential circuit which negates each successive element of its input sequence. If *add* is the combinational circuit which adds two integer signals, then $\llbracket add \rrbracket$ is the sequential circuit which adds each successive pair of integers from an infinite sequence of pairs of integers.

The meaning of higher order functions can be given by a systematic translation, which will be made precise later. The important thing is that the meaning of a μFP function should be given in terms of the corresponding FP function. Consider, for example, the $\triangle_R$ constructor. The input sequence of tuples is separated (using zip) into the tuple of sequences which appear on the wires which are the inputs to the triangle. On the output of the triangle, the tuple of sequences is recombined (using zip^{-1} into a sequence of tuples.

$$
\begin{aligned}
[\![\triangle_R F]\!] &\stackrel{\text{def}}{=} zip^{-1} \circ \triangle_R[\![F]\!] \circ zip \\
&= zip \circ \triangle_R[\![F]\!] \circ zip
\end{aligned}
$$

Proceeding in the same way, we can define constructs in μFP which correspond to each of the constructs of FP. This language describes the way that combinational circuits can be used to process sequences of inputs. In order to describe non-combinational circuits, we will need a non-combinational primitive, $\mathcal{D}$, a *latch* which delays its input by a single time step. The fact that we use doubly infinite sequences allows us freely to manipulate *anti-delays*, $\mathcal{D}^{-1}$, that is, circuits which predict their inputs. The output of an anti-delay at any time is the value which will appear at its input at the subsequent time. An anti-latch shifts each value on a time-sequence forward by one time step. Latches and anti-latches can be cancelled, because $\mathcal{D} \circ \mathcal{D}^{-1} = \mathcal{D}^{-1} \circ \mathcal{D} = \text{id}$. Latches and anti-latches apply to all types of signal, so

$$\mathcal{D} \circ \text{app}_L = \text{app}_L \circ (\mathcal{D} \parallel \propto\mathcal{D}) = \propto\mathcal{D} \circ \text{app}_L$$

Of course, whilst one can implement latches, anti-latches cannot be given a sound physical realisation. They can be used to record requirements on the times at which data are made available to a circuit, and they can be manipulated in the same way as latches in the course of a development, provided that they are eliminated before the implementation is complete. Triangles of latches and anti-latches appear naturally in describing data-skew.

14.5   Duality

The choice of the same notation for μFP as for FP was not deliberately made to cause confusion. The definitions of μFP functions are so chosen that equalities such as

$$\text{fst } zip \circ (\backslash_V F \parallel \backslash_V G) \circ \text{snd } zip = \backslash_V (zip \circ (F \parallel G) \circ zip)$$

are true not only in FP but also in μFP. Moreover, although $\mathcal{D}$ had no equivalent in FP, these equalities continue to be true when applied to sequential circuits, for example

$$F \circ \propto\mathcal{D} = \mathcal{D} \circ F \;\Rightarrow\; (/_H F) \circ \text{app}_R^{-1} \circ \triangle_R \mathcal{D} = /_H (F \circ \text{snd } \mathcal{D}) \circ \text{app}_R^{-1}$$

This last (an instance of what [4] calls a 'retiming' lemma) is particularly useful, giving a way of transforming a combinational ripple-through circuit, $/_H F$, and an expensive data-skewing circuit, $\triangle_R \mathcal{D}$, into a pipelined circuit $/_H(F \circ \text{snd } \mathcal{D})$.

It is, of course, very much simpler to prove these equalities to be true in FP than it is to check the corresponding theorems in μFP. Were it possible to say that any equality true in FP applied also in μFP, performing the proofs by induction in the simple combinational world of FP would be an easy way of checking powerful theorems about sequential circuits. Unfortunately, this is not the case.

Consider the conditional (multiplexer) function, cond, defined in FP by

$$\text{cond} \circ [p, [f, g]] = \begin{cases} f & \text{if } p \text{ returns true} \\ g & \text{if } p \text{ returns false} \end{cases}$$

Since it is stateless, the natural definition of cond in μFP would seem to be $[\![\text{cond}]\!] = \propto$cond. It is readily shown that in FP

$$h \circ \text{cond} \circ [p, [f, g]] = \text{cond} \circ [p, [h \circ f, h \circ g]]$$

by an argument by cases on the value of the output of p. This theorem is not true in μFP for all h: indeed, it is only necessarily true in case h is stateless, in particular

$$\mathcal{D} \circ \text{cond} \circ [p, [f, g]] \neq \text{cond} \circ [p, [\mathcal{D} \circ f, \mathcal{D} \circ g]]$$

To rescue the situation, it will be necessary to make precise the 'systematic translation' of μFP into FP. We will divide μFP constants into two classes, such that a theorem true in FP will necessarily be true in μFP provided that only *timeless* constants appear in its statement. It will transpire that cond is not timeless.

14.6 Types

In order to be precise about the translation of μFP into FP, we will need to assign types to FP expressions. We will assume that there exists some set of base types (e.g. INT, BOOL etc.) The exact nature of base types is unimportant, since we are principally concerned with the size and structure of tuples. We will return, later, to the question of what a type is; for the present we will simply record that $x : T$ if x is of type T.

What is interesting about the type of an object is a record of the the way the object is constructed. For any types $T, U, V, \ldots$ there is a product type $[T, U, V, \ldots]$ for which

$$x : T, y : U, z : V, \ldots \Rightarrow [x, y, z, \ldots] : [T, U, V, \ldots]$$

and there is a type which is the type of all (finite and infinite) homogeneous tuples with components of a given type

$$x_i : T \Rightarrow [x_0, x_1, x_2, \ldots] : T^\star$$

These type constructors are related, because

$$x : [T, T, T] \Rightarrow x : T^\star$$

All zeroth order objects in FP can be given types constructed in these ways.

The type of a function which takes inputs of type A and delivers a result of type B will be denoted $A \to B$. Thus, function application can be typed by the rule

$$f : A \to B, x : A \Rightarrow fx : B$$

Type variables bound by the universal quantifier will be used to indicate polymorphic functions in the usual way. For example,

$$\text{id} : \forall \alpha . \alpha \rightarrow \alpha$$

$$\text{dist}_L : \forall \alpha, \beta . [\alpha, \beta^\star] \rightarrow [\alpha, \beta]^\star$$

A polymorphic function can also have any substitution instance of its type. In general, an object has many types, for example

$$\text{id} : \forall \alpha . \alpha \rightarrow \alpha$$

$$\text{id} : \forall \beta . (\beta \rightarrow \beta) \rightarrow (\beta \rightarrow \beta)$$

$$\text{id} : \forall \alpha, \beta . [\alpha, \beta]^\star \rightarrow [\alpha, \beta]^\star .$$

Note that an application may also return a polymorphic function. For example, id applied to id is again id, the polymorphic identity function.

An assignment of types to an expression is made by choosing instances of types of the constants and variables, and ascribing types to each of its sub-expressions, so that each application is exactly well typed. An assignment of types to an equation between two expressions consists of assignments of the same types to each expression.

14.7   Timelessness

Just as the same symbols and names are used to stand for corresponding objects in FP and μFP, so we will use the same names for their types. Thus, if $f : A \rightarrow B$ in FP has a corresponding meaning in μFP we will write also $[\![f : A \rightarrow B]\!]$.

For example, the FP function *first*, which selects the first of a pair of inputs, has type $\forall \alpha, \beta . [\alpha, \beta] \rightarrow \alpha$. We will write the type of the μFP function $[\![first]\!]$, which takes a *stream* of pairs to the *stream* of the first elements of those pairs, in exactly the same way: $[\![first : \forall \alpha, \beta . [\alpha, \beta] \rightarrow \alpha]\!]$. The streams are implicit in the type. We call *first* timeless, because we can express the meaning (i.e. the translation into FP) of the μFP function in terms of the FP function itself.

$$[\![first]\!] = first \circ zip$$

We simply zip the stream of pairs together, to get a pair of streams. The first stream contains the stream of first elements, so we select it using the FP function *first*. Similarly, we can express other μFP plumbing functions and combining forms in terms of their FP equivalents. This allows us to develop laws about μFP by appealing to the corresponding FP laws. This was the strategy used in [5]. Here, we generalise the approach by defining an injection that relates the meaning of an FP expression to the meaning of the corresponding μFP expression. An object or function will be injected according to its type.

We define a family of functions, π, indexed by our range of monotypes, such that for any timeless object x

$$x : T \;\Rightarrow\; [\![x]\!] = \pi_T x$$

Indeed, this will be the definition of timelessness: an object x of type T (strictly speaking, a pair of objects, one in FP, one in μFP) is timeless precisely when the two meanings are related by the injection π_T. We will say that a polymorphic object is timeless if each of its instances is timeless. Thus, a polymorphic function f is timeless iff

$$f : \forall \alpha.T(\alpha) \;\Rightarrow\; \forall \alpha.([\![f]\!] = \pi_{T(\alpha)} f)$$

In fact, because of our choice of polymorphic primitive functions, it is always the case that $\pi_{T(\alpha)} f$ is independent of α. So, $\pi_{T(\alpha)}$ can be used for $\pi_{\forall \alpha.T(\alpha)}$.

π, being an injection, has an inverse, such that $\pi_T^{-1} \circ \pi_T$ is the identity function on the domain of π_T and $\pi_T \circ \pi_T^{-1}$ is the identity function on the range of π_T. At this point we can be more precise about what a type is: we identify a type with its injection function and call types equal if they have the same injections, distinct if they have distinct injections.

To complete the definition of π, we must define it for each of our type constructors. The injection for a pair type is constructed from the injections for the components. First, the components of the pair of streams are injected, and then the resulting pair is transposed, to produce a stream of pairs.

$$\pi_{[A,B]} = \mathsf{zip} \circ (\pi_A \,\|\, \pi_B)$$

and similarly for larger tuples. In particular, the injection for a homogeneous tuple is

$$\pi_{A\star} = \mathsf{zip} \circ \propto \pi_A$$

Because π is an injection, we can conclude that

$$\pi_{[A,B]}^{-1} = (\pi_A^{-1} \,\|\, \pi_B^{-1}) \circ \mathsf{zip}$$

$$\pi_{A\star}^{-1} = \propto \pi_A^{-1} \circ \mathsf{zip}$$

Thus we define the injection for zeroth order objects.

The injection for the type of a function is chosen to make the following diagram commute for timeless f:

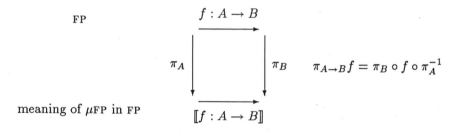

$$\pi_{A\to B} f = \pi_B \circ f \circ \pi_A^{-1}$$

The force of the commutativity of this diagram is that rather than computing the result $[\![f]\!][\![x]\!]$ of applying the μFP function f to x, applying the corresponding FP function to corresponding arguments yields corresponding results.

$$
\begin{aligned}
[\![(fx) : B]\!] &= \pi_B(fx) \\
&= (\pi_B \circ f)((\pi_A^{-1} \circ \pi_A)x) \\
&= (\pi_B \circ f \circ \pi_A^{-1})(\pi_A x) \\
&= [\![f : A \to B]\!][\![x : A]\!]
\end{aligned}
$$

If an equation true in FP is given in which all constants are timeless, and types ascribed to its constants and variables such that all applications are properly typed, then that equation will remain true if interpreted in μFP. We can construct a μFP equation from an FP one as follows. Replace each constant in the FP equation by its projection, and each variable by a projection of a corresponding variable. Because the initial equation is well typed, all of the 'internal' projections cancel as shown in the example above. The only effect is to project each side by the same projection, so the equation is still true of all values of the new variables, provided only that they are of the right type. The result is a μFP equation in which the only requirement on the free variables is that they have the appropriate μFP types. Thus, the laws constructed in this way apply even to circuits which contain the sequential primitive $\mathcal{D}$. So, we can create new μFP laws simply by *promoting* FP laws which contain only timeless constants.

We must decide which of our FP constants are timeless. To do this, we must check that the meanings of the FP and corresponding μFP constants are related by the appropriate injection. For example, since the function $first$ is stateless, we know that

$$
[\![first]\!] = \propto first
$$

We must check that $first$ is timeless, i.e. that

$$
[\![first]\!] = \pi_{[\alpha,\beta]\to\alpha} first
$$

By the definition of π,

$$
\begin{aligned}
\pi_{[\alpha,\beta]\to\alpha} first &= \pi_\alpha \circ first \circ \pi_{[\alpha,\beta]}^{-1} \\
&= \pi_\alpha \circ first \circ (\pi_\alpha^{-1} \parallel \pi_\beta^{-1}) \circ \mathsf{zip} \\
&= \pi_\alpha \circ \pi_\alpha^{-1} \circ first \circ \mathsf{zip} \\
&= first \circ \mathsf{zip} \\
&= \propto first
\end{aligned}
$$

So, $first$ is timeless. Note that here we have used the function first at two different instances. The $first$ in the last line takes a pair of values to a value, whereas the $first$ in the line before takes a pair of streams of values to a stream of values. This use of two instances always happens when we inject a function, and it shows that only polymorphic functions can be timeless. Similarly, the other plumbing functions, such as dist_L, zip and app_R^{-1} can be shown to be timeless.

The structural combining forms such as $\propto$ and $\triangle_R$ are also timeless. The injection gives us a systematic way of giving meaning to the μFP combining forms. Let us take $\triangle_R$ as an example.

$$\triangle_R : \forall \alpha.(\alpha \to \alpha) \to (\alpha^\star \to \alpha^\star)$$

For $\triangle_R$ to be timeless,

$$[\![\triangle_R]\!] = \pi_{(\alpha \to \alpha) \to (\alpha^\star \to \alpha^\star)} \triangle_R$$

$$
\begin{aligned}
(\pi_{(\alpha \to \alpha) \to (\alpha^\star \to \alpha^\star)} \triangle_R)f &= (\pi_{\alpha^\star \to \alpha^\star} \circ \triangle_R \circ \pi_{\alpha \to \alpha}^{-1})f \\
&= \pi_{\alpha^\star \to \alpha^\star} \circ \triangle_R \circ (\pi_\alpha^{-1} \circ f \circ \pi_\alpha) \\
&= \pi_{\alpha^\star} \circ \triangle_R(\pi_\alpha^{-1} \circ f \circ \pi_\alpha) \circ \pi_{\alpha^\star}^{-1} \\
&= \text{zip} \circ \propto\pi_\alpha \circ \triangle_R(\pi_\alpha^{-1} \circ f \circ \pi_\alpha) \circ \propto\pi_\alpha^{-1} \circ \text{zip} \\
&= \text{zip} \circ \propto\pi_\alpha \circ \propto\pi_\alpha^{-1} \circ \triangle_R f \circ \propto\pi_\alpha \circ \propto\pi_\alpha^{-1} \circ \text{zip} \\
&= \text{zip} \circ \triangle_R f \circ \text{zip}
\end{aligned}
$$

So,

$$
\begin{aligned}
[\![\triangle_R f]\!] &= [\![\triangle_R]\!][\![f]\!] \\
&= \text{zip} \circ \triangle_R[\![f]\!] \circ \text{zip}
\end{aligned}
$$

as we had before.

Most of the definitions of the combining forms given in [5] can be calculated in the same way. The only exception is the conditional combining form. We can see why conditional is not timeless by considering the multiplexing circuit cond. cond is not polymorphic; it demands that its first argument be a single boolean. However, when we project cond, we find that we are trying to use it with a stream as its first argument, so that the resulting equation cannot be correctly typed. So, $[\![\text{cond}]\!] = \propto\text{cond}$ is not equal the projection of cond, and cond cannot be timeless. The conditional combining form can be defined in terms of cond, and so is not timeless. In fact, only polymorphic functions can be timeless.

With the exceptions of $\text{cond}, \mathcal{D}$, and $\mathcal{D}^{-1}$, all generic functions mentioned in this paper are timeless. It follows that the theorems quoted in the earlier discussion of FP are true of μFP.

14.8   Statelessness

There is a second, less general, way of promoting laws from FP to μFP. Any FP law can be promoted to μFP, provided all variables are guaranteed to be stateless. Suppose $A = B$ in FP. If A and B are stateless, we know that $[\![A]\!] = \propto A$ and that $[\![B]\!] = \propto B$.

$$A = B \Rightarrow \propto A = \propto B \Rightarrow [\![A]\!] = [\![B]\!]$$

So, $A = B$ in μFP also.

Sometimes, it is appropriate to use a mixture of the two forms of promotion. We noted earlier that the following theorem is true in FP, but is only necessarily true in μFP if h is stateless.

$$h \circ \text{cond} \circ [p, [f, g]] = \text{cond} \circ [p, [h \circ f, h \circ g]]$$

We divide the law into two parts. The first part cannot be made timeless because of the presence of cond.

$$h \circ \text{cond} = \text{cond} \circ [id, [h, h]]$$

The only way in which we can promote this law to μFP is to require h to be stateless. The second part

$$a = b \implies a \circ [p, [f, g]] = b \circ [p, [f, g]]$$

contains only the timeless constants *composition* and *construction*, so it can be promoted to μFP without placing any requirements on the variables. The net result is that the original FP law is true in μFP, provided h is stateless.

14.9   Conclusion

We have introduced the notion of timelessness, and have shown that any FP law which contains only timeless constants can immediately be promoted to μFP. Timelessness is defined by means of an injection which maps the meaning of an FP expression to the meaning of the corresponding μFP expression. With the exception of cond, the multiplexing function, and the sequential constants $\mathcal{D}$ and $\mathcal{D}^{-1}$, all of the plumbing functions and combining forms that we use to describe circuits are timeless. This means that we can reason about complex sequential circuits as though we were still in the simpler FP world. We no longer have to prove each individual μFP law correct.

In those cases where this timeless promotion cannot be used, we have shown that a weaker form of promotion, which demands that all variables be stateless can be applied. Some useful laws can be formed by a hybrid timeless/stateless promotion, and we have given an example of this. The final group of μFP laws comprises those which cannot be proved by promotion. These are laws which apply only to μFP and they will typically refer to the constants $\mathcal{D}$ and $\mathcal{D}^{-1}$, which have no counterparts in FP. The laws in this small group must be proved individually.

We have recently been investigating the use of relations to represent circuits[8]. Essentially, the same development can be performed as is described here, but with relations standing as the first order objects. The relation records the signals at the terminals of a circuit, but abstracts from the causal relationship between inputs and outputs, simplifying the treatment of complex data-flows. In contrast to the approach of [2], by confining the structure of the design to the same second- and higher-order functions described here we preserve the simple equational reasoning about circuits.

Tools are being developed to lighten the clerical burden of applying rewriting rules[3], and to record the transformation process as a means of documenting design decisions.

Acknowlegements

We would like to thank Wayne Luk for comments on an earlier draft, and John Hughes for some stimulating late night discussions about types. Part of the work on this paper was done while the second author was a visiting scientist at the IBM Almaden Research Center, San Jose.

References

[1] J. Backus, 'Can programming be liberated from the von Neumann style?', *Commun. ACM*, vol. 21(8), pp. 613–641, 1978.

[2] M.J.C. Gordon, 'Why higher order logic is a good formalism for specifying and verifying hardware', Technical report No. 77, Cambridge University Computing Laboratory, 1985.

[3] G. Jones and W. Luk, 'Exploring designs by circuit transformation', in W. Moore, A. McCabe and R. Urquhart (eds.), 'Systolic Arrays', Bristol: Adam Hilger, 1987, pp. 91–98.

[4] C.E. Leiserson and J.B. Saxe, 'Optimising synchronous systems', *J VLSI & Comput. Syst.*, 1983, vol. 1(1), pp. 41–67.

[5] M. Sheeran, 'μFP, an Algebraic VLSI design language', D.Phil. thesis, University of Oxford, 1983.

[6] M. Sheeran, 'Designing regular array architectures using higher order functions', in *Proc. Int. Conference on Functional Programming Languages and Computer Architecture*, Springer-Verlag LNCS 201, 1985, pp. 220-237.

[7] M. Sheeran, 'Design and verification of regular synchronous circuits', *IEE Proceedings*, vol. 133, Pt. E, No. 5, pp. 295–304 September 1986.

[8] M. Sheeran and G. Jones, 'Relations + Higher Order Functions = Hardware Descriptions', in *IEEE Proc. Comp Euro 1987*, pp. 303–306

[9] U.C. Weiser and A.L. Davis, 'A wavefront notation tool for VLSI array design', in H.T. Kung, R.F. Sproull and G.L. Steele (eds.), 'VLSI systems and computations' (Computer Science Press, 1981), pp. 226–234.

PART III
Communication Issues

Introduction

Whereas heuristic development of architectures optimized for a particular problem is common, a more systematic approach not only simplifies design, analysis and verification but also provides a common basis for comparing alternative realizations. Such systematic design tools will emerge first from within special purpose architectures implementing deterministic algorithms. Communication and scheduling, which provide the global coordination of otherwise independently executing tasks on distinct processors, are particularly important in such design tools. Given techniques to better understand the interplay between communications and computations, communication and synchronization can be actively used in designing an optimum implementation of a concurrent computing system (rather than passively reacting to specific local behaviors of computing sites during execution of an algorithm).

Engstrom and Cappello's description of the SDEF systolic programming environment, in addition to discussing a specific topic important in its own right, provides a view of the kinds of software aids that would be particularly useful in general concurrent computing environment. This set of software tools addresses the broad range of design issues that arise in any system design.

Schwartz provides a timely review of techniques for deterministic scheduling of a class of recursive/iterative problems. Such problems are common in signal processing. Here, the careful inclusion of delays along paths through a graph representation of the problem provides useful insight not only into the interplay between computation and communication but also into the general issue of ordering the sequence of computing tasks.

Bianchini and Shen also discuss the issue of deterministic scheduling for real time processing problems. In contrast to the case treated by Schwartz (with communications and computations integrated), they consider the case of a global communications environment separated from the localized computing nodes of the computation environment. Such separation of computing and communication issues, though still assumed deterministic, is an important step toward general purpose computing environments.

Moving still further toward general purpose concurrent computing environments, **Jesshope** considers communications from the perspective of data structures, rather than the specific computing steps. This abstraction allows communications to be represented from a programming perspective within an asynchronous communications environment. Jesshope's *active data model* is suitable for MIMD environments and is applied to the Reconfigurable Array Processor being developed at Southampton University.

From a better understanding of the communication environments sought for parallel computing, network design can move beyond structures emphasizing ease of routing and transmission bandwidth to more complex communication environments relaxing limits on parallel computation. The granularity of concurrently executing tasks and the granularity of information propagating across networks have a significant influence on performance issues facing network design. **Dally** addresses the issue of achieving communication environments able to handle fine-grained programming systems. In this case, software overhead delays are reduced by use of hardware to speed up communications. The communication network and its performance (only a 5 μsec delay overhead for message passing, translation and context switching) is illustrated using the J-machine being implemented at MIT. At the workshop, **A Ligtenberg and W. Mooij** discussed a mesh-connected network with distributed, adaptive routing applied to VLSI signal processing arrays.

Implementing communication environments and protocols for complex, parallel computing systems is difficult, given the global nature of such communications within an environment of local computing sites. Software tools which can emulate and simulate networks and parallel systems are important in any real network design. Furthermore, software tools to aid program development (and able to model the communication environment) will be necessary for users. **Berman, Cuny and Snyder** describe a set of software tools, drawing on techniques developed for the CHiP computer and from the Simple Simon programming environment, useful as support for parallel programming environments.

Chapter 15

The SDEF Systolic Programming System [1]

Bradley R. Engstrom [2]
Peter R. Cappello [2]

Abstract

SDEF, a systolic array programming system, is presented. It is intended to provide
1) systolic algorithm researchers/developers with an executable notation, and 2) the
software systems community with a target notation for the development of higher
level systolic software tools. The design issues associated with such a programming
system are identified. A spacetime representation of systolic computations is described briefly in order to motivate SDEF's program notation. The programming
system treats a special class of systolic computations, called atomic systolic computations, any one of which can be specified as a set of properties: the computation's
1) index set (S), 2) domain dependencies (D), 3) spacetime embedding (E), and
nodal function (F). These properties are defined and illustrated. SDEF's user interface is presented. It comprises an editor, a translator, a domain type database, and
a systolic array simulator used to test SDEF programs. The system currently runs
on a Sun 3/50 operating under Unix and Xwindows. Key design choices affecting
this implementation are described. SDEF is designed for portability. The problem
of porting it to a Transputer array is discussed.

[1] This work was supported by the Office of Naval Research under contracts N00014-84-K-0664
and N00014-85-K-0553.
[2] Dept. of Computer Science, University of California, Santa Barbara, CA

15.1   Introduction

15.1.1   Systolic arrays

Systolic Arrays were first reported by Kung and Leiserson [18]. As originally conceived, systolic arrays are special-purpose peripheral processor arrays implemented with VLSI technology. Such arrays use only a small number of processor types, and have regular, nearest-neighbor interconnection patterns. These characteristics reduce the cost of both their design and operation. Kung and Leiserson point out [18],

> The important feature common to all of our algorithms is that their data flows are very *simple* and *regular*, and they are *pipeline algorithms*.

15.1.2   Programmable systolic arrays

While research concerning special-purpose systolic arrays still is ongoing, the view of systolic arrays has broadened to include arrays of general-purpose processors. These arrays, which share the regular interconnection structure of their special-purpose counterparts, are programmable. Examples of general-purpose systolic arrays include the Transputer[3], the Warp [2], and the Matrix-1 [12]. General-purpose systolic arrays have spurred development of systolic programming languages. Relative to algorithmic and hardware development, work on tools for systolic software development is just beginning. An explanation for this is given by Snyder [34].

> Because systolic algorithms are commonly thought of as being directly implemented as hardware arrays, writing systolic programs would appear to be an activity without need for a programming environment. But the appearance is deceiving. There are many times when one indeed does program systolic algorithms: when the systolic array is programmable, during the design process (for simulation purposes) of hard-wired array implementations, when a systolic algorithm is used on a general purpose parallel computer, or when one is engaged in research on systolic algorithms.

Many efforts are being made to meet the need for systolic programming environments. Occam[4] is a concurrent programming language based on Hoare's model of *communicating sequential processes*. Occam produces code for Transputer arrays, also developed by INMOS. A Transputer is a general-purpose processor which

---

[3] Transputer is a trademark of INMOS, Ltd.
[4] Occam is a trademark of INMOS, Ltd.

may be connected to up to four other Transputers using on-chip data links. At Carnegie-Mellon University, the Warp project has developed a language, W2 [3], and its compiler and run-time system in support of a high-speed programmable systolic array. W2 is syntactically similar to Pascal, but also provides interprocessor communication primitives based on message passing. Occam and W2 are significant achievements in systolic array programming. Poker [33] uses several program abstractions that unify parallel programming. The Poker environment has been targeted to 1) the ChiP [14], 2) hypercubes [35], and 3) systolic arrays (Hearts [34]). Hearts, a specialization of the Poker programming environment, integrates the process of specifying, compiling, loading, and tracing systolic computations.

The systems mentioned above are intended to facilitate the production of executable software for hardware arrays: they must provide a usable programming environment. Designers of such systems must attend to such issues as the operating system and user interfaces, and the interprocessor communication protocol.

15.1.3   The SDEF system

The SDEF system constitutes a programming environment for describing systolic algorithms. It includes a notation for expressing systolic algorithms, a translator for the notation, and a systolic array simulator with trace facilities.

The translator generates a C [17] program that performs the computation specified by the SDEF description. After being compiled, this C program can be run on the SDEF systolic array simulator. Fig. 15.1 shows the overall structure of the SDEF environment.

An SDEF program specifies both the computation and communication requirements of a systolic algorithm. The SDEF program also specifies how the systolic algorithm is to be 'embedded' in spacetime [5,24]. This approach differs from that used by Occam, W2, and Hearts. These differences are examined in section 15.2.

The goals of the SDEF system are:

- *To increase the productivity of systolic algorithm researchers.*
  SDEF provides a notation for systolic computations that is precise and executable. Algorithms can be communicated succinctly in a form that is suitable for independent testing and use by others.

- *To increase the productivity of systolic algorithm developers.*
  Data communication 1) between array processors, and 2) between peripheral processors and the file system, is described implicitly in an SDEF program. The SDEF translator (and not the user) creates a C program wherein all data communication is made explicit.

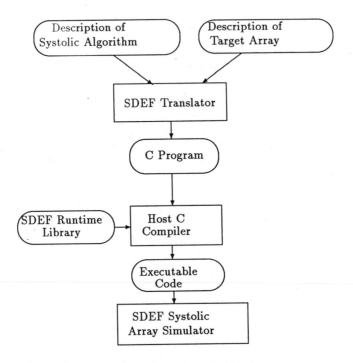

Figure 15.1: Overall structure of SDEF system.

Reliability is enhanced because parts of SDEF programs are reusable, and because SDEF provides extensive trace facilities for testing algorithm descriptions.

- *To support work by others on very high level systolic languages.*
  The SDEF notation is not intended to be the ultimate systolic programming language. Higher level languages are contemplated. Indeed, some are under development. Chen's Crystal [7], a general framework for synthesizing parallel systems, can be specialized to systolic processing [6]. Delosme and Ipsen have started work on a system for producing optimal spacetime embedding of affine recurrence equations [8]. The mapping a systolic computation onto an undersized array has been addressed, for example, by Fortes and Moldovan [10,28,29], Nelis and Deprettere [30], and Navarro, Llaberia, and Valero [29].

Much research conducted into tools for analyzing [4,25,31,5,27,32,7,1], synthesizing [25,5,6,32,13,23], as well as optimizing [22] systolic algorithms. SDEF does not subsume these tools. Where automated, such tools can be connected to SDEF's front-end: SDEF can be used to *express the results* of the analyses, synthesis, and optimizations performed by other tools.

15.1.4   The organization of this paper

Section 15.2 discusses the issues addressed by a systolic programming environment, and examines existing work on such environments. Section 15.3 briefly explains the spacetime representation of a systolic computation. This section also discusses how one spacetime embedding of a systolic computation can be transformed into another. Section 15.4 gives the SDEF notation for describing systolic computations. The advantages and disadvantages of this approach also are discussed. Section 15.5 presents the user's view of the SDEF programming environment, while section 15.6 discusses SDEF's implementation.

15.2   Issues in Systolic Computation Specification

15.2.1   Systolic array programming

Before discussing the general issues of systolic programming, we examine the methods for programming systolic arrays provided by W2 and Hearts. Since W2 is different from Hearts, and both are different from SDEF, they provide a good basis for comparison.

W2

The language W2 was developed to program the Warp processor array. The user views the Warp system as a linear array of identical processors which can communicate with their left and right neighbors. Communication is based on message passing. Figure 2 shows a W2 program taken from [3]. The program evaluates the polynomial $P(z) = c_0 z^9 + c_1 z^8 + \ldots + c_9$ for 100 different values of z. Each of the ten processors has its own coefficient c, which remains constant for each z. The *receive* primitive has four parameters: 1) the direction from which the data is to be read, 2) the name of the channel , 3) the variable to which the data is to be assigned, and 4) the name of an external variable from which to get the data, if the receive is performed by a peripheral processor.

The program presented in Fig. 2 is an example of explicit communication. It is the programmer's responsibility to ensure that data flows correctly, and that sends match receives. Explicit communication is widely used in parallel programming systems. A well known example is the send and receive primitives of Hoare's CSP language.

```
module polynomial (z in, c in, p out)

float z[100], c[10], p[100];

cellprogram (cid : 0 : 9)
begin
   function poly
   begin
      float coeff, xin, yin, ans;
      float temp;
      int i;

/* save the first coefficient and passes the rest.
   Outputs an extra item to conserve the no. of receives and sends
*/
      receive (L, X, coeff, c[0]);

      for i := 1 to 9 do begin
         receive (L, X, temp, c[i]);
         send (R, X, temp);
      end;
      send (R, X, 0.0);

/* Horner's rule: multiplies the accumulated result yin with
   incoming data xin adds the next coefficient */
      for i := 0 to 99 do begin
         receive (L, X, xin, z[i]);
         receive (L, Y, yin, 0.0);
         send (R, X, xin);
         ans := coeff + yin*xin;
         send (R, Y, ans, p[i]);
      end;
   end;

   call poly;
end
```

Figure 15.2: A simple W2 program.

Hearts

Hearts is a derivative of the Poker programming environment. Hearts provides an integrated set of tools to create, run, trace, and debug systolic programs. The Hearts environment provides graphical metaphors to simplify programming. Creating a Hearts program is a five-step process.

1. *Create the communication structure*
   The programmer requests a one- or two-dimensional array of processors. Then, using the mouse and keyboard, the programmer "draws" the communication links between processors, thus specifying the communication graph of the systolic array. Since the interconnection structure for systolic arrays is usually regular, commands are available for specifying an iterative structure.

2. *Write the nodal functions*
   Using a text editor, the programmer writes the sequential code that will run on each processor. The language used incorporates primitives to read and write data to a port (a named communication path). There may be more than one nodal function.

3. *Assign processes to processors*
   After writing one or more nodal functions each processor is assigned the function it is to execute. Any actual parameters to the nodal function are entered at this time.

4. *Assign port names to communication links*
   Communication links between processors are defined in the first step. In order to refer to them in the nodal code, each link is given a name, known as the *port name*.

5. *Assign stream names*
   Input and output at the periphery of the array requires data to be read or written to files. A *stream name* associates file names with input and output ports. It also specifies the index of data within the file that is to be associated with each port. Record boundaries can be located because data item sizes must be fixed.

Both W2 and Hearts use explicit commands to pass messages. In Hearts, a graphical tool is used to specify the communication structure and size of the array. Since the Warp has a fixed architecture, this kind of tool is not needed in the Warp environment. Hearts obtains external data from the underlying Unix file system; W2 uses a shared memory paradigm to access data on the host computer.

The differences in these two systems stem from differences in their goals. The W2 project is working to create a very high speed parallel processing engine. The intent of the Hearts project is to provide a programming environment that facilitates the creation of systolic programs.

15.2.2   Higher level notation

Hearts and W2 are examples of message-based programming: sends and receives are used to pass messages. It is the programmers responsibility to coordinate processes so that sends and receives match. This method is simple to understand, in principle, because the program mirrors the underlying operational mechanisms. It also tends to result in efficient code. Message-based programming however can be error prone, especially when the communication pattern is complex.

Systolic algorithms and applications are becoming increasingly complex. A higher level notation for programming systolic arrays is needed to help programmers cope with this increased complexity. The argument for higher level languages here is essentially the same as that for sequential machines, only more compelling.

15.2.3   Systolic programming issues

Many of the issues in systolic programming have analogues in sequential programming. The complexity of parallel programming increases the importance of some of these issues, such as program re-usability. This section, though not exhaustive, examines some of the important issues in the design of a systolic programming environment. We first mention some *interface issues*.

Operating system: The systolic programming system should provide the array programmer with a natural interface to the operating system of the host computer.

External I/O: Usually in a systolic array, only the peripheral processors access the host or external I/O devices. This complicates the programmed communication of data and program to, and from, the systolic array. The programming system must enable the user to handle gracefully this constraint on external communication. The environment should help the user to ensure that data is ordered and formated correctly.

There also are *program issues*. The reason for creating a programming environment is to make programming simpler (hence more reliable) and faster.

Creation/Modification: Specialized tools are needed to create and modify programs. Programming systems, such as Poker, that provide a complete set of tools are essential.

Error detection: The systolic programming system should be designed to detect as many program errors as possible, as soon as they are committed.

Testing: As systolic programs become more complex, the need for high-level testing and evaluation tools increases. A systolic programming environment should provide specialized facilities for tracing and debugging systolic array programs.

Re-usability: Distinct systolic programs often have one or more components in common, such as their communication pattern. In order to increase programmer productivity, the programming language should provide for the re-use of common program components.

Efficiency: In systolic systems, just as in conventional systems, there is usually a tradeoff between speed and ease of use. The pros and cons of using a high level systolic programming language are generally analogous to those of a high level sequential programming language. High level systolic languages typically incur more overhead than low level languages. It is however generally accepted that the advantages of high level languages justify their cost, except in extremely time critical applications.

15.3 Spacetime Representations

Programming in the SDEF system is based on a spacetime representations of a systolic computations[5,24,28]. This section briefly introduces this idea.

15.3.1 Spacetime representation of systolic computations

The following code fragment computes a matrix-vector product, $y = Ax$, for a 3 × 3 array A, and vectors y and x.

```
for i = 1 to 3 do
    y[i] = 0;
    for j = 1 to 3 do
        y[i] = y[i] + A[i,j] * x[j];  /* inner product step */
    end
end
```

If we "unravel" the **for** loops we can represent the computation with respect to data usage. Such a diagram, for the above code, is given in Fig. 15.3. The figure depicts the data dependence of each inner product step (IPS). There is one IPS process for each entry in the A matrix. It is convenient to associate each process with its corresponding A element index, which we refer to as its *process index*.

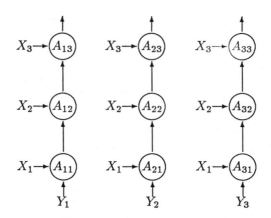

Figure 15.3: Data usage in matrix-vector product.

We can create another representation of matrix-vector product by using a spacetime diagram. Similar to Fig. 15.3, this diagram depicts the data dependence (between inner product processes) in space and time. A spacetime diagram of matrix-vector product is shown in Fig. 15.4. We refer to this particular spacetime representation of matrix-vector product as its *canonical* form. In this design, there are three processors, and three time cycles. The value for y_i is computed in cycle i. The diagram indicates the three processes associated with a y component are all performed during the same cycle. In fact, process(i,1) must complete before process(i,2) starts, which in turn must complete before process(i,3) starts. In this design, which would never be used, the time/cycle thus depends on the size of the matrix. Processor P_1 starts with an initial value of 0 for y_i, then executes the IPS function (i.e., the inner product step) using the values for x_1 and A_{i1} that it received. The result, the new value of y_i, is passed to P_2. The final value of y_i is output by P_3.

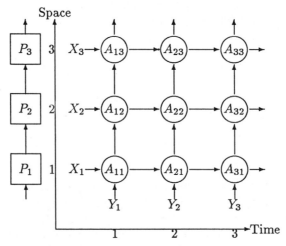

Figure 15.4: Spacetime representation of matrix-vector product.

The position of each node in spacetime (its process index) indicates where and when an IPS process takes place. The representation also shows what, where, and when, data is needed.

One design for a computation can be transformed into another by applying a linear transformation to the indices of each process. The canonical design for matrix-vector product, for example, can be transformed to the Kung and Leiserson[18] design by the following linear transformation.

$$\begin{pmatrix} 1 & 1 \\ -1 & 1 \end{pmatrix} \begin{pmatrix} i \\ j \end{pmatrix} = \begin{pmatrix} Time \\ Space \end{pmatrix}$$

The spacetime representation of the Kung and Leiserson (KL) design is depicted in Fig. 15.5.

15.3.2 Spatial projection

If we project the process graph (embedded in spacetime) onto the spatial subspace (which in this case is a single axis), we obtain the spatial characteristics of the

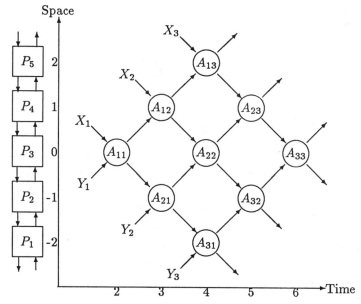

Figure 15.5: Kung and Leiserson design for matrix-vector product.

computation. The set of inner product processes, called the *index set*, in the KL design projects onto a set of five processors (i.e., five points in space). The number of nodes that map to a processor is the number of IPS computations that the processor must perform. The data dependencies, shown as arcs, map to the physical array, indicating the direction that data must flow through the array. For example, the projection of the y data dependence indicates that y values must move upward. Additionally, each processor must have access the entries of the A array that it uses during its IPS computations. Processor P_2 therefore must have access to A_{21} and A_{32}.

15.3.3   Temporal projection

By projecting the process graph onto the temporal axis, we obtain the cycle in which each process is executed. The first IPS process cycle (i.e., the node in the process graph with the smallest time index) occurs in processor P_3. The IPS function requires an x component, a y component, and an element of A. The x and y values must be passed to this center processor by its immediate neighbors. By extending the data arcs in spacetime, we create a schedule for data delivery from the peripheral processors. This is given in Fig. 15.6.

This figure portrays what each processor must do at each cycle. As an example, the actions prescribed for the first three cycles are given in Fig. 15.7.

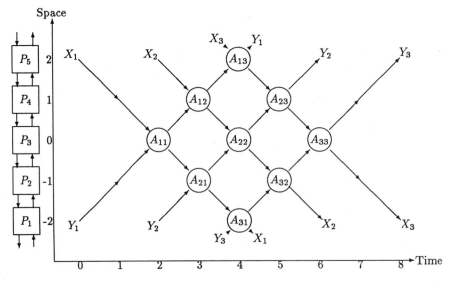

Figure 15.6: Extended space-time diagram showing all data movement.

Cycle	Actions
0	P_5 read the value for x_1 and passes it to P_4. P_1 reads the value for y_1 and passes it to P_2.
1	P_4 read the value for x_1 and passes it to P_3. P_2 reads the value for y_1 and passes it to P_3.
2	P_5 reads x_2 and passes it to P_4. P_3 reads y_1 from P_2, reads x_1 from P_4 and executes an IPS. The new y_1 is passed to P_4, and x_1 is passed to P_2. P_1 reads y_2 and passes it to P_2.

Figure 15.7: Processor actions during cycles 0,1 and 2.

15.3.4 Transforming computations

Many different designs can be obtained from the canonical form using different transformation matrices. The transformation

$$\begin{pmatrix} 1 & 1 \\ 0 & 1 \end{pmatrix} \begin{pmatrix} i \\ j \end{pmatrix} = \begin{pmatrix} Time \\ Space \end{pmatrix}$$

obtains the design presented in Fig. 15.8.

In this design, y values move up, while x values are used in place. The design uses three processors instead of five.

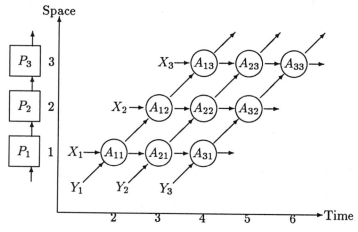

Figure 15.8: Alternate design for matrix-vector product.

A systolic algorithm can be realized by embedding a processes graph in spacetime according to a linear map of its process indices. Different linear maps result in systolic designs with different sets of processor arrays, communication patterns, and relative communication rates. This mathematical mechanism is an important part of the SDEF programming system.

15.4   Specifying a Systolic Computation Using SDEF

SDEF programs are based primarily on the spacetime representation introduced in Section 15.3. In this section, we introduce the four properties used in SDEF to describe a systolic computation. The specification of those properties also is presented, along with a brief discussion of this approach's advantages and disadvantages.

15.4.1   The properties of an atomic systolic computation

SDEF treats a subset of systolic computations. This subset is the one treated in the work of [16,25,26,27,24]. The computational fragments informally correspond to computing uniform recurrence equations[15] inside the nested index loops of a high level language (e.g., 'DO' loops in FORTRAN and 'For' loops in Pascal). We refer to such a systolic computation as an *atomic* systolic computation. Contemporary systolic algorithms are more complex than this[32], but they can be decomposed into atomic components. The task of a contemporary systolic algorithm designer includes *bonding* atomic computations into a *compound* systolic computation. In this paper, we consider only the individual atomic[5] components of a systolic computation. In the Conclusion, we briefly discuss enriching SDEF with a composition feature, which can be used to 'bond' atomic systolic computations into a compound systolic computation.

A notation for a specialized computation should reflect that computation's distinguishing characteristics. SDEF is based on the fact that an atomic systolic computation is characterized by four properties: its S, D, E, and F properties.

The S property: its index set

The index set of a systolic computation is the set of index values over which the computation is defined. These index values define the set of nodes that make up

---

[5] Unless stated otherwise, we hereafter only will refer to atomic systolic computations, and will omit the qualification 'atomic.'

a spacetime representation of a computation as depicted in Section 15.3. It can be thought of, informally, as the indices for one or more arrays, whose elements need to be computed. Two such sets are specified below.

$$S_1 : 1 \leq i \leq j \leq 5$$

$$S_2 : 1 \leq i, j \leq 5$$

Fig. 15.9 depicts the index set S_1. SDEF treats a specific kind of index set: the set of integers in convex polyhedra. Such a set consists of the integer solutions to the convex polyhedron's corresponding linear system $Ax \leq b$. This view includes all index sets that we have seen in practice[19]. We therefore see no reason at this time to use a more general class of index sets.

The D property: its domain dependences

Informally again, each array element must be computed in terms of other array elements. For most systolic algorithms, the computed value of an array element, $a(p)$, depends on array elements whose indices are fixed offsets from p. Such dependences are referred to in the literature as *uniform* data dependences [16,31,32]. Following geometric terminology, they may be called *translation domain* dependences. SDEF works with this type of dependence. Two sets of domain dependences are given below.

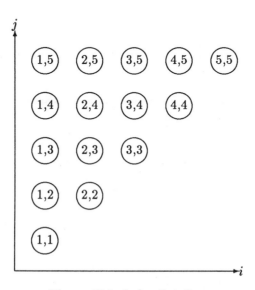

Figure 15.9: Index Set S_1

$$D_1 : x : \begin{pmatrix} -1 \\ 0 \end{pmatrix}, y : \begin{pmatrix} 0 \\ -1 \end{pmatrix}, a : \begin{pmatrix} 0 \\ 0 \end{pmatrix}$$

$$D_2 : x : \begin{pmatrix} -1 \\ 0 \end{pmatrix}, y : \begin{pmatrix} -1 \\ -1 \end{pmatrix}, a : \begin{pmatrix} 0 \\ -1 \end{pmatrix}$$

Fig. 15.10 depicts these two dependence sets. Fig. 15.11 depicts dependence set D_1 applied to the index set S_1. By convention, the data arcs are directed *toward* the process that uses the data. This is the opposite direction of domain dependence vectors, which point to the source of the data, not its destination.

The E property: its spacetime embedding

A process graph can be scheduled on an array of processors in many ways. One topic of systolic array research is concerned with *linear* embeddings of these process graphs into spacetime. Different systolic arrays in the literature often are nothing more than linear transformations of one another in spacetime[25,4,31,24,5,32,27]. A guide to the literature concerned with such manipulations is given by Fortes and Wah[11]. Two examples of spacetime embeddings are given below.

$$E_1 : \begin{pmatrix} 1 & 1 \\ -1 & 1 \end{pmatrix} \begin{pmatrix} i \\ j \end{pmatrix} = \begin{pmatrix} Time \\ Space \end{pmatrix}$$

$$E_2 : \begin{pmatrix} 1 & 1 \\ 1 & 0 \\ 0 & 1 \end{pmatrix} \begin{pmatrix} i \\ j \end{pmatrix} = \begin{pmatrix} Time \\ Space_1 \\ Space_2 \end{pmatrix}$$

Figures 15.5 and 15.8 depict two different linear transformations of Fig. 15.4.

The F property: its process function

In an atomic systolic computation, all processes compute the same function. The function's domain can contain the process index. Apart from the process index, the

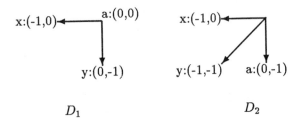

Figure 15.10: Two data dependence sets. Data dependence arcs are directed toward the source of a process's data.

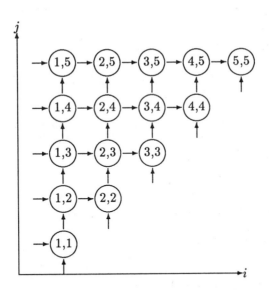

Figure 15.11: Dependence set D_1 applied to index Set S_1.

function's domain equals its co-domain. We give below two different functions that have the same number of inputs and outputs.

F_1(input:x,y,a:output:x',y',a')
integer x, y, a;
{
    x' ← x;
    a' ← a;
    y' ← y + a × x;
}

F_2(input:x,y,a:output:x',y',a')
char x, a;
boolean y;
{
    x' ← x;
    a' ← a;
    y' ← y and (a == x);
}

The property values described above can be combined in a variety of ways. Five distinct designs based on these property values are given below.

1. Upper triangular matrix-vector product: $U = (S_1, D_1, E_1, F_1)$.

2. By changing the index space, we obtain a design for Full Matrix-Vector Product: $M = (S_2, D_1, E_1, F_1)$.

3. By changing the domain dependences, we obtain a design for Polynomial Product (convolution): $C = (S_2, D_2, E_1, F_1)$.

4. By changing the function computed at each vertex, we obtain a design for String Pattern Matching: $S = (S_2, D_2, E_1, F_2)$.

5. By changing the spacetime embedding, we obtain a design for String Pattern Matching that is completely pipelined, operating on a hexagonally-connected array: $P = (S_2, D_2, E_2, F_2)$.

The design (S_2, D_1, E_1, F_1) was first reported by Kung and Leiserson[18]. It is their bidirectional linear systolic array for computing matrix-vector product.

As the above examples illustrate, different systolic computations may share some properties. A good systolic programming environment should exploit this characteristic of systolic computation: it should facilitate the re-use of previously established properties in the specification of a new systolic computation.

Although the S,D,E and F properties are largely independent, there are some weak interdependences that must be noted. These interdependences are:

- The dimension of the index space, the dependence vectors, and the embedding matrix all must agree.

- The number of domains must be equal to the number of arguments to the nodal function.

Distinct properties values thus can be readily substituted, resulting in distinct computations.

The S, D, and E properties are mathematical objects. Specifically, let C be an atomic systolic computation with a dimension of d, and a function arity of a. If C's index space is the set of integers inside an m-sided convex polygon, then its index space is characterized by matrix $S \in \mathbf{Z}^{m \times (d+1)}$, where $\mathbf{Z}$ denotes the set of integers. Each of its a domain dependences is a vector $d_i \in \mathbf{Z}^d$. If the dimension of spacetime into which C is being embedded is l (l is typically 2 or 3), then its spacetime embedding is a matrix $E \in \mathbf{Z}^{l \times d}$.

15.4.2   Specifying the systolic array

SDEF's target array is assumed to be a rectangular grid of orthogonally-connected processors. The array's *size* is specified as an ordered pair of integers (x, y), indicating that there are xy processors, forming an $x \times y$ array.

SDEF assumes that only peripheral processors have an I/O capability. The I/O capability of these processors is expressed in terms of *read* and *write* capabilities for each boundary of the array: left, right, top, and bottom. The capabilities that can be specified are: no capability, read only, write only, or read and write.

An example specification follows. It specifies a rectangular array that can be used to perform a Schreiber design of matrix product for a 5 × 5 matrix.

Size	Top	Bottom	Left	Right
(9,5)	read	none	read, write	read, write

Array specification is unrelated to the specification of S, D, E, and F properties: If a computation is targeted to a physical array, then the computation's spatial projection (set of processors) cannot exceed the size of the physical array.

15.4.3   Specifying the systolic computation

Specifying an S property

As mentioned previously, SDEF accommodates an index set if it is the set of integers inside a convex polyhedron. This set can be described as the integer solutions of a linear system $Ax \leq b$.

To specify such an index set, a user first specifies the *dimension* of the computation (i.e., the number of independent indices). After doing so, the the user can specify the A matrix and b vector that define the boundaries of the convex polyhedron. In SDEF, such linear constraints are referred to as *global constraints*.

In addition, a user specifies *orthohedral bounds*. Orthohedral bounds specify lower and upper limits for an index (i.e., an axis). Below we give the portion of the SDE file that specifies the index set S_1, mentioned earlier in this section. This index set is used in Upper Triangular Matrix-Vector Product (in this case a 5 × 5 matrix).

```
Dimension:    2
Orthohedral Bounds:
    lower upper
    1     5  i
    1     5  j
Global Constraints:
-1 i + j <=  0
```

Orthohedral bounds are a programmer convenience. Since they always can be expressed as linear inequalities, they can be expressed as global constraints.

To obtain the index set S_2, used for Full Matrix-Vector Product, one only needs to remove the global constraint from the specification above.

Specifying a D property

An arc in a systolic computation's cellular process graph represents a domain dependence. Each array variable, referred to as a *domain*, has an associated domain dependence. The programmer must specify each domain's dependence. The dependence set D_1 used in Full Matrix-Vector Product, is specified as follows.

```
Domains:
    name: X   type: int  dependence: i(-1) j( 0)
    name: Y   type: int  dependence: i( 0) j(-1)
    name: A   type: int  dependence: i( 0) j( 0)
```

The domain dependence set D_2, used in Polynomial Product, is like that for D_1 except that the entry for domain A is:

```
    name: A   type: int  dependence: i(-1) j(-1)
```

Specifying an E property

A computation's spacetime embedding is specified as a matrix. The embedding map E_1 used in the Kung and Leiserson Matrix-Vector Product, for example, is specified as follows.

$$\begin{pmatrix} 1 & 1 \\ -1 & 1 \end{pmatrix} \begin{pmatrix} i \\ j \end{pmatrix} = \begin{pmatrix} Time \\ Space \end{pmatrix}$$

Specifying an F property

Function code is written in an SDEF-extended version of C. The idea of building on top of an existing compiler for a sequential machine also has been used in the Poker programming environment [35]. One advantage of doing so is that we may build on the work of others. The C compiler is responsible for producing code for the target machine, and for using its resources (e.g., registers, memory, and buffers) efficiently.

During each 'computation cycle,' this function is invoked with its arguments. It can call any subfunction written by the programmer. SDEF provides two extensions to C. These are 1) *prime* notation, and 2) compiler-generated declarations.

Prime notation specifies the computed value of a domain. For example, the statement

$$Y' = Y + X + 1;$$

means that the computed value of the domain Y is the argument value of Y plus the argument value of X plus 1. Since data locations for Y' and Y are distinct, modification of one does not affect the other. The SDEF translator translates domain references to internal data locations. The information in these data locations is managed by code that is generated by the SDEF compiler. This code ensures that data is moved between data locations and processors in the manner implied by the spacetime embedding. Function code generally is only a small fraction of the program produced by the SDEF translator.

For F_1, mentioned earlier in this section, the user provides the following SDEF code.

```
F1(X,Y,A)
{
    Y' = Y + X * A;  /* compute inner product step */
}
```

Variables Y', Y, X, and A are not declared in the function code; the SDEF compiler inserts declarations for all domains. The type of each domain is specified when the domain is specified. In SDEF notation, a domain needs an explicit assignment statement *only* when it is modified by the function invocation. As another example, the user provides the following code for F_2 (the string pattern matching function).

```
F2(X,Y,W)
{
    Y' = Y && X == W;
}
```

SDEF's extensions to the C language are tailored for ease of expression of systolic function code.

15.4.4 The specification environment

The SDE file

Property specifications are partitioned into 3 files. The A property, the characteristics of the systolic array, are given in the A file. The second file is the SDE file. It contains the specification of a computation's S, D and E properties. Such files are currently created using an SDE file editor. This editor performs error and consistency checking on the data entered. The third file is the F file. It contains the extended C version of the function that executes on the nodes of the process graph. The name of the function is specified in the SDE file.

Another way to create A, SDE, and F files is for a higher level translator to generate them. That is, these files are a *data interface* for higher level systolic specification translators. Such a translator takes as input a higher level specification (higher than that used by SDEF) of a systolic computation, and produces one or more A, SDE, and F files as output. The SDEF translator thus is the back end of such a system.

15.4.5   An example specification

An SDEF *program* for an atomic systolic computation is a specification of the computation's S, D, E, and F property values, as well as a specification of a physical array. We list here the SDE, F, and A files for a Schreiber design 5×5 full matrix product. Fig. 15.12 gives the spacetime diagram for matrix product; Fig. 15.13 presents the SDEF specification for it.

15.4.6   Advantages and disadvantages

The SDEF notation represents a systolic computation as a structure in spacetime. As with all high level notations, the approach has both merits and demerits.

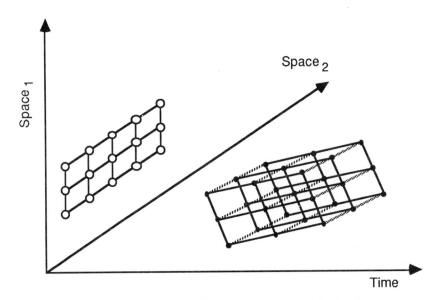

Figure 15.12: Schreiber design for 3×3 matrix product.

SDE File

```
Dimension:    3
Orthohedral Bounds:
    lower upper
     1     5   i
     1     5   j
     1     5   k
Domains:
    name: A    type: float   dependence: i(-1) j( 0) k( 0)
    name: B    type: float   dependence: i( 0) j(-1) k( 0)
    name: C    type: float   dependence: i( 0) j( 0) k(-1)

Function: name: IPS

Embedding:   1 0 1
             1 1 1
            -1 0 1
```

F File

```
IPS(A,B,C)
{
    C' = C + A * B;
}
```

A File

Size	*Top*	*Bottom*	*Left*	*Right*
(9,5)	read	none	read, write	read, write

Figure 15.13: SDEF specification for Full Matrix Product

Having a mathematical basis, an SDEF specification is precise and succinct. Because of their relative independence, the S,D,E and F properties can be re-used freely, creating new algorithms. As shown in Section 15.3, one can change the communication pattern, even the processor topology, by simply changing the embedding matrix. Compare this succinctness to systems in which communication is intermixed with the processor code. Any change in the communication pattern requires modification of the nodal code in one or more places, with each change introducing an opportunity for error. Hand coding of communication however can produce faster, smaller programs that can make use of specialized hardware. This tradeoff is commonly seen in comparing higher-level languages to lower-level languages.

In addition to ease of modification and reusability, SDEF programs exhibit a degree of hardware independence. The run-time system is responsible for creating a virtual, orthogonally-connected machine. The SDEF program is unaware whether it is running on a simulator or on hardware. This increased portability has a drawback: it is difficult for the programmer to take advantage of special hardware capabilities.

The SDEF system is designed to increase the productivity of researchers and developers of systolic algorithms. In this context, the expense of increased run-time overhead is justified. SDEF might be used even when one seeks raw speed. But the programmer may wish to program at a very low level, or at least re-code frequently executed sections.

15.5   The SDEF Environment

The SDEF system facilitates the process of creating and testing systolic algorithms. This section examines tools, other than the translator, that are used to create systolic programs. These tools include support for user-defined domain types and input preparation, and the SDEF simulator for testing SDEF programs.

15.5.1   The SDEF domain type database

An important area in systolic array research is that of accessing external data. In W2, external data appears as elements of arrays in shared memory. In Hearts, data is kept in files composed of fixed size records. By knowing the index of the data, it thus is possible to access the data with a simple file seek.

Our goal in the SDEF system is to support all user-defined data types, and to allow the user to determine the format of data in files. Towards this end, SDEF includes a *domain type* database, and tools for managing it.

For every domain type that can be used in the SDEF system, the domain type database contains 1) a header file that defines the type for use in C programs, and 2) a set of I/O routines for that type. C predefined types do not need a header file, and have default I/O routines. The user however is free to create new I/O routines for C predefined types.

The type of each domain is specified in the SDE file. The SDEF translator includes the header file for each user-defined domain type in the C program it produces (producing an error diagnostic, if an undefined domain type is referenced). The translator also ensures that the correct I/O routine is invoked whenever a domain is read or written. This applies only to external reads and writes; domains are communicated between processors as binary images.

There are many advantages to having user-supplied I/O drivers. The user decides how the input (and output) is to be formated. Integers, for example, can be stored in files in decimal, hexadecimal, octal or even binary images. Sometimes a processor reads data from a device where the format is predetermined. In this case, user-supplied read routines allow acceptance of arbitrary formats. The read and write routines for a type do not need to use the same format. Data can, for instance, be read in binary format and written in ASCII format.

SDEF provides tools for adding to, deleting from, and modifying the domain type database.

15.5.2 Preparing input data

In any systolic computation, the data for each domain is processed in a particular order. This means that external data must be read in a particular order. Since SDEF allows arbitrary formatting of input data, it is not possible to 'seek' a particular domain item, as it is in Hearts. Thus, it is necessarily the user's responsibility for providing data, as required by the user's SDEF program.

The SDEF system however aids the programmer by providing *domain order templates* for each domain. Such templates convey the order of domain items by specifying the processor and cycle in which they are read. These files are computed by the translator, based on the programmer's specification of the index space and the spacetime embedding. The domain order files can be used to ensure that either data in files is formated correctly, or data from a device is generated in the proper order.

15.5.3 The SDEF systolic array simulator

As mentioned in Section 4, SDEF's output is targeted for an orthogonally-connected array. The SDEF simulator provides a means of tracing and testing SDEF programs. It displays a window for each processor in the processor array described by the A file. The window shows the values of domains, and the communication activity for the processor. An example of trace windows during execution is shown in Fig. 15.14. The menu provides the ability to stop or start one or all of the processors. To affect the action of a single processor, the user first *selects* the processor by moving the mouse cursor over the processor's window, and then presses the appropriate mouse button. All simulation control functions are initiated with the mouse.

15.5.4 Running an SDEF program

There are five basic steps used in creating and running an SDEF program. These are:

1. *Create the SDE, F, and A files:* These are the inputs to the SDEF translator. The SDE file is created using a specialized 'SDE' editor; the F and A files are created using a text editor.

2. *Add new domain types to the database:* If any domains in the SDE file have domain types that are not already in the database, then add these domain types to the database. This includes writing a C header file that defines the type, and writing I/O routines to read and write the objects of this type.

3. *Run the SDEF translator:* The SDEF translator produces 1) a C file for the SDEF program, 2) a boot file that contains processor schedules which are loaded at run-time, and 3) domain order files which convey the place and cycle that input data is to be read. The C program automatically is compiled and linked with run-time support routines, producing a program that executes on the SDEF simulator.

4. *Create the input data files:* Using the domain order files created by the translator, the user creates files containing the actual data. The format of the

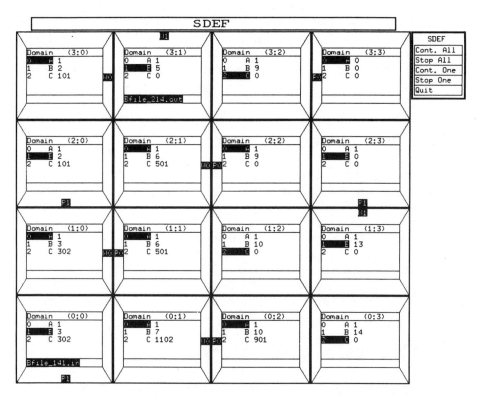

Figure 15.14: SDEF trace windows during execution.

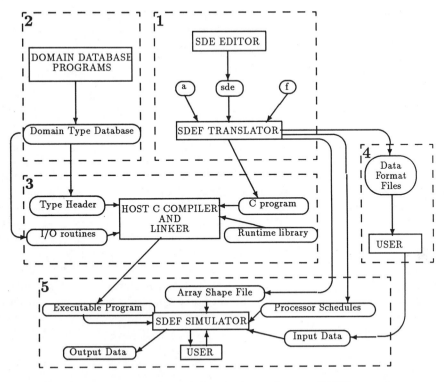

Figure 15.15: Steps in creating and running an SDEF program.

data is determined by the I/O routines for the data type in the domain type database. The order of stream data types is determined by order of use during computation. Statics and initial register values are booted at run-time. Their order is determined by the shape of the processor array, as given in the A file.

5. *Test the program using the SDEF simulator:* Running the program on the simulator allows the user to interactively monitor the program. The SDEF simulator automatically boots internal tables and control files.

A diagram showing the process of creating and running an SDEF program is shown in Fig. 15.15.

15.6 An Implementation of the SDEF System

15.6.1 The Translator

The SDEF translator takes the A, SDE and F files as input and produces a C program to be run on the processors of the systolic array. It is the responsibility of the translator to create sequential code and control data, which, when executed, reproduces the communication and computation structures described by the SDEF program. In this section, we discuss some of the details of generating such a program.

Communication Types

The domain dependences together with the spacetime embedding determine the pattern of communication for a computation. The index set determines the size and shape of the process graph. Figure 15.16 depicts a process communication graph embedded in spacetime (whose spatial projection is a linear array of five processors). Based on the spacetime orientation of its propagation, we classify four types of communication in spacetime. Information can propagate in: (1) time, but not space (memory), (2) space, but not time (broadcasting), (3) both space and time, (4) neither space nor time (information that is used at one point in spacetime). Type (2) is considered to be incompatible with the paradigm of systolic computation. The three types of information propagation that are compatible with systolic computation are illustrated by the three domains (A, W, and X) shown in Fig. 15.16.

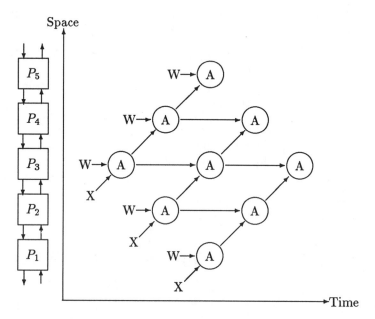

Figure 15.16: A process graph embedded in spacetime whose spatial projection is a linear array of five processors.

We first discuss type (3) communication. This type of communication is called *stream* communication. In the example depicted by Fig. 15.16, values for the X domain propagate between processors from botton to top over time. In this example, I/O is confined to the processors on the ends of the array. All external input values for the X domain thus must be read by processor P_1. External outputs likewise are written by P_5. Note that P_3 is the first processor to perform a computation. Assuming that the nodal function needs all three domains, this processor must wait for an initial X value to be passed from P_1. After a computation completes, results for domain X also need to be passed to P_5 in order to be written externally.

Type (1) communication, exemplified by domain W, is realized with a *register*. Registers are used to realize domain dependences that have no spatial dependence, only a temporal one. The translator detects this type of communication, generating the code to save and restore register values between invocations of the nodal function. Type (4) communication is called *static* 'communication.' Domain A is of this type. In static communication, information propagates in neither space nor time. Such a piece of information can be viewed as a constant embedded in spacetime.

Processor Schedules

A processor's activity is partitioned into *computation cycles*. The process executed by a processor may change from one computation cycle to the next. The SDEF translator generates code which ensures that data arrives at the right place at the right time. To do this, the translator computes a schedule for each processor. This schedule specifies what the processor is to do during each computation cycle. For example, consider domain X in Fig. 15.16. Domain X is used first by P_3. Processor P_3 however does not have access to disk. Processors P_1 and P_2 thus must have, as part of their schedules, instructions to propagate the first X value to P_3.

It may be that the physical array is larger than the spatial projection of the computation (i.e., its set of processors). In this case, the translator generates code for processors outside the computation's spatial projection, if they are used to propagate data from the boundary of the array to processors that participate in the computation. No code is generated for processors that have neither communication nor computation tasks.

At present, the translator requires that the size of the physical array be at least as large as the spatial projection of the embedded computation. That is, the translator does not automatically solve the 'partitioning' problem for the user. Several systematic mapping techniques that solve this problem are under consideration for inclusion into the SDEF system. In the meantime, extant research (see, e.g., [27,28,29,30]) can be incorporated as a front-end to SDEF.

In addition to data propagation, a schedule includes information about whether the data comes from a neighbor or from an external source. It also indicates when a function invocation is to occur using the data obtained by the processor. Whenever a function invocation occurs, the code generated by the translator ensures that the nodal function is passed the correct domain values for all types of domains: stream, register and static. After the nodal function is invoked, the modified domains are propagated in spacetime as required by the user-specified spacetime embedding (E).

Translation Domain Dependencies

The SDEF translator processes dependence vectors that are not simply single steps in time and space. A dependence can, for instance, specify that a domain value be communicated from a point two units away in space and three away in time. SDEF assumes an architecture in which each processor can only send messages to its nearest neighbors. A dependence that requires a movement of two spatial units over a period of three time units is converted to a *sequence* of physically realizable communications called *simple moves*. A simple move is one where data moves 0 or 1 unit in space in exactly one unit of time. Figure 15.17 depicts a spacetime embedding of a process graph, and the resulting embedding after the embedded domain dependences have been realized as simple moves. Each domain is propagated via a sequence of t stages, where t is the time component of its embedded domain dependence vector. In order to realize an embedded domain dependence as a sequence of simple moves, it is necessary and sufficient that $\sum_{i=1}^{d-1} s_i \leq t$, where the domain dependence, after embedding in a d-dimensional spacetime, is of the form $(t\ s_1\ s_2\ \cdots\ s_{d-1})^t$, s_i is a spatial component, and t is its time component.

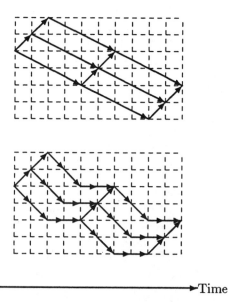

$\longrightarrow$Time

Figure 15.17: An embedding of dependencies and its realization using simple moves.

Even for atomic computations, processors may change their behavior many times during the course of the computation. This is because there is interaction among the domain dependences with respect to their I/O requirements on the physical array. Although, the translator generates one program for all processors, each processor can have as many as $2|D| - 1$ different kinds of computation cycles, where D is the set of domain dependences. This is a tight bound on the number of kinds of computation cycles. For each different kind of computation cycle, the processor has a different *subset* of data domains that it receives and/or sends. Convexity of the systolic computation's index space implies that for each processor there is only one kind of computation cycle in which the nodal function is invoked. All other kinds of computations cycles are used to deliver data to either files, or other processors, in the array.

Processor Data

As depicted in Fig. 15.16, values for the X domain must be read from external storage. Data for stream domains are read at run-time from edge processors. Data values for register and static domains are stored internally by each virtual processor. There are many ways that SDEF could have been designed to provide these values. This is especially problematical since, as in this example, all processors may not have external read/write capability. The translator could have been designed to embed initial values for these domains in the generated code. In this case, the user would have had to retranslate the program to change the data. It also would have meant, given that SDEF generates a single program which runs on all processors, that all processors contain all initialization data, even though any single processor only uses a fraction of the data. To avoid these problems, SDEF programs have an *array initialization* phase (i.e., boot phase) where tables, register values, and static values are loaded from external storage. Each processor therefore needs to store only its own register and static data. During *array initialization*, initial values are read in, and passed to the appropriate processor. Users thus can modify a computation's constants without retranslating. We believe that this design decision is compatible with our goal of increasing programmer productivity.

15.6.2   Virtual Hardware

Architecture and Capabilities

The SDEF simulator provides diagnostic and control facilities. It simulates an array of processing elements which are not too architecturally powerful. Simulating processing elements that are extremely powerful would make the SDEF translator's job too easy, and is unrealistic; real systolic processing elements (e.g., the Transputer) have simple, but focused, computation and communication capabilities. The simulated systolic array is a grid of MIMD processors, each with up to four bidirectional communication channels. Some edge processors are able to read and write externally (i.e., to data files). Fig. 15.18 depicts some typical configurations that fit this model.

The A file provides the translator with information regarding the size of the target processor array. Consider a systolic computation that maps to a 4 × 4 physical array. If the size of the physical array is 5 × 4, then the translator creates schedules for the extra processors, so that they propagate data to the processors that are actually involved in the computation (since it is assumed that only edge processors have access to external data). The A file also enables the translator to handle correctly situations where, for instance, the left side of the array can read data but cannot write data. The translator similarly detects the case when a spacetime embedding of a computation is incompatible with the I/O capabilities of the physical array.

Communication between processors is synchronous, one word per cycle, based on a message-passing protocol. That is, if processor A sends a word to processor B, then A blocks until B receives the word; if B attempts to receive the word before A sends it, then B blocks until the word is sent. The translator has no restrictions on the length of a word. It determines the word size of the underlying machine, generating code accordingly. We have done this to enhance the translator's portability. If a domain data item is larger than the processing element's word size, then the translator generates code for multiple single-word transmissions.

These processing element capabilities may seem too restrictive in light of current hardware projects (e.g., the iWarp) that provide more powerful capabilities. Making our processing elements simple, however, eases the task of porting the SDEF system to real hardware, as it becomes available.

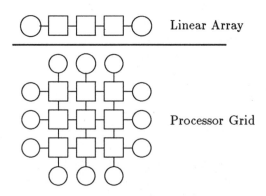

Figure 15.18: Two typical processor and I/O device configurations. The circles in the figure represent I/O devices; squares are processors.

UNIX implementation

The SDEF systolic array simulator uses UNIX processes to simulate processors. The C program produced by the SDEF translator is compiled and linked to a run-time library to produce an executable image. Although the same executable is used for every UNIX process (virtual processor), each processor uses different data, and more, importantly, different schedules. The use of UNIX process facilities means that all of the UNIX process control features, such as suspending and resuming processes, are available for use with the systolic array simulator. All simulator processes are part of a *process group* and thus can be manipulated as a whole. There is also a master process which handles input from the user, and can start and stop virtual processors on command. The disadvantage of using UNIX processes is that there are a limited number of them. In the current implementation, the process limit is over 50. Since systolic algorithms scale, testing can be done on a small array. This UNIX limitation thus has not been too restrictive.

Interprocess communication is done using UNIX sockets, providing buffered byte stream communications between processes. The SDEF synchronous word protocol is created on top of this. External data is read and written using UNIX files.

Each simulator process has a window associated with it that displays internal data, and communication activity. Due to limited screen space, the largest array that can be displayed is a 6 × 6 grid. The use of display windows is optional; the simulator can be used without them.

15.6.3 Portability

Consider what would be necessary to port the SDEF translator and run-time system to an INMOS Transputer system. A Transputer system consists of a host computer connection to an array of processors. Each processor has four I/O channels that can be used to link them. The INMOS system provides a C compiler, and a library of routines used for I/O and inter-Transputer communication.

The SDEF translator would run on the host computer, and does *not* need to be modified. The code that it generates makes no reference to the underlying hardware or operating system. Instead, all system-dependent features are encapsulated in SDEF library routines. For example, one SDEF library routine, *read_word*, reads a word from a particular direction (north, south, east, or west). The code generated by the translator references routines in this *interface library*.

To port SDEF to a Transputer environment, the SDEF run-time system would need to be modified in two places. First, the UNIX calls that spawn processes and establish connections between them would be replaced by calls to the Transputer library routines providing these services. Second, the send and receive calls between processors would be changed from UNIX socket calls to Transputer library calls. Figure 15.19 depicts the paths for creating programs for the SDEF simulator, and for a Transputer array.

The SDEF system is designed to reduce the translator's dependence on particular hardware capabilities. The design accomplishes this by 1) using a commonly available language, C, as the target language for the translator, and 2) encapsulating hardware-specific code. These measures enhance SDEF's portability.

Although the translator is relatively hardware independent, the run-time system needs to be tailored for each implementation; the run-time system is responsible for the display of trace information, and maintaining the view of the array as an orthogonally-connected mesh.

15.7   Conclusions

The SDEF programming system increases the productivity of systolic algorithm researchers. It accomplishes this by providing a notation for specifying atomic sys-

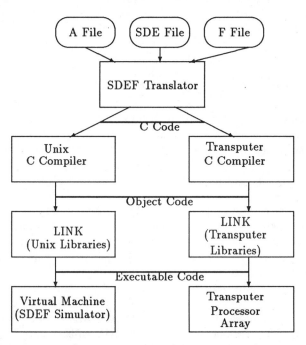

Figure 15.19: One path is for the SDEF simulator; the other, for a Transputer system.

tolic computations that is succinct, precise, and executable. The *communication* requirements are specified in terms of the algorithm's index space (S), domain dependencies (D), and spacetime embedding (E). Systolic algorithms that appear to be quite different (e.g., one operating on a linear array of processors, and another operating on a hexagonally-connected array) often differ *only* in their spacetime embedding (see, e.g., Cappello and Steiglitz[5]). By simply modifying the E property, researchers thus can re-use much of a previously tested algorithm. Re-use applies, as well, to the S, D, and F properties. This notation allows researchers to share their work for independent testing, and to dissect and re-use parts of others' work.

The SDEF programming system increases the productivity of systolic algorithm developers. There are several reasons for this. In a programming language such as Occam or W2, communication constructs are mixed with the program's computation constructs. In SDEF, the communication aspect of the array programming is done with a high level declarative language (the S, D, and E properties). The communication is *implied* by the spacetime embedding, relieving the programmer from issuing explicit send and receive commands within each processor's program. This is perhaps the most significant conceptual difference between SDEF, and W2, Occam, or Hearts. The *reliability* of a communication program is enhanced both by its declarative expression as high level properties, and by the user's ability to re-use previously tested property values. Indeed, the clean separation of communication programming from node computation programming enhances the reliability of both.

Like W2 and Hearts, SDEF provides a set of specialized tools for creating and modifying systolic array programs. The SDEF translator provides high level error detection. For example, it detects spacetime embeddings that cannot be executed due to I/O or size restrictions on the physical array. Users also are provided with a domain *type* database that is extensible. The translator helps users to create domain input files in such a way that they are consistent with the specified index space and spacetime embedding.

The SDEF systolic array simulator is built on top of UNIX's 'process' features. These features consequently are made available through the simulator. The simulator allows users to quickly and easily trace their systolic programs, inspecting their program's actual communication and computation characteristics. Each simulated processor can be started or stopped as desired, allowing observation of each I/O action and domain value.

As with sequential languages there is a tradeoff between ease of use (i.e., human productivity) and run-time efficiency. An SDEF program incurs more overhead than a carefully hand-coded one. But SDEF provides a notation that is hardware-independent and compact, yet executable. Moreover, the SDEF notation can support work by others on very high level systolic programming systems. The S, D, E, and F properties constitute an interface, not only between a human user and the SDEF translator, but also between a higher level system and the SDEF translator. There may be considerable advantage in doing so. For example, the SDEF translator creates the code to support external I/O from a declarative specification of the S, D, and E properties. A higher level language thus need only provide these properties, avoiding much detailed and complicated I/O scheduling. In this way, SDEF facilitates further software tool development for systolic array programming. One such tool might be a spacetime embedding optimizer, such as has been investigated by Li and Wah[22], for example.

Future work is contemplated in two areas. First, the SDEF programming system is designed for portability. All hardware-dependent code is encapsulated into a small call library. The system is especially portable to a Transputer environment. Indeed, one reason that arrays in SDEF are orthogonally-connected is to keep them compatible with the Transputer. The SDEF translator generates C code, and there is a C compiler for the Transputer. The SDEF translator, consequently, does *not* need to be modified to port the SDEF run-time system to an INMOS Transputer array.

Secondly, the programming system can be generalized with respect to the class of systolic computations that it treats. SDEF properties are to an atomic systolic computation, as atomic systolic computations are to a compound systolic computation. They are valuable ways to package its re-usable parts. Incorporating a *composition* capability thus is a natural enhancement to SDEF. Such a capability would permit atomic systolic computations to be bonded together, and reshaped with spacetime embeddings that are appropriate to the context of a complex computation.

References

[1] ALLEN, K. R., AND R. P. PARGAS, "On Compiling Loop Algorithms onto Systolic Arrays," TR #85-11-18, Clemson University, Dept. of Computer Science, 1985.

[2] ANNARATONE, M., E. ARNOULD, R. COHEN, T. GROSS, H-T KUNG, M. LAM, O. MENSILCIOGLU, K. SAROCKY, J. SENKO, AND J WEBB, "Architecture of Warp," *Proc. COMPCON*, Spring 1987.

[3] BRUEGGE, B., C. CHANG, R. COHEN, T. GROSS, M. LAM, P. LIEU, A. NOAMAN, AND D. YAM, "Programming Warp," *Proc. COMPCON*, Spring 1987.

[4] CAPPELLO, P. R., *VLSI Architectures for Digital Signal Processing*, Princeton University, Ph.D. Dissertation, Princeton, NJ, Oct. 1982.

[5] CAPPELLO, P. R. AND K. STEIGLITZ, "Unifying VLSI Array Design with Linear Transformations of Space-Time," in *Advances in Computing research, VLSI Theory*, ed. F. P. Preparata, vol. 2, pp. 23-65, JAI Press, Inc., Greenwich, CT, 1984.

[6] CHEN, M. C., "Synthesizing Systolic Designs," *Proc. Sec. Int. Symp. VLSI Technology, Systems, and Applications*, p. 209-215, Taipai, May 1985.

[7] CHEN, M. C., "A Parallel Language and Its Compilation to Muliprocessor Machines," *J. Parallel and Distributed Computing*, Dec. 1986.

[8] DELOSME, J. M. AND I. C. F. IPSEN, "Systolic Array Synthesis: Computability and Time Cones," Yale/DCS/RR-474, May 1986.

[9] FISHER, A. L., H-T KUNG , L. M. MONIER, Y. DOHI, "The Architecture of a Programmable Systolic Chip," *Journal VLSI and Computer Systems*, 1(2):153-169, 1984.

[10] FORTES, J. A. B. AND D. I. MOLDOVAN, "Parallelism detection and algorithm transformation techniques useful for VLSI architecture design;" *J. Parallel Distrib. Comput.*, May 1985.

[11] FORTES, J. A. B., K. S. FU, AND B. W. WAH, "Systematic Approaches to the Design of Algorithmically Spcified Systolic Arrays," *Proc. Int. Conf. on Acoustics, Speech, and Signal Processing*, pp. 300-303, Tampa, 1985.

[12] FOULSER, D. E., AND R. SCHREIBER, "The Saxpy Matrix-1: A General-Purpose Systolic Computer," *IEEE Computer*, 20(7):35-43, June 1987.

[13] HUANG, C. H. AND C. LENGAUER, "An incremental mechanical development of systolic solutions to the algebraic path problem," TR-86-28, Univ. of Texas, Dept. Computer Science, Austin, Dec. 1986.

[14] KAPAUAN, A., K. Y. WANG, D. GANNON, J. CUNY, AND L. SNYDER, "The Pringle: and experimental system for parallel algorithm and software testing," *Proc Int. Conf. on Parallel Processing*, 1984.

[15] KARP, R. M., R.E. MILLER, AND S. WINOGRAD, "Properties of a Model for Parallel Computations: Determinace, Termination, Queueing," *SIAM J. Appl. Math.*, 14:1390-1411, 1966.

[16] KARP, R. M., R.E. MILLER, AND S. WINOGRAD, "The Organization of Computations for Uniform Recurrence Equations, " *J. of the Assoc. for Comput. Machinery*, 14:563-590, 1967.

[17] KERNIGHAN, B. AND M. RITCHIE, *The C Programming Language*, Prentice-Hall, Inc. Englewood Cliffs, NJ, 1978.

[18] KUNG, H-T AND C. E. LEISERSON, "Algorithms for VLSI Processor Arrays," in *Introduction to VLSI Systems*, Addison-Wesley Publishing Co., Menlo Park, CA, 1980.

[19] KUNG, H-T, *A Listing of Systolic Papers,* Dept. of Computer Science, Canegie-Mellon Univ., Pittsburgh, Dec. 1986.

[20] KUNG, S-Y, K. S. ARUN, R. J. GAL-EZER, AND D. V. B. RAO, "Wavefront array processor: Language, Architecture and Applications", *IEEE Trans. on Computers*, C-31(11), May 1973.

[21] KUNG, S-Y, "On Supercomputing with Systolic/Wavefront Array Processors," *Proc. IEEE*, 1984.

[22] LI, G. J. AND B. W. WAH, "The Design of Optimal Systolic Algortihms," *IEEE Trans. on Computers*, C-34(1):66-77, 1985.

[23] MCCANNY, J. V. AND J. G. MCWHIRTER, "The derivation and utilization of bit level systolic array architectures, " *Int. Workshop on Systolic Arrays*, pp. F1.1-F1.12, Univ. of Oxford, July 1986.

[24] MIRANKER, W. L. AND A. WINKLER, "Spacetime Representations of Computational Structures," *Computing*, Vol. 32, 1984.

[25] MOLDOVAN, D. I., "On the Analysis and synthesis fo VLSI algorithms," *IEEE Trans. Comput.*, C-31:1121-1126, Nov. 1982.

[26] MOLDOVAN, D. I., "On the Design of Algorithms for VLSI Systolic Arrays, " *Proc. IEEE*, 71(1):113-120, Jan 1983.

[27] MOLDOVAN, D. I. AND J. A. B. FORTES, "Partitioning and Mapping Algorithms into Fixed Systolic Arrays," *IEEE Trans. on Computers*, C-35:1-12, 1986.

[28] MOLDOVAN, D. I., "ADVIS: A software Package for the Design of Systolic Arrays," *IEEE Trans. Computer-Aided Design*, CAD-6(1):33-40, Jan. 1987.

[29] NAVARRO, J. J., J. M. LLABERIA, AND M. VALERO "Partitioning: An Essential Step in Mapping Algorithms Into Systolic Array Processors," *IEEE Computer*, 20(7):77-89, June 1987.

[30] NELIS, H. W., AND E. F. DEPRETTERE "Methods and Tools for the Design and Partitioning of VLSI Systolic/Wavefront Arrays," TR, Dept. of Electrical Engineering, Delft University of Technology, 1987.

[31] QUINTON, P., "Automatic synthesis of systolic arrays from uniform recurrent equations," *Proc. 11th Ann. Symp. on Computer Architecture*, pp. 208-214, 1984.

[32] RAO, S. K., "Regular Iterative Algorithms and Their Implementation on Processor Arrays," *Ph.D. Dissertation*, Stanford University, Stanford, October 1985.

[33] SNYDER, L., "Parallel programming and the poker programming environment," *Computer*, 17(7):27-36, July 1984.

[34] SNYDER, L., "A Dialect of the Poker Programming Environment Specialized for Systolic Computation," *Proc. Int. Workshop on Systolic Arrays,* Univ. of Oxford, July 1986.

[35] SNYDER, L. AND D. SOCHA, "Poker on the Cosmic Cube: The First Retargetable Parallel Programming Language and Environment," *Proc Int. Conf. on Parallel Processing,* pp. 628-635, St. Charles, IL, August 1986.

Chapter 16

Cyclo-Static Realizations, Loop Unrolling and CPM: Optimal Multiprocessor Scheduling [1]

D. A. Schwartz [2]

Abstract

Optimal, deterministic scheduling, for a class of iterative/recursive problems is explored. The connection between the program transformation technique of *loop unrolling* and the blocking of a related acyclic graph for the determination of CPM schedules is developed. It is shown that loop unrolling and blocking can increase the parallelism of the realization by allowing the overlapped execution of successive iterations. Optimal overlapping is based on simple bounds for a related cyclic graph. Cyclo-static realizations that can achieve the optimal overlapped schedules are then introduced.

16.1 Introduction

This paper is intended as a tutorial introduction to an optimal technique for the deterministic scheduling of synchronous multiprocessors for a class of iterative/recursive algorithms. Specifically, it presents *cyclo-static* realizations [1]–[3] which are based on an optimal, non-preemptive, periodic scheduling method developed by the author. A detailed example will be presented to show the connection between cyclo-

---

[1] This work supported by JSEP contract DAAL0387K0059
[2] Georgia Institute of Technology, School of Electrical Engineering

static realizations and *loop unrolling* (a technique for exposing parallelism that is used by some compilers for vector or parallel processing computers). Additionally, it is hoped that this will clear up some widely held misunderstandings about the use and application of the *critical path method* (CPM) as a technique for measuring parallelism of algorithms.

After motivating the existence of an optimal periodic schedule, two important bounds on the general optimality and existence of realizations based on some simple graph theoretic concept will be presented. Cyclo-static realization, which can achieve these bounds, will be introduced by presenting it's properties and the fundamentals needed to determine such realizations.

It should be noted that cyclo-static realizations were developed in the context of the application of *fine grain* synchronous processors to real-time filtering and related algorithms in digital signal processing. However, the techniques apply to any grain size and are relevant to many other problems including optimizing the code generated for inner loops of scientific code for computers that support instruction level parallelism (i.e. Multiflow Trace/28).

16.2   CPM

16.2.1   Loop Unrolling/Blocking

The goal of this section is to find a deterministic, minimum time, realization of an iterative/recursive algorithm. It is assumed (initially) that an infinite number of homogeneous, synchronous, processors are available. What is desired is to find a *processor schedule*, which is a specification of what operation each processor performs at a given time instant. We will further constrain the schedule to be *non-preemptive*, in that once an operation starts on a processor, it must run to completion. In other word, you can not perform the first half of a multiplication on one processor and the second half on a different processor or the same processor at a latter time. Subject to these constraints, the secondary goal is to use the minimum number of processors possible to achieve the minimum time. This is equivalent to maximizing processor utilization, assuming that only the required number of processor is present.

To provide an example as a frame work for analysis we will consider the solution of Van der Pol's equation, expressed as two first order ordinary differential equations.

$$\dot{y}_1 \;=\; y_2 \tag{16.1}$$
$$\dot{y}_2 \;=\; \mu(1 - y_1^2)\dot{y}_2 - y_1 \tag{16.2}$$

This type of initial value problem can be easily solved through the conversion to an equivalent integral equation and the application of a numerical integration method.

For this example, a simple *parallel predictor-corrector* will be used. The formula is given by

$$y_{n+1}^p = y_{n-1}^c + 2hf_n^p \tag{16.3}$$

$$y_n^c = y_{n-1}^c + (h/2)[f_n^p + f_{n-1}^c] \tag{16.4}$$

where y_n^p and y_n^c are the predicted and corrected estimates of y at step n, h is the time step size and f_n^p and f_n^c represent $f(x_n, y_n^c)$ and $f(x_n, y_n^p)$, respectively.

This results in the following pseudo-code program to solve Van der Pol's equation.

```
h2 = H*2.0;
h_2 = H/2.0;
time = TIME_START;
while (time < TIME_MAX) {
    tmp1 = yc1 + h2*yp2;
    tmp2 = yc2 + h2*(u*(1.0 - yp1*yp1)*yp2 - yp1);
    tmp3 = yc1 + h_2*(yp2 + yc2);
    tmp4 = yc2 + h_2*(u*(1.0 - yp1*yp1)*yp2 - yp1 +
        u*(1.0 - yc1*yc1)*yc2 - yc1 );
    yp1 = tmp1; yp2 = tmp2; yc1 = tmp3; yc2 = tmp4;
    y(n) = yp1;
    n = n+1;
    time = time + H;
}
```

The program lines in italics are included for clarity, but will be ignored for simplicity in the following analysis.

Fig. 16.1 is a *directed acyclic graph* (DAG) representation of the inner loop of the above program. The nodes of the graph correspond to *atomic* operations of the underlying processing elements, while the branches denote the direction of information flow and the precedence constraints between operations. Examining the graph reveals that the operations corresponding to nodes 12-13-14-15-16-11-17-18 must be performed sequentially. Assume that all numerical operations have a computational *latency* of one unit of time (1 u.t.). Where 1 u.t. corresponds to one clock cycle of the systems synchronous clock (in general they can be an arbitrary integral number of units of time). In this case it must take at least 8 u.t. to perform the sequence of operations 12- $\cdots$ -18. For this graph, this is the longest sequential path and is referred to as the *critical path*. The length or latency of a critical path is the minimum possible time in which the operations of a DAG can be performed with parallel resources since sequential paths must be performed sequentially. The graph in Fig. 16.1 contains 20 nodes of unit latency resulting in a total of 20 u.t. of computational latency. Since the critical path is of length 8 u.t. a lower bound on the number of homogeneous processors needed to achieve the critical path bound

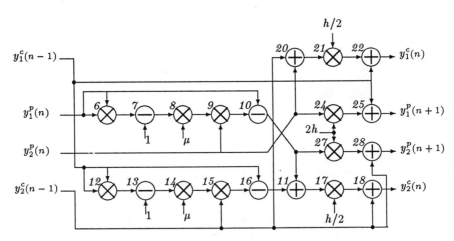

Figure 16.1: A DAG realization of the inner loop of the program for Van der Pol's equation.

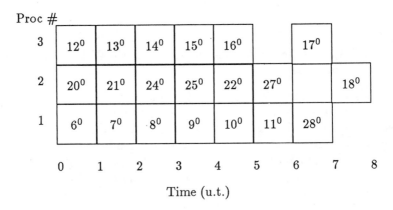

Figure 16.2: Optimal CPM Schedule for DAG in Fig. 16.1.

is[1] $\lceil 20$ u.t./8 u.t.$\rceil = 3$. In general the minimum number of processors, P_{min} is equal to the total latency, D, divided by the length of the critical path, t_{crit}.

$$P_{min} = \left\lceil \frac{D}{t_{crit}} \right\rceil \qquad (16.5)$$

In general P_{min} is not achievable, due to precedence constraints, and a larger number of processors is required to achieve the critical path bound. However, for this graph a three processor solution that achieves the critical path bound exists. One such possible solution (not unique), or schedule, is shown in Fig. 16.2. The schedule is in a form known as a *Gantt Chart*, where the vertical axis of the table corresponds to processor number and the horizontal axis corresponds to units of time. Successive iterations correspond to initiating a shifted copy of the schedule after the completion of the previous iteration. The next iteration would start at time 8. There is no concurrency between iterations.

Before we draw any conclusions from this a few comments about the method of determining the schedule are needed. Given a set of atomic operation, a DAG specifying an algorithm and P processors, the general question of finding the optimal schedule (or assignment of operations to processors) is referred to as the *scheduling problem* and is well known to be an *NP-complete* problem. This implies that the optimal solution is computationally intractable for cases with more than a few nodes. To handle this difficulty a class of heuristic algorithms based on exploiting the knowledge that the operations on the critical path must be scheduled sequentially without other intervening operations have been developed. In general they are referred to as *critical path methods* or more commonly CPM. While CPM is in general suboptimal, it runs in polynomial time and often finds the optimal solution.

For further details on CPM the reader is referred to the literature [4]–[7]. All of the CPM schedules in this paper were generated by a variant of the *list scheduling* algorithm.

At this point we conclude that 8 u.t. is the minimum time in which the DAG in Fig. 16.1 can be computed and that it can exploit a maximum parallelism of three processors. While this is true, it is not true of the underlying algorithm specified in the program code! Consider the program transformation of *unrolling* the inner loop so that two sequential iterations are specified in the loop.

```
while (time < TIME_MAX) {
    yp1_0 = yc1_1 + h2*yp2_1;
    yp2_0 = yc2_1 + h2*(u*(1.0 - yp1_1*yp1_1)*yp2_1 - yp1_1);
    yc1_0 = yc1_1 + h_2*(yp2_1 + yc2_1);
    yc2_0 = yc2_1 + h_2*(u*(1.0 - yp1_1*yp1_1)*yp2_1 - yp1_1 +
        u*(1.0 - yc1_1*yc1_1)*yc2_1 - yc1_1 );
```

---

[1] $\lceil x \rceil$ denotes the 'ceiling' of x, the smallest integer greater than or equal to x.

```
 yp1_1 = yc1_0 + h2*h*yp2_0;
yp2_1 = yc2_0 + h2*(u*(1.0 - yp1_0*yp1_0)*yp2_0 - yp1_0);
yc1_1 = yc1_0 + h_2*(yp2_0 + yc2_0);
yc2_1 = yc2_0 + h_2*(u*(1.0 - yp1_0*yp1_0)*yp2_0 - yp1_0 +
    u*(1.0 - yc1_0*yc1_0)*yc2_0 - yc1_0 );
```
$y(n) = yp1\_0;$
$y(n+1) = yp1\_1;$
$n = n+2;$
$time = time + 2*H;$
}

This program has a corresponding DAG where there are two copies or *blocks* of the DAG depicted in Fig. 16.1. The iteration outputs of the first block feed the corresponding iteration inputs of the next block. This is shown in Fig. 16.3. This *blocked by two* DAG has a critical path of nodes

$$12^0 - 13^0 - 14^0 - 15^0 - 16^0 - 11^0 - 17^0 - 18^0 - 15^1 - 16^1 - 11^1 - 17^1 - 18^1$$

(where the superscript denotes block number). The latency of this critical path is 13 u.t., and $P_{min} = 4$. Fig. 16.4 shows a schedule that achieves the critical path bound with only P_{min} processors. While the latency of the critical path is 13 u.t., the outputs for two successive iterations has been produced. This results in an average latency between successive outputs of $13/2 = 6.5$ u.t., compared to an average latency of 8 u.t. for the original unblocked graph. Note that by unrolling the loop by two we have uncovered more parallelism and decreased the time between successive outputs!

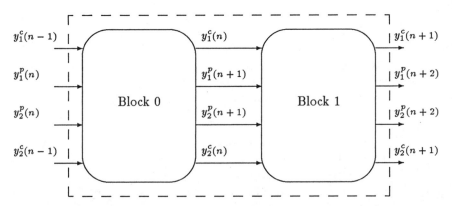

Figure 16.3: A *blocked by two* DAG realization corresponding to an *unrolled by two* inner loop of the program for Van der Pol's equation. (Each block is identical to the DAG in Fig. 16.1).

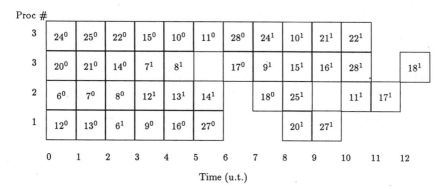

Figure 16.4: Optimal CPM Schedule for blocked by two DAG in Figure 16.3.

This naturally suggests considering unrolling the loop further (increasing the blocking factor). Table 16.1 show the effect of further loop unrolling on the optimal CPM schedule of the blocked graph. From the table it can be seen that all of these CPM schedules achieve the critical path bound with the minimum number of processors. Therefore, all of the CPM schedules are optimal schedules of the corresponding DAG's. Unrolling the loop by two increased the parallelism and the number of processors that could be used in parallel. Further unrolling yielded small increases in parallelism, but not enough to require an additional parallel processor. These increases in parallelism decreased the average latency between successive outputs resulting in faster execution. The average latency appears to be asymptotically approaching a limit and in fact in section 16.3 it will be shown that the limit is an average latency of 5.5 u.t. The lower limit on average latency is referred to as the *iteration period bound*, and it will be shown that it is simple to determine.

Table 16.1: CPM bounds for different blocking factors.

Block size	t_{crit} (u.t.)	P_{min}	P	Avg. latency (u.t.)
1	8	3	3	8.00
2	13	4	4	6.50
4	24	4	4	6.00
8	46	4	4	5.75
16	90	4	4	5.63
25	140	4	4	5.60

Why does blocking result in a lower average latency? Fig. 16.5 shows a portion of the optimal CPM schedule for the loop unrolled by eight case. At time 18 and 23 it can be seen that operations corresponding to three successive iterations are being executed in parallel, and at most times operations corresponding to two successive iterations are present. Within the unrolled portion of the loop it is possible to *overlap* the execution of successive iterations. Successive iterations of a CPM schedule start after the completion of the previous iteration. Thus in a loop with no unrolling there is no opportunity to overlap successive iterations.

All of these gains are not without penalty. There is a significant increase in scheduling/compiling time as the unrolling factor increases and there is the obvious penalty of increased 'code' size.

It is important to note that a monotone decrease in the average latency as a function of the block size does not hold true for all graphs. Many graphs achieve the iteration period bound with a small block size of L. A block size of $M = kL$ would have the same average latency, while a block size of $M \neq kL$ would have a larger average latency. Lee [8] also showed that for some graphs there does not exist a finite block size that can achieve the iteration period bound. Additionally, Lee investigated the question of optimum block size for CPM schedules and concluded that there is no current known solution to the problem. While this is true, practical guidelines will be presented. More importantly, section 16.4 will present *cyclo-static* realizations which achieve the iteration period bound without the penalty of a large 'code' size.

16.2.2   Cyclic Graphs

Up to this point we have been examining *directed acyclic graphs* (DAGs) and have examined the issue of what is the *critical path bound* or minimum average latency. Different block sizes or loop unrolling factors gave rise to different answers to the question. This may appear somewhat odd when we consider that loop unrolling does not modify in any way the purely sequential execution of the underlying algorithm, except that the looping control construct is executed a fewer number of times.

Consider an inner loop that executes a large number of iterations. It can be treated as being effectively a semi-infinite loop. In this case the inner loop can be modeled by a *directed cyclic graph* (DCG, graph with directed loops). Memory locations that are written in one iteration and read in the next iteration are the *iteration/recursion variables* of the loop. These are represented in the cyclic graph by *ideal delays* which are denoted symbolically[2] as z^{-1}. In the first program example, the ideal delays correspond to the variables **yp1, yp2, yc1,** and **yc2**. Similarly, a memory location

---

[2] Z-transform notation from signal flow graphs.

Proc #

#														
3	25^3	27^3	20^3	24^4		10^4	13^4	28^4	8^5	9^5	12^5	13^5	18^4	8^6
3	10^3	18^2	28^3	8^4	9^4	12^4	18^3	7^5	15^4	16^4	11^4	6^6		28^5
2	13^3	14^3	7^4	21^3	22^3	17^3	27^4	20^4	21^4	22^4	25^5	27^5	7^6	15^5
1	17^2	6^4	15^3	16^3	11^3	25^4	6^5	14^4	24^5		10^5	17^4	14^5	20^5
	17	18	19	20	21	22	23	24	25	26	27	28	29	30

Time (u.t.)

Figure 16.5: Portion of an optimal CPM Schedule for a blocked by eight DAG of the program for Van der Pol's equation.

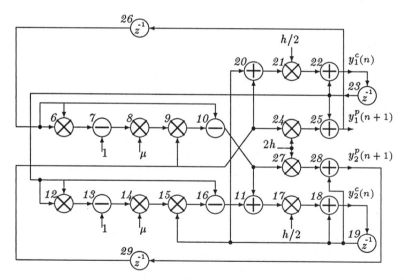

Figure 16.6: A DCG realization of the inner loop of the program for Van der Pol's equation.

that is written in iteration j and then read in iteration $j + n$ corresponds to an *nth* order ideal delay, or z^{-n}.

By treating the inner loop as being semi-infinite we can now model the algorithm by a DCG. Fig. 16.6 illustrates a cyclic graph that models the inner loop of the Van der Pol's equation program. This DCG forms the basis of the blocked DAG's and of the optimal solution that will be presented.

The blocked DAG's can be derived from this DCG by the following algorithm. The first part of the algorithm converts the DCG into a *direct blocked by L* DCG. Direct blocking refers to transforming the original graph that defines a single (set of) output(s) to a form that defines a block of L successive (sets of) outputs. The second part transforms the direct blocked DCG into a blocked DAG.

Let the original DCG be represented as $G = < V, B >$, where V is the set of nodes (or vertices) and B is the set of directed branches of the graph. A branch $(i, j) \in B$ denotes that there is a directed branch from node i to node j. *Direct blocking by L* of graph G results in the DCG G_L

> *Make L copies of G;*
> *The kth copy of the graph is denoted $G^k = < V^k, B^k >$;*
> *let $G_L = < \bigcup_{k=0}^{L-1} V^k, \bigcup_{k=0}^{L-1} B^k >$;*
> **for** *all (nodes) $m \in \{ideal\ delay\ nodes\}$* {
>   **for** k = 0 **to** L-1 {
>   n = *the order of node m^k;*
>   **if** n <= k **then** {
>   *replace node m^k with a dummy node;*
>   *delete the input branch (l^k, m^k);*
>   *add the branch (l^{k-n}, m^k);*
>   } **else** {
>   *replace node m^k with an order $\lceil n/L \rceil$ ideal delay;*
>   *delete the input branch (l^k, m^k);*
>   *add the branch $(l^{(k-n) \bmod L}, m^k)$;*
>   }
>   }
> }

The direct blocked DCG can then be trivially converted to a blocked DAG by transforming all of the ideal delay nodes into a pair of *write, read* nodes. The input branch to the delay element is now the input to a *write* node and the output branches from the delay element now come from a *read* node. Note that the delay element has been split into two nodes without a direct interconnecting branch. Since all loops of a DCG must contain delay elements in order to be realizable the result of the transformation is to break or eliminate all loops. Thus the resulting graph is loop free or *acyclic*.

We now have a method of performing the loop unrolling or blocking of a DAG, and all loop unrolling factors or block sizes have a correspondence to an original DCG. Therefore the question of what is the shortest possible iteration period (or equivalently, what is the maximum parallelism) is a property of the DCG, while the length of the critical path is only a property of a specific block size DAG. This a particularly important point with respect to algorithms that are defined in terms of DCG's, particularly signal flow graphs for digital signal processing algorithms. The literature is full of algorithms that are claimed to be 'better' than other algorithms for highly parallel realizations based on the fact that the critical path of the equivalent unblocked DAG is shorter. However if the DCG had been examined directly, or a larger block size DAG, the 'new' algorithm is often seen to be equivalent or worse than the object of comparison.

16.3   Flow Graph Bounds

By extending the DCG model specified earlier it will be possible to determine the lower bound on the iteration period and the parallelism or minimum number of processors to support a given iteration period. These are properties of realizations based on the underlying cyclic graph. Formally, the iterative/recursive algorithms are defined by a DCG referred to as a *fully specified flow graph* (FSFG), which is very similar to the DAG model that CPM was introduced with.

A FSFG is a shift-invariant flow graph in which the nodal operations are additionally constrained to be the atomic operations of the underlying processors which are to be used in the realization. Thus the atomic operations represent the smallest granularity at which parallelism may be exploited. Specifically, if the underlying processor has a two input (binary) addition operation and a three input summation is required, then the FSFG would represent the three input summation as two separate nodes of two input addition in cascade. Note additionally that if the adder were a static four stage pipelined adder, the adder would be considered indivisible and the four stage pipeline would constitute a single atomic operation (as opposed to four atomic operations).

Consider a FSFG that continuously processes an infinitely long input sequence of data to produce a corresponding infinitely long output sequence of data. It is possible to determine a tightest lower bound on the iteration period (reciprocal of sampling rate), hereafter referred to as the *iteration period bound*. The iteration period bound is a property of the FSFG and of the operational latency of the nodal operations of the graph and is independent of the architecture of the realization. This technique was developed in terms of single time index filters, but applies to any iterative algorithm that might be characterized by a representation, in FORTRAN, containing nested 'DO' loops. However, the interpretation of the iteration period is algorithm dependent in the general case.

The notation with which the iteration period bound is presented is that of Renfors and Neuvo [9]; however this is a reformulation of a result originally due to Fettweis [10]. In a different form, a similar result for the iteration period bound of a SSIMD realization was independently reported by Barnwell and Hodges [11]. The result due to Fettweis was originally applied to asynchronous implementations. Here we are concerned only with synchronous realizations and will extend Fettweis' bound to the synchronous case.

The basic assumptions are that the algorithm to be implemented is represented as a FSFG, and that the *latency* (computation times) for all the nodal operations are known. Nothing is assumed about the communications structure, I/O constraints, or other details of the realization. While specific characteristics of a given architecture may not allow a realization of a specific FSFG at its iteration period bound, the resulting bound reflects the absolute limits on computation rate for the set of all possible architectures based on processing elements with the given atomic latencies.

The *iteration period bound* (IPB), is the minimum achievable latency between iterations of the algorithm. If the algorithm is a digital filter operating on a time series, this translates into the minimum achievable sample period. The iteration period bound is simply computed as follows. First, all the loops in the FSFG are identified. Then the operational delay around each loop, D_ℓ, is computed as

$$D_\ell = \sum_{j \in \ell} d_j \qquad (16.6)$$

where ℓ is the loop index and d_j is the computational delay of the jth node of the ℓth loop. The iteration period bound is then given by

$$T_o = \max_{\ell \in loops} \left\{ \frac{D_\ell}{n_\ell} \right\} \qquad (16.7)$$

where n_ℓ is the total number of delay elements in loop ℓ. Any loop for which $T_\ell = D_\ell/n_\ell = T_o$ is considered to be a *critical loop*.

The iteration period bound is a lower bound for both asynchronous and synchronous systems. In synchronous systems all events must be synchronized with the system clock. Thus in a synchronous system the iteration period bound must be an integral multiple of the system clock (clock period = 1 unit of time (u.t.)). The iteration period bound for a synchronous system is then given by

$$T_{s_o} = \left\lceil \max_{\ell \in loops} \left\{ \frac{D_\ell}{n_\ell} \right\} \right\rceil \qquad (16.8)$$

In this paper, it is implied that the system is synchronous and that the iteration period bound is given by equation (2b). For simplicity, the 's' subscript will be dropped.

Alternatively, the full parallelism implied by equation (16.7) can be achieved by the *direct blocking* of the FSFG by applying the algorithm presented in section 16.2.2.

Let D_c and n_c be the latency and order of delay in the critical loop. Direct blocking by L results in a FSFG where the latency of the critical path is $D_c' = LD_c$, and the order of the delay is unchanged, $n_c' = n_c$. Thus the iteration period bound is increased by a factor of L. However L sets of output are produced resulting in an average iteration period bound that is equal to the iteration period bound for the unblocked case. Direct blocking does not alter the parallelism of the algorithm. However, by choosing L such that n_c divides LD_c, the block iteration period can be made to be an integral number of clock cycles. Thus the full parallelism can be achieved with a synchronous system (this will be shown by an example in section 16.4).

In this computation, ideal delay elements (z^{-1}) are assumed to have zero computational delay. However, extra nodes can be introduced into the FSFG to model the processing delay of an ideal delay, or communications delay, as necessary. If D is the total (sequential) computational delay of all the nodes,

$$D = \sum_{j \in \mathbf{A}} d_j \tag{16.9}$$

where $\mathbf{A}$ is the set of all node indices and dj is the computational delay of the jth node, then the lower bound on the number of processors required to achieve an iteration period of T, is given by

$$P \geq \lceil D/T \rceil \tag{16.10}$$

This result follows very simply. If one iteration is computed every T seconds, and a total of D seconds of computations must be performed, then the average number of parallel operations is D/T. The average number of parallel operations is the lower bound on the number of required processors. Therefore an iteration period of T can never be achieved with fewer than P processors. The ceiling function is required since fractions of processors are not meaningful in a single task environment.

However, this is not a tightest lower bound on the number of processors required to achieve an iteration period of T. For most FSFGs, realizations that achieve strict equality exist. For some FSFG's, as T approaches T_o, the minimum number of processors increases. A precise determination of the tightest lower bound for an arbitrary FSFG and for any T appears to require a combinatorial search based on postulating a lower bound and searching to determine if a compatible realization exists. For cyclo-static realizations where strict equality holds, P is also referred to as the *maximum parallelism* of a graph.

Since T_o is the iteration period bound, the maximum number of processors that can be used with optimal efficiency to realize a FSFG is

$$P_o \geq \lceil D/T_o \rceil$$

It is conjectured that equality holds for most graphs of interest. Therefore, the *processor bound* can be defined as

$$P_o = \lceil D/T_o \rceil \qquad\qquad (16.11)$$

Note: when the equality condition does not hold, P_o is used for the initial postulation of the tightest lower bound on the number of processors. Also note the similarity and connection to equation (16.5).

16.3.1   Optimality

In the context of this work, any realization that achieves an iteration period equal to the iteration period bound is considered to be *rate optimal*, since no faster realization of the underlying graph can exist. Similarly any realization that uses the minimum number of processors, to support an iteration period of T, is considered *processor optimal* (processor efficiency optimal). For some positive Δ, if $T \geq T_o + \Delta$, then the minimum number of processors is $\lceil D/T \rceil$. When $D/T \neq \lceil D/T \rceil$ then on the average there is less than one processor not being 100% utilized. Since no realization (in a single task environment) exists that can achieve an iteration period of T, with less than $\lceil D/T \rceil$ processors, the realization is considered processor optimal even though the processor efficiency may be less than 100%. If $T_o \leq T < T_o + \Delta$, then the minimum number of processors, P, is greater than $\lceil D/T \rceil$. In these cases the processor efficiency, $100\%(PT/D)$, although potentially small, is still considered processor optimal. The difference between 100% and the true processor efficiency is called the *inherent processor inefficiency*. However, as previously mentioned, it appears that for most graphs Δ is is zero. When Δ is non-zero it is usually small relative to T. Therefore choosing $T = T_o + \Delta$ results in ideal processor efficiency and a near rate optimal realization.

If a realization achieves an 'input to output' delay (throughput time) equal to the delay bound, then it is considered *delay optimal*.

Of course, many things besides the structure of the algorithm and the fundamental operational capability of the processors may limit implementations. Clearly, issues such as I/O bandwidths, external resource availability, the number of available processors, and the communications architecture may impact the achievable rate, delay, and processor efficiency of an algorithm. But in its own way, each of these aspects can be addressed and corrected. For a particular multiprocessor system (or a particular VLSI cell library) and a particular FSFG, the bounds described above are fundamental. Hence, if implementations can be developed which achieve these bounds, then it is clear that no other implementations exist which can operate at a higher rate, with less delay, or with higher efficiency. It is this class of optimal implementations which will be generated for cyclo-static systems.

16.3.2 FSFG Bounds, CPM and Inner Loops

The FSFG bounds and optimality criteria are most powerful in producing optimal realizations of filter-like, semi-infinite, recursions. As stated previously, when the inner loop has has a large number of iterations it can be treated as being effectively semi-infinite and therefore compatible with the FSFG bounds. The FSFG model does not support data dependencies, however some simple data dependencies can still be handled. Loop exceptions of the form of conditional loop exits present no difficulties except for a possible efficiency penalty if the loop exits too early. Simple data dependent execution paths can be modeled by executing all paths in parallel with a simple *mask vector* control that only allows the active path to alter the system state. This of course carries obvious penalties in terms of total efficiency, however these may be overridden by the benefits of deterministic schedules.

In terms of directly applying the bounds to CPM realizations, if the critical path is greater than the iteration period bound, increasing the block size factor should be attempted. If the iteration period bound is achieved, then no better solution can exist, so it is optimal. If the critical path increases it is of the case that CPM achieves a minimum critical path for block sizes of $M = kL$, and further block size increases will not result in improvements. If no CPM solution appears to come suitably close to the optimal bounds then another approach is needed. When $T_{s_o} \neq T_o$, a blocking factor that is an integer multiple of the direct blocking factor needed to achieve an integral value for T_o is often effective for achieving maximal parallelism.

It is stated without proof that if all operations are of unit length, then a globally optimal CPM solution is guaranteed to exist. However, it may be necessary to apply a *retiming* technique to the DAG, such as that of Leiserson [12]. If all operations are not of unit length, usually globally optimal CPM solutions will not exist, however they may be nearly optimal.

16.3.3 Example

This section will present examples of the bounds by continuing to explore the Van der Pol's equation program. The FSFG defining the algorithm is shown in Fig. 16.6. As in the examples for the CPM case assume that all operations have a latency of 1 u.t. (for simplicity). The ideal delay elements have an operational latency of 0 u.t. This graph contains a large number of loops, only a representative set of loops, including the critical loop, will be enumerated below. In general it is necessary to systematically examine all simple loops.

loop 1 $22 \rightarrow 23 \rightarrow 22$:
$$T_1 = 1 \text{ u.t.}/1 = 1 \text{ u.t.}$$

loop 2 $12 \rightarrow 13 \rightarrow 14 \rightarrow 15 \rightarrow 16 \rightarrow 11 \rightarrow 17 \rightarrow 18 \rightarrow 19 \rightarrow 20 \rightarrow 21 \rightarrow 22 \rightarrow 23 \rightarrow 12$:

$$T_2 = 11 \text{ u.t.}/2 = 11/2 \text{ u.t.}$$

loop 3 $6 \rightarrow 7 \rightarrow 8 \rightarrow 9 \rightarrow 10 \rightarrow 27 \rightarrow 28 \rightarrow 29 \rightarrow 24 \rightarrow 25 \rightarrow 26 \rightarrow 6$:

$$T_3 = 9 \text{ u.t.}/2 = 9/2 \text{ u.t.}$$

loop 4 $6 \rightarrow 7 \rightarrow 8 \rightarrow 9 \rightarrow 10 \rightarrow 11 \rightarrow 17 \rightarrow 18 \rightarrow 19 \rightarrow 20 \rightarrow 21 \rightarrow 22 \rightarrow 23 \rightarrow 25 \rightarrow 26 \rightarrow 6$:

$$T_4 = 12 \text{ u.t.}/3 = 4 \text{ u.t.}$$

Therefore

$$
\begin{aligned}
T_o &= \max\{T_1, T_2, T_3, T_4, \ldots\} \\
&= \max\{1 \text{ u.t.}, 11/2 \text{ u.t.}, 9/2 \text{ u.t.}, 4 \text{ u.t.}, \ldots\} \\
&= 11/2 \text{ u.t.}
\end{aligned}
$$

and loop 2 is the constraining or critical loop.

There are 20 computational nodes in the graph, each of unit latency. Therefore the total latency, D, is 20 u.t. The lower bound on the minimum number of processors is

$$
\begin{aligned}
P_o &= \lceil D/T_o \rceil \\
&= \lceil 20 \text{ u.t.}/(11/2) \text{ u.t.} \rceil = 4
\end{aligned}
$$

Since T_o is not equal to an integral number of synchronous clock cycles it can not be directly achieved in a synchronous system. Either the iteration period must be rounded up to the next integer, in this case an iteration period of 6 u.t., or it must represent an average iteration period per output. If the FSFG in Fig. 16.6 were direct blocked by two, the total computational latency and the latency of the critical loop would double. In this case the iteration period bound would be given by $T_o = \lceil 22 \text{ u.t.}/2 \rceil = 11$ u.t. However each iteration would produce two sets of outputs for an average iteration period of 11/2 u.t. The parallelism would remain unchanged ($P_o = \lceil 40 \text{ u.t.}/11 \text{ u.t.} \rceil = 4$).

Since the CPM schedules of the DAG's all correspond to direct blocking of the FSFG, whose properties are unchanged by blocking, the above bounds apply to all possible CPM schedules are bounds on specific realizations. Note that the above bounds match those suggested by Table 16.1.

16.4   Cyclo-static Realizations

The fundamental weakness of the CPM method is that it is based on the acyclic graph and not the cyclic graph. Cyclo-static realizations are determined directly

Proc #

Proc											
3	24_1^0		25_1^0	6_1^1		8_1^1	9_1^1		27_1^1		28_1^1
2	8_1^0	9_1^0	10_1^0	27_1^0	7_1^1	28_1^0	24_1^1	10_1^1	25_1^1	6_2^0	7_2^0
1	21_0^1	22_0^1	12_1^0	13_1^0	14_1^0	15_1^0	16_1^0	11_1^0	17_1^0	18_1^0	20_1^1
0	15_0^1	16_0^1	11_0^1	17_0^1	18_0^1	20_1^0	21_1^0	22_1^0	12_1^1	13_1^1	14_1^1
	0 1	2	3	4	5	6	7	8	9	10	

Time (u.t.)

Figure 16.7: Optimal Cyclo-static processor schedule for FSFG in Fig. 16.6.

from the properties of the cyclic graph and allow for the optimal overlapped execution of successive iterations. Cyclo-static realizations apply to iterative/recursive problems and are characterized by a schedule (or program) for one iteration that is periodically repeated. They are an extension of periodic scheduling, or classical multifunction static pipeline scheduling with a constant latency (fixed period). The schedule of an iteration repeats every iteration period. In a static, or classical schedule (i.e. CPM), the initiation of the next iteration of computation can be represented by the initiation schedule of the previous iteration, shifted in time by one iteration period. There is no overlap of iterations. The schedule of the previous iteration completes before the schedule of the next iteration starts. In a cyclo-static schedule the next iteration can be represented by the schedule of the previous iteration, shifted in time by the iteration period and additionally by a fixed shift, or displacement, in the processor space.

A *processor schedule* will refer to a complete description of the operations each physical processor will perform at each time instance. In contrast, a *schedule* will be a list of all operations that must be performed at each time instant independent of the physical processors. In other words, a schedule is a processor schedule without processor assignment. Since the problems under consideration are iterative with a constant iteration period, it is only necessary to specify the processor schedule information for one iteration period. The processor schedule for one iteration can be diagrammed in a manner related to the Gantt charts which were used to represent the processor schedules for the CPM solutions. Fig. 16.7 illustrates a processor schedule[3] for an optimal four processor cyclo-static realization of the Van der Pol's equation program. Again for simplicity (and consistency), all of the operations were chosen to be of unit length, but in general the length of the operations can be any number of clock cycles.

---

[3] Generated by a FSFG compiler developed by H. G. Forren.

Unlike the previous Gantt charts which depicted the schedule of a single iteration of the task, this representation [13] additionally depicts the relationship of successive iterations in a minimal form. As before, the superscript denotes the block number of the direct blocked graph. The subscript denotes the iteration index. This schedule is effectively a time slice of width equal to the iteration period, 11 u.t., that shows all of the concurrent activity related to the execution of the (three) overlapped iterations.

As shown is section 16.3.3, for the Van der Pol's equation program, the iteration period bound was 11 u.t., and the processor bound was four. It was also shown that to achieve the iteration period bound that it was necessary to direct block the FSFG by an integral multiple of two. Since the smallest blocking size results in the lowest cost control structure (or smallest memory requirement), a direct blocked by two FSFG leads to an optimal schedule. Fig. 16.7 shows a schedule that uses the minimum number of processors (four) to support an average iteration period equal to the iteration period bound (11/2 u.t .), since blocking by two produces two sets of outputs per iteration. At time 0 u.t., node 15 of block 1 of the graph[4] is executed on processor 0 and corresponds to an operation of iteration 0. Before the rest of the details needed to interpret the processor schedule are presented it is necessary to present some further background information.

The time dimension of the processor schedule is represented by a discrete time index. Since the processor index and the time index are discrete, the two dimensional space is a lattice which will be denoted as $\mathbf{P} \times \mathbf{T}$. A processor schedule is a pattern in the processor-time space ($\mathbf{P} \times \mathbf{T}$).

If the processors of the system are arranged as a two dimensional array then the processor subspace, $\mathbf{P}$, is two dimensional. In general, the processor subspace is multi-dimensional. The indexing of processors is performed modulo the cardinality of each processor dimension. For example, let $\mathbf{P}$ be one dimensional and of cardinality four (four physical processors), then the processor indexed by zero is the same processor as that indexed by four. For a one dimensional processor subspace, the processor-time lattice is an infinitely long 'cylinder.'

The primary difference between cyclo-static processor schedules and traditional deterministic schedules is that they contain an extra degree of freedom. Successive iterations are not only overlapped and shifted in time, but they are also shifted in processor space by a fixed displacement. The displacement in the lattice between the execution of the same node of the FSFG in successive iterations is conveyed by a vector, $\mathcal{L}$, termed the *principal lattice vector*. Each successive iteration is identical to the previous iteration, but shifted in time by the time component of the principal lattice vector, and shifted in space by the processor component of the principal lattice vector.

---

[4] The first block (iteration) are referred to as block (iteration) 0, not 1.

In terms of the concepts above, a cyclo-static realization can now be more extensively defined. It is a synchronous multiprocessor system that is deterministically scheduled. The scheduling of the system is characterized by its periodicity, in processor space and time, with a period related to the iteration interval. One iteration is a pattern (processor schedule) in the processor-time lattice, $\mathbf{P} \times \mathbf{T}$. The spatial displacement, in the lattice, between successive iterations is denoted by the *principal lattice vector*, $\mathcal{L}$.

The term cyclo-static, connotes an idea similar to cyclo-stationarity of random processes. In a cyclo-stationary process, the statistics of observations separated by an integer multiple of the period of the process are stationary. In a cyclo-static realization, any two operations in the processor-time lattice, separated by an integer multiple of the principal lattice vector, represent the exact same operation of the algorithm for different iterations.

Returning to the processor schedule in Fig. 16.7, the interpretation additionally requires the specification of a principal lattice vector. Since there are four processors and the iteration period is 11 u.t., the set of possible principal lattice vectors are (0,11), (1,11), (2,11) and (3,11), due to the modulo indexing of processors. Of these vectors, (1,11) and (3,11) are equivalent since they are relatively prime with respect to the processor cardinality (four). They both cycle back to their initial positions after four iterations. The vector (2,11) decomposes the processor space into two partitions. The first contains processors {0,2} and the second processors {1,3}. Operations cycle with a period of two iterations within each partition. The vector (0,11) corresponds to a classic static-pipeline schedule with no displacement in the processor space.

Given the processor schedule of Fig. 16.7 and a principal lattice vector of $\mathcal{L} = (1, 11)$, the same operation is successive iterations is displaced 1 in the processor subspace and 11 u.t. in the time subspace. To clarify, at time 0 u.t. node 15 in block 1 is performed for iteration 0. Also at time 0 u.t. node 8 in block 0 is performed for iteration 1. Therefore the operation of node 8 in block 0 for iteration 0 is executed at time 0 u.t.−11 u.t. = −11 u.t., and is performed by processor (2−1) mod 4 = 1. The processor schedule indicated the operation in iteration 1, therefore we *unwrapped* the processor schedule back one iteration and the operation moved $-\mathcal{L}$ in the lattice.

As previously mentioned, for $\mathcal{L} = (1, 11)$, every four iterations, or 44 u.t., the same nodal operation is performed on the same processor. Thus each processor executes the exact same sequence of 44 operations, but skewed in time by successive multiplies of 11 u.t. (the iteration period). This is a generalization of a class of multiprocessor schedules referred to as *skewed single instruction multiple data* (SSIMD) [11]. Similar results apply to the case of $\mathcal{L} = (2, 11)$, except that there is a different skewed sequence of length 22 u.t. in each of the two processor partitions.

The processor schedule thus fully defines the execution scheduling of all operations for all possible principal lattice vectors. Different choices of principal lattice vectors however do result in different interprocessor communications requirements. The case corresponding to the specialized static case, with no processor displacement, requires the minimum communications support. Unfortunately, if all the operations are not of unit length, the fully static schedule usually does not exist. Conditions for its existence can be found in [14].

16.4.1 Determination of Processor Schedules

To find a processor and rate optimal cyclo-static schedule, it is first necessary to determine the flow graph bounds of the defining FSFG. All loops are analyzed to determine the iteration period bound, to identify all critical loops, and to determine the slack time of all other loops. All loops that are not critical have spare time in the computational deadlines of all operations in a loop. This spare time is called *slack time,* and for a loop l_i is:

$$t_{s_i} = n_i T_o - \sum_{j \in l_i} d_j \qquad (16.12)$$

where d_i is the number of unit delays in loop l_i, T_o is the iteration period bound, and d_j is the computational delay of operation j in loop l_i.

For a solution to exist, all operations in critical loops must be scheduled sequentially without gaps. All non-critical loops must be scheduled sequentially with a maximum total gap equal to the slack time of the loop. Non-loop operations can be scheduled at any time after their precedence requirements have been met. Of course, no operation can be scheduled before its precedence requirements have been met. The other key element to finding cyclo-static solutions is the requirement that the processor schedule tile or periodically extend along the direction of the principal lattice vector. The requirement can be simply stated as a constraint that no two points in the processor schedule for one iteration (lattice) may be separated by an integer multiple of the principal lattice vector $(k\mathcal{L})$. More precisely, no two entries in the complete processor schedule of one iteration can be congruent modulo $\mathcal{L}$.

However it is more computationally attractive to separate the processor assignment from the scheduling. Therefore the tiling constraint can gives rise to a simpler constraint referred to as the *processor modulo constraint*. The processor modulo constraint folds the effects of overlapped computations of different iterations to insure that there is never a need for more processors than exist. Since there are no processor assignment, the schedule is partitioned into T (the iteration period) equivalence classes. The maximum size of each class is P ($P \equiv |\mathbf{P}|$), the number of processors.

Let $\mathbf{S}(n)$, a representation of the schedule, be a sequence of parallel operations to be performed at time n, and $|\mathbf{S}(n)|$ is the number of parallel operations at time n. The processor modulo constraint can then be stated as:

$$\sum_{k=0}^{\infty} |\mathbf{S}(kT + t)| \leq P \quad ;\forall t \in \{0, 1, \ldots, T - 1\} \tag{16.13}$$

This is equivalent to considering the processor schedule representation in Fig. 16.7 while ignoring the processor assignment and stating that no column at any time can contain more than P operations. Not surprisingly the different time columns in the processor schedule are precisely the implied equivalence classes that indicate all operations that are in the same time equivalence class.

With the previous constraints and the information on bounds, slack time and precedence requirements a combinatorial optimization based on a constrained depth first search (pseudo branch and bound) can determine admissible processor schedules. This approach has a worst case exponential complexity (in terms of the number of nodes), however many FSFGs have strong constraints on loops or a high density of admissible solutions in the search space resulting in typical acceptable execution times. Based on these simple principles, three compilers that take FSFGs as input and produce cyclo-static processor schedules have been developed at Georgia Tech. Further information on how to find cyclo-static solutions can be found in [1]–[3]. Several examples from filtering are given in [3], as well as extensions to pipelined processing elements and heterogeneous processors. In addition, based on principles that lead to efficient implementations of the cyclo-static compiler, two signal processing computer architectures have been developed [15],[16], with the later currently being implemented.

References

[1] D. A. Schwartz and T. P. Barnwell III, "Cyclo-Static Multiprocessor Scheduling for the Optimal Implementation of Shift-Invariant Flow Graphs," *ICASSP'85*, Tampa, FL, March 1985.

[2] D. A. Schwartz, "Synchronous Multiprocessor Realizations of Shift Invariant Flow Graphs," Elec. Eng., Georgia Inst. of Tech, DSPL-85-2, July, 1985.

[3] D. A. Schwartz and T. P. Barnwell III, "Cyclo-Static Solutions: Optimal Multiprocessor Realizations of Recursive Algorithms," *VLSI Signal Processing II*, Chap. 11, IEEE Press, 1986.

[4] M. J. Gonzalez, "Deterministic Processor Scheduling," *Computing Surveys*, vol. 9, no. 3, Sept. 1977, pp. 173–204.

[5] E. G. Coffman, ed. *Computer and Job-Shop Scheduling Theory*, John Wiley, New York, 1976.

[6] J. Ondáš, "Algorithms for Scheduling Homogeneous Multiprocessor Computers," *Algorithms, Software and Hardware of Parallel Computers*, Eds. J. Mikloško and V. E. Kotov, Springer-Verlag, 1984.

[7] Jan Zeman and G. S. Moschytz, "Systematic Design and Programming of Signal Processors, Using Project Management Techniques," *IEEE Trans. on ASSP*, Dec. 1983, pp. 1536–1549.

[8] E. A. Lee, "A Coupled Hardware and Software Architecture for Programmable Digital Signal Processors," Memorandum No. UCB/ERL M86/54, ERL, Univ. of California, Berkeley, June, 1986.

[9] Markku Renfors and Yrjö Neuvo, "The Maximum Sampling Rate of Digital Filters Under Hardware Speed Constraints," *IEEE Trans. on Cir. and Sys.*, March 1981, pp. 196–202.

[10] A. Fettweis, "Realizability of Digital Filter Networks," *Arch. Elek. Übertragung.*, Feb. 1976, pp. 90–96.

[11] T. P. Barnwell III, C. J. M. Hodges, M. Randolf, "Optimal Implementation of Single Time Index Signal Flow Graphs on Synchronous Multiprocessor Machines," *ICASSP'82*, Paris, France, May, 1982.

[12] C. E. Leiserson, "Optimizing Synchronous Circuitry by Retiming," in *Proc. 3rd Caltech Conf. Very Large Scale Integration*, ed. R. Bryant, Computer Science Press, 1983, pp. 87–116.

[13] H. G. Forren, *personal communication*, 1987.

[14] H. G. Forren and D. A. Schwartz, "Transforming Periodic Synchronous Multiprocessor Programs," *Proc. ICASSP'87*, Houston, TX, April, 1987.

[15] D. A. Schwartz, T. P. Barnwell III, and C. J. M. Hodges, "The Optimal Synchronous Cyclo-static Array: A Multiprocessor Supercomputer for Digital Signal Processing," *Proc. ICASSP'86*, Tokyo, Japan, April, 1986.

[16] S. J. A. McGrath, T. P. Barnwell III, and D. A. Schwartz, "A WE-DSP32 Based, Low-Cost, High Performance, Synchronous Multiprocessor for Cyclo-Static Implementations," *Proc. ICASSP'87*, Houston, TX, April, 1987.

Chapter 17

Network Traffic Scheduling Algorithm for Application-Specific Architectures [1]

Ronald P. Bianchini, Jr. [2]

John Paul Shen [2]

Abstract

For many application-specific and mission-oriented multiple processor systems, the interprocessor communication is deterministic and can be specified at system inception. This specification can be automatically mapped onto a physical system using a network traffic scheduler. An iterative network traffic scheduler is presented which, given the arbitrary topology of the communication network, translates the deterministic communication into a network traffic routing pattern. Previous work has shown the existence of a network traffic scheduling algorithm based on a fluid-flow model that converges to an optimal solution. Issues of iteration complexity and convergence rate of the algorithm are discussed in this paper along with the overall methodology of application-specific architecture design. An upper bound on the traffic scheduling time can be determined. It is further shown that incremental traffic pattern changes can be more efficiently scheduled than total system rescheduling. Such incremental changes can model slowly-changing nondeterministic interproces-

[1] This work was supported by NSF Contract No. DMC-8451501, ONR Contract No. N00014-86-K-0507, and in part by the Semiconductor Research Corporation (SRC).

[2] Carnegie Mellon University, Pittsburgh, PA

sor communication. Hence, the algorithm presented in this paper can function as a (pseudo) dynamic traffic scheduler.

17.1   Introduction

This paper focuses on the automated design and implementation of high-performance embedded computing systems for mission-oriented applications. Such applications require real-time processing of data, generally derived from sensor inputs. Typically, algorithms employed in such applications require intensive numerical computation and extremely high data-flow bandwidth. However, the algorithm, or more specifically the program control flow, is generally deterministic. The physical system is typically characterized by a high-performance parallel architecture with deterministic interprocessor communication. Multiple special-purpose processors are usually used to achieve the required computational power. Examples of such systems include data and signal processing systems for aerospace applications, and vision and speech processing systems for robotics applications.

The paper presents a cost-effective methodology to the design of such special-purpose embedded systems. This methodology, called Application-Specific Architecture Design (ASAD), incorporates the following attributes.

1. **Integration of Algorithm and Architecture:** A key attribute of this methodology involves the simultaneous consideration of both the application algorithm and the supporting hardware architecture. Judicious integration of this type can lead to synergism of software and hardware.

2. **Algorithmically-Oriented Architecture:** Characteristics of the application algorithms are exploited to achieve highly efficient special-purpose systems. System structure is dictated by, or adapted to, the algorithmic structure.

3. **Semi-custom Design:** A semi-custom design approach is taken in which the possible design space has been appropriately constrained. At each level of the design hierarchy, there is a limited number of possible variations. Frequently, generic structures or templates are used as starting points and then customized, to suit specific applications, based on the algorithmic dictates.

4. **Automated Design Tools:** A systematic and rigorous methodology and associated software tools are developed and used to automatically achieve the necessary customization effectively and at low cost. Clearly, such automated design tools must be validated.

The integrated approach to design algorithmically-oriented architectures can produce finely-tuned high-performance systems; while the combination of the semi-custom design approach and the use of automated design tools ensures that these special-purpose systems can be cost-effectively designed and implemented.

An overview of the ASAD methodology is presented in this paper. This methodology employs an architecture compilation approach which translates a target architecture, specified by the system designer, to a host architecture which is the specification of the physical system. The hierarchy of procedures for performing architecture compilation is described. The focus of the remainder of the paper is on the design of the traffic routing network which is used to support interprocessor communication. Recent research on parallel systems has clearly shown that the most difficult problem is in the area of interprocessor interconnection and communication [14]. A network traffic scheduling algorithm is developed which functions as the "compiler" in the automated design of the communication network. This algorithm is proved to converge to an optimal traffic pattern and its computational complexity is shown to be polynomially bounded. Several interesting extensions to this work are also discussed.

17.2   Application-Specific Architecture Design

The ASAD automated system design methodology involves three major components: target architecture, host architecture, and architecture compilation; See Figure 17.1. The target architecture is the input to the design process, it is specified by the system designer to characterize the application algorithm. The host architecture specifies the available hardware system. It has a generic structure or template, which is further specified and customized based on the application algorithm. Varying degrees of customization of the generic host architecture are possible. Architecture compilation is the automated design process. It involves a methodology and associated software algorithms and tools.

17.2.1   Target Architecture

The target architecture specifies the overall desired system function, or the application algorithm. Separate specifications of the computation and communication tasks are used. The computation task is specified by the actual application code. The communication task is specified by a data flow description. This data flow description must be capable of capturing the natural hierarchy of the structure in the application code and the information characterizing the interprocessor communication.

Computation Specification. The computation specification is the actual application code written in a high level language. The application code can contain a hierarchy of constructs such as primitive operators, functional procedures, and

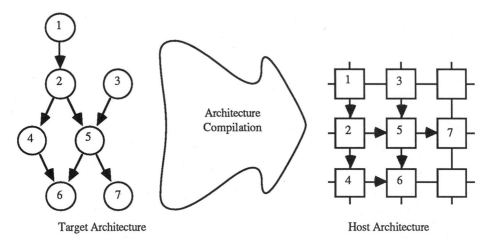

Figure 17.1: ASAD Automated System Design Approach.

reusable subprograms. Examples of primitive operators include floating point operators and inner products operators. An example of a functional procedure is the FFT routine. Reusable subprograms, e.g. DSP filter code, form a higher level of computation specification and are made up of functional procedures. The entire application code is functionally partitioned into *software modules*. Typically, each software module contains one or more functional procedures or reusable subprograms. Each software module is mapped onto a physical module during the architecture compilation process.

Communication Specification. After the algorithm is partitioned and software modules are defined, the communication specification describes the intermodule communication. Intermodule communication is specified by the intermodule dataflow pattern and associated bandwidths, and can be represented graphically. One such graphical representation language is SIGNAL [2]. A graphical representation includes:

- Data-flow pattern - a directed graph

    * graph nodes - represent software modules, $\{m_i\}$

    * directed edges - represent data-dependency between software modules, (s_i, d_i)

- Data-flow bandwidth - a label on each directed edge indicates the volume of data transfer, v_i

A set of notations is presented to facilitate subsequent discussion of the compilation process. Let $M = \{m_i\}$ denote the set of modules. Since there exists an one-to-one correspondence between software modules and physical modules, the same notation can be used for both. Each intermodule communication or data transfer is denoted by an ordered pair (s_i, d_i) where $s_i, d_i \in \{m_i\}$ and denote the source and destination modules, respectively. For every ordered pair (s_i, d_i), the actual data-flow bandwidth of volume of data transfer is denoted by v_i. Hence, each intermodule communication is characterized by (s_i, d_i) and v_i.

17.2.2 Host Architecture

The host architecture specifies the actual hardware structure. It provides a generic structure or template which can be customized to support specific applications. The host architecture has separate structures for computation, performed by the modules, and communication, supported by a communication network; See Figure 17.2. Arbitrary network topologies and module designs can be used, however, a limited number of useful topologies and module types can be stored in a library for repeated use.

Module Specification. All application computation is performed by the *physical modules* of the host architecture. An one-to-one mapping exists between the software modules of the target architecture and the physical modules of the host architecture. Each physical module performs, locally, all the computation of a particular software module. These modules are special-purpose processing elements or functional processor, such as FFT processors [21] and floating-point array processors. The number of modules of each type and the number of types and their physical locations in the network is discussed as part of the customization procedure. Each module has exactly one instruction stream, but can incorporate multiple

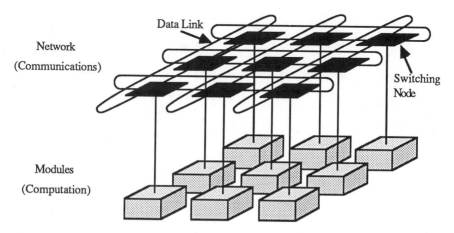

Figure 17.2: Host Architecture Example: Toroid Network of Nine Processor Modules.

data paths to achieve much small-grain parallelism. Each module is connected to the network via a switching node.

Network Specification. The communication network of the host architecture implements the intermodule communication specified by the application algorithm. The network consists of *switching nodes* and *data links*. The switching nodes are identical microcoded engines; see Figure 17.3. Each is an individually programmable, high-speed communication processor. Each switching node has a port to an associated module, and a limited number of additional connecting ports to other switching nodes. The switching nodes collectively function as the network operating system in performing intermodule traffic routing. The network data links are bidirectional high-bandwidth transmission media. Each link is directly connected to two adjacent switching nodes, and in conjunction with the switching nodes supports store-and-forward block-transfer of data between modules.

The network can have arbitrary topology as realized by the data links interconnecting the switching nodes. A generic network structure incorporating a particular topology can function as a template, and can be customized by the microcode downloaded into the switching nodes to achieve a certain traffic routing pattern. Each topology can be designed to cater to a specific class of applications; a small number of such efficient "virtual" network topologies can be prespecified and stored in a library.

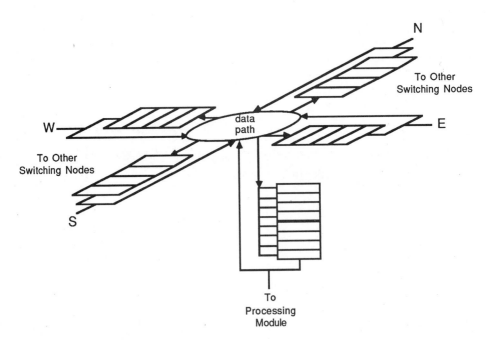

Figure 17.3: A Four-Ported Switching Node Architecture.

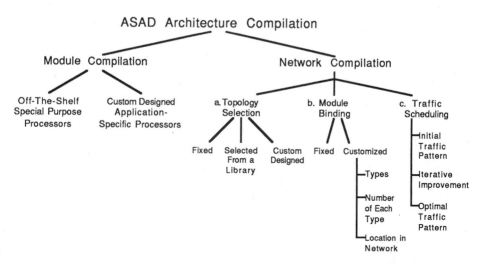

Figure 17.4: ASAD Architecture Compilation.

17.2.3 Architecture Compilation

The architecture compilation process accomplishes the ASAD automated design methodology. It separates the compilation of computation (module compilation) and communication (network compilation). An overview of the ASAD architecture compilation process is shown in Figure 17.4.

Module Compilation. The module compilation task is the automated synthesis of physical modules from software modules specified in the target architecture. A physical module can be a processor selected from a library of off-the-shelf special-purpose processors, or a custom designed application-specific processor. A separate project called Application-Specific Processor Design (ASPD) [7] is underway at CMU to develop the procedure and associated software tools needed for the automated design of application-specific processors. The ASPD paradigm is encompassed in the overall ASAD methodology. Features of ASPD include: code analysis, parallelism extraction, data-path synthesis, data-memory synthesis, control-path synthesis, and microcode generation. Results from the ASPD work are being published elsewhere.

Network Compilation. The network compilation task customizes a generic physical network to implement the intermodule communication of the target architecture, and is the focus of this paper. The network compilation procedure is divided into topology selection, module binding, and traffic scheduling. The network topology can be fixed, i.e. predetermined or already existing, judiciously selected from a library, or custom designed. Module binding involves the mapping of software modules to physical modules, including specifying physical module types, number of each type, and locations in the network. This mapping problem is analogous to the placement problem of computer-aided physical design of integrated circuits [8]. In our initial investigation [3,4], module binding appears to be less critical in terms of impacting overall system throughput than traffic scheduling. Furthermore, for many practical systems, good module binding results have been obtained [16] or can be easily generated using well-known algorithms [8]. Hence, module binding is not considered further in this paper.

Traffic scheduling involves the mapping of the intermodule communication onto the physical communication network, given a specific binding of the modules; See Figure 17.5. This mapping problem is similar to the routing problem of computer-aided physical design of integrated circuits [8]. The input to the traffic scheduling process includes the description of intermodule communication in terms of source-destination module pairs and their bandwidths or traffic volumes, and the binding of the source and destination modules to physical modules. The output of the traffic scheduling procedure includes a traffic routing pattern, i.e. the specific paths of intermodule communication, and associated down-loadable microcodes for the switching nodes. The current design of the traffic scheduler employs a multiple-commodity fluid-flow [11] model to achieve its goal of maximum total network throughput.

17.2.4   Scope of Paper

The balance of this paper focuses on the network compilation task, or more specifically the traffic scheduling problem described above. Traffic scheduling constitutes the most important and interesting aspect of the network compilation task of architecture compilation. This entails the mapping of the intermodule communication specified by the target architecture onto the physical communication network specified by the host architecture, as illustrated in Figure 17.5. In a previous publication

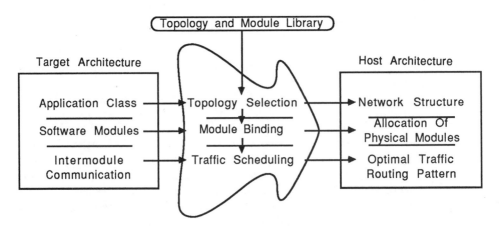

Figure 17.5: Network Compilation Overview.

[5], the authors presented an algorithm for traffic scheduling which converges to an optimal traffic routing pattern. This original algorithm is summarized here. A modified algorithm is then presented, and shown to be equivalent to the previous algorithm in terms of converging to an optimal solution. The further constraint added to the modified algorithm permits the rigorous analysis of its computational complexity. The algorithm complexity is discussed, and the algorithm is shown to be quite efficient. Several extensions to this work are ongoing and are summarized.

17.3 Traffic Scheduling Algorithm

The previously published traffic scheduling algorithm based on a fluid-flow model [5] is summarized in this section. An additional constraint is placed on the algorithm to form an equivalent algorithm which is more amenable to complexity analysis. Criteria for determining optimality of traffic routing patterns are presented and both algorithms are shown to generate optimal solutions.

17.3.1 Original Algorithm

The network traffic scheduling algorithm is a form of policy iteration [17]. The final state is initially predicted, and improvements are made in each iteration to eventually reach the correct final state. This successive reroute improvement technique has been proven to converge to an optimal traffic routing pattern [5]. An optimal traffic pattern is defined as any traffic pattern that permits the greatest total volume of traffic flow in the network, i.e. there are enough saturated links such that no additional traffic flow can be added.

The basic policy iteration algorithm for network traffic scheduling involves:

1. **Initial Guess:** Generate an initial traffic pattern using a simple algorithm, e.g. shortest path.

2. **Iterate:**

   a. **System Characterization** : Calculate all path costs between all source and destination nodes.

   b. **System Improvement** : For a particular source-destination pair, reroute traffic from a high cost path to a low cost path. Average the traffic across both paths to obtain equal path cost, and hence balance the load on both paths. Cost is proportional to the utilization of a path.

   End when no further traffic reroutes are possible.

3. **Clean Up**: Proportionally raise all traffic volumes to just saturate the highest utilized links.

For the reroute operation of reducing the utilization of highly saturated links, an ideal path reroute is illustrated in Figure 17.6. Figure 17.6a shows three paths between the traffic source, **S**, and destination **D**. Each path is utilized to varying degrees reflecting the utilization of the highest utilized link in each path. Path cost is directly proportional to the path utilization. For Figure 17.6a the path utilizations are from top to bottom: 0.5, 0.9, 1.0. Traffic is rerouted in Figure 17.6b to produce three paths of equal utilization: 0.3 units of traffic are added to the top path, and 0.1 and 0.2 units of traffic are removed from the two lower paths. In the final network, all three paths are equally utilized at 0.8. The traffic volume on the three paths can then be proportionally increased to saturate all three paths and obtain maximal traffic flow between **S**, and **D**.

An optimal traffic pattern results when a cutset of saturated network links is formed. An example of an optimal traffic pattern is shown in Figure 17.7 involving two source-destination pairs. The example traffic pattern is optimal since no additional traffic can be sent across the cutset links shown, as each link in the cutset is saturated. The following properties regarding a cutset must hold in order to obtain an optimal traffic pattern:

1. All links in the cutset must be saturated to their capacities, i.e. having the maximal utilization of 1.0.

2. The removal of links of the cutset separates the network into two disconnected subnetworks.

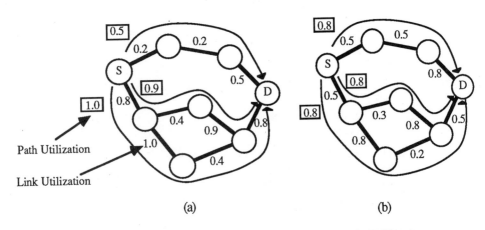

(a)   (b)

Figure 17.6: Rerouting Traffic to Average Path Utilization.

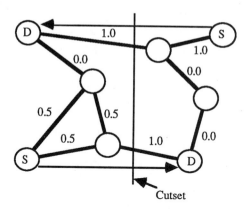

Figure 17.7: Cutset of an Optimal Network Traffic Pattern.

3. Any traffic path involving a cutset link must have its source and destination nodes in different subnetworks, i.e. traffic cannot cross the cutset twice.

As in the example shown, the above properties hold and hence prohibit additional traffic from being scheduled across the cutset. The network is then saturated, and the traffic pattern is effectively optimal.

The proof of convergence to an optimal traffic pattern for the above algorithm is briefly described. Details of the proof can be found in [5]. The problem is formatted as a transportation network problem [18]. An objective function is defined to measure the quality of the system state, or traffic pattern. The objective function defined is the sum of the costs associated with the network links, which is a function of the link utilizations. The convergence proof of the algorithm is dictated by the following arguments:

1. The objective function is shown to be a monotonically increasing function, thus having one absolute minimum and no other local minimum.

2. It is shown, by construction, that the traffic scheduling algorithm will reduce the value of the objective function at each iteration.

3. It is further shown that the algorithm will only stop iterating when a minimum point is reached.

Items 1, and 2 above imply that the algorithm is always approaching the minimum state at each iteration, and thus cannot cycle around a set of suboptimal states. Item 3 above guarantees that the minimum being approached is the absolute minimum in the solution space, and thus the algorithm is guaranteed to converge to an optimal traffic pattern possessing the minimum cost.

17.3.2   Modified Algorithm

A modified algorithm is presented which more easily permits the determination of
the algorithm convergence rate. Both algorithms perform the same function, except
for how traffic reroutes are chosen. In the modified algorithm, an additional constraint is used when selecting possible traffic reroutes. The constraint is as follows:
a reroute is selected only if the high-cost path contains one of the highest utilized
links in the network. This constraint reduces the number of possible reroutes and
guarantees that each traffic reroute reduces the volume of traffic on one of the highest utilized links. This constraint is instrumental in determining the convergence
rate of the algorithm.

It must be shown that the modified algorithm also converges to an optimal solution.
This can be shown by comparing it to the original algorithm, which is guaranteed to
converge to the optimal solution [5]. Two lemmas are needed. To simplify further
discussion, the original algorithm is referred to as **algo**, and the modified algorithm
is referred to as **algm**.

Lemma 1: Two network traffic patterns will yield the same total volume of network
traffic flow, if the most utilized links of the two traffic patterns have the same
utilization value.

An informal argument for Lemma 1 follows. In any traffic pattern, the highest
utilized link determines the limiting factor in proportional traffic increases. The
traffic can be uniformly increased to just saturate this link, and no further. This is
shown by the example in Figure 17.8. In the figure, both traffic patterns have at least
one maximum utilized link with utilization 0.5 (1/2 of the available bandwidth).
The interprocessor communication can be realized with up to twice the currently
specified traffic volume. Thus both networks can support the same total volume of
traffic flow, namely twice the currently specified volume. In this case, it so happens
that both traffic patterns are also optimal, since at twice the currently specified
volume, a saturated cutset is formed, indicating that no additional traffic flow can
be added to either network.

Lemma 2: The modified traffic scheduling algorithm **algm** supports the same
maximum total network traffic volume as that supported by the original algorithm **algo**.

The proof of lemma 2 is outlined below:

1. Run **algm** to produce an optimal traffic pattern S_1.

2. Run **algo** with solution S_1 as the initial guess to produce the optimal traffic
   pattern S_2.

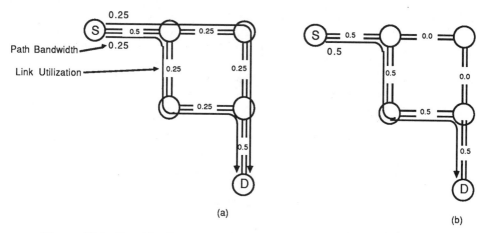

Figure 17.8: Two Traffic Patterns With The Same Total Traffic Flow.

3. Show that by the definition of **algm**, S_1 and S_2 must have the same highest utilized links.

4. Then by Lemma 1, solutions S_1 and S_2 must support the same maximum total volume of traffic flow.

Step 3 of the above outline is shown by examining the differences between solutions S_1 and S_2. The first reroute attempted by algorithm **algo** on solution S_1 cannot move traffic from any highest utilized link of the network, since if such a reroute existed **algm** would have accomplished it. The same is true for the ith reroute taken by **algo**. At the $(i-1)^{th}$ iteration, traffic is rerouted between paths that are utilized less than the highest utilized link in the network, so that neither path utilization is raised above or equal to the highest utilized link by the reroute. Therefore, the same set of highest utilized links exist at both iterations i and $i-1$; and if a reroute away from one such link is not possible at iteration $i-1$, then it is not possible at iteration i. It must be the case that the set of highest utilized links of both solutions S_2 and S_1 are the same. Hence, proving Step 3, and Lemma 2.

Lemmas 1 and 2 lead to the following theorem.

Theorem 1: The modified traffic scheduling algorithm **algm** is guaranteed to converge to an optimal network traffic pattern.

To prove Theorem 1, it must be shown that the added constraint will not prohibit **algm** from reaching an optimal traffic pattern. This is based on the fact that **algo** has been proven to converge to an optimal traffic pattern. It is shown by Lemmas

1 and 2 that the two algorithms may produce two different optimal traffic patterns, but both patterns yield the same maximum total traffic flow. By Lemma 2, both algorithms produce the same highest link utilization. According to Lemma 1, if the highest link utilization is the same, then both traffic patterns must support the same total volume of traffic in the network. Since it has been shown that **algo** converges to an optimal traffic pattern, this leads to the fact that **algm** also converges to an optimal traffic pattern.

The difference between solutions from the two algorithms are illustrated by Figure 17.8. An algorithm of type **algo** will tend to produce the traffic flow pattern as that shown in Figure 17.8a, and an algorithm of type **algm** will tend to produce the traffic flow pattern as that shown in Figure 17.8b. Basically **algm** will tend to reduce the highest utilized links, thus reaching the optimal solution without further smoothing the lesser utilized links. On the other hand **algo** tends to be less efficient in that it performs further traffic smoothing after an optimal solution is obtained. The example shown illustrates that **algm** and **algo** can produce different traffic patterns but the same maximal traffic flow.

The basic policy iteration algorithm for network traffic scheduling used for the remainder of the paper is **algm** which is described below:

1. **Initial Guess**: Generate an initial traffic pattern using a simple algorithm (e.g. shortest path).

2. **Iterate**:

   a. **System Characterization** : Calculate all path costs between all source and destination nodes.

   b. **System Improvement** : For a particular source-destination pair, reroute traffic from a high cost path to a low cost path. The high cost path must include one of the network's highest utilized links. Average the traffic across both paths to obtain equal path cost for both paths.

   End when no further traffic reroutes are possible.

3. **Clean Up**: Proportionally raise or lower all traffic volumes to just saturate the highest utilized links.

17.3.3   Algorithm Implementation and Illustration

The above described traffic scheduling algorithm has been implemented. The implemented scheduler assumes an external host on which the scheduler runs. The results from the scheduler execution are then downloaded into the network of switching nodes. The implemented scheduler is written in C and runs on a VAX 11/785.

This scheduler is now illustrated. Although the scheduler can accommodate a network with any arbitrary topology, the simple nine node toroid network topology is used here as an example host architecture. The network consists of nine processing modules interconnected by a three by three toroid structure. The target architecture communication specification consists of four software modules, labeled S_1, D_1, S_2 and D_2, and a single unit of data flow between each of the two source-destination pairs. This is represented by the intermodule communication description $< (s_i, d_i), v_i >$ as:

- (S_1, D_1), 1 unit/cycle>, and

- (S_2, D_2), 1 unit/cycle>.

The module binding of the four software modules to physical modules is as shown in Figure 17.9. The initial network traffic pattern, at iteration 0, is shown in Figure 17.9a.

The traffic scheduler optimizes the use of the network by utilizing multiple paths between each traffic source and destination pair. The optimal network traffic pattern is represented by traffic paths, and the volume of traffic on each path as shown in Figure 17.9b. The solution is provably optimal by observing the following properties of the cutset of saturated links shown: all cutset links are saturated, the network is separated into two subnetworks by the cutset, and no traffic crosses the cutset twice. Since the traffic sources are contained in one subnetwork, and since all cutset links are saturated, additional traffic cannot be scheduled across the cutset to the destination nodes. The final solution supports traffic volumes at three times the originally given bandwidths, or the target architecture requirements can be supported by network links having bandwidth capacity of 1/3 unit/cycle.

This scheduler consists of approximately 600 lines of C code. An optimal traffic pattern was found for the above example using less than a second of a VAX 11/785 elapse time. The execution trace of the network traffic scheduler, showing the iteration number and the network link utilizations is shown in Figure 17.10. To reduce the size of the printed results, the iterations are displayed only after the utilization of all the highest utilized links have been reduced. The scheduler produced the results after seven such iterations. The cutset is formed by the links with identification numbers: 1, 2, 4, 5, 11, and 17.

Execution times for the network traffic scheduler on several example systems have been quite reasonable. Examination of classical solutions, e.g. linear programming formulations [5,10], has shown that the complexity for such a problem is typically exponential. The complexity of the network scheduling algorithm presented in this paper can be shown to be polynomially bounded by the number of network nodes.

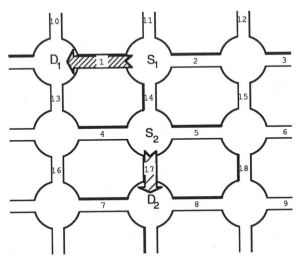

Initial Traffic Routing Pattern

(a)

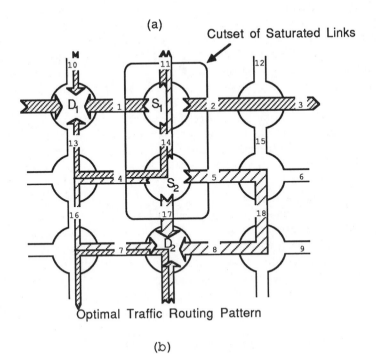

Optimal Traffic Routing Pattern

(b)

Figure 17.9: Network Traffic Scheduler Example.

Iteration Number	Network Link Utilization																	
	1	2	3	4	5	6	7	8	9	10	11	12	13	14	15	16	17	18
0	1.00	0.00	0.00	0.00	0.00	0.00	0.00	0.00	0.00	0.00	0.00	0.00	0.00	0.00	0.00	0.00	1.00	0.00
2	0.50	0.50	0.50	0.50	0.00	0.00	0.50	0.00	0.00	0.00	0.00	0.00	0.00	0.00	0.00	0.50	0.50	0.00
7	0.24	0.31	0.24	0.43	0.37	0.37	0.31	0.43	0.24	0.24	0.37	0.06	0.24	0.31	0.12	0.31	0.37	0.12
9	0.34	0.31	0.24	0.37	0.37	0.37	0.24	0.34	0.15	0.15	0.34	0.06	0.24	0.37	0.12	0.24	0.37	0.12
12	0.34	0.32	0.24	0.35	0.35	0.34	0.26	0.34	0.21	0.18	0.34	0.12	0.21	0.35	0.14	0.23	0.35	0.12
19	0.34	0.34	0.24	0.34	0.34	0.32	0.27	0.34	0.24	0.21	0.34	0.14	0.19	0.34	0.13	0.22	0.34	0.13
32	0.33	0.33	0.26	0.33	0.33	0.26	0.33	0.33	0.25	0.24	0.33	0.13	0.17	0.33	0.11	0.25	0.33	0.13

Traffic Increased by 3x to Saturate Network:

	1	2	3	4	5	6	7	8	9	10	11	12	13	14	15	16	17	18
	1.00	1.00	0.79	1.00	1.00	0.78	0.99	0.99	0.74	0.73	1.00	0.41	0.53	1.00	0.33	0.75	1.00	0.41

Figure 17.10: Network Traffic Scheduler Execution Trace.

17.4 Complexity Analysis of the Algorithm

The overall algorithm complexity is determined by the complexity of each iteration, and the iteration count. The *iteration complexity*, or the complexity of each iteration, is determined by the number of computation steps required to perform each traffic reroute. The *iteration count* is an upper bound on the number of iterations needed to reach the optimal traffic pattern, which is an indication of the algorithm convergence rate. The following variables are used in this section:

N: the number of switching nodes, and hence modules, in the network.

L: The number of links in the network.

T: The total volume of traffic to be routed.

17.4.1 Iteration Complexity

Each iteration of the algorithm requires the following three steps:

1. **Determine lowest cost paths.**

2. **Determine critical paths.**

3. If possible, **reroute traffic** from a critical path to a lowest cost path.

These computation steps are illustrated in the algorithm flow chart shown in Figure 17.11. First, the *lowest cost paths* are found. These are the paths between all source-destination pairs containing links of lowest utilization. Traffic can be moved onto these paths to alleviate congestion on higher utilized links. Secondly, critical paths originating from each node are found. A *critical path* is the highest cost path between any two nodes that contains traffic. A critical path must already contain traffic, since traffic will be moved from certain critical paths to alleviate network congestion. The third step tests for potential traffic reroutes. A reroute is permitted between two nodes, **S** and **D**, if the critical path between **S** and **D** contains at least one of the network links with the highest utilization, and its path cost is greater than the least cost path between **S** and **D**. The complexity of each of the three iteration steps is addressed independently.

The worst-case complexities of the three steps can be shown to be $L log_2 L$, $NL log_2 L$, and $2N$, respectively. Hence, the total iteration complexity is the sum of the com-

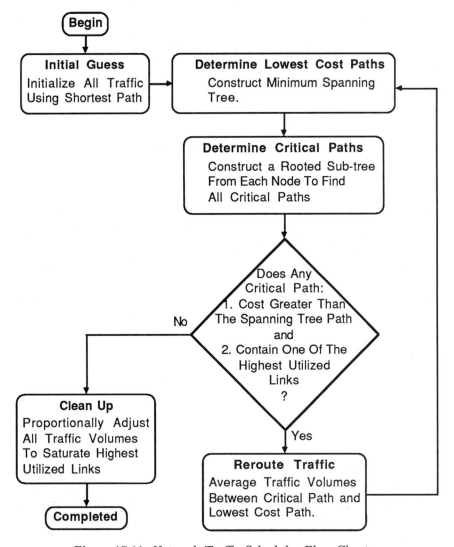

Figure 17.11: Network Traffic Scheduler Flow Chart.

plexities of the three steps in the iteration. Therefore, the worst-case complexity for each iteration of the network traffic scheduling algorithm is as stated in the theorem below.

Theorem 2: The worst-case computation complexity of each iteration of the modified algorithm **alg$_m$** is of the order:

$$L log_2 L + N \cdot L log_2 L + 2 \cdot N$$

where **N** is the number of switching nodes and **L** is the number of links in the network.

17.4.2 Iteration Count

To determine the convergence rate of the iterative procedure, an upper bound is determined for the number of iterations required for convergence. To do this the worst possible case scenario for convergence is determined, i.e. assuming starting with the worst possible initial traffic pattern and requiring the final traffic pattern to be the best possible. The rate at which the algorithm converges for this case is determined. Once this is obtained, the maximum possible number of iterations required between any two traffic patterns is also determined, and is presented in the following theorem.

Theorem 3: The maximum number of iterations required for the convergence of the network traffic scheduling algorithm **alg$_m$** is **L $\cdot$ T**, where **L** is the number of links in the network, and **T** is the total volume of traffic to be routed.

Since the algorithm is guaranteed to reroute at least one unit of traffic in each iteration, and based on Theorem 3, the rate of convergence of the algorithm is proportional to the product **L $\cdot$ T**. For a given network, **L** is fixed, so the number of iterations required to converge to the optimal traffic pattern is linearly proportional to **T**, the total volume of traffic to be routed. Details of the proofs for Theorems 2 and 3 can be found in [6].

17.4.3 Traffic Segmentation

The iteration count presented above assumes that the network structure can support the entire system traffic volume of **T**. Specifically, the total traffic volume of **T** dictates that a certain volume of traffic **t $\leq$ T**, be routable by each switching node. The assumption is that each switching node is capable of supporting any arbitrary **t** units of traffic within a single network traffic routing cycle. A network traffic routing cycle is defined as a single pass through the switching node traffic routing microcode. This assumption may not be valid for a particular physical realization of the network. The capability of the switching node architecture, as shown in Figure 17.3, to support **t** units of traffic is directly bounded by the size of the microcode memory. Specifically, a switching node with memory size **m** can only support the routing of at most **m** units of traffic per cycle assuming each microinstruction routes a unit of traffic. Thus the total volume of traffic that can be routed in each cycle is bounded by the size of the microcode memory in each switching node.

In the case that **m** < **t**, the **t** units of traffic must be segmented and routed using multiple cycles. The minimum number of segments, hence routing cycles, required is **s** = **t**/**m**]. The total volume of traffic is then completely routed after **s** network traffic routing cycles.

With limited microcode memory size in the switching node, there will also be a limit on the volume of traffic, $\mathbf{T_{max}}$, that can be routed in one cycle. Once that limit is reached, the traffic must be segmented. For **T** ≤ $\mathbf{T_{max}}$ the maximum iteration count is as before, i.e. **L** × **T**. If **T** is greater than $\mathbf{T_{max}}$ then the maximum iteration count for traffic scheduling is now **L** × $\mathbf{T_{max}}$. Of course, with traffic segmentation, multiple network traffic routing cycles are needed to route the **T** units of traffic. However, the scheduling of the traffic routing pattern only needs to be performed for each routing cycle or effectively for $\mathbf{T_{max}}$ units of traffic. Since L and $\mathbf{T_{max}}$ are both fixed for a particular physical realization of the network, the iteration count is bounded by the constant **L** × $\mathbf{T_{max}}$.

Theorem 4: The physical upper bound on the iteration count is bounded and strictly a function of the physical parameters of the network. This bound is **L** × $\mathbf{T_{max}}$, where **L** is the number of network links and $\mathbf{T_{max}}$ is the maximum total volume of network traffic which can be supported, by the switching node microcode memory, in one cycle.

17.4.4 Algorithm Complexity

The total complexity of the traffic scheduling algorithm is the product of the complexity of each iteration and the iteration count, i.e. *IterationComplexity* × *IterationCount*. This leads to the following theorem:

Theorem 5: The computational complexity of the network traffic scheduling algorithm **alg$_m$** is

$$O([\{\mathbf{L}\} + \{\mathbf{N} \times \mathbf{L}log_2\mathbf{L}\} + \{2 \times \mathbf{N}\}] \times [\mathbf{L} \times \mathbf{T}])$$

where **N** is the number of switching nodes in the network, **L** is the number of links in the network, and **T** is the total volume of traffic to be routed.

Based on Theorem 5, the following observations can be made. N and L are closely related and determined by the network topology, and are independent of T. The relationship between N and L is determined by the degree of connectivity, d, of the switching nodes. Assuming that each switching node has d connecting ports, the relationship between N and L is

$$L = \frac{N \times d}{2}.$$

In many practical application-specific systems, the degree of connectivity, d is a relatively small number, typically from four to eight. Hence, considering that L is a constant multiple of N, L can be replaced by N in the algorithm complexity expression. Therefore replacing L with N and keeping only dominant terms, the complexity of the traffic scheduling algorithm becomes:

$$O(\{N^2 log_2 N\} \times \{N \times T\}), \text{ or}$$

$$O(N^3 log_2 N \times T).$$

The algorithm complexity, as shown above, reflects the worst possible case complexity and is a function of the total volume of traffic being routed and the number of network nodes. This worst case complexity appears to be strongly influenced by the network size. Fortunately, in most practical systems N is typically less than 100.

The complexity result presented above leads to interesting observations concerning the computation time of the network traffic scheduling algorithm. For example, for a given network the number of nodes is a constant, and the traffic scheduling time becomes linearly dependent on the total volume of traffic being routed. More importantly, once the total volume of traffic being routed becomes large enough to force traffic segmentation, the traffic volume that must be dealt with in the traffic scheduling algorithm becomes a constant, T_{max}. The computation time for the traffic scheduling algorithm effectively becomes a constant. Hence, for a given physical network with a fixed topology and switching node architecture the computation time for the traffic scheduling algorithm is linearly dependent on the total volume of traffic being routed. When the traffic volume exceeds the routing capacity of the switching node and traffic segmentation occurs, then the computation time becomes effectively a constant.

The algorithm presented can be viewed as a solution for a type of multiple-commodity network flow problems [11]. Currently, the single-commodity network flow problem is well understood, and algorithms with polynomial complexity have been developed. For example, Sleator and Tarjan [22] have shown that an algorithm of complexity $NlogN$ can be implemented. Solutions for the multiple-commodity network flow problem exist, but require greater complexity, and are not as extendible to a parallel implementation. Such solutions include linear programming [10], which imposes a high number of variables, in the problem formulation, and a high degree of complexity in computation time. Another type of solution is a matrix formulation of the problem, as in [15]. Such solutions require a high complexity overhead, and a central global matrix, and thus limit their usefulness for a parallel architecture.

17.5   Extensions and Summary

Currently, the final refined version of the traffic scheduling algorithm **algm** has been implemented, and a number of actual example systems have been scheduled. Several interesting and promising extensions to this work are being pursued and are summarized below.

17.5.1   Switching Node Design and Implementation

To facilitate evaluation of the concepts developed herein, an exerimental hardware effort is currently underway. A VLSI implementation of the network switching node [9] has been designed and simulated. The final layout is being prepared for submission to MOSIS for fabrication. The VLSI chip that has been designed is a bit-sliced microcoded engine for traffic routing. A complete switching node would consist of the VLSI switch chip, a memory chip and a sequencer or counter chip. All of the interfacing to other chips and the bus for downloading are included on the VLSI switch chip. The VLSI switch chip architecture, shown in Figure 17.12, consists of a module interface bus, and a network interface with connectivity of four ports. Output on each network port is queued by a flush through FIFO shift register. The output to the module interface bus is queued by an address-in-shift-out register. This structure facilitates re-ordering of data from network paths of different lengths before it is sent to the module. Data is loaded from the network into the register location that is addressed by the microcode, and is shifted out to the processor port. Both queue structures are illustrated in Figure 17.13. The VLSI switch chip also provides a microinstruction register, and a global bus interface. The global bus is connected to all the network switching nodes, and facilitates downloading of the microcode. Downloading is accomplished by addressing the desired switching node via the global bus. The microcode for the switching node is then supplied on the global bus and loaded into the microstore.

17.5.2   Parallel Traffic Routing Algorithm

In the discussion so far, a host external to the application system is assumed for performing the traffic scheduling task. It would be desirable to have the network in the application system perform self scheduling without requiring an external host. The traffic scheduling task can be distributed among and performed by the network switching nodes themselves in a cooperative fashion, and communication

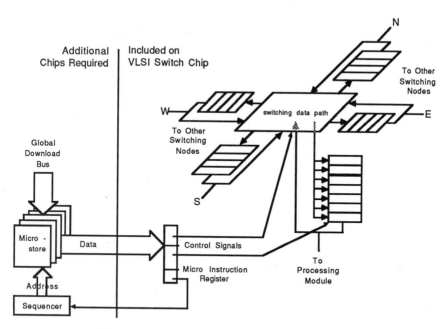

Figure 17.12: Switching Node Architecture.

between the nodes during traffic scheduling can occur on the network links. The hardware structure to support the additional traffic scheduling task at the switching node is shown in Figure 17.14. The present scheduling algorithm would need to be parallelized to run on the switching nodes of the network. Such a network scheduler is feasible and a preliminary version of the algorithm has been implemented [5]. The control flow of the parallel algorithm is analogous to the present algorithm.

The preliminary version of the parallel traffic scheduler is implemented in Concurrent C [3]. AT&T Bell Labs has given Carnegie-Mellon University permission for its use [12,20]. Concurrent C offers extensions to allow the C programming language to simulate concurrent programs. Concurrent C provides mechanisms for process creation, termination, and inter-process synchronization and interaction based on the *rendezvous* concept [1]. Concurrent C is implemented on a VAX 11/785 under the Unix operating system. A parallel traffic scheduler written in Concurrent C has been implemented and generates the exact same solutions as the serial algorithm for several examples. Further work is in progress to port the parallel traffic scheduler to parallel hardware. Currently, the parallel traffic scheduler runs on the Intel Hypercube simulator, and work is in progress to move it to the actual Hypercube machine.

---

[3] Concurrent C was implemented at AT&T Bell Labs

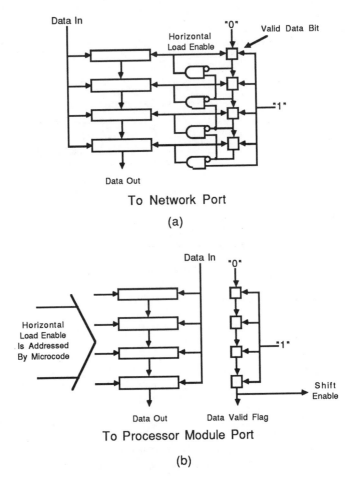

Figure 17.13: Switching Node Queue Structures.

17.5.3   Incorporation of Pseudo-Dynamic Traffic

The initial constraint placed on the network traffic of application-specific systems is that intermodule communication and hence traffic flows be static and deterministic at network compile time. This constraint permits the network compiler to optimally allocate the network resources prior to system run-time. It is desirable to relax this constraint to permit changing traffic requirements. The pseudo-dynamic traffic model permits incremental changes of the traffic requirements. This can be accomplished by incorporating concurrent system operation, i.e. traffic routing and traffic scheduling. While the network switching nodes are performing traffic

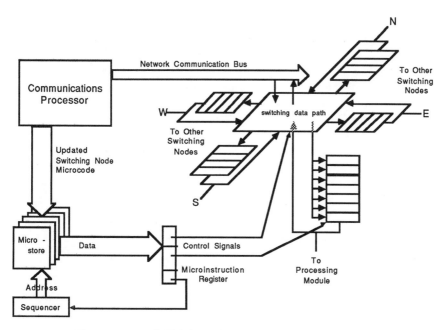

Figure 17.14: Self-Scheduling Network Switching Node.

routing in the foreground, the traffic scheduling algorithm can be executed in the background in response to incremental changes in the traffic requirements. At the appropriate time the new traffic routing microcode generated by the traffic scheduling algorithm can be loaded into and hence update the switching node microstore. In the above model, the network traffic pattern is permitted to change incrementally. The frequency of incremental changes is limited by the scheduling time of the additional traffic. It has been shown that the scheduling time depends on the volume of traffic being scheduled; in fact the traffic scheduling time for a given network is linearly proportional to the traffic being scheduled, as stated in Theorem 6. As a result, less traffic being routed during the incremental traffic changes permit faster scheduling times and a higher frequency of pseudo-dynamic traffic changes. The structure of the proposed self-scheduling switching node architecture, Figure 17.14, is designed to support the pseudo-dynamic traffic model. The incremental traffic scheduling computation occurs in the computation processor, and utilizes the least utilized links of the network for system messages during scheduling.

17.5.4 Incorporation of Reconfiguration for Fault Tolerance

Due to the nature of application-specific system requirements, fault tolerance issues must be addressed. Such embedded, mission-oriented, real-time processing systems, require recovery from system faults without the intervention of a system operator.

Fault tolerance in such a system includes fault detection and fault recovery. A number of projects are currently underway in the CMU Center for Dependable Systems which are focusing on fault detection within processing modules [19,23]. System wide fault detection has also been addressed by other recent research efforts [13]. Our work focuses on the fault recovery task at the network level. Fault recovery can be accomplished via reconfiguration around faulty resources. Such reconfiguration can be accomplished by traffic rescheduling, with the assumption that redundant nodes and links exist in the network. In the same manner that pseudo-dynamic traffic permits incremental changes in traffic requirements, fault recovery via reconfiguration incorporates incremental changes in resource availability. As resources are determined to be faulty, the network is rescheduled to avoid them. As with pseudo-dynamic traffic, the rescheduling time is proportional to the degree to which the network resources must be rescheduled.

17.5.5   Summary

A methodology called Application-Specific Architecture Design (ASAD) for the automated design of special-purpose mission-oriented computing systems is presented. This methodology proposes an architecture compilation approach, which involves systematic procedures for the automated design of computation modules (module compilation) and the communication network (network compilation). The core of the network compilation procedure is a network traffic scheduler. An efficient network traffic scheduling algorithm, guaranteed to converge to an optimal traffic pattern, has been developed and is presented in this paper. The computation complexity of the algorithm is presented. Several extensions to this work are currently being pursued. These extensions include the VLSI circuit design of a network switching node, parallelizing of the algorithm to facilitate network self-scheduling, adaptation of the algorithm to accommodate pseudo-dynamic traffic scheduling, and implementation of network reconfiguration strategies for fault tolerance. More interesting Results are forthcoming.

References

[1] M. Ben-Ari, *Principles of Concurrent Programming,* Prentice Hall, Englewood Cliffs, NJ, 1984.

[2] A. Benveniste, P. Bournai, T. Gautier, P. Le Guernic, *SIGNAL: A Data Flow Oriented Language for Signal Processing,* Technical Report 378, Institut de Recherche en Informatique et Systems Aleatoires, March, 1985.

[3] R. P. Bianchini, Jr and M. S. Schlansker, A High Performance Interconnect for Concurrent Signal Processing, In *VLSI Signal Processing,* pp. 39-49, IEEE, November, 1984.

[4] R. P. Bianchini, Jr. and J. P. Shen, Automated Compilation of Interprocessor Communication for Multiple Processor Systems, In *IEEE International Conference on Computer Design,* IEEE, October, 1986.

[5] R. P. Bianchini, Jr. and J. P. Shen, Interprocessor Traffic Scheduling Algorithm for Multiple-Processor Networks, *IEEE Transactions on Computers,* C-36(4):396-409, April, 1987.

[6] R. P. Bianchini, Jr. and J. P. Shen, *Complexity Considerations of a Network Traffic Scheduling Algorithm,* Technical Report, Dept. of Electrical and Computer Engineering, Carnegie Mellon University, Pittsburg, PA, 15213, July, 1987.

[7] M. Breternitz Jr and J. P. Shen, *ASPD: Application-Specific Processor Design,* Technical Report, Department of Electrical and Computer Engineering, Carnegie Mellon University, Pittsburg, PA 15213, May, 1987.

[8] M. A. Breuer (editor), *Design Automation of Digital Systems,* Prentice Hall, Englewood Cliffs, NJ, 1972.

[9] T. Cobourn and M. Johnson, *Data Switch Chip,* Technical Report, Department of Electrical and Computer Engineering, Carnegie Mellon, Pittsburg, PA 15213, May, 1987.

[10] G. B. Dantzig, *Linear Programming and Extensions,* Princeton University Press, Princeton, NJ, 1963.

[11] L. R. Ford and D. R. Fulkerson, *Flows in Networks,* Princeton Universty Press, Princeton, NJ, 1962.

[12] N. H. Gehani abd W. D. Roome, *Concurrent C,* Technical Report, AT&T Bell Laboratories, Murray Hill, NJ, 1985.

[13] S. H. Hosseini, J. G. Kuhl and S. M. Reddy, A Diagnosis Algorithm for Distributed Computing Systems with Dynamic Failure and Repair, *IEEE Transactions on Computers,* C-33(3):223-233, March, 1984.

[14] K. Hwang, F. A. Briggs, *Computer Architecture and Parallel Processing,* McGraw-Hill Book Company, New York, 1984.

[15] J. L. Kennington and R. V. Helgason, *Algorithms for Network Programming,* J. Wiley, New York, NY 1980.

[16] E. A. Lee and D. G. Messerschmitt, Static Scheduling of Synchronous Data Flow Programs for Digital Signal Processing, *IEEE Transactions on Computers,* C-36(1):24-35, January, 1987.

[17] E. Minieka, *Optimization Algorithms for Networks and Graphs,* Marcel Dekker, Inc., New York, New York and Basel, 1978.

[18] G. F. Newell, *Traffic Flow on Transportation Networks,* The MIT Press, Cambridge, Massachusetts, 1980.

[19] M. A. Schuette and J. P. Shen, Processor Control Flow Monitoring Using Signature Instruction Streams, *IEEE Transactions on Computers,* C-36(3):264-276, March, 1987.

[20] B. Smith-Thomas, *Managing I/O in Concurrent Programming: The Concurrent C Window Manager,* Technical Report, University of North Carolina at Greensboro, Greensboro, N.C. 27412, 1985.

[21] E. E. Swartzlander, Jr and G. Hallnor, High Spewed FFT Processor Implementation, In *VLSI Signal Processing,* pp. 27-34, IEEE, November, 1984.

[22] R. E. Tarjan, *Data Structures and Network Algorithms,* Society for Industrial and Applied Mathematics, Philadelphia, PA, 1983.

[23] K. D. Wilkin and J. P. Shen, Embedded Signature Monitoring: Analysis and Technique, In *International Test Conference,* September, 1987.

Chapter 18

Implementations of Load Balanced Active-Data Models of Parallel Computation

Chris Jesshope [1]

Abstract

VLSI has encouraged the use of large scale parallelism in computer systems. This paper introduces an active-data model of parallelism applied to arbitrary data structures. An implementation of this model is described, and its limitations are sought. This implementation on the RPA computer system uses a fine-grain, SIMD-like, array-in-memory system, hosted by the INMOS transputer. Process or algorithmic parallelism can therefore be exploited at the top level of the system by replication of this basic unit. This synergism will be explored through the consideration of objects implementing the active-data model as a means of exploiting efficient and portable systems.

18.1   Introduction

In order to produce a self-consistent paper, I have included, as background to the active-data model, enough details about a processor array development at Southampton University to describe a detailed implementation. More details about the chip and system architecture, and its micro-programming environment can be found in Jesshope (1987), Jesshope et. al. (1987) and Jesshope and Stewart (1986).

---

[1] Dept. of Electronics and Computer Science, The University of Southampton, England

The Reconfigurable Processor Array (RPA), is a fine-grain, SIMD-like, array-in memory system, hosted by the INMOS transputer. The design differs from other similar SIMD designs, such as the ICL and AMT DAPs and the MPP, in that each processor has more autonomy; the action of each processing element (PE) is modified by local state information. Instruction decoding is adapted, depending on which of the 64 states a PE is in. This state can be considered equivalent to a preset field in a classical micro-coded controller. However as there are many (16-16K) PEs, the preset modification to the control word is distributed across the system and allows data dependencies, providing a MIMD flavour to some operations.

The design and implementation of a Transputer/RPA prototype is well advanced and the design of a custom CMOS chip containing 16 bit-slices is nearing completion, a single bit-slice test chip has already been fabricated on the Southampton, 3 micron, in-house, n-well, CMOS process. A photograph of this is shown in Figure 18.1. A micro-programming system based on a low level simulation of the RPA has now been in use for over a year, and many operations such as IEEE floating point have been implemented and timed on this system. For a single 64 chip RPA system, floating point performance (IEEE single precision) ranges from 6 to 25 MFLOPS and integer performance (16 bit) from 16 to 500 MIPS. These figures are for a 1024 PE system and greater performance can be achieved either by increasing the number of PEs in the system or by replication of this basic transputer/RPA unit.

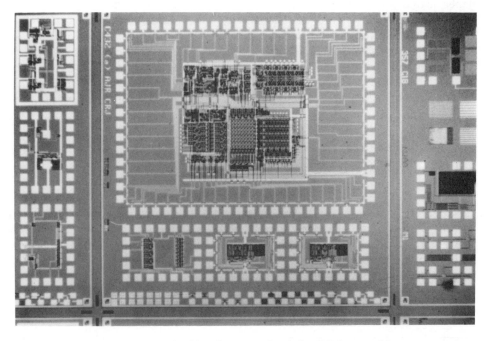

Figure 18.1: A chip photograph of the RPA test chip.

18.2   The Active-Data Model

There is no doubt that in the most general case, parallelism introduces additional complexity into the programming task. The use of a multiplicity of instruction streams compounds the ability of the programmer to introduce bugs into her programs. There is nothing new in this, as virtual concurrency has been used for at least a decade now as a means of controlling the complexity of the interactions between the various components of operating systems. It is also well known that operating systems are notorious for their ability to conceal errors, waiting for the appropriate conjunction of circumstances to manifest their ailment.

Deadlock is a prime example of an error introduced through concurrency. Deadlock is the ability of two or more processes to mutually block each other, for example, by each waiting for another to perform an action, which is precluded by the action of another process. The classical example is a ring of processes each trying to read from another, but none willing to write first; were one process willing to write before read, the a write-read-write sequence would ripple around the ring of processes.

Such problems arise because of the asynchronous nature of communication or synchronization between instruction streams in this process based view of concurrency. There are other models however, which avoid the problems associated with this process view. The alternative viewpoint is to consider the data structure and to perform concurrent operations on all elements of the structure. This is the model adopted for obvious reasons by SIMD and vector-based pipelined computers. However it has a much wider applicability than has been exploited to date and it provides an ideal framework for the exploitation of asynchronous parallel systems.

Because this model has not been well developed, it tends to be dismissed as too restrictive and inefficient. It will be shown here however, that properly developed the model can be applied to a wide range of applications and implemented on many diverse architectures. Indeed, it will be shown that the model developed is far from being restrictive, in that it may efficiently simulate a MIMD environment on a SIMD machine, where the processes are replaced by active data and load balancing is achieved by the equitable distribution of that active data. Programming over such a model can be made more attractive and less prone to error than its sequential counterpart. This is because the semantic content of the instruction has been raised without introducing issues such as synchronization and deadlock.

One of the reasons that parallelism has not been exploited as much as it could have been over the last decade, is the issue of mapping the problem domain over the machine domain. To date, this has been the responsibility of the user, who has had little assistance. No model of parallel computation can be considered global unless it describes attributes of a computation, rather than a particular computer system. One of many counterexamples to this is DAP Fortran, which describes computations based on synchronous data operations for a particular SIMD array processor. In order to provide portability and to make a model general, virtualization schemes must be considered which abstract the user away from the physical realization of the computer system. This can be compared to virtual machine architectures, as found in UCSD Pascal, where the system is implemented over a virtual stack based machine. However with parallel systems, it is the structure of the underlying architecture, rather than the instruction set that must be hidden.

The active-data model of concurrency provides for simultaneous operations across any data structure or object. The operations on the individual elements of the structure take place as if simultaneously, hence even if implemented over MIMD hardware, we have a synchronous model, without deadlock. This model is best illustrated by considering the following example, which contains both an expression of, and an assignment to the same structure or object:

$$A \leftarrow e(A, B)$$

In this example the expression over A and B would be evaluated for all elements of A, before an assignment was made back to the structure A. This model is expressed in APL (Iverson 1962), CmLisp (Hillis 1975) and the array extensions of FORTRAN 8X.

This model of concurrency can be readily applied to any replicated system, indeed no assumption has yet been made about the control strategy, and implementations of this model could equally well be made over MIMD or SIMD architectures. The major problem however, is one of mapping data onto the available processors and maintaining a high utilization of those resources.

The virtualization scheme proposed for this active-data model is based on the data structure and more importantly its activation, which provides a distributed data-driven control structure. The model uses the abstraction of a single virtual processor per data structure element, although at any given time only few virtual processors may be allocated to a real resource (the real processors in the system). This scheme is analogous to virtual memory for the storage of large structures. Indeed the model may contain a memory virtualization to provide processing over structures too large for the memory requirements, thus virtual processors may themselves reside in backing store.

In practice not all elements of a data structure will require updating, and in a sequentially based language the selection is performed by control operations such as indexing arrays and list traversal. Both selection techniques may also be augmented by conditional control structures in the programming language, such as If and Case statements. The active-data model is based on concurrently processing the active (selected) data structure elements. Typically the procedure for writing programs in the active-data model can be decomposed into the following sequence:

- define elements of structure for updating, and

- evaluate the method to update the structure.

This partioning is illustrated schematically in Figure 18.2, where both stages can be thought of as different methods applied to a data object. In relating this to conventional sequential programming models, the activation would translate into some control structure and the method or code would represent the body of the construct.

In general, activation may be by:

- sub-structure selection, a row from a matrix for example;

- structure replication, creating a matrix by the repetition of a row for example;

- association, a class of elements identified by some key;

- propagation or discrete function evaluation, using mappings between sets; and

- the state of associated structures or expressions, such as spatially defined conditional operations.

Although the first two create deterministic sets of activated elements (assuming static arrays), and can be explicitly defined at compile time, the latter operations are data dependent and mappings from active data structure elements to physical resources can not be determined at compile time.

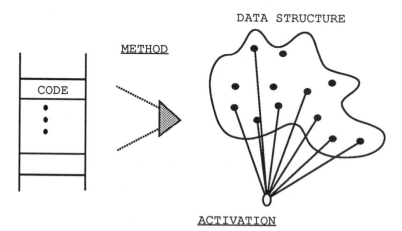

Figure 18.2: A diagram illustrating the partioning of the active-data model into activation and methods.

It is possibly to extend this model to one of an encoded activation stage, which could select an appropriate method for each encoding. For example, consider an activation similar to that shown in Figure 18.2, but which coloured a subset of the elements of the structure black and the remainder white. It is possible to use this coding as a method selector. This of course is nothing more than the distributed implementation of the If ... Then ... Else construct. The important point to note it that the activation may be decoupled from the method and an encoding or tag associated with the data structure elements. Sequencing of operations may thus be data driven, an effective technique for implementing load balancing. For example, provided that both sets more than cover the available resources, then an efficient execution of both methods is possible, even on a SIMD system. This represents the implementation of multiple instruction streams over a SIMD model, through sequencing and load balancing. Schemes for an equitable distribution of the two sets over the available resources are considered in section 18.3.

Of course it is possible to bring to bear the whole armoury of structured program design to this methodology, with a hierarchy of data structures and appropriate classes of methods. Figure 18.1 for example could represent a user selecting a group of files for processing from some graphical interface, in a database application, with the code being built from lower level methods, including further activations based on data association or discrete function evaluation.

In order to consider the problems of load-balancing, an abstract machine and some implementation details must be considered. The abstract machine used here reflects the duality of the model; the separation of activation and methods. It comprises a set of processors for performing routing operations, to activate and distribute data, and another set of processors to process the data (these would also provide local activation of data).

Figure 18.3 illustrates the abstract machine. The array of communication processing elements (CPEs) are joined by a communication network and each CPE is also connected, by shared memory, to a processing element (PE), which performs the computation required by the model. Both PE and CPE may require other memory, however a number of data queues are maintained within the shared memory:

Network queue: This is for packet routing only, and provides a queue of packets waiting to be forwarded through the network by the CPE, this queue may be filled by the CPE or the PE and is emptied only by the CPE.

Work queues: A number of queues are maintained to hold active data for currently-active methods. These hold data packets waiting to be processed by the PE, these queues may be filled by the CPE or the PE, but are only emptied by the PE.

The network may be of any topology, but simple routing strategies exist for regular networks. Implementation restrictions are likely to define its topology. On each cycle, the CPE should receive and process at least one data packet from each di-

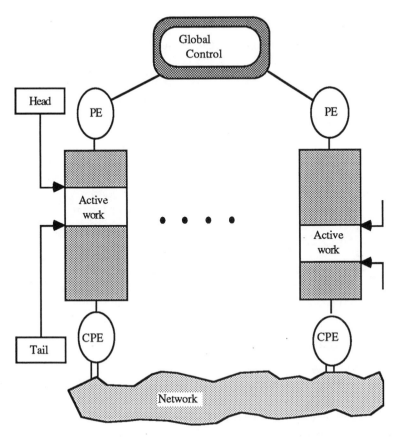

Figure 18.3: A schematic diagram of the active-data abstract machine.

rection in the network. The processing required simply differentiates which queue
the data should be placed in, the network queue, for further forwarding, or one of
the work queues if the packet has arrived at its destination. The destination may
be defined by an address held within the packet, or by some state held locally in
the CPE. This latter situation occurs in some load-balancing algorithms.

A second network, the global control, links the PEs in a fan-in/fan-out tree. This is
to provide broadcast and reduction operations, which are central to the active-data
model. The requirement for these control mechanisms is illustrated by the simple
example of operations over sets, where the most frequent operations involve both of
these global communications operations. The set is one of the most powerful data
structures in many symbolic applications. It is used extensively in load-balancing
algorithms. Examples of set operations are:

- is_in: requires association of an object with every member of the set, and

- max: requires the reduction over the set using the greater than operation.

- sum: requires the reduction over the set using the add operation.

18.3   Load Balancing

This section describes and enumerates different load balancing techniques and flags the requirements for efficient implementation. Both static and dynamic load balancing will be considered and related to the activation methods described above. These techniques have direct analogue with process load balancing, but are more efficient to implement and moreover are more closely represenative of the semantics of parallel algorithm development.

Put simply, load balancing is the process of evenly distributing entries in the work queues, either based on local or global information concerning the distribution of work. The techniques can be classified as follows:

18.3.1   Static deterministic

We can assume that for static structures, an even distribution of data structure elements over the available resources will be made at compile time. During execution however, sub-structures may be activated from the evenly distributed source structure. In the worst case for example, it is possible that all activated data may be associated with a single processor. A more realistic situation is illustrated in Figure 18.4, where a row has been selected from a matrix, mapped over the available resources. The two axes represent processors and memory, or alternatively parallel and sequential execution.

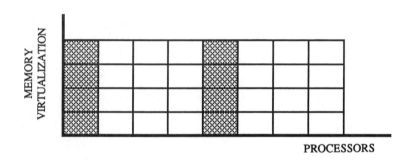

Figure 18.4: Load balancing on a static deterministic activation.

It is possible with a statically distributed regular structure to redistribute the elements of a selected sub-structure to obtain a more evenly distributed load. Indeed this is the major technique of current SIMD application implementations. For certain classes of structure and transformation, there are reported techniques which can exploit the deterministic nature of this re-mapping, such as the parallel data transforms described by Flanders (1982). Alternatively, providing a distributed address space by means of a packet routing protocol will handle any permutation of data.

18.3.2   Static non-deterministic

A second class of load balancing is associated with the non-deterministic distribution of active data structure elements. Such distributions arise from activation by conditional operations for example. Here it will not be known a-priori, where the activated elements will be found. Provided that it can be expected or determined that a reasonably even distribution of active elements will result, the use of work queues as defined in the abstract machine, will allow automatic load balancing in such circumstances. This is illustrated in Figure 18.5, which shows the activation over a regular data-structure. It can be seen here, that given active element queues, only two passes of the method would be required. In current SIMD implementations, such as the ICL DAP, this technique is not supported by the language, because the data structures within the language only form one layer in the sequential space. In fact this space is explicitly programmed by the user application.

The implementation of the queue model on a machine like DAP is not trivial, as the distributed queue structure requires local addressing, which the DAP does not support directly. It can of course be simulated over a memory block in logarithmic time, using the bit address as an activation mechanism, and this is described below. The implementation is not efficient and there is a trade-off therefore in the complexity of the method implemented; for a very simple method the technique would not be appropriate.

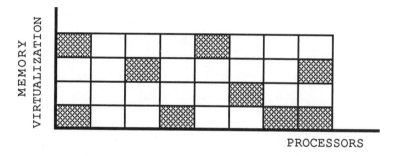

Figure 18.5: Load balancing on a static non-deterministic activation.

18.3.3 Dynamic non-deterministic

The final class of load balancing is the most general and would provide load balancing in all eventualities. The situation in which this would be used is where an uneven and non-deterministic distribution of active-data structure elements would result. For example this may occur in discrete function evaluation or marker propagation. The situation is illustrated in Figure 18.6. What is required is a redistribution of active data structure elements, which can be achieved either by local or global algorithms. In a local scheme, packets of data would be redistributed based on local comparisons over work queue lengths. However this is likely to result in a great deal of unrequired activity in the communications network.

A global scheme may be implemented by taking a histogram of the distribution of the load, from which a desired average load could be established. From this desired load, a routing strategy can be adopted, which forwards packets of information, not on address, but on local load compared to desired average load. For example if the average load is broadcast to all processors, through the global control network, then if the processor's local load is to within some bounds greater than the average load, then it would emit packets; if it were less than the local load, it would accept packets; otherwise it would simply forward packets. The direction of forwarding would need to producing a randomizing effect, but also adapt to traffic density.

The problem in load balancing non-regular or linked structures is in maintaining the integrity of pointers across the array. The migration of virtual processors can of course be achieved by addressing virtual nodes, rather than physical nodes and having a routing strategy which adapts to any load balancing migration. Alternatively data may be diffused temporarily to achieve a balanced load for processing a particular method, and then by attaching return addresses or 'springs' to the data

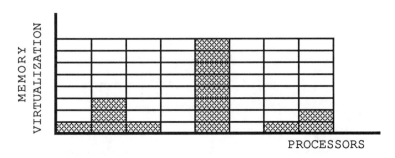

Figure 18.6: Load balancing on a dynamic non-deterministic activation.

packets when farmed out, the original connected structure can be recreated, by an addressed-packet, final-routing phase.

18.4 Implementation Issues

Little reference has yet been made to the implementation of the abstract machine. There are however, a number of issues that must be taken into consideration, when choosing an implementation.

First consider the MIMD/SIMD issue. Obviously the MIMD implementation has advantages in that it may run different methods on different processors at the same time. This implementation would be ideal for a data driven object oriented paradigm, where it can be conceived that the packets flowing within the system may contain method selectors. However, it has already been shown that by exploiting load balancing techniques, multiple methods may also be efficiently implemented on SIMD implementations. The methods are executed in sequence and provided that there are sufficient active elements for each method, then a full utilization of the available resources may be achieved. There are also positive advantages for SIMD implementations; in many methods there is a requirements for broadcast and reduction communications within the PE structure. In a SIMD machine, these operations would be implemented as a part of the global control scheme. Data within the global control word may by used as an associative key, and a reduction by logical sum tree of a value in each PE is a common SIMD control mechanism; the root of the tree being sensed by the global controller. A MIMD implementation would have to source the broadcast data from either a nominated processor or could perhaps source it from any processor. Moreover the realization of broadcast and reduction is likely to be provided by the slow action of distributed communicating processes, rather than by electronic (or optical) signals, as could be implemented on a SIMD machine.

At Southampton we have been investigating the implementation of this model of parallelism over both MIMD and SIMD structures, indeed the aim of this research is to provide a fusion of such architecture in complex systems, so as to have a common system target architecture. The SIMD implementation has been over the RPA, and the MIMD implementation over transputer networks. The SIMD implementation is described here in detail, as it has been a more interesting and difficult piece of research.

The SIMD implementation issues are found at the lower levels of systems code, and have been implemented in micro-code over the RPA simulator (Jesshope and Stewart 1986). An exploratory implementation of a packet routing communications schemes has been made, which has involved packet forwarding protocols and queue management. The difficulties encountered and solutions found are outlined here.

18.4.1 The RPA PE

The RPA processing element (PE), like most other SIMD PEs selects an input from one of a number of neighbouring PEs, in this case from one of the four orthogonal neighbours. However unlike most other arrays, the RPA PE is able to locally choose the direction selected. This local asymmetric behavior is of great benefit in mapping regular data structures over the RPA array, and is described in more detail in (Jesshope et al 1987, Jesshope 1987). It will be shown here that this also provides for an efficient utilization of the communications structure, when implementing the non-deterministic packet routing communications system.

In order to better understand the implementation, a brief description of the PE is required. The RPA PE has a 2-bit source and 2-bit result bus. The result bus can be connected, using a local control field, to the source bus of one of the four neighbours. Two bits of data may be passed into and out of each PE in a single cycle, over the same leg if necessary (ie north selects south and south selects north). The storage in each PE comprises two eight bit stacks (capable of pushing two bits per cycle) and 64 bits or RAM organized as an 8 × 8 block, with byte-wide, parallel-to-serial conversion provided by two shift registers between the source and destination busses. These shift registers also contain parallel comparator circuits (giving $<$, $>$, and $=$ as two bits enabled onto the source bus). It is also possible to locally address the word store, using the top three bits of the bitstack to select one of the eight bytes of RAM.

18.4.2 Packet Routing Protocol

As is seen above, the basics of implementing an active-data model of concurrency are two-fold:

- a packet addressing scheme to provide a "shared" memory address space for each data-structure under consideration. This space may be augmented include a coding over the data-structure or object value space;

- a mechanism for maintaining a number of stacks or queues of active-data.

For the packet switched communications scheme, the address of the data is provided at the source, as a part of the packet, but in the RPA hardware, data must be selected at its destination. We therefore have to implement a protocol to invert the sense of the direction control from a 'send to' to a 'receive from' address, and resolve any contention that may result. This protocol is implemented in micro-code and must establish as many channels between PEs as possible. This is achieved by a sequence of polling operations. For obvious reasons we have chosen a local but deterministic algorithm, which is efficient to implement. Although it does not provide the optimal solution, it provides a good compromise between the number of channels implemented and the overhead required to establish those channels, both of which contribute to the effective bandwidth over those channels.

Absolute data packet addressing is used, because the PE can provide rapid routing information using the eight bit comparitor. Although we anticipate building a 32×32 RPA, this scheme would support arrays of up to 128×128.

The packet routing is performed on a set of addressed packets of data, distributed in stacks across the RPA array. Each packet is forwarded by the protocol described below, so that it will eventually arrive at its addressed location. Obviously the RPA is only able to configure communications channels to its immediate neighbours (at full parallelism) and thus long distance communications are implemented by multiple cycles of the same algorithm. Although not implemented, it is possible to run packet routing and data injection as two 'concurrent' methods, such that when the packet routing load became light, more packets could be injected into the routing queues.

In each cycle, each processor is capable of transmitting and receiving a packet of data from one of its immediate neighbours. Provided that both events occur, buffering is performed in internal RAM, for packet sizes up to 32 bits (16 bit address and 16 bit data).

Because of the reconfiguration registers, the direction from which a packet is received is calculated from a locally determined protocol. Each processor is polled by its four nearest neighbours. The net effect of this protocol is to establish a set of closed arcs across the array, over which data is transmitted. This transmission will advance as many packets as possible towards their destination.

There is of course the potential for contention, where more than one processor wishes to send a packet to the same destination, using different direction codes. There are two strategies to avoid this contention; a backtracking algorithm can be implemented to reroute one or more of the contending packets, so that buffering is provided in the network, or alternatively, a priority scheme can be implemented so that one of the contending packets will always win, in which case buffering may be required at one or more of the transmitting neighbours. We have adopted the latter strategy, as it is deterministic (packet routing can be interrupted on stack overflow) and in which no packet is forwarded to a more distant location from its destination.

The buffering required can use the implementation of packet queues in globally addressed external memory, which is described below. The algorithm to establish a set of channels is given below, for one cycle, and Figure 18.7 illustrates the result of such a polling process. Four successive cycles rotate the direction priority order, so as to achieve an equal, average-net-flow across the array.

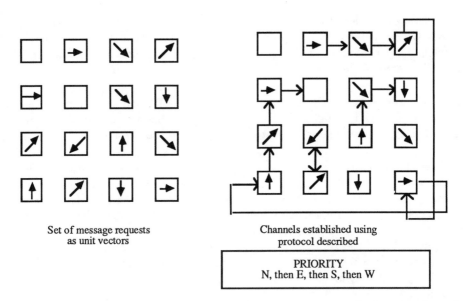

Set of message requests
as unit vectors

Channels established using
protocol described

PRIORITY
N, then E, then S, then W

Figure 18.7: An illustration of a cycle of the routing protocol, showing first the requests as a set of unit vectors in each PR and also the channels established using the protocol.

In the following description of the protocol, a set is mapped over the array of PEs, with one member of the power set mapped onto each PE. To implement the protocol we define the following signals:

- R_i, $i \in \{n, s, e, w\}$, the set of PEs requiring channels in $n, s, e,$ and w respectively.

  $\exists\ R_i \cap R_j \neq \emptyset,$

- A_i, $i \in \{n, s, e, w\}$, the set of PEs having their requests acknowledged.

  $\forall i, j\ R_i \cap R_j = \emptyset,$

- S_i, $i \in \{n, s, e, w\}$, the set of PEs enabled to receive from a given direction, and W, the set of PEs currently able to accept a channel.

A function $\mathit{Shift}(X, \mathit{dir})$ is required to describe the protocol, which shifts the set X of signals in the direction $\mathit{dir} \in \{n, s, e, w\}$.

Protocol

North channel:

$$A_n \leftarrow R_n$$

$$S_s \leftarrow Shift(A_n, n)$$
$$W \leftarrow -S_s$$
$$R_e \leftarrow R_e - A_n$$
$$R_w \leftarrow R_w - A_n$$

East channel:

$$A_e \leftarrow Shift(W, w) \cap R_e$$
$$S_w \leftarrow Shift(A_e, e)$$
$$W \leftarrow W - S_w$$
$$R_s \leftarrow R_s - A_e$$

South channel:

$$A_s \leftarrow Shift(W, n) \cap R_s$$
$$S_s \leftarrow Shift(A_s, s)$$
$$W \leftarrow W - S_n$$
$$R_w \leftarrow R_w - A_s$$

West channel:

$$A_w \leftarrow Shift(W, e) \cap R_w$$
$$S_e \leftarrow Shift(A_w, w)$$

The implementation of the above protocol is very efficient, as set operations such as intersection and difference can be performed at 10 billion operations per second on the 1024 RPA array. This scheme requires between 14 and 22 microseconds to forward up to 1024 32 bit packets through the array, assuming a 100 nanosecond clock cycle. The minimum figure is the protocol and transmission, on top of which there are several possible buffering stages, for injection, for contention and for unloading. This represents in the worst case, a rate of 1.5 Mbits per second per processor, or a total communication bandwidth over a 1024 PE RPA of 1.5 Gbits per second. Larger packets will amortise the protocol over more data bits and will asymptote to a peak rate of 6 Mbits per second per processor, which assumes no buffering.

18.4.3   Stacks and Queues on SIMD Machines

It is the implementation of the buffering techniques that causes concern for the implementation of an active-data model on a SIMD machines. Some inefficiency can be incurred in a software implementation of packet routing, because of the high communications bandwidth per unit processing power, compared with more complex processors, such as the transputer. The high relative bandwidth comes about because the processor is narrow, frequently 1 bit, and therefore processing operations on a single processor are slow requiring many bit-serial operations.

It is this same reason that makes the implementation of stacks and queues problematic. Because the processor is simple, it would be uneconomic to provide relative addressing at the individual processor, or at least to provide relative addressing to the external RAM; the RPA does support local relative addressing to its internal RAM. External relative addressing would have required an address but from each PE on the 16 PE RAP chip, which would have made the pinout of that chip intolerably large. Even 8 bit relative addressing would require an additional 128 pins, more than doubling the current pin count, and would be particularly restrictive in practice. Instead we have opted for a software solution. Indeed the problem is similar to that of floating point normalization when performing array valued floating point addition. For in this problem each processor calculates a local normalization shift distance, which must be used to shift its local mantissa, moreover this must be achieved with a common instruction and a common address.

The complexity of address decoding is logarithmic, so it should be possible to provide a solution which normalizes floating point numbers, or aligns the heads of stacks and queues in logarithmic time. Indeed because the stack is accessed from only one position, it should be possible to maintain a stack which can be read from or written to in constant time.

The assumptions above are only true if all data can be moved in constant time. However, because of the bit serial nature of the processor, the complexity of these operations also contains a term in the "word" length. For example a b bit mantissa can be aligned to an arbitrary bit position in:

$$t = O(b \log_2 b)$$

common-address memory references. This $b \log_2 b$ complexity may seem an excessive overhead until it is recognized that addition on that same b bit mantissa will require $O(b)$ single bit operations.

The same analysis applies to the alignment of a queue, if the queue is contained in a b bit block, the tail can be aligned on an arbitrary bit position in:

$$t = O(b \log_2 b)$$

common-address memory references. The queue however, will be maintained in increments of the packet size p and thus the complexity will be:

$$t = O(b \log_2(b/p))$$

common-address memory references. Unfortunately if p is large, then b is also likely to be large. But again we must remember that each packet in the queue must be processed by the local processor, either bit serially, in arithmetic operations, or bit serially through a communications channel. But in this case, unlike the floating point normalization operation, on a single cycle of the packet routing engine, one packet of data is added to the queue, which in turn will require the movement of the entire b bit block of memory. Thus if the queue length were q packets, i.e.

$$q = b/p,$$

then the constant of proportionality for adding a single packet onto the queue is q, and for aligning the queue is $q \log_2 q$. The stacks or queues are of course distributed, so we must consider a little more detail of the mechanics of the operations, in order to illustrate the optimizations.

Adding to an active-data queue during a routing phase of the active-data model is essentially a stack push, see Figure 18.8 for details of the implementation; the local length field must be updated and the packet must be added at the address *lowmem* . To accommodate the new packet, the existing stack must also be moved up in memory by the size of a packet, p bits. Remember that these are conditional operations, as the packet routing is an non-deterministic operation, and the arrival of a packet will be flagged at each processor. It will be seen later that a global maximum length of queue/stack is required.

In Figure 18.8, an example of this operation is illustrated, and it is assumed that PEs 1, 3 and 5 of Figure 18.8a will receive packets of data from the network. In this case the result of the stacking operations will be as shown in figure 18.8b. It should be noted that the global information *maxlength* is mandatory, to predict stack overflow. To access data from the stack or queue requires the reverse of the procedure above for the stack, but in the case of the queue, because data is removed from the opposite end of the structure, a realignment on the head of the queue is required, prior to the stack push operations. This is illustrated in Figure 18.9.

Clearly it is possible to move the data in place, between *lowmem* and *lowmem* + *maxlength* $* p$. At each stage i,

$$1 < i < \log_2 \text{maxlength},$$

then blocks of data of length,

$$(\text{maxlength} - \text{maxlength}/2^i) * p$$

may if activated, require shifting up in memory by

$$p * \text{maxlength}/2^i$$

bits. The activation criterion for this shift is whether the value

$$maxlength - length$$

has a bit set in the i th position of its binary representation. This flag is available from a once-only global/local subtraction of lengths. The total operation count to implement this alignment over a distributed queue of p bit packets, whose maximum length is maxlength, roman is therefore as follows:

$$p * \log_2(maxlength - 1)$$

binary data move operations across the array.

Obviously the linear operations of push and pop on stacks should be more frequent than those of realignment, to minimize the logarithmic dependencies of realignment. This argument also provides for greater flexibility in exploiting asynchronous load

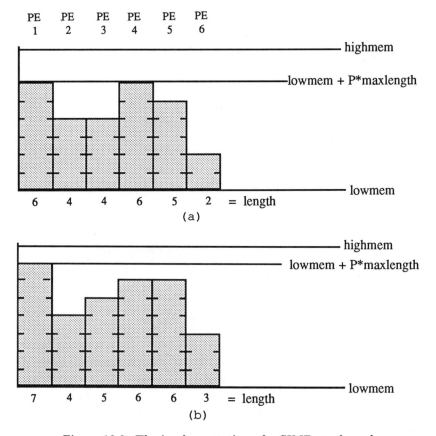

Figure 18.8: The implementation of a SIMD stack push.

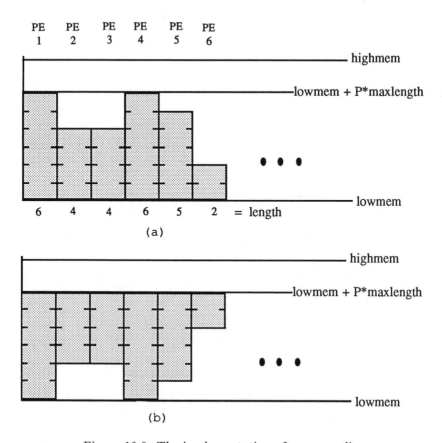

Figure 18.9: The implementation of a queue align.

balancing within the active-data model. A counter to this argument, is that the incremental cost of flowing a packet through a heavily loaded system, incurs the overhead of the greater sack length. However, because the active-data model is data driven, packet routing can be suspended at any time in order to flush out queues by processing the active data. Indeed with hardware implementations (indirect addressing is mandatory here) there is no reason why asynchronous concurrency should not be used between processing and packet routing.

There are optimizations that may be performed on the stack based operations, that incur a near incremental cost in push and pop operations, but they involve a periodic realignment stage. In the scheme implemented, a number of overflow slots are provided for the stack. The amount of shifting required now depends on the number of overflow slots that are required on each push or pop operation. For example, if four slots are provided, on the first pass, one p bit block write is required, plus a flag write to signify the slot vacancy. On the second pass, some push or pop operations will fill the first slot, and some may use the second slot.

Depending on the sparsity and the activation of the push and pop operations, at some stage later, it will be found that all slots are full and a realignment will be required, thus emptying the overflow slots

This alignment with n overflow slots would require

$$p * maxlength * \log_2 n$$

common-memory references. However it should be noted that the maximum cost of the push operation is $4 * p$ shift operations and if the distribution of packets is even, the average cost should be near $2 * p$ bit shifts. For a given expected queueing density, it should be possible to optimize the operation of the stack and queue, to provide the right balance between stack and queue operations and between queue length and occupancy slot optimizations. What is being traded is dense operations that move the whole queue, with sparse operations on a single packet.

Simulations on various simple data transformations over the RPA range (Fong 1987) has shown that even with little data on average per processor, queue lengths can build up alarmingly. An illustration is given in Figure 18.10, for a perfect-shuffle permutation across the 2-D RPA array, with initially one data packet per processor. It can be seen that the maximum buffering requirement is for 10 packets.

18.5   VLSI: The Future and the Sequential Mould

Replication is likely to be the major mechanism for increasing computer performance in the coming decade. This trend is already being observed in commercial computer designs, in the class of computers known as near supercomputers, such as AMT, DAP, Meiko computing surface, Alliant FX/Series, Intel IPSC, NCube and Sequent Balance, all of which contain multiple processors. The reason for this surge in interest in this general class of architecture are the needs of VLSI, which are twofold; economic VLSI designs require:

- regular layout with regular interconnect patterns, and

- the economies of scale.

Both of the properties above are found in memory chips and this contributes directly to their low cost. Unfortunately memory does not contribute to increased processing performance. Indeed it perpetuates the what I call the sequential mould of programming, which actively discriminates against the successful exploitation of replication.

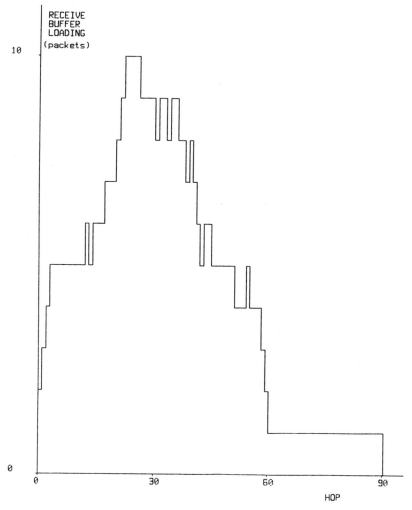

Figure 18.10: The accumulated statistics of a perfect shuffle transformation of 1024 data packets on a 1024 PE array, with the data evenly distributed initially.

The sequential mould states that: "A machine may have unused resources, providing that those resources are memory and NOT processors". Nobody minds running a 5 Mbyte application on a 10 Mbyte machine, but to use only 500 out of 1000 available processors would be considered a heinous crime. This paper has presented an abstract model of concurrency, which attempts to alleviate the insecurity felt by programmers in the sequential mould, when not all processors in the system are active. It does this by proving dynamic load-balancing over a distributed data model of concurrency.

Of course, the introduction of such a schemes necessarily introduces overheads, which must use a finite amount of the machine's resources. The overheads in this model are in general small, compared with the analogous model of dynamic process-based load balancing. Indeed this is the key thesis of this paper; the data abstraction of processing is inherently more suitable for efficient utilization of highly replicated computing resources than the process based abstraction.

References

[1] P. M. Flanders, *A unified approach to a class of data movements on array processors,* IEEE Trans. Comput., vol. C-31, pp. 405-408, (1982).

[2] J. L. C. Fong, *Microprocessor array routing simulator,* 3rd year report, University of Southampton, Dept. of Electronics and Computer Science, 1987.

[3] W. D. Hillis, *The Connection Machine* , MIT Press, 1985.

[4] K. E. Iverson, *A Programming Language* , Wiley, 1962.

[5] C. R. Jesshope, A. Rushton, A. Cruz, and J. Stewart, *The structure and application of RPA a highly parallel adaptive architecture,* in *Highly Parallel Computers* , Elsevier Science Publishers, North-Holland, pp. 81-95, 1987.

[6] C. R. Jesshope, *The RPA as an intelligent transputer memory system,* in *Systolic Arrays* , Adam Hilger, (Moore, McCabe and Urquhart, Eds), pp. 283-293, 1987.

[7] C. R. Jesshope and J. M. Stewart, *MIPSE - a micro-code development environment for the RPA computer system,* Software Engineering 86 , Peter Peregrinus, (Barnes and Brown, Eds), pp. 184-196, 1986.

Chapter 19

A Fine-Grain, Message-Passing Processing Node [1]

William J. Dally [2]

Abstract

This paper describes a processing node for a fine-grain message-passing concurrent computer. The node consists of a processor, a communication unit, and a memory. To reduce the overhead of message passing and task switching to $5\mu s$, the node incorporates a send instruction, a fast communication system, hardware message buffering and dispatch, and a general translation mechanism. These mechanisms work together to implement a fine-grain programming system.

19.1   Introduction

The natural grain size of many parallel algorithms is about 20 instructions. To fully exploit the concurrency in such algorithms, we must be able to efficiently execute tasks of this length. The message transmission and reception overhead of existing systems is in excess of 200 instruction times. With such a large overhead, these systems must execute tasks at the artificially large grain size of about 1,000

---

[1] The research described in this paper was supported in part by the Defense Advanced Research Projects Agency under contracts N00014-80-C-0622 and N00014-85-K-0124 and in part by a National Science Foundation Presidential Young Investigator Award with matching funds from General Electric Corporation.
[2] Massachusetts Institute of Technology, Cambridge, MA

instructions. If we can reduce the overhead and operate at the natural grain size, we can effectively apply 100 times as many processing elements to the problem.

At MIT we are developing the J-Machine [12], a fine-grain message-passing concurrent computer that will efficiently execute programs with 20-instruction tasks. The J-Machine consists of a number (initially 4096) of possibly different processing nodes that communicate over a high speed network. The network is implemented using a Network Design Frame (NDF) [10], a communication controller integrated into the pad-frame of a chip. The center of the NDF chip is used to implement a node-specific processor. The primary processing node of the J-Machine is an NDF wrapped around a Message-Driven Processor (MDP) [9], a symbolic processing element.

A J-Machine constructed from MDP/NDF processing nodes combines features of both message-passing and shared memory machines. Like the Caltech Cosmic Cube [25], the Intel iPSC [18], and the N-CUBE [21], each node of the J-Machine has a local memory and communicates with other nodes by passing messages. The J-Machine can exploit concurrency at a much finer grain than these early message passing computers. Delivering a message and dispatching a task in response to the message arrival takes $5\mu s$ on the J-Machine as opposed to 5ms on an iPSC. Like the BBN butterfly [3] and the IBM RP3 [22] the J-Machine provides a global virtual address space. The same IDs (virtual addresses) are used to reference on and off node objects. Like the InMOS transputer [17] and the Caltech MOSAIC [20] the MDP/NDF is a single chip processing element integrating a processor, memory, and a communication unit.

The next section introduces the problems associated with fine-grain concurrent computation by means of a concurrent factorial program. In Section 19.3 the mechanisms provided by the MDP/NDF to support this style of programming are presented. The SEND instruction, message delivery, message buffering, scheduling and dispatching, and the MDP translation instructions are described. This section also discusses some of the features we didn't implement and the reasons why they were abandoned. Section 19.4 illustrates how the MDP/NDF mechanisms work together to execute the factorial program from Section 19.2.

19.2   The Problem

Concurrent programming is often considered harder than sequential programming because of partitioning, communication, and synchronization. If a machine is programmed at a very low level, with the programmer explicitly specifying the partition and the communication, concurrent programming can indeed be a difficult task, and the programs produced are rarely portable. However, with suitable programming abstractions [11], concurrent programming need be no harder than sequential programming.

```
(Integer) factorial
  ^ 1 rangeProduct: self '

(Integer) rangeProduct: n
  | mid |
  self = n ifTrue: [^self].
  mid <- self + n // 2.
  ^ (self rangeProduct: mid) * (mid+1 rangeProduct: n) '
```

Figure 19.1: A Factorial program in Concurrent Smalltalk. This program executes an average of 8 assembly instructions in response to a message.

Consider the factorial program shown in Figure 19.1. The program calculates $n!$ by recursively dividing the interval of integers to be multiplied. To compute $n!$, the depth of the recursion is $\log_2 n$. This program is patterned after one by Theriault [29] p.33. and is written in Concurrent Smalltalk (CST) [5], a concurrent programming language based on Smalltalk-80 [14]. A description of the programming language is beyond the scope of this paper.

The computation graph for this program is shown in Figure 19.2. Each node of this graph represents a context object created to hold the state of one activation of the **rangeProduct:** method. A message from the parent node in the tree creates the context, sends two **rangeProduct** messages, and then suspends awaiting the replies. This initial activation executes 16 *working*[3] instructions for internal nodes. Leaf nodes execute only 6 instructions to send their argument back. The first reply message resumes the context, saves its argument, and suspends (4 instructions). The second reply performs the multiply and sends a reply up the tree (8 instructions[4]). The average number of working instructions executed in response to a message is ≈ 8.

Operating at a fine-grain, in addition to exploiting more concurrency, also simplifies communication. Using the natural partition of the program, communication is implicitly specified by interactions between the named objects in the program. Synchronization is also implicit. The arrival of a message containing the required data schedules execution. For example, in Figure 19.1 the expression (self rangeProduct: mid) causes a rangeProduct: message to be sent. The arrival of this message schedules the execution of the method.

---

[3] These instruction counts consider only useful code. Message reception and address translation overheads have been factored out.

[4] If the multiply exceeds the machine precision additional time is required to perform a bignum multiply.

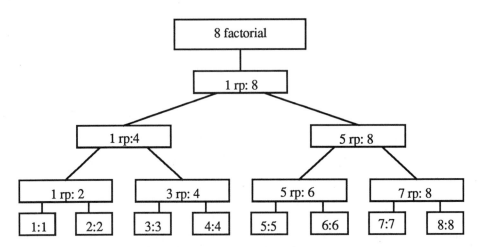

Figure 19.2: Computation graph for the factorial program.

In some concurrent programming systems, communication is made difficult by non-uniform naming: local objects are referenced differently than non-local objects. In the Cosmic Kernel [28], for example, local objects may be referenced through a pointer, while global objects require an explicit message send and receive. Providing a global address space allows objects to be referenced via a single mechanism (the virtual address) regardless of their location, and relieves the programmer of the bookkeeping required to keep track of node numbers. Programs become both easier to write and more portable.

The logical partition of most programs is fine-grained (8 word objects, 20 instruction methods). Unifying this logical partition with the physical partition of a program simplifies programming and results in greater concurrency. To efficiently execute the resulting fine-grain program, a machine must have fast communication, low message reception overhead, and low scheduling overhead. To simplify communication between named objects, a machine must support a global address space. The mechanisms used by the J-Machine to meet these requirements are described in the next section. While we intend to use these mechanisms to execute object oriented [27] or actor [1] [2] programs, the same mechanisms are required to support shared memory programming systems such as Multilisp [15].

19.3   Mechanisms

The following sequence of actions is involved in sending and receiving a message.

1. The originating node *translates* the ID (address or name) of the destination object into a destination node number.

2. The originating node *sends* the message.

3. The network *transmits* the message to the destination node.

4. The destination node *receives* the message.

5. The destination node *decides* whether to execute or buffer the message.

6. If the decision is to buffer, the node *buffers* the message.

7. Eventually, the message is executed and the node *translates* the receiver ID and message selector into a receiver address and a method address.

8. The method is executed.

9. The method suspends and transfers control to the next message.

On a 4K node J-Machine, this sequence of operations takes $\approx 5\mu s$[5]. The mechanisms implemented in the hardware of the system to accelerate this operation include.

1. A send instruction (2).

2. A fast communication mechanism, the NDF [10] (3).

3. A message unit that controls the reception and buffering of messages (4 and 6).

4. A scheduling mechanism that (5) decides when to preempt execution and (9) selects a message to be executed when a method suspends.

5. A general translation mechanism (1 and 7).

Send Instruction

The MDP injects messages into the network using a send instruction that transmits one or two words (at most one from memory) and optionally terminates the message. The first word of the message is interpreted by the network as an absolute node address (in x,y format) and is stripped off before delivery. The remainder of the message is transmitted without modification. A typical message send is shown in Figure 19.3. The first instruction sends the absolute address of the destination node (contained in R0). The second instruction sends two words of data (from R1 and R2). The final instruction sends two additional words of data, one from R3, and one from memory. The use of the SENDE instruction marks the end of the message and causes it to be transmitted into the network. In a Concurrent Smalltalk message, the first word is a message header, the second specifies the receiver, the third word is the

---

[5] This estimate assumes that the message is not buffered, that all caches hit, and it ignores time spent actually executing the method.

```
SEND    R0           ; send net address
SEND2   R1,R2        ; header and receiver
SEND2E  R3,[3,A3]    ; selector and continuation - end msg.
```

Figure 19.3: MDP assembly code to send a 4 word message uses three variants of the SEND instruction.

selector, subsequent words contain arguments, and the final word is a continuation. On our register-transfer simulator, this sequence executes in 4 clock cycles.

Early in the design of the MDP we considered making a message send a single instruction that took a message template, filled in the template using the current addressing environment, and transmitted the message. Each template entry specified one word of the message as being either a constant, the contents of a data register, or a memory reference offset from an address register (like an operand descriptor). The template approach was abandoned in favor of the simpler one or two operand SEND instruction because the template did not significantly reduce code space or execution time. A two operand SEND instruction results in code that is nearly as dense as a template and can be implemented using the same control logic used for arithmetic and logical instructions.

Message Communications

Communication between nodes is performed by the NDFs on the nodes along the route. The NDF performs routing, buffering, and flow control to deliver messages in a 2-D mesh connected network. These functions are performed entirely within the NDF. No memory bandwidth or CPU time on intermediate nodes is used by message delivery. NDFs are connected by 9-bit communication channels that are expected to operate at 50MHz for a throughput of (450Mbits/sec). The propagation delay through this self-timed router is 20ns. Wormhole routing [7] [8] [19] is used to give an idle-network latency of $20D+80L$ ns, where D is the distance in channels and L is the message length in 36-bit words. In a 4096 node machine, for example, the average distance is 42 and the worst case distance is 126 for respective latencies of $1.3\mu s$ and $3\mu s$ respectively. In [7] it is shown that these latencies increase only slightly as network traffic is increased to within 25% of saturation.

Figure 19.4 shows how the network of NDFs delivers a message in the J-Machine. The message is injected into the network at node (1,5). The source NDF converts the absolute destination address (4,1) into a relative address (3,-4). The message is then routed in the positive X direction with the head *flit* (flow control digit) containing the relative X address decremented at each stage. Message delivery is pipelined with each flit occupying one stage of the pipeline. At node (3,5), the relative X address is decremented to zero. Upon seeing this zero, Node (4,5) strips the X address and begins routing the message in the negative Y direction.

TIME

NODE	0	1	2	3	4	5	6	7	8	9	10	11	12	13
(1.5)	3	-4	A	B	C									
(2.5)		2	-4	A	B	C								
(3.5)			1	-4	A	B	C							
(4.5)				0	-4	A	B	C						
(4.4)					-3	A	B	B	C	C				
(4.3)						-2	-2	A -2	A A	B	C	C		
(4.2)										-1	A	B	C	
(4.1)											0	A	B	C

Figure 19.4: The NDFs deliver the message A B C from (1,5) to (4,1) using worm-hole routing. Each column reflects the contents and location of the message at a particular time. Buffering compresses the message when blockage occurs.

At node (4,3) the head of the message blocks for two cycles (because the channel to (4,2) is in use). The tail of the message continues to advance compressing the message in the NDF buffers. After two network cycles[6] the blockage is removed and the entire message proceeds to the destination (4,1). The relative Y address (now 0) is stripped and the remainder of the message is delivered to the node.

A block diagram of the logic that performs this routing is shown in Figure 19.5. The NDF contains two priority levels that implement logically separate networks sharing the same set of physical wires. Figure 19.5 shows only one level. Each level consists of four direction data paths, one for each of the cardinal directions (+X, -X, +Y, -Y). A message from the processor (P channel) has its address converted from absolute to relative by subtracting the local node address. A two-way switch then selects the proper direction by examining the sign bit of the resulting relative X address. Each direction data path has two inputs (the direction input and the preceding dimension) and two outputs (the direction output and the next dimension). It arbitrates between the two inputs and performs a zero check on the head flit to select the appropriate output. If the direction output is selected, the head flit is incremented/decremented. The head flit is stripped if the dimension output is selected.

Partitioning the router into separate direction data paths significantly reduces both the area and delay as compared to previous designs based on a central crossbar

---

[6] The network is completely self-timed – there is no clock. The channel cycles are aligned here for purposes of illustration only.

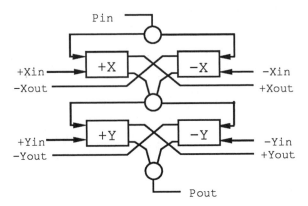

Figure 19.5: The NDF consists of separate dimension data paths that forward a message in its current direction or switch it to the next dimension.

switch [6]. In the most common case (a message continuing in the current direction) a flit sees only a single 2 way switch between the input and output. Additional switching is only required when a message switches dimensions.

Absolute integer node addressing was selected to simplify passing node numbers through the network – viz. with absolute addressing a node's address is the same on each node. Relative integer addressing is used internally by the NDF. Rather than converting from absolute addressing at the input to the network, we considered using relative addresses everywhere and adjusting them on-the-fly as they passed through the network. The incrementers and decrementers required to perform this adjustment are already in the NDF data paths. However, it proved cumbersome to detect which flits passing through the network contained network addresses in a particular dimension. We also considered representing addresses as polynomials over GF(2) and using a Galois incrementer [4] (the combinational equivalent of a linear feedback shift register) to adjust addresses. This approach had the advantage of requiring only a single gate delay to perform an increment/decrement however handling polynomial addresses proved difficult for some parts of the system software.

Message Reception

Message reception overhead is reduced to $\approx 1\mu s$ by buffering, scheduling, and dispatching messages in hardware. The MDP maintains two message/scheduling queues (corresponding to two priority levels) in its on-chip memory. As messages arrive over the network, they are buffered in the appropriate queue. The queues are implemented as circular buffers. It is important that the queue have sufficient

```
MOVE    [1,A3],R0   ; get method id
XLATE   R0,A0       ; translate to address descriptor
RES     2           ; transfer control to method
```

Figure 19.6: MDP assembly code for the CALL message.

performance to accept words from the network at the same rate at which they arrive. Otherwise, messages would backup into the network causing congestion. To achieve the required performance, special addressing hardware is used to enqueue or dequeue a message word with wraparound and full/empty check in a single clock cycle. A queue row buffer allows enqueuing to proceed using one memory cycle for each four words received. Thus a program can execute in parallel with message reception with little loss of memory bandwidth.

The MDP schedules the task associated with each queued message. At any point in time, the MDP is executing the task associated with the first message in the highest priority non-empty queue. If both queues are empty, the MDP is idle – viz., executing a background task. Sending a message implicitly schedules a task on the destination node. The task will be run when it reaches the head of the queue. This simple two-priority scheduling mechanism removes the overhead associated with a software scheduler. More sophisticated scheduling policies may be implemented on top of this substrate.

Messages become *active* either by arriving while the node is idle or executing at a lower priority, or by being at the head of a queue when the preceding message *suspends* execution. When a message becomes active, a handler is dispatched in one clock cycle. The dispatch forces execution to a physical address specified in the message header. This mechanism is used directly to process messages requiring low latency (e.g., combining and forwarding). Other messages (e.g., remote procedure call) specify a handler that locates the required method (using the translation mechanism described below) and then transfers control to it. For example, the call handler is shown in Figure 19.6. The first instruction gets the method ID (offset 1 into the message). To facilitate access to the message arguments, hardware initializes register A3 to contain an address descriptor (base/length) for the current message. The next instruction translates the method ID into an address descriptor for the method. If the translate faults, because method is not resident or the descriptor is not in the cache, the fault handler *fixes* the problem and reschedules the message. If the translation succeeds, the final instruction (resume) transfers control to the method.

An early version of the MDP had a fixed set of message handlers in microcode. An analysis of these handlers showed that their performance was limited by memory accesses. Thus there was little advantage in using microcode. The microcode was eliminated, the handlers were recoded in assembly language, and the *message opcode* was defined to be the physical address of the handler routine. Frequently used handlers are contained in an on-chip ROM. This approach simplifies the control structure of the machine and gives us flexibility to redefine message handlers to fix bugs, for instrumentation (e.g., to count the number of sends), and to implement new message types.

The message queue originally allocated storage from the heap for each incoming message. This eliminated the need to copy messages when a method suspended for intermediate results. However, the cost of allocating and reclaiming storage for each message proved to be prohibitive. Instead, we settled on the preallocated circular buffer. When a method suspends for intermediate results, message arguments are copied into a context object. The overhead of this copying is small since the context must be created anyway to specify a continuation and to hold live variables. The fixed buffer also provides a convenient layering. Priority zero messages are sent when the memory allocator runs out of room and priority one messages are sent when the priority zero queue fills.

Translation

The MDP is an experiment in unifying shared-memory and message-passing parallel computers. Shared-memory machines provide a uniform global name space (address space) that allows processing elements to access data regardless of its location. Message-passing machines perform communication and synchronization via node-to-node messages. These two concepts are not mutually exclusive. The MDP provides a virtual addressing mechanism intended to support a global name space while using an execution mechanism based on message passing.

The MDP implements a global virtual address space using a very general translation mechanism. The MDP memory allows both indexed and set-associative access. By building comparators into the column multiplexer of the on-chip RAM, we are able to provide set-associative access with only a small increase in the size of the RAM's peripheral circuitry.

The translation mechanism is exposed to the programmer with the ENTER and XLATE instructions. ENTER Ra,Rb associates the contents of Ra (the key) with the contents of Rb (the data). The association is made on the full 36 bits of the key so that tags may be used to distinguish different keys. XLATE Ra,Ab looks up the data associated with the contents of Ra and stores this data in Ab. The instruction faults if the lookup *misses* or if the data is not an address descriptor. XLATE Ra,Rb can be used to lookup other types of data. This mechanism is used by our system code to cache ID to address descriptor (virtual to physical) translations, to cache ID to node number (virtual to physical) translations, and to cache class/selector to address descriptor (method lookup) translations.

```
MOVE     [1,A3],R0          ; Receiver ID
XLATE    R0,A2              ; Receiver descriptor
MOVE     [0,A2],R0          ; Receiver object header
AND      R0,CLASS_MASK,R0 ; Isolate class
OR       R0,[2,A3],R0       ; combine w/ selector
XLATE    R0,A0              ; get method
RES      2
```

Figure 19.7: Send handler translates class and selector into address descriptor for method and transfers control.

Tags are an integral part of our addressing mechanism. An ID may translate into an address descriptor for a local object, or a node address for a global object. The tag allows us to distinguish these two cases and a fault provides an efficient mechanism for the test. Tags also allow us to distinguish an ID key from a class/selector key with the same bit pattern.

Most computers provide a set associative cache to accelerate translations. We have taken this mechanism and exposed it in a pair of instructions that a systems programmer can use for any translation. Providing this general mechanism gives us the freedom to experiment with different address translation mechanisms and different uses of translation. We pay very little for this flexibility since performance is limited by the number of memory accesses that must be performed.

19.4 Fine-Grain Programming

To illustrate how these mechanisms work together to execute a concurrent program recall the factorial example from Section 19.2. When the factorial message is received it is immediately buffered. When the message becomes active, control is dispatched to the SEND handler. Shown in Figure 19.7, this handler forms a key from the class of the receiver and the message selector and looks up the associated method using the XLATE instruction. After executing 7 instructions, control is transferred to the factorial method for class Integer.

MDP assembly code for the factorial method is shown in Figure 19.8. This code performs no computing. It reformats its message for the first rangeproduct and transmits it. Thus it is a good indicator of the overhead associated with sending a message; this 12 instruction sequence takes 18 clock cycles (estimated) to send a five word message. The total time from factorial message in to rangeProduct: message out is 30 clock cycles ($1.5\mu s$ on a 20MHz MDP).

```
        MOVE    1,R1           ; receiver is 1
        CALL    ID_TO_NODE     ; translate to node number in R0
        SEND    R0             ; send address
        DC      SEND_HEADER    ;
        SEND2   R0,R1          ; send header and receiver
        DC      RANGE_PRODUCT  ; send selector
        SEND    R0             ;
        SEND    [1,A3]         ; send argument
        SENDE   [3,A3]         ; continuation
        SUSPEND
```

Figure 19.8: Assembly code for the factorial method passes its continuation in a
rangeProduct message.

This code uses continuation passing to return its result. The factorial code does not
create a context to await the result of the rangeProduct:. Instead, it passes the
continuation (where to reply to) from its message in the rangeProduct: message.
The rangeProduct method then returns to the original sender.

The rangeProduct: code (not shown) creates a context to await the results of its
two message sends[7]. The ID of this context is included in the continuation field
of each of these messages. After the messages are sent, a SUSPEND instruction is
executed to pass control to the next message in the queue while awaiting the replies.

When the first reply arrives, it stores its value in the context and tests for the
presence of the other reply value. Finding that value absent, execution is again
suspended. The arrival of the second reply reactivates the context, stores its value,
and (finding the other result present) performs the multiply and sends the result
in a reply message to its continuation. The scheduling of operations during this
combining is performed by dataflow. As soon as both operands are present, the
operation is performed.

19.5 Conclusion

The MDP/NDF processing node efficiently executes fine-grain concurrent programs
by providing mechanisms that reduce the overhead of message-passing, translation,
and context switching to $\approx 5\mu s$. Reducing overhead to a time comparable with the
natural grain size of many concurrent programs allows the programmer to exploit

---

[7] A four-instruction inline sequence allocates this context off a free list.

all of the concurrency present in these programs rather than grouping many grains together – reducing the concurrency to improve the efficiency.

The MDP provides very general hardware mechanisms that can support many different concurrent programming models including conventional message-passing [28], actors [1] [2], futures [15], communicating processes [16], and dataflow [13]. All of these programming models require the same execution mechanisms: communication, synchronization, and translation. Specializing a machine for a particular model of computation results in only a small increase in performance.

At the time of this writing the NDF design is complete and a chip has been submitted for fabrication. Instruction and register transfer level simulations of the MDP have been written and used to test the architecture. Transistor level design, and artwork design for the MDP are underway.

There are many promising directions for future research. The mechanisms described here efficiently execute concurrency at a grain size of $5\mu s$. Many numerical programs, however, have potential concurrency at the level of single operations. Architectures must be developed that can exploit this concurrency without incurring the overhead of message delivery or synchronization.

Another critical problem is the development of (communication, processor, and memory) resource management policies for concurrent operating systems. It is quite easy to write a program with sufficient concurrency to swamp any concurrent machine. A concurrent operating system must provide a means to *throttle back* such massively concurrent applications to match the concurrency to the available resources.

Acknowledgement

The following MIT students have contributed to the work described here: Linda Chao, Andrew Chien, Stuart Fiske, Soha Hassoun, Waldemar Horwat, Jon Kaplan, Michael Larivee, Paul Song, Brian Totty, and Scott Wills.

I thank Tom Knight, Gerry Sussman, Steve Ward, Dave Gifford, and Carl Hewitt of MIT, and Chuck Seitz and Bill Athas of Caltech for many valuable suggestions, comments, and advice.

References

[1] Agha, Gul A., *Actors: A Model of Concurrent Computation in Distributed Systems*, MIT Press, 1986.

[2] Athas, W.C., and Seitz, C.L., *Cantor Language Report*, Technical Report 5232:TR:86, Dept. of Computer Science, California Institute of Technology, 1986.

[3] BBN Advanced Computers, Inc., *Butterfly Parallel Processor Overview*, BBN Report No. 6148, March 1986.

[4] Blahut, Richard E., *Theory and Practice of Error Control Codes*, Addison-Wesley, 1983, pp. 65-90.

[5] Dally, William J., *A VLSI Architecture for Concurrent Data Structures*, Kluwer, Hingham, MA, 1987.

[6] Dally, William J. and Seitz, Charles L., "The Torus Routing Chip," *J. Distributed Systems*, Vol. 1, No. 3, 1986, pp. 187-196.

[7] Dally, William J. "Wire Efficient VLSI Multiprocessor Communication Networks," *Proceedings Stanford Conference on Advanced Research in VLSI*, March 1987, pp. 391-415.

[8] Dally, William J. and Seitz, Charles L., " Deadlock-Free Message Routing in Multiprocessor Interconnection Networks," *IEEE Transactions on Computers,* Vol. C-36, No. 5, May 1987, pp. 547-553.

[9] Dally, William J. et al., "Architecture of a Message-Driven Processor," *Proceedings of the 14th Symposium on Computer Architecture*, June 1987, pp. 189-196..

[10] Dally, William J., and Song, Paul., "Design of a Self-Timed VLSI Multicomputer Communication Controller," To appear in, *Proc. ICCD-87*, 1987.

[11] Dally, William J., "Concurrent Data Structures," Chapter 7 in [26].

[12] Dally, William J., "The J-Machine: A Concurrent VLSI Message-Passing Computer for Symbolic and Numeric Processing," to appear.

[13] Dennis, Jack B., "Data Flow Supercomputers," *IEEE Computer*, Vol. 13, No. 11, Nov. 1980, pp. 48-56.

[14] Goldberg, Adele, and Robson, David, *Smalltalk-80, The Language and its Implementation*, Addison-Wesley, Reading, Mass., 1983.

[15] Halstead, Robert H., "Parallel Symbolic Computation," *IEEE Computer*, Vol. 19, No. 8, Aug. 1986, pp. 35-43.

[16] Hoare, C.A.R., "Communicating Sequential Processes," *CACM*, Vol. 21, No. 8, August 1978, pp. 666-677.

[17] Inmos Limited, *IMS T424 Reference Manual*, Order No. 72 TRN 006 00, Bristol, United Kingdom, November 1984.

[18] Intel Scientific Computers, *iPSC User's Guide*, Order No. 175455-001, Santa Clara, CA, Aug. 1985.

[19] Kermani, Parviz and Kleinrock, Leonard, "Virtual Cut-Through: A New Computer Communication Switching Technique," *Computer Networks,* Vol 3., 1979, pp. 267-286.

[20] Lutz, C., et al., "Design of the Mosaic Element," *Proc. MIT Conference on Advanced Research in VLSI,* Artech Books, 1984, pp. 1-10.

[21] Palmer, John F., "The NCUBE Family of Parallel Supercomputers," *Proc. IEEE International Conference on Computer Design, ICCD-86,* 1986, p. 107.

[22] Pfister, G.F. et al., "The IBM Research Parallel Processor Prototype (RP3): Introduction and Architecture", *Proceedings ICPP*, 1985, pp. 764–771.

[23] Seitz, Charles L., "System Timing" in *Introduction to VLSI Systems,* C. A. Mead and L. A. Conway, Addison-Wesley, 1980, Ch. 7.

[24] Seitz, Charles L., et al., *The Hypercube Communications Chip,* Display File 5182:DF:85, Dept. of Computer Science, California Institute of Technology, March 1985.

[25] Seitz, Charles L., "The Cosmic Cube", *Comm. ACM,* Vol. 28, No. 1, Jan. 1985, pp. 22-33.

[26] Seitz, Charles L., Athas, William C., Dally, William J., Faucette, Reese, Martin, Alain J. , Mattisson, Sven, Steele, Craig S., and Su, Wen-King, *Message-Passing Concurrent Computers: Their Architecture and Programming,* Addison-Wesley, publication expected 1987.

[27] Stefik, Mark and Bobrow, Daniel G., "Object-Oriented Programming: Themes and Variations," *AI Magazine,* Vol. 6, No. 4, Winter 1986, pp. 40-62.

[28] Su, Wen-King, Faucette, Reese, and Seitz, Charles L., *C Programmer's Guide to the Cosmic Cube,* Technical Report 5203:TR:85, Dept. of Computer Science, California Institute of Technology, September 1985.

[29] Theriault D.G., *Issues in the Design and Implementation of Act2*, MIT Artificial Intelligence Laboratory, Technical Report 728, June 1983.

Chapter 20

Unifying Programming Support for Parallel Computers [1]

Francine Berman [2]
Janice Cuny [3]
Lawrence Snyder [4]

Abstract

Highly parallel computation requires programming support tools distinct from those that have been developed for sequential computation. In this paper, we discuss characteristics of parallel programming support that we have found to be useful. In addition, we describe a compatible collection of existing tools – taken from the Poker Parallel Programming Environment [1], the Prep-P Mapping Preprocessor [2] and the Simple Simon Programming Environment [3] – that could form the basis of an integrated, comprehensive parallel environment.

20.1   Introduction

Highly parallel computation requires programming support tools distinct from those that have been developed for sequential computation. Parallel programmers must, for example, specify large numbers of nearly homogeneous processes together with their interactions and points of synchronization. They must map logical process

---

[1] Sponsored by ONR contracts N00014-86-K-0218 & -0264 and N00014-84-K-0647 and NSF grant DCR-8416878.
[2] University of California at San Diego, San Diego, CA
[3] University of Massachusetts, Amherst, MA
[4] University of Washington, Seattle, WA

structures onto disparate hardware structures and they must debug programs in the absence of reproducibility and consistent global states. Focusing on nonshared memory architectures, we discuss characteristics of the support needed for these activities. In addition, we describe a compatible collection of existing support tools that could form the basis of an integrated, comprehensive parallel environment.

The existing tools that we use as our examples come from the Poker Parallel Programming Environment [1], the Prep-P Mapping Preprocessor [2] and the Simple Simon Programming Environment [3]. Poker provides a model of unified facilities for specifying parallel programs: sequential code segments, processor assignments, interconnections, and the distribution of external data streams are all described via a consistent graphics interface. Prep-P is a preprocessor that automatically maps logical interconnection structures of varying sizes and topologies onto fixed size architectures. It produces multiplexed code for processes assigned to single processors. Simple Simon is a rudimentary programming environment for prototyping support tools that has been used in the development of a preprocessor for code specification, a graph editor and a "pattern-oriented" parallel debugger.

Taking a programming environment to be the integrated collection of all language and operating system facilities needed to support the programmer, we concentrate on aspects specific to highly parallel computation. The discussion is divided into three sections: program specification; mapping; and debugging. In each section, we describe general characteristics of parallel support environments that we have found to be useful and we discuss the extent to which these characteristics are present in existing collections of tools. In the final section, we show how the tools from our examples could be integrated into a single, comprehensive programming environment for parallel computation.

20.2   Program Specification

As experience with parallel programming increases, many aspects of program specification will likely change. At this point, however, we have found the following general characteristics of specification support to be useful:

- *Explicit parallelism.* At least in the near future, explicit parallelism (in which the programmer is responsible for the parallel decomposition of his algorithm) will be the most efficient method of exploiting the capabilities of many MIMD architectures.[4]

- *Database perspective.* It is convenient to view a parallel program, not as a single, monolithic structure, but as a database with parts that are accessed and manipulated independently.

---

[4] Alternatives may arise in the future from work on programming restructuring [4] or from the design of radically different architectures [5].

- *Scalability.* Because programs are often developed "in the small" and then scaled for massive parallelism, it must be possible to specify a "family" of parallel programs parameterized by size.

- *Automatic distribution and specialization of code segments.* Often the processes of a highly parallel program are nearly homogeneous and therefore, an environment should provide convenient mechanisms for the specification, distribution and specialization of common code.

- *Explicit description of process interconnection structures.* Explicit descriptions provide a natural medium for understanding parallelism, a basis for graphical displays, structural information potentially useful in mapping [2,6] and redundancy for automatic error detection.

- *Graphics interface.* Programmers will need sophisticated graphics to cope with the potentially overwhelming amount of information present in parallel programs.

Examples of program specification tools can be found in both Poker and Simple Simon.

20.2.1 Program Specification in Poker

Poker, originally designed for the CHiP architecture [7], was the first parallel programming environment and it has many of the above characteristics. A Poker program is a relational database; five different specification modes are provided, each corresponding to a different database view. We describe them using the example of an odd-even transposition sort on a linear array.

Poker views may be used in any order but typically a programmer begins with the "switch setting" view where he gives an explicit description of his communication graph by drawing it on a grid of processors. As shown in Figure 20.1, he does this graphically by connecting processors (PE's) with line segments; it is also possible to restore previously saved embeddings and to access a library of commonly used embeddings. The embeddings are not automatically scalable. Next the programmer might describe his set of process code bodies employing a standard text editor.

Figure 20.2 shows the code for two process types needed in the transposition sort; the parameterized code is written in a standard sequential language (here C) that has been extended to include interprocess I/O directed to logical ports. The use of parameterized process types makes the code specification scalable. Two additional views - one for assigning processes to processors and the other for defining logical port names – are used to connect code with the embedded communication graph. Both have a graphical interface. Processes and their parameter values are assigned

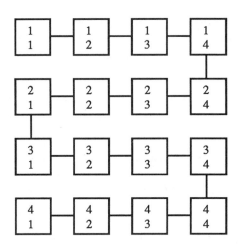

Figure 20.1: Poker Switch Settings View.

```
/* code for an even numbered process*/
code even:

trace value, num2sort, index: ;/* what to trace*/
ports left, right; /* name of the ports*/

/* parameters from Code Names
 *n - number of PE's involved in the sort
 *v - this PE's initial value
 *c - used to determine boundary PE's */

main(n,c,v)
int n,v; char c;
{
  int num2sort, value;
  int index, otherwise;

  if (c == 'M' || c = 'E' || c = 'S') {
    num2sort = n; value = v; /* set local variables */
    othervalue = 1000; /* set othervalue to maximum */
    /* do the sort */
    for (index = 0; index < num2sort / 2 + 1; index++) {
      /* even index */
      if(c != 'S'){   /* if not the starting PE */
        left <- value;
        othervalue <- left;
        if (othervalue > value) value = othervalue;
      }
      /* odd index */
      if (c != 'E'){   /* if not the end PE */
        right <- value;
        othervalue <- right;
        If (othervalue < value) value = othervalue;
      }
    }
  }
}
```

```
/* code for an odd numbered process*/
code odd;

trace value, num2sort, index: ;/* what to trace*/
ports left, right; /* name of the ports*/

/* parameters from Code Names
 *n - number of PE's involved in the sort
 *v - this PE's initial value
 *c - used to determine boundary PE's */

main(n,c,v)
int n,v; char c;
{
  int num2sort, value, inuse;
  int index, otherwise;

  if (c == 'M' || c = 'E' || c = 'S') {
    num2sort = n; value = v; /* set local variables */
    othervalue = 1000; /* set othervalue to maximum */
    /* do the sort */
    for (index = 0; index < num2sort / 2 + 1; index++) {
      /* even index */
      if(c != 'E'){   /* if not the ending PE */
        right <- value;
        othervalue <- right
        if (othervalue > value) value = othervalue;
      }
      /* odd index */
      if (c != 'S'){   /* if not the starting PE */
        left <- value;
        othervalue <- left
        If (othervalue < value) value = othervalue;
      }
    }
  }
}
```

Figure 20.2: Poker Process Code. The programs, written with an extension to C, give the process codes needed for an odd/even transposition sort.

1 odd 1	1 even 2	1 odd 3	1 even 4
16	16	16	16
800	100	150	300
'S'	'M'	'M'	'M'

2 even 1	2 odd 2	2 even 3	2 odd 4
16	16	16	16
900	50	400	775
'M'	'M'	'M'	'M'

3 odd 1	3 even 2	3 odd 3	3 even 4
16	16	16	16
600	375	250	750
'M'	'M'	'M'	'M'

4 even 1	4 odd 2	4 even 3	4 odd 4
16	16	16	16
450	425	850	25
'E'	'M'	'M'	'M'

Figure 20.3: Poker CodeNames View. Code names (here either EVEN or ODD) are assigned to each processor along with values for any parameters.

to processors using the "code names" facility, as shown in Figure 20.3; tedious replication is avoided by buffering capabilities that enable the user simply to deposit common values throughout regions of the array. Logical port names are linked with physical communication channels using the "port names" facility. As shown in Figure 20.4, logical port names can be associated with one of eight physical ports spaced around the periphery of each processor; again replication of buffered values is possible.

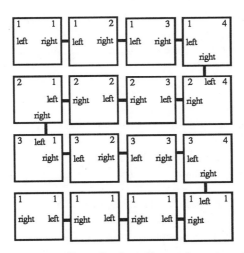

Figure 20.4: Poker Port Names View. Each processor is assumed to have eight ports evenly spaced around its periphery; port names are added by the programmer in the appropriate locations.

The final aspect of a Poker program specification (not shown in our example) is the description of external input/output streams. Poker provides a tool for defining "input names" in which the records of external files are to be partitioned into data streams on input and the manner in which data streams are to be composed into files on output.

20.2.2   Simple Simon Specification Tools

The Simple Simon environment has been used to prototype two tools – a macro preprocessor for code specification and a graph editor – that could easily be incorporated into Poker-like programming environment.

The macro preprocessor, called the Generic Code Preprocessor (GCP), is used to avoid redundant code specification and unnecessary runtime tests for nearly homogeneous processes. In the above program, for example, each process would execute one of four possible instruction sequences depending on whether it was an EVEN or ODD type process and whether or not it was on an end of the array. Note that the first distinction was made at compile time (each process type has its own code) and the second distinction was made at runtime (by the test on the parameter c). GCP code for these processes is shown in Figure 20.5; it consists of a single program. During expansion, individual code segments are constructed for each of the four different processes using the interconnection information stored in the program database.

The Simple Simon graph editor will be a general facility for describing process interconnection structures as families of labeled graphs, independent of their physical realization. The description of graph families permits scaling, and independence from physical realization permits greater portability, allowing graphical displays that match the programmer's mental picture of his structure.

The programmer describes a graph family by drawing its smallest member and then modifying that drawing into the next larger family instance; all remaining members are automatically generated by iterating the modifications. Ultimately, input to the editor will be graphical although currently it is textual (mirrored graphically) as shown in Figure 20.6. To make significant changes to the layout of a graph, a "graph assistant" [3] can be invoked. It has a number of heuristics that are employed at the user's direction to create aesthetic displays as shown in Figure 20.7.

Existing program specification tools already exhibit many of the desirable characteristics that we have listed. If these tools were combined around a common database, they would form a coherent basis for the initial development of parallel programming environments. The resulting environments, however, would be very low level; we expect that each of these tools will be replaced as, with more experience in parallel computation, we develop more appropriate abstractions. Work in other two areas – mapping and debugging – has proven to be even more challenging.

```
/* odd/even transposition sort */
#include "/user/simon/include/simon.h"

PID(arglist)
argstruct arglist;
{
    int argc; char **argv;
    int num2sort, value, othervalue, idx;

    getargs(arglist, argc, argv); /* coerce argc/argv from argument list */
    getiarg(argv, 1, value); /* get sort value from argv[1] */

    /* open necessary channels to neighbors */
#   ifdef left
        copen(left, "rw");
#   endif
#   ifdef right
        copen(right, "rw");
#   endif

    go(); clock_on();

    num2sort = M;   /* N is the number of processes */

    for (idx = 0; idx < num2sort/2+1; idx++)
    {
        /* even phases */
#       ifdef ODD
#          ifdef right
            PUT("odd<->even", right, param(value));
            GET("odd<->even", right, param(othervalue));
            if (othervalue < value) value = othervalue;
#          endif
#       endif
#       ifdef EVEN
#          ifdef left
            PUT("odd<->even", left, param(value));
            GET("odd<->even", left, param(othervalue));
            if (othervalue > value) value = othervalue;
#          endif
#       endif

        /* odd phases */
#       ifdef EVEN
#          ifdef right
            PUT("even<->odd", right, param(value));
            GET("even<->odd", right, param(othervalue));
            if (othervalue < value) value = othervalue;
#          endif
#       endif
#       ifdef ODD
#          ifdef left
            PUT("even<->odd", left, param(value));
            GET("even<->odd", left, param(othervalue));
            if (othervalue > value) value = othervalue;
#          endif
#       endif
    }
    clock_off()
}
```

Figure 20.5: Simple Simon GCP code for the even/odd transposition sort; this code generates four different process types.

...
(setq tree (list (add-vertex N)))

...
(setq left tree)
(setq right (copy tree E))

...
(setq root (list (add-vertex N)))
(connect left root)
(connect right root)
(setq tree (append root left
right))

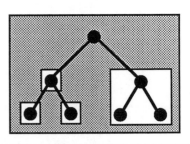

...
(setq left tree)
(setq right (copy tree E))
(setq root (list (add-vertex N)))
(connect left root)
(connect right root)
(setq tree (append root left right))

Figure 20.6: Graph Editor Example. The family of complete binary trees is described. The operations needed to draw the smallest family member (here a three node tree) are shown in the first three pictures; the "modifications" that will be automatically iterated to draw larger family members are shown in the fourth picture.

20.3 Mapping

The term mapping has been used to cover many aspects of parallel program implementation [8] but we use it here to refer to the process of fitting a logical interconnection structure onto a physical machine that may differ in size (number of processors) or topology (interconnection of processors). We find the following to be useful characteristics for its support:

- *Automatic mappings.* Mapping is an extremely difficult problem that requires detailed knowledge of the target machine. As a result, it will be beyond the

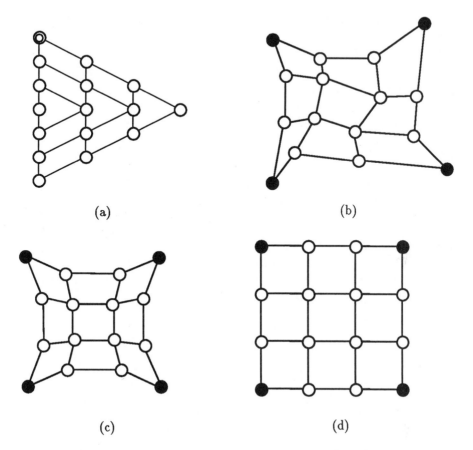

Figure 20.7: Graph Assistant Example. (a) The original layout of a 4 X 4 mesh; (b) the mesh after a heuristic to minimize "tension" on the edges has been invoked, enabling the user to identify and place the corners; (c) the graph after repeated use of the tension heuristic; and (d) the graph after repeated use of a heuristic to equalize edge lengths.

capabilities of many programmers and programming environments will have to provide automatic mappings.

- *General mappings.* Because we are most interested in general purpose architectures, we expect that programmers will employ a variety of interconnection structures and so will need mapping support for arbitrary graphs.

- *Interactive.* Because it is not possible to automatically provide efficient mappings for all graphs, it will be useful to provide feedback to the programmer, allowing him to make modifications where needed.

- *Division into subproblems.* Since mapping is so complex, the problem can fruitfully be decomposed into three subproblems: algorithm partitioning (called *contraction*), partition layout (placement) and channel routing.

Prep-P is a preprocessor which automatically maps arbitrary (bounded-degree), logical interconnection structures onto target CHiP machines, multiplexing code for processes assigned to a single processor. Input to Prep-P is a graph represented by an adjacency list; it is currently specified in a simple graph description language but the graph editor could easily be used.

Prep-P first contracts the interconnection graph to the appropriate size for the target machine, partitioning the process set so that each partition executes on a different processor. An initial partitioning is randomly generated and then a local neighborhood search is performed using a cost function that takes into account the expected traffic along an edge. The contraction produced by Prep-P for a even/odd transposition sort program is shown in Figure 20.8. Efficient library contractions of common interconnection structures are also available.

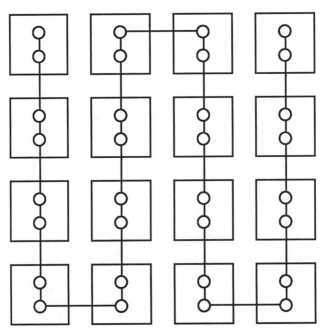

Figure 20.8: Prep-P contraction of a 32 node linear array into 16 partitions.

Prep-P next places the partitions of processes onto a grid of processors using an adaptation of the Kernighan and Lin divide-and-conquer algorithm for circuit placement [9]. Starting with a random subdivision of the graph into four (equal size) subgraphs, pairwise exchanges of partitions between the subgraphs are considered as in the Kernighan and Lin algorithm; the cost function is based on the number of edges crossing subgraph boundaries. The resulting subgraphs are assigned to the quadrants of a grid in such a way that the cost of routing (based on the number of edges and the distance between quadrants) is minimized. This procedure is then iterated for each (subgraph, quadrant) pair until the remaining subgraphs are singletons.

In the last step, Prep-P connects the layout on the target parallel architecture, performing the switch setting function for CHiP machines. It uses a breadth first search to find an available route between each pair of adjacent processors, choosing the pairs in increasing order of the (Manhattan) distance between their components. The final mapping of the even/odd transposition sort produced by Prep-P is shown in Figure 20.9.

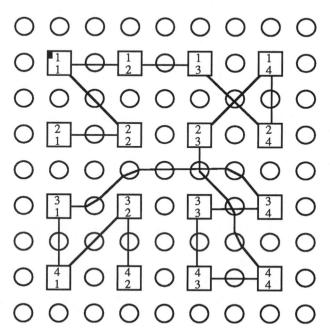

Figure 20.9: Final Prep-P mapping (placement and routing) of the layout in the previous figure.

There is a great deal left to be done in the area of mapping. Several mapping techniques were evaluated during the design of Prep-P [10] but other techniques might be considered. It might be possible, for example, to use program information (such as computation loads,bandwidth, or memory constraints) to generate more efficient mappings. In addition, mapping strategies for architectures other than CHiP machines must be developed. Finally, interactive tools that would enable the programmer to understand, evaluate and possibly modify system generated mappings would be useful. (It may be possible to apply recent work on debugging for performance [11,12] in the development of such tools.)

20.4   Parallel Debugging

Debugging support is critically important and quite difficult. Highly parallel computation is not amenable to existing debugging techniques since parallel programs do not have the consistent global states, manageable quantities of potentially relevant information or reproducibility that have formed the basis for sequential debugging paradigms. Instead, their behavior is best understood in terms of the flow of data and control resulting from interprocess communication, leading us to identify the following characteristics of debugging support:

- *"Pattern-Oriented."* The flow of interprocess data and control in a highly parallel program is often very structured: fine grain, tightly coupled processes communicate across regular interconnection networks resulting, at least logically, in patterned data and control flows. Since the programmer often understands the activity of his program in terms of these patterns, their identification will form the basis for highly parallel debugging paradigms.

- *Animation.* Patterns of activity are best understood visually and so we expect that animation will play an important role.

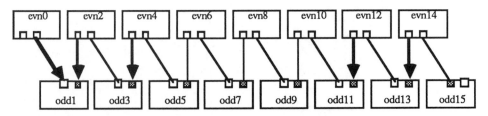

Figure 20.10: Snapshot of Belvedere's animation. Active ports and channels are highlighted and individual messages are shown as arrowheads moving across channels.

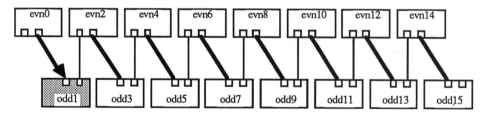

Figure 20.11: Snapshot of Belvedere's animation of a high-level abstract event (an ODD processor swap) from the perspective of process ODD1. Note that all processes and channels involved in the swap are highlighted but only the activity in which ODD1 participates (the message transmission) is shown.

- *Integration with sequential tools.* While many errors in parallel programs result from interprocess interactions, the sequential code segments are also suspect. As a result, parallel debuggers should interface with standard sequential debugging tools.

Simple Simon has been used to prototype a "pattern-oriented" debugger, called *Belvedere.*[5] Belvedere is a trace-based, *post-mortem* debugger. It treats the event trace of a system as a relational database, providing animations of user-selected events.[6] Displays for animation are determined by positional information stored in the program database (normally obtained with the use of the editor or the graph assistant). Figure 20.10 shows a sample snapshot of an animation.

Because complex, asynchronous programs often present patterns that are difficult to interpret, Belvedere has adopted the notion of abstract, user-defined events [13,14] and includes facilities to impose user-defined perspectives on those events, as shown in Figure 20.11. Viewing the behavior of a program from a single perspective may well be misleading (the use of perspectives often sequentializes concurrent behavior, for example) and so to gain accurate insights into his program's behavior, the user will have to view it from a variety of perspectives. This is easy to do in Belvedere and we expect that it will provide a valuable tool for determining discrepancies between actual and intended behaviors.

Belvedere exhibits two of the three desirable characteristics of parallel debuggers to some extent but much remains to be done. Belvedere, for example, animates the user-defined communication patterns that occurred during a program's execu-

---

[5] *Belvedere* comes from the Latin *bellus* meaning "beautiful" and *vidēre* meaning "view."

[6] While most existing systems do not provide such detailed information, its presence will allow us to determine which types of debugging information are most useful, perhaps providing insight for future hardware designs.

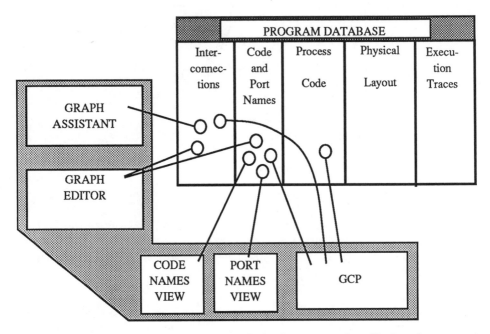

Figure 20.12: Program specification tools in the proposed, unified environment.

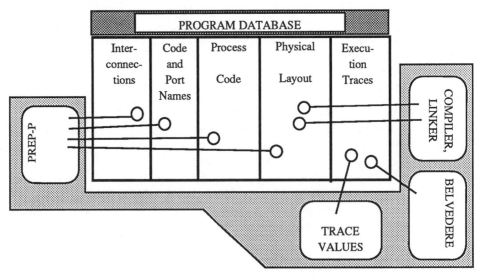

Figure 20.13: Program mapping, compilation and execution in the proposed, unified environment.

tion; incorrect programs, however, might not exhibit the expected patterns and so it is necessary to develop animations of "near misses" (patterns that are almost matched). In addition, Belvedere is not at all integrated with sequential tools; it may be possible, however, to incorporate something like "instant replays" [15] to interface to sequential tools.

20.5   A Comprehensive Parallel Programming Environment

To support highly parallel computation, parallel programming environments must provide integrated facilities for all aspects of a programmer's task. We believe that the tools outlined above could form the basis for such a unified, comprehensive environment.

Any comprehensive environment would have to be integrated around a common database. Using the Poker model, such a database could include descriptions of logical interconnection structures, process code, sets of ports, process names, and parameter values, mappings from the logical interconnections to hardware configurations, I/O descriptors, and traced event streams.

Program specification tools could be used to modify the database as shown in Figure 20.12. The programmer would describe his logical interconnection structure using the graph editor. Graphical views and animations of the program would be governed by display information provided by the editor, possibly obtained with the use of the graph assistant. Associated port names, code names and parameters would be specified either with the graph editor's labeling facilities or with the use of the corresponding Poker views. Process code segments, entered with a standard text editor, would be automatically replicated and specialized by the Simple Simon macro-preprocessor. I/O streams would be described using the Poker Input View.

Program compilation would begin the Prep-P mapping Preprocessor. After mapping, the program could be compiled, loaded and executed using existing Poker facilities as shown in Figure 20.13. The execution would produce the trace of events used in debugging. Debugging could proceed through the use of Belvedere, possibly interfaced with more traditional debugging aids such as the Poker trace value facility.

The integration of these tools into a single environment would provide the programmer with the range of tools that he will need in programming highly parallel architectures. Such an environment would, in addition, form a framework for the development and evaluation of more sophisticated support facilities.

Acknowledgements

We would like to thank the members of the Poker, Prep-P and Simple Simon Projects for their many contributions to this work.

References

[1] Lawrence Snyder, "Parallel Programming and the Poker Programming Environment," *Computer* 17(7), pp. 27-37 (1984).

[2] Francine Berman, "Experience with an Automatic Solution to the Mapping Problem," in *The Characteristics of Parallel Algorithms*, Leah Jamieson, Dennis Gannon and Robert Douglass (eds.), MIT Press (1987).

[3] Janice E. Cuny, Duane A. Bailey, John W. Hagerman, and Alfred A. Hough, The Simple Simon Programming Environment: A Status Report. COINS Technical Report 87-22 (May 1987).

[4] David A. Padua, David J. Kuck and Duncan H. Lawrie, "High-Speed Multiprocessors and Compilation Techniques," *IEEE Trans. on Computers* C-29(9), pp. 763-776 (September 1980).

[5] Arvind and Rishiyur S. Nikhil, "Executing a program on the MIT Tagged-Token Dataflow Architecture," *Parallel Architectures and Languages Europe, Lecture Notes in Computer Science 259*, J.W. de Bakker, A.J. Nijman and P.C. Treleaven (eds.), Springer-Verlag, pp.1-29 (June 1987).

[6] Duane A. Bailey and Janice E. Cuny, "An Approach to Programming Process Interconnection Structures: Aggregate Rewriting Graph Grammars," *Parallel Architectures and Languages Europe, Lecture Notes in Computer Science 259*, J.W. de Bakker, A.J. Nijman and P.C. Treleaven (eds.), Springer-Verlag, pp.112-123 (June 1987).

[7] Lawrence Snyder, "Introduction to the Configurable, Highly Parallel Computer," *Computer* 15(1), pp. 47-56, 1982.

[8] Workshop on Performance Efficient Parallel Programming. Zary Segall and Lawrence Snyder (eds.), Technical Report Carnegie-Mellon University (1986).

[9] B. Kernighan and S. Lin, "An Efficient Heuristic Procedure for Partitioning Graphs," *Bell System Technical Journal* 49(2) (February, 1970).

[10] Francine Berman and Patricia Haden, A Comparative Study of Mapping Algorithms for an Automated Parallel Programming Environment. Technical Re-

port CS-088, Department of Computer Science, University of California, San Diego.

[11] Zary Segall and Larry Rudolph, "PIE - A Programming and Instrumentation Environment for Parallel Processing," Technical Report CMU-CS-85-128, Carnegie-Mellon University, 1985.

[12] Karsten Schwan, Michael Kaelbling, and Rajiv Ramnath, "A Testbed for High-Performance Parallel Software," Technical Report OSU-CISRC-TR-85-5, The Ohio State University, 1985.

[13] Peter C. Bates, "Debugging Programs in a Distributed System Environment," University of Massachusetts, COINS Technical Report 86-05 (January 1986).

[14] Peter C. Bates and Jack C. Wileden, "High-level debugging of Distributed Systems: The Behavioral Abstraction Approach," *Journal of System Software* 3, pp. 255-244 (1983).

[15] Thomas J. LeBlanc and John M. Mellor-Crummey, Debugging Parallel Programs with Instant Replay. Butterfly Project Report 12, Computer Science Department, University of Rochester (September 1986).

PART IV
Fault Tolerance and Reliability

Introduction

Within a highly parallel computing environment, with several processors cooperating in the execution of a problem, failure of a single processor, memory, etc. may be difficult to detect merely by observing the results of the computation. Issues of fault tolerance, testing and reliability become far more important than in conventional sequential computations where failures tend to be obvious (i.e. catastrophic).

Dahbura provides an overview of *system-level diagnosis*. The principles which have evolved in this field are particularly relevant, given the distributed nature of concurrent computation. The general problem considered here concerns a large number of processors, with each processor capable of directly communicating with a subset of all other processors. Identification of the occurrence of failure and isolation of the offending unit are the fundamental questions addressed.

Rucinski and Polaski address a related question, self-testing and reconfiguration of 2-D VLSI processor arrays. This reconfiguration is necessary since manufacturing defects need to be tolerated. Many reconfiguration schemes have been proposed, both for VLSI and WSI. Testing to determine faulty circuitry, however, is seldom given the emphasis deserved (since isolation of a fault is a precondition for reconfiguration). Rucinski and Polaski provide a generalization of reconfiguration schemes and describe a testing mechanism through the use of *drones*.

Software modeling support is critical in evaluating reliability, just as software tools are critical in other areas of parallel computer design and analysis. **Veeraraghaven and Trivedi** describe a powerful software modeling tool (SHARPE), illustrating its use with several examples. The hierarchical modeling approach used here provides the rich description detail needed at the lowest levels while reducing the analysis information for interpretation of higher level behavior.

Software tools to assist the design of reliable parallel processing systems are also a critical need. **Sharma and Fuchs** address the issue not only of detecting errors but also recovery of the computation from errors. Recovery is a particularly complex issue, requiring error containment to avoid corruption beyond definable boundaries and storage of previous states to allow roll back of the computation. These issues are considered from the perspective of multiprocessor systems.

Kumar provides reliability and performance results for a class of fault-tolerant multistage networks called Augmented Shuffle-Exchange Networks (ASENs). His comparison of ASENs to a variety of Multistage Interconnection Networks (MINs) provides a useful overview of those other networks.

Faults in the processor, memory or communication environment do not leave the system necessarily inoperable since there are other processor, memory and communication elements still operating. This allows use of a concurrent processing environment, despite faults, at a reduced level of performance. **Koren and Koren** address the issue of performance degradation in the presence of faults, drawing on multiprocessors and multi-stage networks to illustrate the issues and analytical techniques.

Multi-stage networks provide a versatile communications environment for general purpose parallel programming environments, providing short path lengths and simple routing algorithms. They also provide an efficient, regular structure of switching elements. Extensions of the simple multi-stage network are described by **Rau and Fortes**, providing networks with fewer stages, networks with minimum link complexity for one-to-one connections using uniform switches and fault tolerant versions of the minimum link complexity networks.

The last two papers consider fault tolerant VLSI/WSI systems. **Belfore et al.** consider "inherently fault tolerant" systems, illustrating the general objective with the example of a holographic plate. Broken into several pieces, each piece of the plate retains the full image of the original plate (with decreased resolution). The simulation of neural networks, using the traveling salesman problem as an example, is used to illustrate the principles of such fault tolerance. **Kim and Reddy** address error detection and array reconfiguration within the tight architecture of a systolic array, with LU-decomposition the systolic algorithm.

Chapter 21

System-Level Diagnosis: A Perspective for the Third Decade

Anton T. Dahbura [1]

Abstract

This paper gives an overview of twenty years of achievement in system-level diagnosis and examines the antinomy of this flourishing theoretical research area that has yet to have any apparent practical impact in an era of unforetold technical advances. The potentially important role of system-level diagnosis is discussed relative to future multicomputer systems.

21.1   Introduction

Two decades ago this year, a paper by Preparata, Metze, and Chien concerning a "connection assignment problem for diagnosable systems" appeared in *IEEE Transactions on Electronic Computers* [1]. That work, which introduced the now well-known PMC model for identifying faulty computing elements in distributed systems, was to spawn literally hundreds of journal papers and conference reports on what is now called system-level fault diagnosis.

Unfortunately, the practical implications of this ever-growing body of knowledge remain few. Of the myriad commercial and military fault tolerant computers that

---

[1] AT&T Bell Laboratories, Murray Hill, NJ

have emerged recently, not a single one relies upon anything resembling system-level diagnosis for identifying faulty components. Furthermore, the small number of experimental systems which intend to demonstrate viability of the system-level diagnosis techniques are too simplistic to be of widespread value. In spite of this, research activity in the area of system-level diagnosis remains healthy. Indeed, breakthroughs concerning some of the more fundamental theoretical problems posed early on have been made only recently. New researchers, models, problems, and some solutions are appearing on the scene; moreover, funding for such research appears to be strong.

In this paper, twenty years of achievement in system-level diagnosis research will be reviewed relative to the progress accomplished in the design of scientific, commercial, and military fault-tolerant computers. It is argued that, while many of the problems defined and solutions presented in the system-level diagnosis literature are primarily of academic interest, the underlying philosophy upon which the field is based will become a natural, and indeed, critical consideration for maintaining system dependability as the demand for more powerful computational devices drives the technology towards more massive, complex, radical, and diverse designs.

In Section 21.2, a survey of the system-level diagnosis area is presented. In Section 21.3, some fault detection and diagnosis techniques used in fault-tolerant systems are described. Section 21.4 outlines certain assumptions inherent in system-level diagnosis work which makes the concepts difficult to apply to present-day systems. Section 21.5 discusses the potential role of system-level diagnosis in future multicomputer dependability considerations.

21.2   A Survey of System-Level Diagnosis

21.2.1   The PMC model

The overwhelming majority of theoretical work in the area of system-level fault diagnosis has its origins in a 1967 paper by Preparata, Metze, and Chien [1] in which a model is proposed for considering system-level faults in multiprocessor systems. In this so-called *PMC model*, a set of n independent processors, or *units*, is assembled such that each unit has the capability of communicating with a subset of the other units. These units function according to specified programs, accept data from other units or external sources, process and store data internally, and provide data to other units or to external sources.

In general, such systems are subject to the occurrence of *faults*, that is, disruptions of the specified logical behavior of the system. The disruptions to be considered are those within one of the units which cause the digital output of the unit to be different from that which is specified for a given sequence of inputs. Accordingly, a *permanently faulty* unit is one which exhibits a constant logical behavior which is different from its specified logical behavior, and a *fault-free* unit is one which constantly exhibits its specified logical behavior. It is assumed that faulty units can incorrectly claim that fault-free units are faulty and that faulty units are fault-free. The assumption that the units are independent implies that a fault which occurs in one unit does not affect the behavior of any of the other units.

Each unit is assigned a particular subset of the other units for the purpose of testing. It is assumed that a test performed by one unit on another takes place over a direct, bidirectional communication link between the two units and consists of a sequence of generated stimuli and corresponding responses to the stimuli. Furthermore, the tests are assumed to be *complete* in the sense that if the unit which performs a test is fault-free then the result of the test accurately reflects the condition of the unit under test. It is assumed that a bounded subset of these units is permanently faulty, implying that the outcomes of tests performed by such units are unreliable. Given such a system, two natural issues arise:

1) *Diagnosability Problem:* Given a collection of units and a set of test assignments for each unit, what is the maximum number t_p of arbitrary units which can be faulty such that the set of faulty units can be *uniquely* identified on the basis of any given collection of test results?

2) *Diagnosis Problem:* Given a testing assignment for a system such that at most t_p processors can always be uniquely identified on the basis of any given collection of test results, does there exist an efficient procedure for identifying the faulty units?

Indeed, much of the published system-level diagnosis effort considers at least one of the two aforementioned problems. The remaining work, discussed later in this section, deals with variants of the PMC model.

21.2.2 Formal concepts and definitions

The multiprocessor system proposed by Preparata, Metze, and Chien consists of n units, denoted by the set $U = \{u_1, u_2, \ldots, u_n\}$. Each unit $u_i \in U$ is assigned a particular subset of the remaining units to test, and it is assumed that no unit tests itself. The complete collection of tests in S, called the *connection assignment* of S, is represented by a directed graph $G = (U, E)$, where each unit $u_i \in U$ is represented by a vertex and each edge $(u_i, u_j) \in E$ if and only if u_i tests u_j. An *outcome* a_{ij} is associated with each edge $(u_i, u_j) \in E$, where a weight of $0(1)$ is assigned to a_{ij} if u_i evaluates u_j to be fault-free (faulty). Since the faults to be considered here are permanent, the outcome a_{ij} of a test is reliable if and only if u_i is fault-free. The set of test outcomes of S is called the *syndrome* of S. If $a_{ij} = 0(1)$ then it will be said that u_i has a 0-link (1-link) to u_j and that u_j has a 0-link (1-link) from u_i. A seven-unit system and a syndrome are given in Figure 21.1.

Definition 1: A system S is t_p *-diagnosable* [1] if, given a syndrome, all (permanently) faulty units in S can be correctly identified, provided that the number of faulty units present does not exceed t_p.

Based on Definition 1, the Diagnosability Problem can be restated as the problem of determining whether a given system S is t_p -diagnosable for a given t_p.

21.2.3 PMC results

Preparata, Metze, and Chien were primarily concerned with the characterization of t_p -diagnosable systems. They noted that for non-negative integers t_p and n, if S is t_p -diagnosable then $n \geq 2t_p + 1$ and that if $n \geq 2t_p + 1$ then it is always possible to construct a t_p -diagnosable system by allowing each unit to test every other unit. They also showed that if S is t_p -diagnosable then each unit $u_i \in U$ must be tested by at least t_p other units, implying that at least nt_p tests are required in a t_p -diagnosable system. They defined a family of t_p -diagnosable systems, called

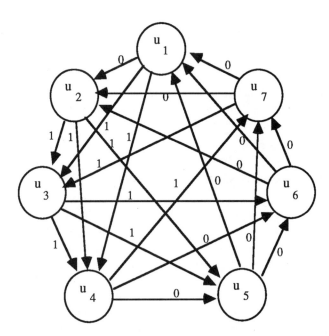

Figure 21.1: A connection assignment for a multiprocessor system and a syndrome

D_{1,t_p} systems, which contain the minimum number of tests, nt_p :

Definition 3: A system S is a D_{1,t_p} system if $t_p < n/2$ and for $m = 1, 2, \ldots, t_p$:

$$E = \{(u_i, u_j) : j - i = m \bmod n\}.$$

The system shown in Figure 21.1 is a $D_{1,3}$ system and is therefore t_p -diagnosable for $t_p = 3$.

21.2.4 Characterization work

The first complete characterization of t_p-diagnosable systems was provided by Hakimi and Amin [2] in 1974.

Theorem 1: A system S is t_p -diagnosable if and only if

1) $n \geq 2t_p + 1$,

2) each unit is tested by at least t_p others, and

3) $|\Gamma(X)| > p$, for all $X \subset U$ such that $|X| = n - 2t_p + p$ and $0 \leq p \leq t_p - 1$, where $\Gamma(X) = \{u_j : (u_i, u_j) \in E, u_i \in X, \text{ and } u_j \in U - X\}$.

For the special case in which no two units test each other, Hakimi and Amin showed that condition 2) of the Hakimi/Amin characterization is necessary and sufficient. Thus, if it is known that no two units test each other in an arbitrary system, the Diagnosability Problem can be solved in $O(nt_p)$ operations by testing condition 2) of Theorem 1. Finally, they showed that if the directed connectivity $\kappa(G)$ [47] of G is at least t_p and $n \geq 2t_p + 1$ then S is t_p -diagnosable. For the system shown in Figure 21.1, as an example, no two units test each other and each unit is tested by three others; furthermore, $\kappa(G) = 3$ in this case. Both properties imply from the above that for this system, $t_p = 3$.

Several others have given equivalent characterizations of t_p -diagnosable systems, including Allan, Kameda, and Toida [3], etc.

21.2.5 Diagnosis algorithms

The years following the work by Hakimi and Amin enjoyed a flurry of activity in the development of an efficient algorithm to solve the Diagnosis Problem for t_p -diagnosable systems, starting in 1975 with the first such algorithm, by Kameda,

Toida, and Allan [4]. The so-called KTA approach is to guess whether an arbitrary unlabeled unit in S is fault-free or faulty. Based on this guess, other units in S are implicitly labeled, and whenever a unit is implied to be faulty by a guess, the number of remaining unlabeled units which can be faulty is reduced by one. If a contradiction is encountered, the algorithm must backtrack. For t_p -diagnosable systems, the number of nodes in the search tree is linear in t_p and $O(|E|)$ operations are required at each node, so that the resulting algorithm is of complexity $O(t_p|E|) \equiv O(n^3)$. The published algorithm suffers from a technical error, pointed out later by Corluhan and Hakimi [5], which does not detract from its validity [6].

In 1977, Madden [7] gave an $O(n^3)$ diagnosis algorithm for t_p -diagnosable systems, largely based on the KTA algorithm, which linked the problem to that of finding a *minimum vertex cover set* [47] of an undirected graph, that is, a set K^* of vertices of minimum cardinality such that every edge in the graph has at least one end in K^*.

A 1978 paper by Fujiwara and Kinoshita [8] showed that, given an *arbitrary* testing assignment for a graph and a collection of test results, as opposed to the special class of t_p -diagnosable systems, it is extremely unlikely (that is, it is NP-complete [48]) that an efficient solution exists for the problem of identifying a set of units of minimum cardinality which, if faulty, could account for the given test results.

In 1984, Dahbura and Masson [6] presented an $O(n^{2.5})$ diagnosis algorithm for the class of t_p -diagnosable systems. The algorithm is based on the concept of an *L-graph,* defined as follows:

Definition 4: Given a t_p -diagnosable system S and a syndrome, the *L-graph* of S is an undirected graph $G_L = (U_L, E_L)$ such that $U_L \equiv U$ and

$$E_L = \{(u_i, u_j): u_i \in L(u_j)\},$$

where $L(u_i)$ is the set of all units in U that may be deduced to be faulty under the assumption that u_i is fault-free.

The set $L(u_i)$ is called the *implied faulty set* of u_i. The L -graph of S is undirected because of the symmetric properties of the implied faulty set: $u_i \in L(u_j)$ if and only if $u_j \in L(u_i)$. Furthermore, G_L can contain self-loops, since it is possible that $u_i \in L(u_i)$. The L -graph for the syndrome of the system shown in Figure 21.1 is given in Figure 21.2.

The Dahbura-Masson fault diagnosis algorithm is based on the following fundamental theorem of t_p -diagnosable systems:

Theorem 2: Given a syndrome for a t_p -diagnosable system S, the set of faulty units in S is the unique minimum vertex cover set of G_L.

Let $K^* \subseteq U_L$ be the unique minimum vertex cover set of G_L and let $M^* \subseteq E_L$ be a *maximum matching* [47] of G_L, that is, the set of edges of maximum cardinality in G_L such that no vertex in U_L is incident to more than one edge in M^*. Dahbura and Masson showed that L -graphs generated by t_p -diagnosable systems belong to the class of *external* graphs; that is, for such graphs $|K^*| = |M^*|$. Based on this, an $O(n^2)$ technique was given for determining K^* from M^*. It was shown that since $O(n^{2.5})$ operations are required to generate G_L, $O(n^{2.5})$ operations are required to compute M^*, and $O(n^2)$ operations are needed to compute K^* from M^*, the complexity of the algorithm is $O(n^{2.5})$. For the L -graph shown in Figure 21.2, the minimum vertex cover set K^* and a maximum matching M^* are highlighted. Note that $|K^*| = |M^*| = 3$ and therefore, u_3, u_4, and u_5 are the faulty units in this example.

21.2.6 Special classes of diagnosable systems

The Dahbura/Masson algorithm is the most efficient designed to date for t_p -diagnosable systems; however, more efficient algorithms are known for special cases of t_p -diagnosable systems. Among these is the $O(nt_p)$ algorithm given by Meyer and Masson [9] for D_{1,t_p} systems. Other $O(nt_p)$ algorithms for generalized D_{1,t_p} systems were given by Mallela [10] and by Chwa and Hakimi [11]. All of these sys-

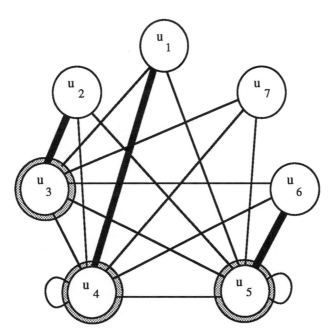

Figure 21.2: The L-graph for the system shown in Figure 21.1. A maximum matching and the minimum vertex cover set are highlighted.

tems have highly regular connection assignments. Hakimi and Chwa [12] gave an $O(|E|)$ algorithm for the class of t_p -diagnosable systems in which u_i tests u_j if and only if u_j tests u_i. Dahbura, Masson, and Yang [13] characterized a special class of t_p -diagnosable systems for which the faulty units can be identified in a straight-forward manner. For this class of so-called t_p -*self-implicating* systems, they gave a diagnosis algorithm which is also linear in the number of tests in S. The notion of self-implication extends to all systems such that $\kappa(G) \geq t_p$, which includes D_{1,t_p} systems. The system shown in Figure 21.1 is, therefore, a self-implicating system.

21.2.7   Sullivan's diagnosability algorithm

Neither the characterization of t_p -diagnosable systems given by Hakimi and Amin nor later versions appearing in the literature suggest polynomial-time algorithms for the Diagnosability Problem. However, Sullivan [14] in 1984 gave an $O(|E|n^{3/2})$ diagnosability algorithm which is based on *network flow* [47]. In Sullivan's procedure, a directed graph $G' = (U', E')$ is constructed, where $U' = U \cup \{u_s\}$ and $E' = E \cup \{(u_s, u_i) : u_i \in U\}$. Each edge (u_s, u_i) is given capacity $1/2$ and all other edges are given infinite capacity. Finally, each vertex $u_i \in U$ is given capacity $c(u_i) = 1$.

Theorem 3:   A directed graph G is t_p -diagnosable if and only if for all $u_i \in U$ the value of the maximum flow in the network G' with source u_s and sink u_i is greater than t_p.

Sullivan shows that an $O(|E|n^{1/2})$ algorithm for finding maximum network flow, performed with each of the n vertices of G' as the network sink, computes the optimal value for which S is t_p -diagnosable. For example, the network G' for the system of Figure 21.1 is given in Figure 21.3. A maximum flow of 3.5 is given when u_1 is the network sink. By the symmetry of the system, it must be concluded that $t_p = 3$ for this system.

21.2.8   Alternative diagnosis strategies

The second major area of system-level diagnosis work is that of defining different strategies for performing diagnosis. The first alternative to the previously described approach was suggested by Preparata, Metze, and Chien [1], who observed that in a diagnosable system, it may be desirable to identify some, but perhaps not all, of the faulty units in the system at the same time. The advantage of this approach, called *sequential diagnosis*, or *diagnosis with repair* (as opposed to the more traditional *one-step diagnosis*, or *diagnosis without repair*), is that far fewer total tests are needed to perform the diagnosis. In sequentially diagnosable systems, a round of tests are applied and at least one faulty unit is guaranteed to be identified. After each round, faulty units which are identified are replaced by spares; this process continues until no faulty units remain. Preparata, Metze, and Chien defined a simple class of t_p -sequentially diagnosable systems and also explored the sequential diagnosability properties of single-loop systems. A few other papers have defined sequentially diagnosable systems with extremely sparse connection assignments [15]-[18]; however, neither useful characterizations nor efficient diagnosis algorithms have been discovered for the general case.

Another approach to take in the diagnosis of multiprocessor systems is less conservative in nature and is known as t/s -diagnosis [19]; that is, given that there are at most t faulty units in a system, t/s -diagnosis isolates the faulty units to within a set of at most s units. Thus, there exists the possibility of incorrectly diagnosing fault-free units as faulty, but all faulty units are correctly identified. Karunanithi and Friedman explored t/s -diagnosis for single-loop and D_{1,t_p} systems and gave sequential and one-step diagnosis procedures for each. Since then, most of the work on t/s -diagnosis has concentrated on special classes of t/s -diagnosable systems such as t/t -diagnosable systems [20]; again, the advantage is that far fewer total tests are required for diagnosis.

In 1981, Hakimi and Nakajima [21] introduced another diagnosis strategy for diagnosable systems, called *adaptive diagnosis*. Here, testing occurs alternately with

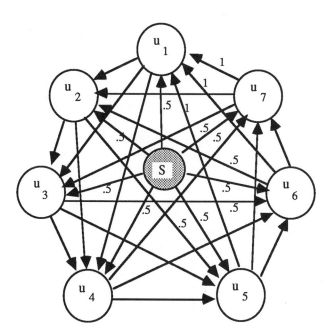

Figure 21.3: A diagnosability graph G' for the system of Figure 21.1

diagnostic processing: the next test to be applied is a function of the test results collected so far. The idea is to first identify a unit which is fault-free and then to assign that unit to test the rest of the units in the system. Later work showed that correct diagnosis is achievable in at least $1.5n - 1.5$ total tests for $n \geq 3$ [22].

Another promising diagnosis technique introduced by Nair, Metze, and Abraham [23], allows for identification of faulty units concurrently with system operation, and is known as *roving diagnosis*. Here, a time-varying subset of units is dedicated to performing system tasks while the remainder of the units test one another. Saheban and Friedman [24] have refined the concurrent diagnosis approach for scheduling optimal testing assignments among the subset of units undergoing testing.

Smith [25] introduced an approach to the diagnosis problem which is called *universal diagnosis,* since it is independent of the connection assignment of the system. He defined two simple algorithms which may lead to the replacement of both fault-free and faulty units: the first scheme identifies as faulty any unit which fails at least one test; in the second scheme, the unit which fails the most tests is identified as faulty. Rounds of testing and replacement proceed until no unit fails any tests. The principal advantages of the method are that the algorithms are computationally efficient and the number of fault-free units replaced is usually low.

21.2.9   Other models of diagnosable systems

The third major direction in which system-level diagnosis has focused is in defining new models which either are based on or are generalizations of the original PMC model. Maheshwari and Hakimi [26] associated with each unit in the system an a *priori* probability of failure. The objective is then to identify the *most likely* set of faulty units given a syndrome. A class of diagnosable systems for the probabilistic model is defined as follows:

Definition 5 [26]: A system S is *probabilistically diagnosable (ρ -diagnosable)* if for any syndrome there exists a unique allowable fault set F whose a *priori* probability of failure is at least ρ, assuming that the probability of occurrence of the set of faulty units in S is at least ρ.

Maheshwari and Hakimi also gave a technique for converting the a *priori* probabilities to positive-valued weights: the greater the weight of a unit, the lower its a *priori* probability of failure and vice-versa. An example of a ρ -diagnosable system ($\rho = .0096$) is shown in Figure 21.4. An $O(n^3)$ diagnosis algorithm for the weighted model was given in [27]; the diagnosability problem for ρ -diagnosable systems was shown by Sullivan to be NP-complete [28]. Other models which include probabilistic parameters have been considered by Blount 1977 [29], by Simoncini and Friedman 1978 [30], and by Barsi 1981 [31].

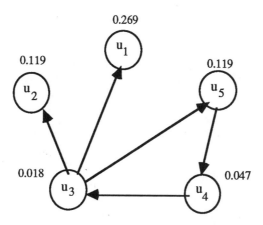

Figure 21.4: A probabilistically diagnosable system ($\rho = .0096$)

One variation of the PMC model which has commanded a considerable amount of interest is the model introduced by Barsi, Grandoni, and Maestrini [32]. In the BGM model or *asymmetric invalidation* model, it is assumed that a faulty unit is not as devious so as to claim that a faulty unit which it tests is fault-free. Thus, a unit which is passed by any other unit must be fault-free. For this simplified problem, Barsi, Grandoni, and Maestrini gave a complete characterization of diagnosable systems, showed that for completely connected connection assignments, $t_p = n - 2$, and proved that D_{1,t_p} systems are optimal for this model.

Kuhl and Reddy [33] first suggested the idea of distributing the diagnosis task among the processors themselves. Thus, the goal is for each fault-free processor to ultimately deduce the set of faulty units. A fault-free unit immediately knows the status of the units which it tests; for a unit which it finds to be fault-free, it can consequently "trust" it and request its list of faulty processors. This iterative process proceeds until all faulty units are made known to the fault-free units. Kuhl and Reddy showed that the necessary and sufficient condition for successful distributed diagnosis is that for the connection assignment, $\kappa(G) \geq t_p$; in the original work and later refinements, the technique was modified to include link failures.

A generalization of the PMC model which has received widespread attention is to incorporate the possibility of *intermittently* faulty units into the system. In Mallela and Masson's pioneering work [34],[35], a fault-free unit may declare an intermittently faulty unit to be fault-free because of the intermittent nature of the fault. They defined a t_i -*diagnosable system* to be one in which, assuming at most t_i intermittently faulty units, a fault-free unit will never be diagnosed as faulty. Therefore, the diagnosis can possibly be incomplete, but never incorrect. They showed that in a system which is known to be t_p -diagnosable, the value of t_i can be as great as t_p, but never greater, or as small as $\lfloor (2t_p)/3 \rfloor$, but never less. The system shown in Figure 21.1, for instance, is t_i -diagnosable for $t_i = 2$. Yang and Masson [36] gave an $O(|E|)$ diagnosis algorithm for t_i -diagnosable systems.

Mallela and Masson later generalized their results on intermittent faults by defining the class of t_h/t_{hi} -*hybrid diagnosable systems,* in which at most t_h units may be faulty, of which at most t_{hi} may be intermittently faulty and the remaining faulty units must be permanently faulty. They reported several interesting interrelationships among t_h, t_{hi}, and t_p.

In 1980, Malek [37] (and independently by Chwa and Hakimi [38]) introduced an *undirected* graph model for t_p -diagnosable systems, in which each system task is assigned to two units to be performed; these agreements and disagreements among the units are the basis for identifying the set of faulty units. Dahbura, Sabnani, and King [39] extended this model to allow faulty units to perform some of their tasks correctly, thereby modeling intermittently as well as permanently faulty units. They also showed that a simple, but possibly imperfect, diagnosis algorithm almost always correctly identifies the subset of faulty units in this model. Finally, Maeng and Malek [40] varied the model by requiring that each comparison of two units be performed by a third unit instead of by a centralized processor.

The works described above are by no means intended to be a comprehensive chronicle of system-level diagnosis research; rather, they are a sampling of the numerous variants of the original idea suggested by Preparata, Metze, and Chien. The following sections describe the fault detection and localization techniques used in practice in fault-tolerant multiprocessors, point out the limitations of system-level diagnosis theory relative to these actual systems, and discuss the role system-level diagnosis will play in the design of future multiprocessor systems.

21.3 Fault Detection and Localization in Fault-Tolerant Systems

The vast majority of commercial, military, and scientific multiprocessor systems developed in recent decades has relied on a number of related techniques for detecting and diagnosing faults at the processor level. Of particular interest here are those systems which are designed to meet high availability, long life, and/or critical computations demands [41].

Even with careful design and manufacturing procedures, faults will eventually plague a system due to the inherent and unavoidable unreliability of its components. In fault-tolerant design, redundancy is used to negate the effects of such faults.

Although there are several steps involved in a fault-tolerant system's response to a fault, ranging from fault confinement to repair and reintegration, the two stages of interest here are *fault detection and diagnosis*. Note that, in practice, fault detection is assumed at the processor level or below, while diagnosis normally implies at the board level or below; in contrast, system-level *diagnosis* is almost always assumed to be at the processor level.

Redundancy can take the form of extra time and/or extra hardware, and fault detection and diagnosis can be performed off-line and/or on-line. Off-line strategies have the disadvantage of degrading or even halting system throughput while the system is under test; furthermore, system integrity is left unmonitored during run-time. On-line methods reduce the severity of these problems but typically the cost in terms of added redundancy is greater.

Since the dawn of the computer age, systems which have been built with fault-tolerance as a design goal have used at least one, and usually more, of the following related techniques: 1) massive replication of hardware and/or software with voting; 2) error detecting and/or correcting codes including self-checking components; 3) validity checks of software and diagnostics of hardware; and 4) watchdog timers and sanity messages. Note that this is by no means an exhaustive list, but does reflect the most common strategies used. Details of the aforementioned strategies and examples of systems which utilize each are not included here but can be found in [41]-[43].

21.3.1 Massive replication of hardware and/or software with voting

In most fault-tolerant systems, massive replication can be found at virtually every level of hardware and software, including gates, chips, boards, processors, intercommunication structures, memory, I/O, and even power supplies. The most general form of massive replication is known as *N-modular redundancy (NMR)*, in which like computations are performed by N different processing elements, which submit their results to a voter. The voter then decides which of these results to propagate to the rest of the system, thereby *masking* the effects of any faults. Furthermore, a disagreement detector used in conjunction with the voter is able to diagnose which of the N elements are faulty so that they may be taken off-line and possibly be replaced by spare elements.

NMR is effective for detecting a wide variety of permanent and transient non-overlapping faults, although its cost is N times that of an equivalent simplex system in addition to the cost of the voting hardware. There is also overhead associated with potential synchronization problems among the components. The voter and common clock are, of course, single points of failure for the NMR cluster; circumventing these problems is non-trivial without adding even more hardware to the system. The problem of common-mode failures can be alleviated somewhat by a technique called *N-version programming*, in which each of the N elements performs a different algorithm to compute the same result [42].

The special case of N-modular redundancy where $N = 2$ is called *duplication* and is probably the most popular form of massive replication used to date. In a duplicated system, the results of the two processing elements are submitted to a *comparator* which then determines whether the results agree. In the case of a disagreement, other techniques must be relied upon to diagnose the fault. In some present-day multiprocessor systems [43], a duplicated system is duplicated once more; when a fault is detected, the disagreeing pair is taken off-line and the remaining pair resumes normal operation. Duplication in its simplest from suffers from the same comparator-failure and common-mode-failure weaknesses as the general NMR approach.

Finally, there are two variations of massive replication which are worthy of mention here. First, certain computations may be performed more than once on the same processor. This strategy, known as *time redundancy*, reduces the amount of hardware needed but increases the computation time considerably. Also, while time redundancy can be effective for detecting transient failures, it is less so for failures which are permanent unless different algorithms are used for the different versions of the computations. Secondly, the voting procedure implemented in the SIFT computer [42] takes place in software; each processor compares its own results to those computed by other N processors, where both N and the elements in an NMR cluster vary depending on the criticality of the process.

21.3.2   Error detecting and/or correcting codes including self-checking components

Error detecting and correcting codes are the most sophisticated fault-tolerant technique in use today, thanks mostly to many years of intensive research in the area, and can be found at all levels of fault-tolerant system design, although they are most frequently associated with protecting the integrity of a system's memory elements or communication channels among components within the system. Some of the advantages of coding techniques are: 1) they are a relatively efficient use of redundancy, both in terms of the extra hardware and coding/decoding time required; 2) a wide variety of codes is available for different error models and levels of error detection and correction; and 3) the error detection and correction is performed on line so that the effects of a fault can be removed or contained immediately. Examples of common codes include parity checks, M-of-N codes, and checksums, to name but a few [42]. Coding techniques are also applied at the logic level to create what are known as *self-checking circuits* [41], in which incorrect outputs take the form of non-codewords and can be detected immediately by a self-checking checker. Such circuits introduce added overhead at the chip level but have the advantage of removing many single points of failure.

21.3.3 Validity checks of software and diagnostics of hardware

Validity checks (also known as *consistency checks* [42]) are normally performed in software and are simple tests which are used to ensure that the operation of the applications software and/or operating system is reasonable. An example of a validity check is a routine which checks that a memory address to be accessed is valid.

A different kind of test is that which is used to verify that the operation of the hardware of a component is correct. This is normally performed off-line and often requires additional hardware. Run-time diagnostics are most often designed at the functional level of a component; that is, the basic functions of the element under test are checked, as opposed to establishing the correct operation of the gates or circuits of which the higher level element is composed, although *built-in self-test (BIST)* techniques are the exception to this [41].

21.3.4 Watchdog timers and sanity messages

Watchdog timers are extra hardware or software mechanisms which monitor in an on-line fashion whether anticipated interactions among system components take place within a pre-specified time period. The technique is relatively inexpensive and therefore fairly common at all levels of system design; however, the fault coverage of watchdog timers is quite limited and the integrity of the timers themselves must be considered.

A variation of watchdog timing which has recently gained popularity is the concept of a *sanity message,* in which a processor periodically sends a communication to some part of the rest of the system; the message can be a simple "I am alive" signal or can be the result of a more thorough exercise of the processor's circuitry.

Of course, no single method of those described above is suitable for all systems and neither is one method sufficient for meeting the dependability needs of a system; instead, cost, technology, and application associated constraints tend to make a certain combination of methods to be that of choice for a particular design. In many cases, the fault detection and diagnosis techniques are applied in a hierarchical fashion; for example, there may be numerous watchdog timers, parity checks, and self-checking circuits within chips which have been duplicated on circuit boards in triplicated processors which communicate on a duplicated bus, and so forth. In such systems, there can also be a hardware *maintenance subsystem* which manages error reports and takes consequent action. This adds to the overhead required for maintaining reliability in a system and can create more single points of failure.

21.4   Unrealistic System-Level Diagnosis Assumptions

It is apparent from the overviews given in the previous two sections that the direction fault-tolerant system design has taken has little resemblance to the concepts developed in system-level diagnosis. One reason for this divergence is that the problems which have been defined in the system-level diagnosis area are based on simplifying assumptions which, while making for both interesting and profound mathematical questions, have not closely adhered to the constraints encountered by designers of fault-tolerant systems. In this section, the unrealistic assumptions inherent in system-level diagnosis research are described. In the next section, the prospect for the future of system-level diagnosis as a viable technique for fault-tolerant systems is examined.

From a practical standpoint, system-level diagnosis concepts are of limited value in present-day fault-tolerant system design due to the following limiting assumptions:

21.4.1   The existence of tests

In much of the literature on system-level diagnosis, it is assumed that components have the ability to *test* one another. Furthermore, the original PMC model assumes that the tests are *complete* in the sense that a fault-free unit u_i which tests a faulty unit u_j determines that u_j is faulty.

Run-time testing in actual multiprocessor systems is usually impractical because: 1) the complexity of existing processors prevents meaningful testing within a reasonable time; 2) testing must be performed off-line, thereby degrading system throughput; and 3) often, off-the-shelf processors are used by fault-tolerant multiprocessor manufacturers, so that acquiring sufficient information to design tests for such processors is infeasible.

To overcome the PMC testing assumption, the comparison models described in Section 21.2 have been developed [37]-[40]. Also, some of the probabilistic models mentioned in Section 21.2 include incomplete tests [30],[31].

21.4.2   Permanently faulty units

For the first ten years of research in system-level diagnosis, only permanently faulty units were considered. Furthermore, in the earliest work on the comparison approaches to diagnosis, it was assumed that faulty units perform *all* assigned jobs incorrectly.

More recent work, starting with the introduction of intermittently faulty units to the PMC model by Mallela and Masson [34],[35], has focused on removing these assumptions. In particular, one recent approach using the comparison model [39] handles both intermittent faults and permanent faults which are manifested externally in the form of transient errors.

Another related assumption is that the fault-free or faulty status of a unit is unchanged during the time test or comparison results are accumulated and diagnosis is performed. Relaxing the model to handle intermittently faulty units alleviates this as well.

21.4.3   The number of faults is bounded

While most system-level diagnosis results are valid when large numbers of processors are faulty simultaneously, in practice it is the case that fault detection and diagnosis strategies are concerned primarily with handling single faults. As evidenced in the previous section, the fault-tolerant strategies employed in practice are quite costly and incomplete at best, even for single faults. The underlying assumption is that the occurrence of multiple failures is so unlikely that the added overhead to safeguard against them using mainstream techniques is not warranted, especially when single faults are likely to be detected and repaired in time to prevent them from accumulating.

On the other hand, a scenario in which multiple faulty processors are commonplace is Wafer-Scale Integration (WSI), in which production-time yields are relatively low. In WSI, a self-diagnosing wafer may contain mostly faulty processing elements, in which case the bounds imposed by system-level diagnosis may be too low. Systems which operate in hostile or volatile environments may also experience multiple fault components simultaneously.

Several of the efforts in system-level diagnosis are aimed at removing the bounded-fault assumption altogether [25],[39]; the approach taken is to define diagnosis algorithms which almost always, but not in every case, identify the set of faulty units, but which do not depend on a preset bound to function properly.

21.4.4   Test scheduling

In most system-level diagnosis work, the coordination among tests and system job assignments is not considered. It is assumed that test results are acquired in an unspecified manner in order to effect diagnosis. However, in practical systems it is desirable that test scheduling not adversely affect system throughput.

For this reason, the concurrent and adaptive fault diagnosis techniques described previously have been introduced [21]-[22],[33]. One way in which such techniques can be implemented with a minimal impact on throughput is to use the *spare capacity* available in most multiprocessor systems; it has been shown in [44] that for moderately loaded systems, a sufficient percentage of jobs can be duplicated using the system's spare capacity to provide a basis for fault detection and diagnosis with virtually no degradation of system response time.

21.4.5   Worst-case approach to diagnosis

Typically, strategies for fault diagnosis have been designed to handle every possible fault situation, especially those in which faulty units collude by incorrectly diagnosing other faulty units as fault-free and fault-free units as faulty. In practice, the added complexity required for diagnosing such improbable scenarios is generally unwarranted. The approaches described in [25],[39] are devised specifically for diagnosing when the solution is relatively straightforward; analysis shows that this is almost always the case.

21.4.6   Hardware faults

System-level diagnosis, like the fault detection and diagnosis techniques described in the previous section with the exceptions of N-version programming and software validity checks, are primarily intended for treating hardware faults, as opposed to design flaws in software or operator errors. Although software and operator faults have been known to account for a significant percentage of errors in some existing systems [46], the numbers vary sharply from one system to another. Which of these fault classes will predominate in future multiprocessor architectures is uncertain.

21.4.7   Centralized diagnosis

In most system-level diagnosis research, it is assumed that there is a central observer which can reliably collect test or comparison data and perform some diagnosis algorithm reliably. While many fault-tolerant systems in use today depend on a maintenance subsystem to handle error reporting and recovery routines, such added hardware may actually reduce the overall reliability of the system. Furthermore, some processors within a system may not be easily observable, making such an alternative an expensive one. For these reasons, Kuhl and Reddy [33] have proposed the *distributed* diagnosis approach described in Section 21.2.

21.5 Fault Detection and Localization Strategies for Today and Tomorrow

This decade is witness to the birth of the next age of automatic computational devices: multiprocessor systems, including distributed and parallel computing. The underlying motivation for the development of such systems is that technological progress is inherently application-driven; present technical resources stimulate new ideas for applications which require new capabilities and thus, the state-of-the-art is impelled forward. This impetus, of course, is forced to occur within the fundamental constraints imposed by nature.

There is no doubt that there will always be demand for physically smaller yet more powerful computers. This demand can be met through advances in *components, software,* and *architectures.*

Most of the improvements in the size and speed of computational devices thus far is due to the development of smaller, faster components. While this trend will continue for the foreseeable future, the cost-effectiveness of gains in this direction is likely to be reduced. More efficient software will yield gains in speed and in memory requirements; however, there are underlying limitations which are fairly well understood regarding the complexity of problem-solving [48]. Although breakthroughs are inevitable for specific instances, dramatic developments in the area as a whole are uncertain.

The third area which has not yet realized its full potential is that of *architectures.* Until now, the classical Von Neumann architecture has been that of choice. But now that improvements in software efficiency and component speed are becoming difficult to realize, it is worthwhile to explore alternative architectures, especially since the technology is available to assemble large numbers of simple processors to operate together in a system. Indeed, systems that are composed of multiple processing devices have just begun to appear and are relatively rudimentary in concept and small in scale.

For such comparably small systems, the simple fault detection and localization techniques described in Section 21.3 are best suited. Furthermore, customers are presently willing and able to absorb the overhead, in terms of increased cost and size, in order to obtain the needed dependability for their applications. However, as the scale of such systems increases, it is contended that such brute force methods will cease to be expedient since 1) the increased scale of future systems and the radically different architectures taking form will present new "reliability bottlenecks" which present conventional fault detection and diagnosis techniques alone will be unable to handle effectively; and 2) the framework will exist for sensible alternatives such as self-monitoring and self-diagnosis.

System-level diagnosis inherently requires that a number of processor-level errors be detected before diagnosis occurs. In some applications, this cannot be tolerated since corrupted data are totally unacceptable. This stringency is not, however, without price. When action is taken to diagnose the error, system resources which would have been used for system throughput are now taken off-line for testing. If the error report had been due to a transient fault, such action is for naught. Also, had the error been due to a fault in a different processor, localizing the faulty processor could be more costly in terms of system resources than waiting for more error reports to occur. Thus, system-level diagnosis is not necessarily more expensive to invoke than traditional techniques.

In multiprocessor systems of tomorrow, as in today's computer systems, users will tend to be unwilling to sacrifice performance for reliability. Therefore, strategies for fault detection and localization must not noticeably degrade system throughput. Unless the processors which compose the system are extremely simple, it is unlikely that thorough diagnostics will be practical. Tests must be run in the system's spare capacity and will have relatively low coverage. The alternative is to perform comparisons of tasks which the processors execute; a reasonable percentage of such tasks could be performed in the system's spare capacity as shown in [44]. In addition, the comparison approach is better than testing for handling transient errors and software errors.

For systems in which multiple faulty components are unlikely to exist simultaneously, the system-level detection and diagnosis strategies must be designed to handle a small number of faults with an extremely high success probability and yet still be able to handle greater numbers with a lower, yet non-zero, success probability. In this way, faulty processors will be identified promptly, before more accumulate.

In actual systems, worst case scenarios are highly unlikely; therefore, it will be more effective to focus on relatively benign situations which assume that faulty processors are not necessarily malicious.

The strategy described in [39] obeys the criteria described above and, furthermore, can be implemented both in the spare capacity of a system and in a distributed manner [45]. Such an approach could be used in a multiprocessor system *along with*, not as a replacement for, some of the fault detection and diagnosis techniques described in Section 21.3 such as error detecting codes and validity checks.

21.6 Conclusions

The multiprocessor family of the future is likely to be highly diverse. While it is improbable that system-level diagnosis is best for all classes of systems, the same could be said for any other approach. Clearly, more work is needed to devise system-level diagnosis strategies within a framework of realistic assumptions, but the message is clear: in systems composed of large numbers of independent elements, the elements themselves have the potential to monitor one another to detect, diagnose, and isolate faulty members. As in all research areas, sound theoretical work in this direction must continue. In parallel, encouragement should be given to applied research for investigating strategies which can be shown to be useful in designing the dependable systems of tomorrow.

References

[1] F.P. Preparata, G. Metze, and R.T. Chien, "On the connection assignment problem of diagnosable systems," *IEEE Trans. Electron. Comput.*, vol. EC-16, pp. 848-854, Dec. 1967.

[2] S.L. Hakimi and A.T. Amin, "Characterization of the connection assignment of diagnosable systems," *IEEE Trans. Comput.*, vol. C-23, pp. 86-88, Jan. 1974.

[3] F.J. Allan, T. Kameda, and S. Toida, "An approach to the diagnosability analysis of a system," *IEEE Trans. Computers*, vol. C-23, pp. 1040-1042, Oct. 1975.

[4] T. Kameda, S. Toida, and F.J. Allan, "A diagnosing algorithm for networks," *Information and Control*, vol. 29, pp. 141-148, 1975.

[5] A.M. Corluhan and S.L. Hakimi, "On an algorithm for identifying faults in a t -diagnosable system," in *Proc. 1976 Conf. on Inf. Sci. and Sys.*, The Johns Hopkins University, pp. 370-375, April. 1976.

[6] A.T. Dahbura and G.M. Masson, "An $O(n^{2.5})$ fault identification algorithm for diagnosable systems," *IEEE Trans. Comput.*, vol. C-33, pp. 486-492, June 1984.

[7] R.F. Madden, "On fault-set identification in some system level diagnostic models," Science Institute, Univ. of Iceland, technical report, Oct. 1978, and personal correspondence, Jan. 1983.

[8] H. Fujiwara and K. Kinoshita, "On the computational complexity of system diagnosis," *IEEE Trans. Comput.*, vol. C-27, pp. 881-885, Oct. 1978.

[9] G.G.L. Meyer and G.M. Masson, "An efficient fault diagnosis algorithm for symmetric multiple processor architectures," *IEEE Trans. Comput.*, vol. C-27, pp. 1059-1063, Nov. 1978.

[10] S. Mallela, "On diagnosable systems with simple algorithms," in *Proc. 1980 Conf. on Inf. Sci. and Sys.*, Princeton Univ., pp. 545-549, March 1980.

[11] K.Y. Chwa and S.L. Hakimi, "On fault identification in diagnosable systems," *IEEE Trans. on Comput.*, vol. C-30, pp. 414-422, June 1981.

[12] S.L. Hakimi and K.Y. Chwa, "Schemes for fault tolerant computing: a comparison of modularly redundant and t-diagnosable systems", *Information and Control*, vol. 49, pp. 212-238, June 1981.

[13] A.T. Dahbura, G.M. Masson, and C.L. Yang, "Self-implicating structures for diagnosable systems," *IEEE Trans. Computers*, vol. C-34, pp. 718-723, Aug. 1985.

[14] G. Sullivan, "A polynomial time algorithm for fault diagnosability," in *Proc. 25th Ann. Symp. on Found. of Comp. Sci.*, IEEE Comput. Soc. Publ., October 1984, pp. 148-155.

[15] P. Maestrini, "Complexity aspects of system diagnosis," in *Proc. 17th Ann. Allerton Conf. Commun., Contr., Comput.*, Allerton House, Monticello, IL, Oct. 1979, pp. 329-338.

[16] A. Kavianpour and A.D. Friedman, "Efficient design of easily diagnosable systems," in *Proc. 3rd USA-Japan Comput. Conf.*, 1978, pp. 251-257.

[17] P. Ciompi and L. Simoncini, "Analysis and optimal design of self-diagnosable systems with repair", *IEEE Trans. Comput.*, vol. C-28, May 1979, pp. 362-365.

[18] U. Manber, "System diagnosis with repair," *IEEE Trans. Comput.*, vol. C-29, Oct. 1980, pp. 934-937.

[19] S. Karunanithi and A.D. Friedman, "Analysis of digital systems using a new measure of system diagnosis", *IEEE Trans. Comput.*, vol. C-28, Feb. 1979, pp. 121-133.

[20] C.-L. Yang, G.M. Masson, and R.A. Leonetti, "On fault isolation and identification in t_1/t_1 -diagnosable systems," *IEEE Trans. Comput.*, vol. C-35, July 1986, pp. 639-643.

[21] S.L. Hakimi and K. Nakajima, "On adaptive system diagnosis," *IEEE Trans. Comput.*, vol. C-33, pp. 234-240, March 1984.

[22] P.M. Blecher, "On a logical problem," *Discrete Math.*, vol. 43, 1983, pp. 107-110.

[23] A.D. Friedman and L. Simoncini, "System-level diagnosis," *IEEE Computer*, March 1980, pp. 47-53.

[24] F. Saheban and A.D. Friedman, "Diagnostic and computational reconfiguration in multiprocessor systems," in *Proc. ACM Ann. Conf.*, Dec. 1978, pp. 68-78.

[25] J.E. Smith, "Universal system diagnosis algorithms," *IEEE Trans. Comput.*, vol. C-28, May 1979, pp. 374-378.

[26] S.N. Maheshwari and S.L. Hakimi, "On models for diagnosable systems and probabilistic fault diagnosis," *IEEE Trans. Comput.*, vol. C-25, March 1976, pp. 228-236.

[27] A.T. Dahbura, "An efficient algorithm for identifying the most likely fault set in a probabilistically diagnosable system," *IEEE Trans, Comput.*, vol. C-36, April 1986, pp. 354-356.

[28] G. Sullivan, "The complexity of system-level fault diagnosis and diagnosability," Ph.D. thesis, Yale University, 1986.

[29] M.L. Blount, "Probabilistic treatment of diagnosis in digital systems," in *Proc. 1975 Symp. Fault Tolerant Comput., IEEE Comput. Soc. Publ.*, June 1975, pp. 72-77.

[30] L. Simoncini and A.D. Friedman, "Incomplete fault coverage in modular multiprocessor systems," in *Proc. ACM Ann. Conf.*, Dec. 1978, pp. 210-216.

[31] F. Barsi, "Probabilistic syndrome decoding in self-diagnosable digital systems," *Digital Processes*, vol. 7, 1981, pp. 33-46.

[32] F. Barsi, F. Grandoni, and P. Maestrini, "A theory of diagnosability of digital systems," *IEEE Trans. Comput.*, vol. C-25, pp. 585-593, June 1976.

[33] J.G. Kuhl and S.M. Reddy, "Fault-diagnosis in fully distributed systems," in *Proc. 1980 Int. Symp. on Fault Tolerant Comput.*, IEEE Comput. Soc. Publ., June 1980, pp. 100-105.

[34] S. Mallela and G.M. Masson, "Diagnosable systems for intermittent faults," *IEEE Trans. Comput.*, vol. C-27, pp. 560-566, June 1978.

[35] S. Mallela and G.M. Masson, "Diagnosis without repair for hybrid fault situations," *IEEE Trans. Comput.*, vol. C-29, June 1980, pp. 461-470.

[36] C.-L. Yang and G.M. Masson, "A fault identification algorithm for t_i - diagnosable systems," in *Proc. 1985 Int. Symp. on Fault Tolerant Comput.*, IEEE Comput. Soc. Publ., June 1985, pp. 78-83.

[37] M. Malek, "A comparison connection assignment for diagnosis of multiprocessor systems," in *Proc. 1980 Int. Symp. on Fault Tolerant Comput.*, IEEE Comput. Soc. Publ., June 1980, pp. 31-36.

[38] K.-Y. Chwa and S.L. Hakimi, "Schemes for fault-tolerant computing: a comparison of modularly redundant and t-diagnosable systems," *Info. and Control*, vol. 49, 1981, pp. 212-238.

[39] A.T. Dahbura, K.K. Sabnani, and L.L. King, "The comparison approach to multiprocessor fault diagnosis," *IEEE Trans. Comput.*, vol. C-36, March 1987, pp. 373-378.

[40] J. Maeng and M. Malek, "A comparison connection assignment for self-diagnosis of multiprocessor systems," in *Proc. 1981 Int. Symp. Fault Tolerant Comput.*, IEEE Comput. Soc. Publ., June 1981, pp. 173-175.

[41] D.K. Pradhan, *Fault-Tolerant Computing Theory and Techniques.* Englewood Cliffs, NJ: Prentice-Hall, 1986.

[42] D.P. Siewiorek and R.S. Swarz, *The Theory and Practice of Reliable System Design.* Digital Equipment Press, 1982.

[43] Eds, "Computers that are 'never' down," *IEEE Spectrum*, vol. 22, April 1985, pp. 46-54.

[44] A.T. Dahbura, K.K. Sabnani, and W.J. Hery, "Performance analysis of a fault detection scheme in multiprocessor systems," in *Proc. 1987 ACM SIGMETRICS Conf. on Meas. and Model. of Comp. Sys.*, ACM Press, May 1987, pp. 143-154.

[45] A.T. Dahbura and K.K. Sabnani, "A distributed algorithm for system-level diagnosis," *AT&T Conference on Interconnection and Communication Issues in Future Systems*, May 1986, pp. 43-45.

[46] J. Gray, "Why do computers stop and what can be done about it?", in *Proc. 5th Symp. on Rel. in Dist. Software and Database Sys.*, IEEE Comp. Soc. Publ., Jan. 1986, pp. 3-12.

[47] J.A. Bondy and U.S.R. Murty, *Graph Theory with Applications.* New York: Elsevier North Holland, Inc., 1976.

[48] M.R. Garey and D.S. Johnson, *Computers and Intractability.* San Francisco, CA: W.H. Freeman, 1979.

Chapter 22

Self-Diagnosable and Self-Reconfigurable VLSI Array Structures

Andrzej Rucinski [1]
and John L. Pokoski [1]

Abstract

The issue of self-testing and self-reconfiguration of two-dimensional VLSI array structures are discussed in this paper. Analysis and synthesis are facilitated by separating redundant circuits into two separate categories. ST-redundancy which represents built-in circuitry used solely for self-diagnosis, and SR-redundancy which represents extra resources used for fault-tolerance. The amount of redundancy can vary, e.g. a BIST method for self-testability employs approximately 10% ST-redundant circuitry while 200% of SR-redundancy plus a voter which is ST-redundant represents a signal-fault tolerant device. The concept is implemented as a double-layered VLSI array architecture. The developed model is very general. Some important arrangements like a centralized host system, a fault-tolerant distributed multiprocessor array developed by Kuhl and Reddy, or Koren's self-reconfigurable VLSI array are special cases of $S(M, N, t_o)$ systems, as introduced here. In one special case, 100% of ST-redundancy allows the testing to be performed at a polynomial time regardless of the array size.

---

[1] University of New Hampshire, Durham, NH

435

22.1   Fault-Tolerant VLSI Array Structures

With the advent of VLSI and WSI technologies and the increasing demand for computational parallelism, the rapid development of array structures: e.g. "smart memories", computational and micro-computer arrays has been observed. Since the assumption that VLSI and WSI circuits can be perfect is not acceptable, faults [SAMI83], yield [KORE81], and reliability [KORE87] of these structures has been extensively studied in the literature. Generally, fault-tolerance can be achieved by masking the faults [KIM85], and by detecting faults and reconfiguring the array [PRAD86].

Three basic aspects of fault-tolerant VLSI array structures are: architecture, redundancy, and reconfiguration. Here, we review briefly some of the techniques that are relevant to these aspects and to the subject of this paper.

There are two major architectural types of VLSI arrays: node-switched arrays and arrays with segregated switches.

In the first type, the processing elements perform internally all the switching necessary to establish connections. Architecture of this kind has been studied in [KORE81]. Drawbacks include: complicated switching mechanism, non-transparent fault-tolerance, and an extremely low utilization rate of good PE's.

VLSI arrays in which connections are established by using external switches are analyzed in [ROSE83]. Perhaps the best-known example of this class is the BLUE CHIP built at Purdue University [SNYD82]. Greater flexibility of possible topologies is provided since the communication and computational spaces are separated. This advantage is offset by the necessity of having extra circuitry and, hence, chip area. Additionally, configuring strategies are non-trivial and result in eccentric layouts. A hybrid approach has been presented in [PRAD86].

To be able to compute in the presence of faults, a two dimensional array must employ redundancy. Time redundancy techniques which enable self-testing at the expense of processing speed have been developed by the research team at the Technical University of Milano, Italy [SAMI83]. Another technique that uses alternate logic which performs concurrent diagnosis and computation [ARBA87] seems to be very attractive for systolic structures where some PE's are active and others are idle at any given time. This "natural" redundancy has resulted in the interstitial fault-tolerance of systolic arrays [KUHN83]. The method allows the array to tolerate any single fault, however it requires massive redundancy. The combination of time and space redundancy techniques has been analyzed by Agraval [AGRA85].

The motivation for the reconfiguration of VLSI arrays is quite natural and twofold: yield improvement and reliability improvement.

Yield improvement is performed at production time by incorporating some redundancy on a chip or wafer, performing an external test on a chip, and finally making hard metal wiring between non-faulty processing elements. It can be accomplished by putting an extra metalization layer or using laser programming techniques. This approach of static reconfiguration enhances the yield because defective wafers can be accepted. The complexity of the reconfiguration algorithms is not critical and no on-chip controlling circuitry is required. Mangir and Avizienis [MANG82] have shown that the yield improvement in planar, two-dimensional VLSI arrays saturates when the redundancy reaches 10%. Leighton and Leiserson [LEIG85] developed structuring procedures for some topologies minimizing the length of the longest wire in the system. Kung and Lam [KUNG84a, KUNG84b] proposed a simple method to bypass faulty PE's using extra registers. Koren and Breuer [KORE84] confirmed the previous yield saturation hypothesis by analyzing reconfigurable VLSI arrays with duplicate PE's and self-testing capability. Self-testing through duplication may actually produce a lower yield than the non-redundant design. Generally, the tradeoff between the number of PE's, yield and tolerated faults has been analyzed by Green and Al Gamal [GREE84]. A basic assumption of the above papers is that the probability of intercommunication links and controlling/self-checking hardware failing is negligible. This is reasonable due to their simplicity and the small area they occupy. However, to investigate the problem thoroughly, the assumption of a reliable hard core of the system has to be relaxed. Additionally, the essence of restructuring is to change and influence a wafer. This may create new manufacturing defects which cannot be tolerated and repaired any more. An excellent review of fault-tolerant techniques for yield enhancement can be found in [MOOR86].

Array reliability can be improved by applying dynamic reconfiguration or soft wiring which is driven by a host at compile-time or at run-time. This approach allows the reconfiguration-controlling algorithm to be fairly complex when the host is powerful. This approach has been pioneered by Manning [MANN77], who showed how a VLSI array can be repaired by eliminating all flawed cells using the "loading arm" strategy. However, there are serious observability and error latency problems due to the limited input/output accessibility of VLSI arrays. Aubusson and Catt [AUBU78] simulated growing fault-free spirals for square and hexagonal arrays. The Diogenes approach introduced by Rosenberg [ROSE83] favors redundancy in an array's communication links rather than in its PE's. The resulting fault-tolerant array utilized 100 percent of the available cells, but the throughput slows down due to prolonged wiring spanning logically adjacent PE's. Fortres and Raghavendra considered two aspects of reconfigurability when both the algorithm and the processor array can be reconfigured [FORT84].

The other method for reliability improvement in array structure assumes that an array is able to test and reconfigure itself (e.g. at run-time). This dynamic self-configuration falls into the soft wiring category. An array equipped with self-testing and self-structuring capabilities does not add to host computation overhead, but extra circuitry is required. The simplicity of a structuring algorithm is critical in order to maintain the balance between the probability of survival and the added silicon area. The classic example of this approach is Koren's algorithm [KORE81] which is suitable for an environment where each PE is able to test all its neighbors. A desired structure (different topologies are considered) grows gradually, and defective PE's are simply not incorporated.

Structuring algorithms in VLSI arrays have been intensively studied by Negrini, Sami and Stefanelli [SAMI83]. Their model of fault-tolerant array assumes that very little PE redundancy is available, usually a single spare row and/or column. However, the connecting scheme is relatively complex and extra multiplexors are added. Presented algorithms for two-dimensional VLSI arrays are simple [SAMI83] and the utilization ratio is high. Reconfiguration is achieved by means of simple rules which are implemented as an overhead circuit accompanying each PE. Fussel and Varman [FUSS84] analyzed the problem of structuring pipelines in defective arrays. They assume that two-way pipelining is possible and a desired structure is wrapped around a previously grown spanning tree of fault-free nodes. In [CHOI84] the fault-diagnosis method called Token Triggered Comparison with Duplicated Data (TTCDD) for reconfigurable systolic arrays is presented. Fault-detection is performed on-line using test-tokens and operation duplication. The formal approach, based on tessellation automata theory, to a fault-tolerant array of processor is given in [WALT81, GOLL84]. Fault-tolerance is achieved by masking any single cell failures caused by erroneous state transition or by erroneous outputs.

There are major problems that are not addressed satisfactorily in the referenced literature and that have plagued VLSI and WSI ambitions for almost 20 years:

- redundancy limit on a wafer

- diagnostic hard-core requirement

- VLSI testing

The first problem relates to the yield of two-dimensional wafers. It has been shown [MANG82, SAMI83, KORE84] that above a given level, added chip area and interconnection complexity lead to decreasing reliability and yield. Unfortunately, all techniques presented result in a relevant increase in chip area.

The second issue is the presence of a hard-core in an array. Both hard and soft wiring require that source circuitry is reliable and cannot be tested. This assumption has been justified to some extent [KUNG84a], but in the case of increasingly complex wafers it might not be valid.

The last problem is VLSI testing. Testing can be performed externally and internally (self-testing). Both techniques are very difficult to apply to VLSI and WSI arrays. It may not be possible to apply a test to all the PE's from an external source because of pin limitations and also due to the large number of PE's. The observability of the test results is limited because of the presence of faulty PE's. On the other hand, the built-in test circuitry increases silicon area which is costly and may violate the yield saturation rule. Unfortunately, the problem of efficient testing in VLSI array structures is often overlooked and not considered thoroughly, despite the fact that it is a necessary condition to apply any restructuring technique. An exception can be found in [CHOI84].

We are suggesting the following remedies to the discussed problems:

- three-dimensional (multilayered) VLSI

- system-level diagnosis

- concurrent testing

The diagnosis process in array structures can be divided into two consecutive steps: the detection phase and the location phase. Both steps can be performed either externally or internally. In external detection, tests are generated by an external tester. The most difficult case of VLSI testing occurs when each PE has to be isolated and individually tested. However, systolic algorithms for testing array structures have been developed [VERG86]. The systolic approach greatly reduces the testing time. If each PE is tested by its neighbors, the detection phase (self-detection) is performed internally.

Two techniques used in the location phase which are performed externally by an independent observer are mentioned here: the "classic" Preparata-Metze-Cheng (PMC) model has been studied extensively in the literature; the systolic approach [SOMA84] is more suitable for VLSI computational arrays. The location phase algorithms can also be driven internally by PE's themselves and this strategy represents pure self-testing. Each PE can employ either global knowledge or local knowledge. The distributed fault-tolerance for multiprocessor systems developed by Kuhl and Reddy [KUHL80] fits into the first category. The diagnostic validity is assured by the "mutual trust" principle. The method is flexible and distributed, but the broadcasting is a serious problem. Since the principles of locality and simplicity are not observed, this approach cannot be easily adopted for VLSI systems. The second strategy with locally knowledgeable PE's is incorporated in self-testable VLSI and WSI fault-tolerant restructurable array structures [KORE81]. Knowledge is reduced to the limit and each PE is unaware of the diagnostic situation beyond its vicinity.

The paper presents a general class of fault-tolerant VLSI arrays which are self-diagnosable and self-reconfigurable. The diagnosis is to be performed periodically and used as an easy maintenance aid to pinpoint errors caused by latent faults. It can also be helpful in avoiding a possible catastrophic failure of the array caused by the superposition of many undetected faults.

In the next section the general concept of fully self-diagnosable array structures will be discussed. We will also present a brief summary of research results which have been either already published or submitted for publication. Section 22.3 addresses directions of future research to be performed.

22.2   Double-Redundancy Double-Layer Scheme

Processing elements in the systems we are considering must perform the control tasks of diagnosis and reconfiguration as well as the computational tasks, thus reducing the time for useful computation. One solution might be to separate the tasks into separate PE's. This could be done in a completely distributed two-layered manner as introduced in [RUCI85, RUCI86]. This section will briefly discuss the two-layered approach show its generalizability, and present results of our research.

The technique assumes the use of auxiliary, low-cost processors for error diagnosis and reconfiguration. In one sense the concept is similar to that of watchdog processors. However, in our case the array of PE's is positioned under an umbrella of DRONE processors which are also arranged in an array-type structure. Each layer implemented as a single WSI wafer takes advantage of multilayered VLSI technology by incorporating extra I/O ports along the Z axis. An example of a double-layer structure is indicated in Figure 22.1. A system is shown there as a graph with each node representing a processor. $M = 9$ denotes the number of DRONES, $N = 36$ denotes the number of working processors, and $t_o = 4$ denotes the self-diagnosability level of the system (i.e. all faulty processors can be identified by the array provided that the number of faulty processors does not exceed t_o). In this particular case each DRONE is responsible for four working PE's. Each layer in this pyramidal structure has a different topology. The DRONE layer constitutes an infinite grid and incorporates a system-level diagnosis. If either a DRONE or its working PE group fails, the information about failure is spread through the DRONE layer, and both elements are eliminated from the system. The presented scheme employs two separate redundancies for fault-tolerance. In order to perform self-testability at a minimal time, an array must employ source redundancy called ST-redundancy (redundancy for self-testability). Self-reconfigurability in a faulty environment requires other extra resources - SR-redundancy. For example, a BIST method for self-testability employs approximately 10% of redundant circuitry [BUEH82]. On the other hand, 100% of SR-redundancy enables a device to be a single fault detectable. 200% SR-redundancy plus a voter which is a ST-redundant circuit makes a device a single-fault tolerant. To achieve both self-diagnosis and self-reconfiguration both redundancy types are needed.

The ratio between the number of DRONES and working processors, $k = M/N$, $1/N \leq k \leq 1$ which is called a distribution level, measures the amount of ST-redundancy. There are two extremal cases: a single DRONE in the system repre-

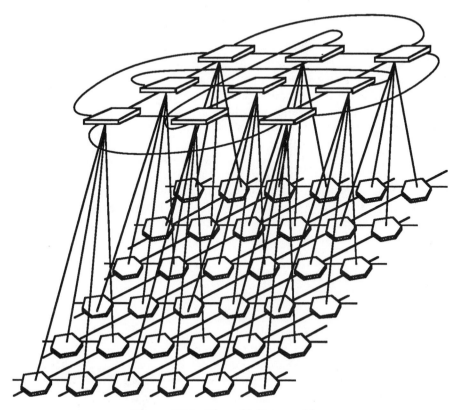

Figure 22.1: The self-diagnosable array.

sents the minimal amount of self-diagnostic circuitry, namely $1/N$ % . On the other hand, the maximal amount of ST-redundancy is when each PE is guarded by its own DRONE. In that case there is a 100% ST-redundancy in the system.

Finally, let us compare $S(M, N, t_o)$ systems with other existing self-diagnosable and self-reconfigurable approaches. There are two parameters characterizing a $S(M, N, t_o)$ system: the distribution level (resources) and the knowledge level of each DRONE (robustness). Thus, the space of distributed fault-tolerant arrays determined by these two factors is a triangle as shown in Figure 22.2. Horizontal lines represent systems having a constant ratio of DRONEs and working PE's (ST-redundancy is constant). The plot is a frame-work which helps to compare important classes of self-testable systems. The distribution level also determines the number of working PE's being tested by each DRONE. Testing can be performed sequentially by isolating and testing one working PE at a time. This approach resembles the "roving emulator" technique [BREU83]. Systolic testing is also possible. In the latter, a whole subarray of N/M elements of working PE's is tested simultaneously. Each vertex of the space represents a well-known testing scheme. The vertex $S(1, N, t_o)$ corresponds to a fault-tolerant array with a single DRONE, i.e. a central host which has to have, by definition, global knowledge. If the host is dedicated to the array it can be interpreted as a BIST-type circuit for self-testability. On the opposite side, there is a system $S(N, N, t_o)$ where each DRONE knows only about its neighbors. This is a double-layered version of Koren's approach [KORE81]. The bottom-right vertex represents a double-hardware system where each DRONE is equipped with global knowledge. So, this is a modified version of the distributed fault-tolerant multiprocessor system proposed by Kuhl and Reddy [KUHL80]. Clearly, the area to the left of a line between $S(N, N, t_o)$ and $S(1, N, t_o)$ is not practical since the total knowledge possessed by the drones does not cover the full array. The area near the lower-left corner seems to be more suitable for VLSI. Although this implies multiple DRONES, use of chip real estate is efficient since each DRONE can be simple. A low knowledge level preserves the principle of locality and the resulting high distribution of knowledge makes an array less susceptible to catastrophic failures.

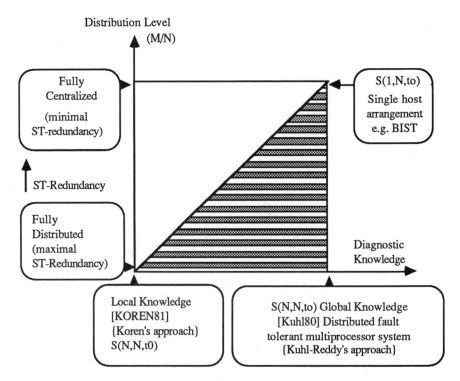

Figure 22.2: The space of distributed fault-tolerant $S(M, N, t_o)$ arrays (N is given)

However, limited knowledge results in the necessity of developing very simple re-structuring algorithms in flawed arrays. Thus, the reconfiguration flexibility is relatively small which, in turn, can result in a low utilization ratio of available, fault-free PE's. This disadvantage can be offset by increasing the knowledge level. On the other hand, the overall cost can be reduced by moving towards a more centralized scheme. This strategy, however, requires more robust, complex DRONES. deteriorates testing performances and results in wasted subarrays of working PE's if the corresponding DRONE becomes faulty. The above coherent taxonomy generalizes the concept of extra circuitry for self-diagnosis (ST-redundancy). The model is valid for both node-switched and segregated switches arrays.

We have developed a formal basis which is used for both testing and reconfiguration. The $S(M, N, t_o)$ systems are described as tesselation automata [WALT81] but this "classic" description has been substantially modified to handle both aspects of fault-tolerance. We also have analyzed the efficiently of a double-layered approach. Some preliminary results show there is an optimal ratio of distribution, i.e. an optimal value of k, to achieve peak performance.

As far as the interconnection schemes for ST-redundancy are concerned, we consider only locally planar and regular arrays. Such structures can be directly related to the structures of crystallographic groups [DURB85]. There are only seventeen unique crystallographic groups, the number of which can be reduced to seven if groups of different orientation of individual elements are treated identically. Thus, there are only seven unique, regular, locally planar interconnection schemes possible for distributed computers.

As the number of faulty processors increases, the self-diagnosability level t_o of a $S(M, N, t_o)$ system generally deteriorates. However, this trend can be reversed by elimination of some "pseudo-faulty" processors from the systems. This diagnostic reconfiguration phenomenon is analyzed along with the corresponding algorithm in [RUCI87a].

Basic, fully distributed strategies for different topologies, e.g. pipes, meshes and binary trees have been presented in [RUCI85]. More complex structures were analyzed in [RUCI86]. The approach assumed that primitive structures mentioned above could be embedded into the same physical array of working PE's independently and then integrated to form a uniform system. Recently, we have developed structuring strategies suitable for any distribution level k. Those strategies have been analyzed in terms of systems efficiency. The structuring strategy has been further refined. The system can be reconfigured at a minimal time with a reduced amount of changes. The impact of catastrophic failures is also greatly reduced.

Moreover, this method of fault-tolerance is superior over the NMR-method in terms of the SR-overhead [RUCI87b].

22.3   Conclusions and Future Developments

The presented general class of fault-tolerant VLSI arrays allows the testing to be performed at a polynomial time regardless of the array size. The concept of ST- and SR-redundancies enables an easy and elegant classification and comparison of fault-tolerant systems with different overhead for fault-tolerance. Some existing solutions, like Koren's [KORE81] or Kuhl-Reddy's [KUHL80] approaches are special cases of the presented scheme. The interconnection schemes of both layers may be different. It means that the concept of ST-redundancy can be applied to any structural graph which represents a computer architecture. However, since $S(M, N, t_o)$ systems employ a massive redundancy, this research is useful to the military community for signal processing applications in either unmanned space or airborne applications. It seems that typical commercial applications do not require such stringent reliability requirements.

Further research will deal with three separately treated problems: self-testing, self-reconfiguration and recovery. It seems to be reasonable to concentrate on the self-testing issue first. It includes the following: occurrence of faults and errors due to faults, propagation of errors, and finally their detection and location. Initially, when an array is fault-free, there are $\Omega(Mt_o)$ tests performed to maintain the t_o self-diagnosability level. However, if a new communication graph due to faults is created, a new testing graph has to be devised as well. Two diagnosing reconfigurations policies will be considered:

- allowing a specific minimum level of performance after identification and confinement of faults;

- achieving a maximum level of performance after identification and confinement of faults, see [RUCI87a].

The problem deals (i) with the reduction of the testing overhead, which has a twofold nature. It is dynamic, due to new diagnostic situations, and it is system-level oriented. If the first characteristic is considered, it should be noted that the self-diagnosibility level of a graph depends on its connectivity [KUHL80], i.e. on its weakest link. Thus the initial $\Omega(Mt_o)$ testing procedure is more powerful than necessary to maintain the current achievable level of self-diagnosability. The second characteristic has a more general aspect. A promising approach to reduce the number of tests is to use Hamiltonian testing graphs and/or the Byzantine general agreement strategy. Even a slight test reduction would be very beneficial for a system-level diagnosis since the PE testing is time-consuming. This, in turn, would allow us to increase the test frequency, thus improving system reliability.

At this stage of research on $S(M, N, t_o)$ systems, it is desired to develop at least a simulation model to verify the feasibility of the presented concept. Perhaps it would lead to a conclusion that testing time overhead was not an acute problem. However, if not, then other possible ways to reduce the testing overhead can be considered. Systolic testing is one of them.

The systolic mechanism can be used in the following manner: the whole array is partitioned into subarrays, the number of which corresponds to the number of DRONES. Then each subarray is tested systolically in parallel with each DRONE functioning as a testing device. Thus, the testing time is significantly less than for sequential testing, and fewer interconnections are required as compared to independent parallel tests, yet high fault-coverage is still preserved. A large partition has the advantage of fast testing with few DRONES, while a small partition implies simpler diagnosis and less likelihood of multiple faults invalidating the test. This approach seems to be especially promising in a case of multiple faults when the test outcomes can be invalidated by other faults along sensitizing paths.

References

[ABRA87] J. A. Abraham et. al., "Fault tolerance techniques for systolic arrays," *IEEE Computer* , July 1987, pp. 65-74.

[AGRA85] P. Agraval, "A novel fault tolerant distributed system architecture," *Proc. Int. Conf. Comp. Design: VLSI in Computers* , ICCD 1985, pp. 760-763.

[AUBU78] R. C. Aubusson, I. Catt, "Wafer scale integration: a fault-tolerant procedure," *IEEE Journal of Solid State Circuits,* vol. SC-13, no. 3, June 1978, pp. 339-344.

[BREU83] M. A. Breuer, A. A. Ismael, "Roving emulation as a fault detection mechanism," *Proc. 13th Symp. Fault-tolerant Comput.,* June 1983, pp. 206-215.

[BUEH82] M. G. Buehler, M. W. Sieviers, "Off-line, built-in test techniques for VLSI circuits," *IEEE Computer* , June 1982, pp. 69-82.

[CHOI84] Y. H. Choi, S. M. Han, M. Malek, "Fault diagnosis of reconfigurable systolic arrays," *Proc. Int. Conf. Comp. Design: VLSI in Computers,* ICCD 1984, pp. 451-455.

[DURB85] J. R. Dubin, *Modern Algebra: An Introduction,* John Wiley & Sons: New York, 1985.

[FORT84] J. A. B. Fortes, C. S. Raghavendra, "Dynamically reconfigurable fault-tolerant array processors," FTCS-14, June 1984, pp. 386-392.

[FUSS84] D. S. Fussell and P. J. Varman, "Designing systolic algorithms for fault tolerance," *Proc. Int. Conf. Comp. Design*, ICCD 1984, Oct. 1984, pp. 623-628.

[GOLL84] N. S. Gollakota, F. G. Gray, "Reconfigurable cellular architecture," *Proc. 1984 Parallel Proc. Conf.*, pp. 377-379.

[GREE84] J. W. Green, A. El Gamal, "Configuration of VLSI arrays in the presence of defects," *Journal of the ACM*, vol. 31, no. 4, Oct. 1984, pp. 694-717.

[KIM85] J. H. Kim, S. M. Reddy, "A fault-tolerant systolic array design using TMR method," *Proc. Int. Conf. Comp. Design: VLSI in Computers*, ICCD 1985, pp. 769-773.

[KORE81] I. Koren, "A reconfigurable and fault tolerant VLSI multiprocessor array," *Proc. 8th Symp. Comp. Architecture*, 1981, pp. 425-442.

[KORE84] I. Koren and M. A. Breuer, "On the area and yield considerations for fault-tolerant VLSI processor arrays," *IEEE Trans. Comp.*, vol. C-33, Jan. 1984, pp. 21-27.

[KORE87] I. Koren and D. Pradhan, "Modeling the effect of redundancy on yield and performance of VLSI systems," *IEEE Trans. Comp.*, vol. C-36, March 1987, pp. 344-355.

[KUHL80] J. Kuhl, S. Reddy, "Distributed fault-tolerance for large multiprocessor systems," *Proc. 7th Ann. Symp. Comp. Architecture*, May 1980, pp. 23-30.

[KUHN83] R. H. Kuhn, "Interstitial fault-tolerance - a technique for making systolic arrays fault-tolerant," *Proc. 16th Ann. Conf. Syst.*, Jan. 1983, pp. 215-224.

[KUNG84a] H. T. Kung, M. S. Lam, "Fault-tolerance and two-level pipelining in VLSI systolic arrays," *Proc. MIT Conf. Advanced Research in VLSI*, Jan. 1984, pp. 76-83.

[KUNG84b] H. T. Kung, M. S. Lam, "Wafer scale integration and two level pipelined implementations of systolic arrays," *Journal of Parallel and Distributed Processing*, vol. 1, no. 1, 1984.

[LEIG85] F. T. Leighton, C. E. Leiseron, "Wafer scale integration of systolic arrays," *IEEE Trans. Comp.*, vol. C-34, no. 5, May 1985, pp. 448-461.

[MANG82] T. E. Mangir and A. Avizienis, "Fault tolerant design for VLSI: effect of interconnect requirements on yield improvement of VLSI designs," *IEEE Trans. Comp.*, vol. C-31, No. 7, July 1982, pp. 609-615.

[MANN77] F. B. Manning, "An approach to highly integrated, computer maintained cellular array," *IEEE Trans. Comp.*, vol. C-26, 1977, pp. 536-552.

[MOOR86] W. R. Moore, "A review of fault-tolerant techniques for the enhancement of integrated circuit yield," *Proc. IEEE*, vol. 74, no. 5, May 1986, pp. 684-698.

[PRAD86] D. K. Pradhan (ed.) *Fault-Tolerant Computing, Theory and Techniques*, Prentice Hall, Englewood Cliffs, NJ 1986.

[ROSE83] A. L. Rosenberg, "The Diogenes approach to testable fault-tolerant arrays of processors," *IEEE Trans. Comp.*, vol. C-32, no. 10, Oct. 1983, pp. 902-909.

[RUCI85] A. Rucinski, J. L. Pokoski, "A fault-tolerant distributed multiprocessor system for systolic algorithms," *Proc. Int. Conf. Comp. Design: VLSI in Computers*, ICCD 85, Oct. 1985, pp. 754-759.

[RUCI86] A. Rucinski, J. L. Pokoski, "Polystructural, reconfigurable, and fault-tolerant computers," *Int. Conf. Distr. Comp. Systems*, May 1986, pp. 175-182.

[RUCI87a] A. Rucinski, J. L. Pokoski, "Distributed diagnostic reconfigurability in array structures," *30th Midwest Symp. on Circuits and Systems*, August 1987.

[RUCI87b] A. Rucinski, J. L. Pokoski, "Efficiently self-reconfigurable VLSI arrays," to be presented at the *5th Int. Workshop on Integrated Electronics and Photonics in Communications*, October 1987.

[SAMI83] M. G. Sami and R. Stefanelli, "Reconfigurable architectures for VLSI implementation," *Proc. Nat'l Comp. Conf.*, NCC 1983, May 1983, pp. 565-577.

[SNYD82] L. Snyder, "Introduction of the configurable, highly parallel computer," *IEEE Computer*, Jan. 1982, pp. 47-56.

[SOMA84] A. K. Somani, V. K. Agarwal, "System level diagnosis in systolic systems," *Proc. Int. Conf. Comp. Design: VLSI in Computers*, ICCD 1984, pp. 445-449.

[VERG86] A. Vergis, K. Steiglitz, "Testability conditions for bilateral arrays of combinational cells," *IEEE Trans. Comp.*, vol. C-35, no. 1, Jan. 1986, pp. 13-22.

[WALT81] S. M. Walters, F. G. Gray, R. A. Thompson. "Self-diagnosing cellular implementation of finite-state machines," *IEEE Trans. Comp.*, vol. C-30, no.12, Dec. 1981, pp. 953-959.

Chapter 23

Hierarchical Modeling for Reliability and Performance Measures [1]

Malathi Veeraraghavan [2]
Kishor Trivedi [2]

Abstract

In this paper we model three aspects of fault-tolerant multiprocessor systems and study their influence on both performance and reliability measures in a combined way. These include concurrency, contention and fault-tolerance. Hierarchical modeling allows complex systems to be analyzed easily by splitting the overall model into layers. SHARPE is a powerful tool which allows the use of eight different model types that can be hierarchically combined to obtain a solution for some measure of the whole system. SHARPE also allows general distributions and hence instead of assuming exponentially distributed random variables, we can model more realistic cases. Examples presented here are small for ease of explanation; however, larger problems have also been solved within this framework. This paper thus lays the ground work for modeling existing systems with real data using these techniques.

[1] This work was supported in part by the Air Force Office of Scientific Research under grant AFOSR-84-0132, by the Army Research Office under contract DAAG29-84-0045 and by the James B. Duke Foundation under a Graduate Fellowship to the first author.
[2] Duke University, Durham, NH

23.1   Introduction

Distributed and multiprocessor systems employ concurrency and fault-tolerance extensively [4].   The need to share a limited set of resources among a competing group of users leads to contention. It is necessary, therefore, to be able to include these three aspects of system behavior in the models of multiple processor systems. The objective of this paper is to perform a unified analysis of these important aspects of a few systems. The modeling tool used for most of this analysis is called SHARPE (Symbolic Hierarchical Automated Reliability and Performance Evaluator), an overview of which is given in the appendix.

Traditionally, concurrency and contention are analyzed for their effects on system performance, while fault-tolerance for its effects on system dependability. In this paper, we study the effects of these aspects on different performability measures which combine both performance and reliability. For example, while analyzing a task graph that employs concurrency for performance, we need to take into account the possibility of (unreliable) processors, on which the the task graphs execute, failing. Similarly, while a fault-tolerant system is usually analyzed for its dependability, its total computational power, as it goes through a series of states by reconfiguration upon failures of components, is also an important parameter.

Secondly, by using hierarchical modeling and allowing the random variables that represent different system parameters to have general distributions we obtain more realistic models that could be used for analysis of real existing systems. This has been possible due to the versatility of SHARPE. It uses model hierarchy both for complex model specification and complex model solution. This aspect of SHARPE avoids the pitfalls of Markov models (namely, large state space and stiffness) while at the same time, it circumvents the independence assumption of combinatorial models such as reliability block diagrams and fault trees.

In this paper, we illustrate the techniques of hierarchical modeling of complex systems for combined reliability and performance measures through a few examples Section 23.2 models the use of concurrency in the execution of a job. An example involving contention for memories and buses in a multi-processor system is modeled in Section 23.3. In Section 23.4 we model a gracefully degradable fault tolerant array processor for both reliability and performance measures.

23.2   Concurrency

As an example of concurrency, we consider the task graph shown in Figure 23.1 adopted from [3]. In [3] task execution times were assumed to be deterministic; however, here we treat individual task execution times as random variables. The number associated with each node in Figure 23.1 is taken as the mean time to

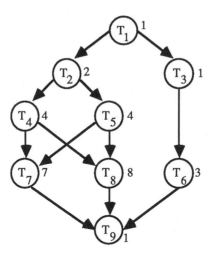

Figure 23.1: An example task graph.

completion of the task associated with that node in seconds. Times to completion of individual tasks are assumed to be exponentially distributed random variables, unless otherwise mentioned. Acyclic series-parallel graph models along with Markov models (to handle cycles in graphs) are used to solve this problem.

In the four subsections in this section, we first analyze the task graph in order to assess the effect of concurrency on pure performance measures under the assumption that the underlying processors on which the task graph executes do not fail. Next we allow for the possibility of either a task failing or the server (processor) on which the task executes failing. In the third subsection, we illustrate how complex systems that use this task graph could be analyzed easily using hierarchical modeling. Finally, we study the effect of different distributions for task completion times on performance measures.

23.2.1   Performance Measures for Completely Reliable Systems

Performance models for the execution of this task graph on systems with one, two and three processors are developed. The uniprocessor system performs these tasks $(T_1, \ldots, T_9)$ sequentially. Two different scheduling schemes, schedule 1 and schedule 2, are used for the two-processor case and only one scheduling scheme is studied for the three processor system. Graph models representing these four cases are shown in Figures 23.2a, 23.2b, 23.2c and 23.2d respectively. The word "max" in Figure 23.2 used at branch points shows that the task at which the branches meet again can be executed only upon completion of all required tasks on those branches. Thus

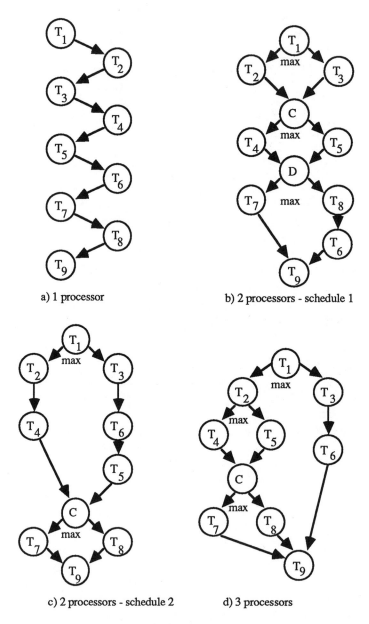

a) 1 processor

b) 2 processors - schedule 1

c) 2 processors - schedule 2

d) 3 processors

Figure 23.2: Series-parallel graph models.

Table 23.1: Performance measures for uni- and multi-processor systems.

Case	Speedup	DFP(25)
One processor	1.0	0.6398
Two processors - schedule 1	1.3124	0.37469
Two processors - schedule 2	1.3527	0.34783
Three processors	1.4518	0.29123

the completion time of the task graph in Figure 23.2d is given by

$$T_1 + max(T_3 + T_6, T_2 + max(T_4, T_5) + max(T_7, T_8)) + T_9$$

In Figure 23.2d, one processor is scheduled to do tasks T_3 and T_6 while the other two processors are scheduled to tasks on the branch starting at T_2. So the three processors could be simultaneously scheduled to tasks T_6, T_4 and T_5. Note that dummy nodes (such as C and D) with zero completion times are introduced wherever necessary in order to obtain series-parallel graphs without altering the problem.

Two performance measures are used in this analysis, speedup and dynamic failure probability. "Speedup" (S_k) is defined by:

$$S_k = ET_1/ET_k$$

where ET_1 and ET_k are the mean execution times of the task graph on uniprocessor and multiprocessor system with k processors, respectively. Speedup gives us a measure of the benefits of using multiprocessor systems and hence could be used in optimization studies. The second measure "Dynamic Failure Probability at Time d" $(DFP(d))$ is defined as the probability of not completing the job by time, d. Since SHARPE computes the distribution function of the completion time of task graphs, $DFP(d)$ is easily computed. This is a useful measure since it gives us the probability of not meeting a fixed deadline, and thus we are able to assess the timeliness property, an important characteristic of real-time systems.

The results of the analysis obtained from SHARPE are given below in Table 23.1. For the deterministic case with two processors using schedule 1, the speedup given in [3] is 1.8, while for the random case, the speedup is 1.3124 as shown in Table 23.1. As expected, the three-processor case has the best speedup, and for the two-processor case, schedule 2 has better speedup than schedule 1. We use a deadline of 25 seconds for the dynamic failure probability measure. We note that apart from speedup, the use of multiple processors significantly reduces the probability of not meeting the deadline.

23.2.2   Combined Measures for Unreliable Systems

We now relax the assumption of perfect processors. Two failure possibilities are considered separately, task and server failures.

Table 23.2: Combined measure for one- and three- processor systems.

Case	$DFP_{TF}(0.1, 25)$
One processor	0.86045
Three processors	0.72541

Task Failures

Let p_{fi} denote the failure probability of task i. The distribution function assigned to the time to complete task i is then $(1 - p_{fi})(1 - e^{-\lambda_i t})$, a defective distribution. The combined performance-reliability measure used here is "Dynamic Failure Probability with Task failures"$(DFP_{TF}(p, d))$ where p is the probability that a task fails, i.e. $p_{fi} = p \,\forall\, i$. It is defined as the probability that the job is not completed by time d when each task is allowed to fail with probability p. Table 23.2 shows the value of this measure for one- and three- processor systems with $p = 0.1$ and $d = 25 \; seconds$. This table has been included so that the reader could compare these values with those of Table 23.1 to see the effect of allowing task failures.

Fig. 23.3 shows plots of the variation of $DFP_{TF}(p, d)$ with respect to d for both one and three processor systems and for three values of p, 0, 0.1 and 0.5. When $p = 0$ DFP_{TF} reduces to DFP as defined in the previous section. Note that for $p = 0.5$, the values of DFP_{TF} remains the same for one and three processors for the time range considered.

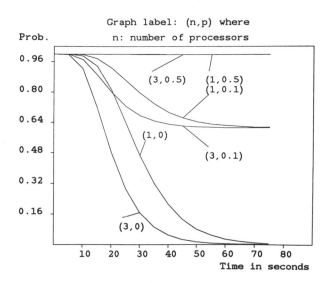

Figure 23.3: DFP_TF(p,d) vs. d

Table 23.3: Omission failure probability.

p	OFP - 1 processor	OFP - 3 processors
0.0	0	0
0.1	6.1258e-01	6.1258e-01
0.5	9.9805e-01	9.9805e-01

Another useful measure that can be obtained from this analysis is the probability that the job never completes also called the omission failure probability ($OFP(p)$). This can happen since tasks are allowed to fail and not recover. The omission failure probability is obtained directly in SHARPE by requesting the mass at infinity in the overall execution time distribution for the task graph. This obviously depends on the value of p as shown below in Table 23.3.

Server Failures

A server that is performing a task (each node in Figure 23.1) is now allowed to fail. A server failure-recovery-task completion sequence is modeled as a Markov chain (Figure 23.4) as failure, recovery and task completion times are all assumed to be exponentially distributed. The task graph of Figure 23.1 is assumed to be executed on one- and three-processor systems where processors are allowed to fail as per Figure 23.4. The time to completion of each task in the graph models of Figures 23.2a and 23.2d is now represented by the time to absorption into state "COMP" of Figure 23.4. Note that τ in Figure 23.4 is replaced by the task completion rate (1/mean task completion time) as obtained for each task from Figure 23.1, while γ and μ are the same for all tasks as they represent the processor failure and recovery rates respectively. The combined measure obtained here is called "Dynamic Failure Probability with Server failures" ($DFP_{SF}(\gamma, \mu, d)$), where γ is the failure rate of each processor and μ is the recovery rate of each processor. $DFP_{SF}(\gamma, \mu, d)$ is the probability of not completing the job by time d, in the presence of server failures

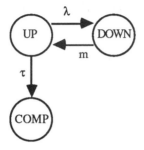

Figure 23.4: Markov model of a server failure.

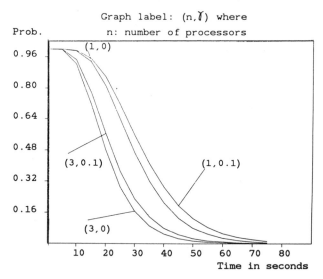

Figure 23.5: DEP_SE(γ,1,d) vs. d.

and recoveries with rates γ and μ respectively. The recovery rate μ is *1 /sec* for all servers.

Figure 23.5 shows the variation of $DFP_{SF}(\gamma, 1/sec, d)$ with respect to time d for two different values of server failure rate(γ) for one- and three-processor systems. For $\gamma = 0$, the measure $DFP_{SF}(\gamma, \mu, d)$ reduces to $DFP(d)$ defined in section 23.2.1. For large values of d, the number of processors in the system does not affect the dynamic failure probability which reduces to 0.

The difference between the model in the last section and this one is that in the previous section once a task fails, the job (whole task graph) fails since no recovery mechanism was included for a task failure. Here, however, the system can recover from a server failure and hence the job always completes. Note that the words "server" and "processor" have been used interchangeably in this example. [3]

---

[3] This example also illustrates both the use of hierarchical modeling, the sublayer being the Markov model (Figure 23.4) and the top layer the graph model (Figure 23.2), as well as the the ability to have distributions other than exponential for each graph node. It has however been included in this section because it primarily models unreliable servers.

23.2.3 Illustration of Hierarchical Modeling

Two examples are given here explicitly to show the use of hierarchical modeling. To analyze complex systems, we split the models into layers, analyze each layer separately and then pass parameters calculated from the lower to the upper layers. First, as the tool SHARPE allows only acyclic series-parallel graphs, we use this technique to model a cyclic graph which vastly increases the types of jobs that can be analyzed for the use of concurrency. The second example is the analysis of an M/G/1 queue. In a real system jobs of different types arrive at a multiprocessor system at varying rates. Here we model a hypothetical system where all jobs are of the type shown in Figure 23.1 and they arrive at a constant rate. This is done to illustrate how after independent analysis of a task graph we can use the results for analyzing systems that use these graphs.

Cyclic Graphs

The task graph in Figure 23.1 is modified to make it cyclic and then an analysis for pure performance measures (no task or server failures), speedup and $DFP(d)$, defined in section 23.2.1, is done. Figure 23.6 shows the modified graph where tasks T_3 and T_6 have been replaced by a cycle. The word "prob" in Figure 23.6 stands for "probabilistic", i.e. upon completion of task T_6, task T_3 is executed with probability p and with probability $(1 - p)$ control leaves the cycle. To analyze this, we first

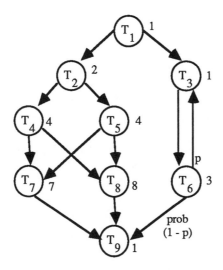

Figure 23.6: An example task graph with a cycle.

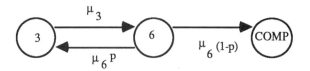

Figure 23.7: Markov model for the cycle in task graph.

model the subgraph with a cycle as a Markov chain (Figure 23.7) and then use the time to absorption into state "COMP" in Figure 23.7 as the time to completion of the T_3, T_6 sequence in the upper layer graph model of the hierarchy. The task graph shown in Figure 23.6 is run through one and three-processor systems using the same scheduling schemes as in section 23.2.1. The rates used in Figure 23.7 are $\tau_3 = 1$ and $\tau_6 = 1/3$, the same as the inverse of the mean time to completion of tasks T_3 and T_6 in Figure 23.1. The value used for p in Figure 23.7 is *0.1*. Results (speedup and $DFP(25\ seconds)$) obtained by this analysis is shown in Table 23.4. By comparing these results with those in Table 23.1 we see that introduction of the cycle increases the probability of not completing the job in 25 *sec* for both one- and three-processor systems.

M/G/1 queue

We assume that task graphs of the form in Figure 23.1 arrive in a Poisson manner with rate $\lambda = 0.03\ /sec$. The system is then analyzed as an M/G/1 queue. The performance measures evaluated are the traffic intensity (mean service time/mean interarrival time), mean response time (waiting time + service time) in seconds and mean service time in seconds. The server is modeled first as a three-processor system (scheduling same as in Figure 23.2d), then as a two-processor system with schedules 1 and 2 (Figures 23.2b and 23.2c) and finally as a uniprocessor. There are no task or server failures. The lower layer of the hierarchy is the solution of the graph models in Figures 23.2a, 23.2b, 23.2c, and 23.2d. The mean time to completion of the job (task graph) is the mean service time. The service times for all four systems (time to complete the job) are non-exponentially distributed random variables. The upper layer of the hierarchy are simple expressions, the P-K formulas [9]. SHARPE has

Table 23.4: Performance measures for a cyclic graph.

Case	Speedup	DFP(25)
One processor	1.0	0.92491
Three processors	1.4421	0.69265

Table 23.5: Performance measures for an M/G/1 queue.

M/G/1 Queue ($\lambda=0.03$)	ρ	Mean response times	Mean service times
Three processors	0.64065	44.386	21.355
Two processors-schedule1	0.70867	57.597	23.622
Two processors-schedule2	0.68759	52.913	22.920
One processor	0.93009	271.79	31.003

the ability to evaluate expressions, which can be regarded as a basic model type. The mean response times along with traffic intensities are shown in Table 23.5. This illustrates an example of how we could hierarchically combine the results of submodels to analyze a complex model.

23.2.4 Effect of Different Distributions for Task Completion Times

We then test the effect of assigning distributions other than exponential for the completion times of tasks. The task graph used is that in Figure 23.1 and no task or server failures are considered. The servers used are the one and three processor systems with scheduling as in Figures 23.2a and 23.2d. Performance measures are speedup and $DFP(25\ seconds)$ as defined in section 23.2.1. Results are shown in Table 23.6. For each case in Table 23.6, only one node's distribution is explicitly mentioned. All other nodes in the graph models are assumed to have exponentially distributed times to completion unless otherwise mentioned in the Table 23.6. The mean task completion times have been maintained the same while

Table 23.6: Effect of varying distributions of some task times.

Case	Speedup	DFP(25) 1 processor	DFP(25) 3 processor
Exp. distr. for T1	1.4518	0.63980	0.29123
Hypoexp. distr. for T1	1.4518	0.63986	0.29080
Hyperexp. distr. for T1	1.4518	0.63902	0.29295
Exp. distr. for T3	1.4518	0.63980	0.29123
Hypoexp. distr. for T3	1.4519	0.63986	0.29121
Hyperexp. distr. for T3	1.4507	0.63902	0.29145
Exp. distr. for T7	1.4518	0.63980	0.29123
Hypoexp. distr. for T7	1.4753	0.65706	0.27037
Hyperexp. distr. for T7	1.4152	0.59617	0.28928
Exp. distr. for T9	1.4518	0.63980	0.29123
Hypoexp. distr. for T9	1.4518	0.63986	0.29080
Hyperexp. distr. for T9	1.4518	0.63902	0.29295

changing distributions and the values used for the mean times are as shown in Figure 23.1. The distributions used are exponential, hypoexponential with two stages and hyperexponential with two phases with branch probabilities of 0.25 and 0.75.

The results in Table 23.6 show that if the node that is assigned these various distributions (hypoexponential and hyperexponential) is one that cannot be performed concurrently with any other task (T_1 and T_9 in Figure 23.1), then speedup is insensitive to the distributional assumption. Otherwise, the hypoexponential case gives the most speedup since this distribution has a smaller coefficient of variation than either the exponential or the hyperexponential. Changing the distributions of completion times of nodes T_1, T_3 and T_9 have the same effect on $DFP(25)$ for the one processor system while we get different values for this parameter by changing the distribution of the completion time of node T_7. This is because tasks T_1, T_3 and T_9 have the same mean time to completion of unity while T_7 has a mean time to completion of 7. Hence its contribution to the distribution of the total completion time is different.

23.3   Contention

The second aspect of multiprocessor systems modeled here is bus and memory contention. We use a single bus system with three processors and three memories to illustrate hierarchical modeling for performance and reliability measures under the assumption of various processes having general (non-exponential) distributions.

In the two subsections that follow we first analyze the single bus system for a measure of contention for the bus under the assumption that the bus never fails. Next, we allow bus failures and analyze the single bus system for combined reliability and performance measures. In this analysis we use both hierarchical modeling and non-exponential distributions.

23.3.1   Performance Measures

In this section we model the single bus system for pure performance measures which give us a quantitative idea of the extent of contention. A continuous time semi-Markov model is used to represent the three processor, three memory, single bus system. The random variable that represents the processing time between any two accesses is assumed to be exponentially distributed, while the access times (time spent by a processor accessing a memory module) are allowed to have general distributions. Memory references are assumed to be nonuniform. Figure 23.8 shows the semi-Markov model used for this analysis. A state of the chain is a 3-tuple with each element representing the state of a processor. An element of a tuple

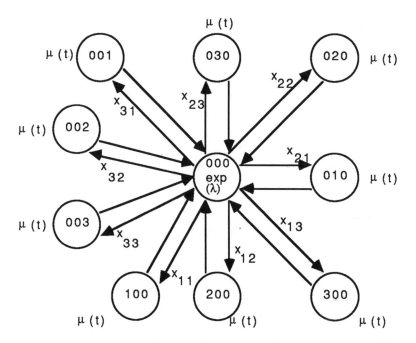

Figure 23.8: Semi-markov model of a three processor, three memory, one bus system.

could be $0, 1, 2$, or 3 where 0 stands for the processor either processing or waiting while any other number represents the memory module being accessed by that processor. Processors compete for the access to a single bus. λ stands for the rate of processing between memory accesses. λ is the same for all processors. $\mu(t)$ is the rate of memory access completion, which is the same for all processors. It is time-dependent since memory access times are allowed to be generally distributed. The assumption that mean memory access time and mean interaccess processing time are the same for all processors could be easily relaxed for this example.

We first define the parameters of this model. Let X represent the access matrix whose component x_{ij} represents the probability that processor i requests memory module j. Matrix X is a 3×3 matrix as the system is assumed to have three processors and three memories. The mean access time is assumed to be 10 μsec while the mean processing time between accesses $(1/\lambda)$ is taken as 2 μsec. By keeping the interaccess processing times exponentially distributed while allowing the access times to be generally distributed we get a semi-Markov chain. SHARPE has the capability to solve irreducible semi-Markov chains. The access matrix X was chosen to be

$$\begin{pmatrix} 0.6 & 0.2 & 0.2 \\ 0.3 & 0.4 & 0.3 \\ 0.5 & 0.3 & 0.2 \end{pmatrix}$$

Table 23.7: *BUSBUSY* for a single bus, three processor and three memory system.

Distribution of access times	BUSBUSY
Exponential	0.9375
Erlang (2 stages)	0.9375

Row one of the matrix X shows the probabilities with which processor one chooses memory modules one, two and three.

The performance measure used here is the probability of finding the bus busy, *BUSBUSY*. Though this is not a direct measure of contention, it gives us an upper bound for the probability that there is contention for the single bus. The model chosen allows us to get only this simple measure instead of a more meaningful value like the average number of active processors because of the nonrestrictive assumptions like nonuniform access and nonexponential processes.

Table 23.7 shows the value of *BUSBUSY* for different access time distributions with same mean access rates. The fact that the performance measure is insensitive

to the distribution of access times is easily explained [5].

23.3.2   Combined Measures and Hierarchical Modeling

The single bus system with three processors and memories described in the previous section is considered here. All the input parameters to the model are the same except that access times are allowed to be only exponentially distributed. As the critical component in this system is the bus we now allow the possibility of a bus failure and model such a system for a combined reliability-performance measure using hierarchical modeling.

Figure 23.9 shows a Markov model for the failure process of the bus. Times to failure and repair of the bus are assumed to be exponential. The bus is allowed to fail only when it is busy. When the bus is busy, two processes compete, the access process (constant rate of μ) and the failure process (constant rate γ). Upon completion of access the system goes back to the state (000) shown in Figure 23.8. Two levels of hierarchy are used. The Markov model in Figure 23.9 is the lower layer. The upper layer is the semi-Markov model shown in Figure 23.8 where the general distributions marked on the arcs from the various states to state (000) are the same and is given by the distribution of the time to absorption to state ACC in the Markov model of Figure 23.9. We first solve the Markov model (Figure 23.9) for the distribution of the time to absorption into state ACC and then use this for the distributions on the arcs of the upper level semi-Markov model (Figure 23.8).

Table 23.8: Bus failure allowed with rate γ.

γ (/hour)	$BUSBUSY(\gamma)$
0.0	0.93750
0.00001	0.93741
0.0001	0.93662

The bus repair rate (τ) is assumed to be 0.1 /hour. The access rate (μ) is the same as in the previous section $(1 \times 10^5$ /sec$)$ and the interaccess processing rate λ is 5×10^5 /sec.

The combined reliability-performance measure is $BUSBUSY(\gamma)$ which is the probability that the bus is up and busy when a bus failure rate of γ is allowed. Table 23.8 shows different values of $BUSBUSY(\gamma)$ for different failure rates. As the bus failure rate increases the probability of the bus being down or non-functional increases, thus reducing the probability of finding the bus up and busy.

23.4 Fault Tolerance

The architecture analyzed is a dynamically-reconfigurable fault-tolerant array processor [2]. Markov reward models are used for this analysis. The processor consists of an $(n_1 \times n_2)$ array. When a fault occurs in any one of the processors the system is reconfigured as another array of smaller dimensions. Two methods of degradation are described in [2]. The first is a Successive Row Elimination (SRE) and its complement Successive Column Elimination (SCE). The other method is to alternatively remove rows and columns (ARCE - Alternate Row Column Elimination).

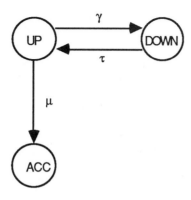

Figure 23.9: Markov model for bus failure-repair-access cycle.

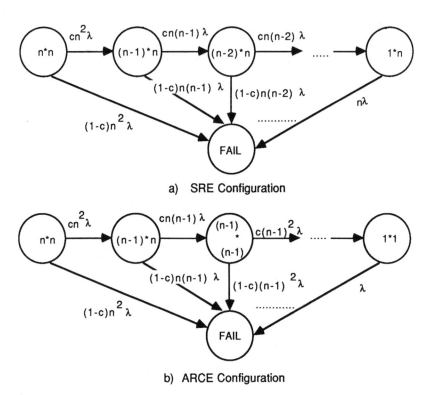

Figure 23.10: Markov models of the two configurations of the degradable 10*10 array (note: n=10).

Figure 23.10 shows the Markov reward models of these two reconfiguration techniques. A 10×10 array is considered for this analysis. λ is the constant failure rate (exponential time to failure) of any processor, it being the same for all processors. Coverage of a failure is represented by c, i.e. the probability with which a fault is detected and the system successfully reconfigured by removing the row or the column carrying the faulty processor. We first model these two systems for pure reliability measures and then define certain performability measures and compare the two degradation methods. Finally we introduce one repairman and analyze the cyclic Markov chains with absorbing states that model the two repairable configurations for their reliability. Hierarchical modeling and Markov reward models have been used in the analysis.

23.4.1 Reliability Measures

We use two different models for the failure handling process.

1. A Markovian model allowing only permanent faults.

Table 23.9: Failure Model Definitions and Coverage Values.

Failure model	Input		Coverage
	$\lambda(t)$	ptr	c
Model #1	0.0001	0	8.99999975e-01
Model #2	0.0011	0.9	8.99999537e-01

2. A Markovian model allowing both permanent and transient faults (transient fault rate is assumed to be 10 times the permanent fault rate).

Figure 23.11 shows a general fault model which for different values of certain parameters represents the above two models. Table 23.9 shows the values of these input parameters for the different models. The failure rate λ is measured in /hour. The assumptions common to these two models are as follows:

1. State *det* represents the detection state. Mean detection time is $1/\delta$ (= 5 *msec*) and detection time is exponentially distributed.

2. State *rec* is the reconfiguration state. Mean reconfiguration time is $1/\mu$ (= 1 *sec*) and reconfiguration time is exponentially distributed.

3. The probability of reconfiguration is shown as q in Figure 23.11. It is assumed to be 0.9 for all three models. The system is assumed to fail in case of imperfect reconfiguration.

4. Failure rate λ and *ptr*, the probability that a fault is transient, are the two parameters that are varied to accommodate the two models.

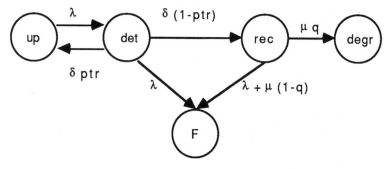

Figure 23.11: General model of a failure handling process.

Table 23.10: Reliability Measures for the two Configurations.

Failure model	SRE Output		ARCE Output	
	MTTF	F(1000)	MTTF	F(1000)
#1	1.52778726e+03	4.83061042e-01	4.40990369e+03	4.97713852e-01
#2	1.38889363e+02	9.99935299e-01	4.00897818e+02	8.48011708e-01

5. State *degr* represents the system in a reconfigured but degraded state.

6. Near coincident faults are assumed to lead to a system crash, i.e. a fault in states *det* or *rec* are assumed to lead to failure.

We use hierarchical modeling to incorporate this submodel of the failure handling process (Figure 23.11) into the model representing the overall system degradation shown in Figure 23.10. The model in Figure 23.11 is first solved for the probability of being absorbed in state *degr*. This is less than one since the system could finally reach state F instead. Coverage c, defined earlier is the probability of reaching state *degr*. So we first obtain the value of c by solving the model in Figure 23.11 and then use this value in Figure 23.10 to obtain various system-wide dependability measures.

1. MTTF - Mean Time To Failure is the expected value of the time taken to reach state $Fail$ in Figure 23.10.

2. Unreliability, $F(t)$, at a fixed time, t, in hours.

Table 23.9 shows the values of c for the two different models of the failure handling process and Table 23.10 shows these two reliability measures for the two configurations SRE and ARCE and for the two failure models.

23.4.2   Performability Measures

To obtain some measure of the performance of the system, we use the notion of rewards. State i in the Markov models of Figure 23.10 is assigned a reward rate r_i. This is a measure of the useful work done by the system when in that state. Based on these state reward rates we define the following combined measures of performance and reliability.

1. Reliability, $(R(t))$, is the probability that in the interval $(0,t)$, the system has not failed.

2. Accumulated reward, Y, until system failure is the total reward accumulated before the system crashes.

3. Expected instantaneous reward rate, $E[X(t)]$, is the expected reward rate at time t.

4. Expected accumulated reward, $E[Y(t)]$, is the expected accumulated reward in the interval $(0,t)$.

5. Expected uptime, $E[U(t)]$, is the mean time spent by the system in up states in the interval $(0,t)$.

6. Time averaged accumulated reward, $E[Y(t)/t]$, is the average of the total reward accumulated by time t.

7. Expected interval availability, $E[A_I(t)]$, is the time averaged expected uptime.

Defining equations for these above measures are given below.

$$R(t) = \sum_{i \epsilon\ up\ states} P_i(t)$$

$$Y = \sum_k r_k H_k$$

where r_k and H_k are the reward rate attached to and holding time of state k, respectively.

$$E[X(t)] = \sum_{i \epsilon\ all\ states} r_i P_i(t)$$

$$E[Y(t)] = \sum_i r_i \int_0^t P_i(\tau)d\tau$$

$$E[U(t)] = E[Y(t)] \quad with \quad r_i = 1,\ i\epsilon\ up\ states\ and\ r_i = 0,\ i\epsilon down\ states$$

$$E[Y(t)/t] = E[Y(t)]/t$$

$$E[A_I(t)] = E[U(t)]/t$$

Let the states in Figure 23.10a be numbered 0 (corresponding to $n*n$) to 9 (corresponding to $1*n$) and the states in Figure 23.10b be similarly numbered from 0 (corresponding to $n*n$) to 19 (corresponding to $1*1$). Let the reward rate attached to state k be denoted by r_k; it is equated to the the the number of processors that are functional in that state. For example, state 0 of Figure 23.10a is given the reward rate 100 as $n = 10$, since as stated before we are modeling a 10×10 array and state 15 of Figure 23.10b is given the rate 6 as there are two rows and three columns left in this state. Reward rates of the $FAIL$ states is zero as no useful work is done when the system is non-functional.

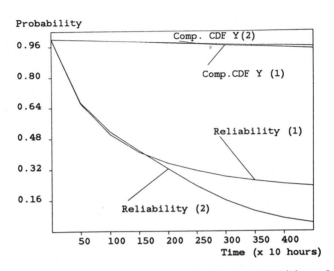

Figure 23.12: Reliability curves for the ARCE(1) and SRE(2) configurations.

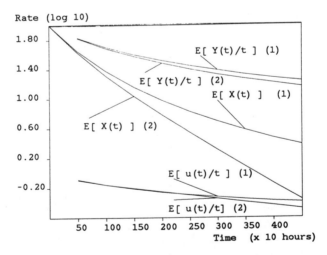

Figure 23.13: Reward rate curves for the ARCE(1) and SRE(2) configurations.

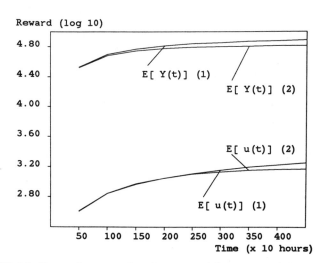

Figure 23.14: Reward curves for the ARCE(1) and SRE(2) configurations.

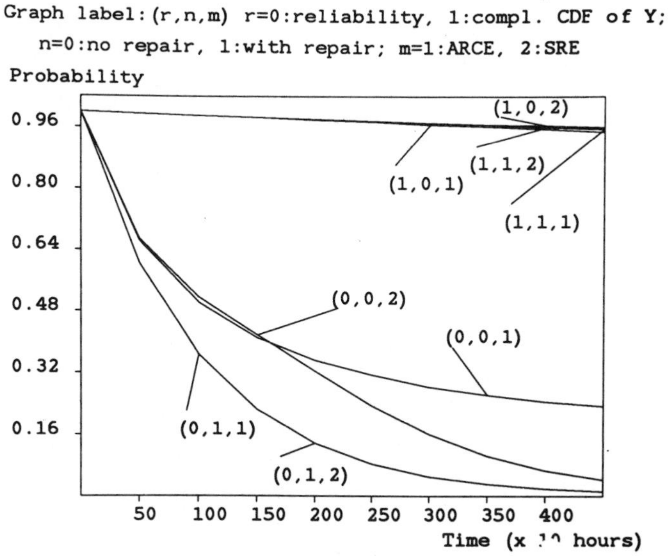

Graph label: (r,n,m) r=0:reliability, 1:compl. CDF of Y;
n=0:no repair, 1:with repair; m=1:ARCE, 2:SRE

Figure 23.15: Reliability curves for the two configurations.

For the two configurations, the reward rates are then given by:

$$r_k = n(n - k) \quad 0 \leq k \leq (n - 1) \quad for \quad SRE, \quad and$$

$$r_k = (n - \lceil \tfrac{k}{2} \rceil)(n - \lfloor \tfrac{k}{2} \rfloor) \quad 0 \leq k \leq (2n - 1) \quad for \quad ARCE.$$

We analyze the two models in Figure 23.10 for these seven measures using the failure handling Model #1. Three graphs are shown comparing different measures for the two configurations. In Figure 23.12, reliability and complementary CDF of Y are compared for the two architectures. At smaller values of time, the SRE configuration has better reliability than the ARCE. When performance is also considered the complementary CDF of Y shows small differences for the two configurations, SRE being the better one. However for other performance measures, $E[X(t)], E[Y(t)/t]$ and $E[A_I(t)]$ as shown in Figure 23.13, ARCE configuration gives better values. We note that measures that ignore degraded levels of performance underestimate system effectiveness.

This is further corroborated by Figure 23.14, which compares $E[Y(t)]$ and $E[U(t)]$. $E[Y(t)]$ is always greater than $E[U(t)]$ since it represents the total work done and up states are assumed to produce work represented by a reward rate greater than unity.

23.4.3   Repairable Systems

Incorporation of a repair model into these two configurations is relatively easy. An exponentially distributed repair time with a constant rate $\gamma = 1$ per hour (from non-failure states only) is added to the Markov models of Figure 23.10. Only one repairman is assumed. The failure handling process is modeled by Model #1 in this analysis.

Figure 23.15 shows the reliability and the complementary CDF of Y of the systems with and without repair for the two configurations. It is interesting that the addition of local repair actually degrades the system reliability. The primary reason is that repair tends to keep the system in the high reward states where the coverage leakage to the failure state is high. Thus if we consider the complementary distribution of Y instead of reliability, this effect should disappear. In Figure 23.15, we see that the reliability curves for both configurations with repair are the same but lie considerably below the curves corresponding to the reliability of the system without repair. The curves showing the complementary CDF of Y with repair for the two configurations are also the same. However it is greater than the corresponding curve without repair only for the ARCE configuration. This is because there are more states in the model for the ARCE system than in the SRE system which allows repair to have a visible influence. The relative values of λ and γ make the complementary CDF of Y with repair for the SRE configuration take on smaller values than the corresponding values for the system without repair.

23.5 Conclusions

We have demonstrated different techniques to model complex systems for performance and reliability measures under non-restrictive assumptions using the hierarchical modeling power of SHARPE. Important aspects of multiple processor systems such as concurrency, contention, fault-tolerance and degradable performance have been modeled in a natural way. The capability to model such diverse aspects of system behavior is achieved by the availability of eight different model types in SHARPE(precedence graphs, reliability block diagrams, fault trees, acyclic Markov chains, acyclic semi-Markov chains, irreducible Markov chains, irreducible semi-Markov chains and cyclic Markov chains with absorbing states). We plan to extend these model types by allowing product-form queuing networks and generalized stochastic Petri nets as model types. We exploit the concept of model decomposition in two ways. First, by allowing a hybrid and hierarchical combination of submodels, we use the most appropriate model type for a given subsystem, thus simplifying the task of model construction. Second, and perhaps even more important, this hierarchy is further exploited in order to obtain computational efficiency in model solution by separately solving submodels and combining submodel results into a solution for the overall model. The interested reader may consult [8] for further details and more examples.

APPENDIX: Description of SHARPE

References [6] and [7] give a detailed description of SHARPE. We present a brief description of the main aspects here.

The user has four main modeling options – a graph model, a fault tree (FT) /reliability block diagram (RBD) model, a Markov/semi-Markov model and a Markov reward model. The graph model can be used to study the performance of concurrent programs. The FT/RBD model is used only for reliability/availability problems. The Markov models are used for program performance, system performance or system reliability/availability evaluation. Markov reward models are used for the combined evaluation of performance and reliability.

For the combinatorial models (graphs, fault trees and reliability block diagrams), the user inputs a description of the model – the connectivity structure and the distribution attached to each node. In the graph model, nodes could represent different tasks in a program. In an RBD model, they could stand for components in a system. SHARPE solves these models for the distribution function of either the time to completion of a program (a performance problem) or the time to failure of a system (a reliability problem). This distribution function is a symbolic expression in time, t.

The solution technique adopted for these combinatorial models is efficient. In particular, costly conversion to and subsequent solution of a Markov chain model are not done. Currently there is a restriction to series-parallel combinatorial models only. A non-series-parallel model would have paths that have common elements. As the paths are no longer independent, one of three techniques needs to be used – conditioning, inclusion-exclusion or sum of disjoint products. We have chosen the last method [1] to solve non-series-parallel models combinatorially to extend the applications of SHARPE. The implementation of this technique is in progress. It should be noted, however, that non-series-parallel combinatorial models can often be solved using the current capability of SHARPE to combine Markov and combinatorial submodels [8] and [7].

The Markov models need the state transition rates and the initial state probabilities as user inputs. These models are not restricted by various types of dependencies like the combinatorial ones. However, they may have a number of states that is exponential in the number of system components. They are also usually not compact enough for the user to describe easily. However, SHARPE models can be hierarchical in the sense that output from one submodel can be used as input to another. Therefore, the efficiency of combinatorial methods is retained where they are applicable while the power and the flexibility of Markov models are available whenever necessary.

The model types

SHARPE provides eight model types:

1. series-parallel reliability block diagrams

2. fault trees without repeated components

3. series-parallel acyclic directed graphs

4. acyclic Markov (reward) models

5. irreducible Markov (reward) chains

6. cyclic Markov (reward) chains with absorbing states

7. acyclic semi-Markov (reward) chains

8. irreducible semi-Markov (reward) chains

When modeling reliability, each component is assigned a distribution function for its time to failure. This could also be a simple probability. The analysis of the system yields the distribution function for the time to failure of the whole system. If availability is to be modeled, each component is assigned a probability that it is unavailable at time t and the system analysis yields the instantaneous unavailability function for the whole system.

The CDF of the time to failure of a system or the time to completion of a program can be printed as a closed-form expression in time t. This is particularly convenient since the CDF can be evaluated at any specific instant of time t. In addition, the mean and the variance of the time to failure (time to completion) are also computed. In the case of irreducible Markov and semi-Markov chains, the steady-state probabilities are computed. From this the expected steady-state reward rate is also computed. Furthermore, a SHARPE user can define functions and variables that contain outputs available from one or more submodels.

Distribution functions

We evaluate distribution functions as symbolic functions in time t. This requires that the class of functions we choose for the distribution function be closed under the operations of convolution, order statistics and weighted averages as these are the operations performed in solving all eight model types. The class of exponential polynomials (exponomials) satisfies these criteria as well as being valid distribution functions. An exponomial is represented as

$$\sum_{j=1}^{n} a_j t^{k_j} e^{b_j t}$$

where k_j is a non-negative integer and a_j and b_j are real or complex numbers. The need to allow complex numbers arises in the analysis for the CDF of time to absorption in a cyclic Markov model with absorbing states. Each exponomial is stored as a set of triples (a_j, k_j, b_j) inside SHARPE. The class of exponential polynomials includes commonly used distribution functions such as the exponential, Erlang, hypoexponential and the hyperexponential.

Masses at zero and infinity are allowed for these distribution functions. A mass at zero represents a discrete probability value at zero. This could represent, for instance, the possibility of a latent fault. A mass at infinity would make the distribution defective. An example could be a program that does not run to completion due to software/hardware faults.

23.6 Acknowledgements

Thanks are due to LTC Jim Blake for his suggestions and Beth Rothmann for her help with the figures.

References

[1] A. Grnarov, L. Kleinrock and M. Gerla. A new algorithm for network reliability computation. In *Proc. of Comp. Network Symposium*, December 1979.

[2] J. A. B. Fortes and C. S. Raghavendra. Dynamically reconfigurable fault-tolerant array processors. In *Proc. IEEE Int. Symp. on Fault-Tolerant Computing, FTCS-14*, pages 386–392, 1984.

[3] K. Hwang and F. A. Briggs. *Computer Architecture and Parallel Processing.* McGraw Hill, 1984.

[4] J.F. Meyer. Performability modeling of distributed real-time systems. In *Int'l Workshop on Applied Mathematics and Performance Reliability Models of Computer Communication Systems*, Univ. of Pisa, 1983.

[5] Sheldon M. Ross. *Stochastic Processes.* John Wiley & Sons, 1983.

[6] R. Sahner and K. S. Trivedi. Performance and reliability analysis using directed acyclic graphs. *IEEE Transactions on Software Engineering*, pp. 1105-1114, October 1987.

[7] R. Sahner and K. S. Trivedi. Reliability modeling using SHARPE. *IEEE Transactions on Reliability*, R-36(2):186–193, June 1987.

[8] R. A. Sahner and K. S. Trivedi. *SHARPE: Symbolic Hierarchical Automated Reliability and Performance Evaluator, Introduction and Guide for Users.* Technical Report, Duke University, Durham, NC, September 1986.

[9] Kishor S. Trivedi. *Probability & Statistics with Reliability, Queuing & Computer Science Applications.* Prentice-Hall, 1982.

Chapter 24

Applicative Architectures for Fault-Tolerant Multiprocessors [1]

Madhumitra Sharma [2]
W. Kent Fuchs [2]

Abstract

This paper proposes functional programming frameworks for the design of highly reliable multiprocessor systems. In contrast to imperative programming environments, a functional environment offers elegant, relatively simple, and efficient solutions to concurrent error detection and recovery problems in multiprocessors. Specific fault tolerance mechanisms for upset exposure, fault containment, secure task assignment, and recovery are developed for a class of applicative multiprocessor architectures. Verification of abstract behavioral characteristics of applicative tasks is used for exposing faults during the execution of tasks. The fault containment mechanism is based on isolation of stack and heap segments of tasks. A protocol for secure task assignment is defined between system components. The architecture permits incremental, distributed, and asynchronous backups of system state. Finally, recovery is accomplished, even in the worst cases, by re-execution of a small number of tasks.

---

[1] This research was supported in part by the National Aeronautics and Space Administration (NASA) under Contract NASA NAG 1-613 in cooperation with the Illinois Computer Laboratory for Aerospace Systems and Software (ICLASS), by the Joint Services Electronics Program (U.S. Army, U.S. Navy, and the U.S. Air Force) under Contract N00014-84-C-0149, by the National Science Foundation under Grant No. US NSF DCR84-10110, by the U. S. Department of Energy under Grant No. US DOE-DE-FG02-85ER25001, and by an IBM Donation.

[2] University of Illinois at Urbana-Champaign, Urbana, IL

475

24.1   Introduction

Functional programming frameworks are proposed in this paper for the design of highly reliable multiprocessor systems. In contrast to imperative programming environments, a functional environment offers elegant, relatively simple, and efficient solutions to concurrent error detection and recovery problems in multiprocessors. Functional languages are also known to offer features appropriate for development of large parallel software systems including elegant data and procedure abstraction techniques, easily identifiable parallelism, and the relative ease of proving program correctness [Vegd84]. Together, these characteristics make them viable candidates for designing highly reliable multiprocessors. This is illustrated in this paper through the development of mechanisms for efficient and inexpensive concurrent error detection and recovery in a class of applicative multiprocessor architectures.

A general approach to overall system structuring for fault tolerance has been described as being based on *(1) the use of idealized fault tolerant components, (2) recursive structuring, and (3) atomic actions* [Rand84]. Such overall system structuring helps in reducing the complexity of fault tolerance mechanisms and thus leads to higher reliability. The applicative architecture presented herein is well structured with respect to the above criteria. It employs abstract components defined such that each component can appropriately handle faults within itself and those in others it interacts with. Thus, they qualify as *idealized fault tolerant components*. Further, the architecture is *recursively structured* since units within the system are identical and the functionality of each such unit matches that of the system. Finally, the proposed design is based on *applicative tasks* , which are even more suitable for recovery purposes than are *atomic actions*. When an atomic action fails, the machine state has to be restored to that which existed before the atomic action began execution. An applicative task, on the other hand, can simply be re-executed from the task definition because the corrupted machine state can be ignored by future computation. Since no state restoration is required, the functional approach is also free from the *Domino Effect* .

Applicative systems exploiting large-grain parallelism are considered in this paper. While studies on fault-tolerance aspects of such systems have not been reported in literature, researchers have studied fault tolerance issues in machine architectures exploiting fine-grain parallelism - namely, dataflow and reduction architectures.

Several fault tolerance techniques for dataflow machines have been presented by Hughes [Hugh83], Leung and Dennis [Leun80], Misunas [Misu76] and Srini [Srin85]. These techniques, which include checkpointing the entire system state, encoded interconnection, and triple-modular redundancy, do not take advantage of the functional environment characteristics. Grit has outlined fault tolerance mechanisms for reduction-based multiprocessors [Grit84] and more recently, recovery issues for similar systems have been studied by Lin and Keller [LiKe86]. The target systems in these studies exploit parallelism at a very *fine grain* . Functional programming is, however, also very attractive for exploiting parallelism at higher levels. Parallel systems with a coarser grain of parallelism incur less overhead and may be considered more pragmatic. Large-grain parallelism has been employed in systems such as the Rediflow multiprocessor [Kell84], AMPS [Kell79], and in LISP systems such as those proposed in [Hals84] and [Gabr84], which permit side effects.

The work in [Grit84] and [LiKe86] is most closely related to this paper. The specific differences are again primarily due to the granularity at which parallelism is exploited. In Grit's paper, each task reduces a sub-expression and returns a result value to its parent. In architectures considered in our paper, data structures such as trees and linked lists are allowed. Consequently, a node does not simply receive a single result value from its child, but may receive a pointer to an entire tree or list. Elements of the data structure might have been computed by various other nested functions. This consideration significantly changes the problem of identifying the set of tasks to be re-executed for recovery from errors. Also, while Grit's is a message passing model, our architecture uses a shared memory system (in the sense of a single global address space). Pointer based data structures necessitate shared memory. Finally, in the design of fault tolerance mechanisms, Grit has assumed the processors to be *fail-stop* ; that is, a processor becomes silent when a fault occurs. In our study, processors are capable of corrupting the system and mechanisms have been designed to minimize, detect and recover from such corruption. Lin and Keller's work [LiKe86] is devoted to recovery issues alone. They, too, consider processors to be fail-stop. The discussion is at an abstract "process" level and does not consider hardware or software implementation issues like organization of stacks and heaps and synchronization between tasks.

We begin with a definition of the system architecture in Sec. 24.2. Sec. 24.3 describes a set of mechanisms for upset exposure and fault containment. In Sec. 24.4, we propose a task assignment protocol called the *Dialogue* . The Dialogue defines messages between processor and memory modules and serves to establish a mutual watchdog relationship between them. Sec. 24.5 discusses implementation of replication schemes within the functional framework. Secs. 24.6 and 24.7 are devoted to recovery issues. In Sec. 24.6, it is shown that applicative architectures permit distributed, asynchronous, and incremental backups of the system state for recovery. Finally, recovery procedures are presented in Sec. 24.7.

24.2 System Architecture

In this section, an abstract system architecture for an applicative multiprocessor system is defined. This definition will serve as the target architecture for fault tolerance techniques described subsequently. In specifying the architecture, we maintain a level of generality and abstraction such that a large class of implementation strategies may be covered.

24.2.1 Multiprocessor Organization

The multiprocessor consists of a pool of processing elements (PEs), a pool of memory modules (MMs), and an interconnection network (IN). Since no assumptions are made regarding the interconnection network, a large class of machine architectures are covered. Thus, widely differing configurations such as shown in Figures 24.1a and 24.1b are permitted. Figure 24.1a represents a pool of processors sharing a multiport memory system via a multistage interconnection network as in the CEDAR multiprocessor. Figure 24.1b, on the other hand, shows a configuration where processor-memory modules are connected together with an interconnection network as in hypercube architectures.

24.2.2 Programming Environment

Programs are written in a functional language like *pure LISP* and are executed concurrently on the multiprocessor. Parallelism is achieved by evaluating function arguments (and, possibly, the function definition itself) in parallel. In order that the system be efficient, not all expressions are evaluated as separate tasks. Instead, processes are spawned only for *major* function applications. *Major* functions may be identified either by a compiler or by the programmer.

It should be noted that single assignment languages are permitted. It shall also become evident that programs are required to be functional only externally, that

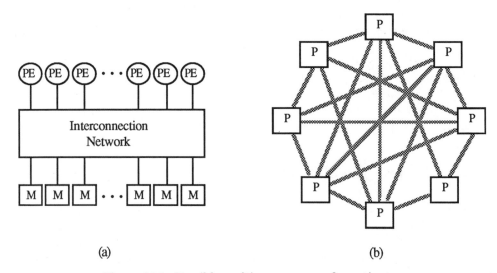

(a) (b)

Figure 24.1: Possible multiprocessor configurations.

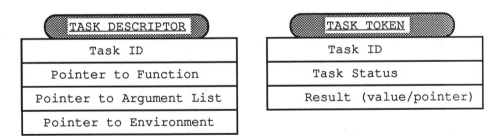

Figure 24.2: Task descriptors and tokens.

is, as seen by the system memory and other processors. A task could internally use side-effects as long as they are limited to PE registers or local memory.

24.2.3 Operation

A request for the application of a *main* function to a given set of arguments initiates a computation. The *main* function application is evaluated by a recursive spawning of subtasks that evaluate subexpressions. The subtasks may execute on the same PE as the current task or may be assigned to another one. After creating children tasks, a parent suspends itself until results from the children become available. When the children return results, the parent continues execution and finally dies, producing results for its parent. Such operation results in a dynamically growing and collapsing task tree.

24.2.4 Data and Control Structures

Task Nomenclature

Ancestors and successor tasks for a given task have to be identified at run time for the purpose of recovery. We define task-ids so that this identification becomes trivial. A task-ID of "1" is assigned to the root task. Assuming a binary task tree, the children of a task N are assigned IDs of $(2N)$ and $(2N+1)$. The nomenclature is easily extended to allow an arbitrary number of children for each task. For example, if a task has three children, IDs of $(4N)$, $(4N+1)$ and $(4N+2)$ are assigned to the children.

Task Descriptors

A *Task Descriptor (TD)* is illustrated in Figure 24.2. It defines an applicative task completely. A task creates a subtask simply by placing a TD for the subtask on a task queue.

Task Tokens

Task tokens (Figure 24.2) are used for synchronization between a parent task and its children. When a task is created, the parent also creates a corresponding token. The child returns its result via the token. The status field in the token is used to indicate one of five states of the task:

X: Task does not exist.

C: Task has been created but has not yet been scheduled for execution.

A: Task is active, i.e., it is currently being executed by some PE.

S: Task is suspended. It spawned subtasks and is waiting for results.

D: Task is complete. The result field in the token contains the value of, or pointer to, the result.

Task tokens are organized in a tree. Each token includes pointers to tokens for subtasks. Thus, given the task-ID, and given that the location of the root token is fixed, the token of any task can always be located.

Heap Segments

As in LISP, data cells for each task are allocated from a heap. The heap is implemented as a distributed set of subheaps - one in each memory module. Heaps are broken down into *Heap Segments* - each segment being a contiguous set of cells. Each activation of a task uses a new heap segment.

Control Stack

Each task also gets a *Stack Segment* to be used as a control stack for calls to **non-task functions** (functions that are executed sequentially within a task and not as independent tasks). When a task spawns subtasks and suspends itself, the corresponding stack segment also becomes dormant. When it resumes, it continues to use the same stack segment. Finally, when the task completes and returns its results to the parent, the stack segment is purged.

24.2.5   Information Transfer Relationships Between Tasks

A task is assumed to spawn all of its subtasks at one time and immediately suspend itself. Thus, in the absence of faults, each task is active only twice in its lifetime - once, when it first comes up for execution after creation, and next, when it resumes execution after a suspension. The state A, therefore, consists of two substates, $A1$

and $A2$. $A1$ corresponds to the first activation of the task and $A2$ to the second. This assumption is necessary only for simplicity in illustrating the fault tolerance mechanisms. Extension of the techniques to allow an arbitrary number of activations can be made easily.

Notation and Definitions :

- T_i denotes a task with task-ID $= i$.

- T_i^a represents the a'th activation of a task whose task-id is i. a can be 1 or 2 corresponding to states A1 and A2 of the task. Each T_i^a is referred to as a "task".

- H_i^a is the Heap Segment for task T_i^a.

- Σ_i is the Stack Segment for task T_i.

- $A(T_i^a)$ is defined as the set of all ancestors of the task T_i^a.

- $S(T_i^a)$ is the set of all successors of the task T_i^a.

- $A_T(T_i^a)$ is defined as the set of all tasks in $A(T_i^a) \cup S(T_i^a)$ that are executed before T_i^a.

- $S_T(T_i^a)$ is defined as the set of all tasks in $A(T_i^a) \cup S(T_i^a)$ that are executed after T_i^a.

The arrows in Figure 24.3 indicate the order of execution of the tasks and the flow of information between them. For example, for the task T_1, $A(T_1) = \{T_0\}$, $S(T_1) = \{T_3, T_4\}$. Similarly, for the task T_1^2, $A_T(T_1^2) = \{T_0^1, T_1^1, T_3, T_4\}$ and $S_T(T_1^2) = \{T_0^2\}$. Thus, a task T_i^a can reference any data produced by tasks in $A_T(T_i^a)$ and conversely, results produced by T_i^a can be referred to only by tasks in $S_T(T_i^a)$.

24.3   Upset Exposure and Error Containment

Hardware mechanisms described in this section limit error propagation between tasks and, at the same time, permit concurrent detection of errors due to a large class of failures in the system. The approach is based on verification of abstract behavior of applicative tasks and very little extra hardware is required.

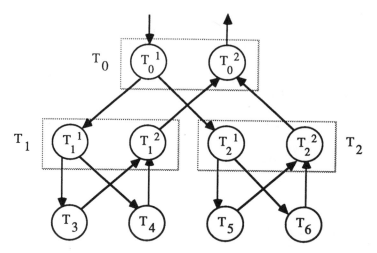

Figure 24.3: Information transfer relationship between tasks.

24.3.1   Motivation

In a functional environment, each data (heap) cell is written only once. All subsequent references to the cell are read references. Further, the cell is written as soon as it is allocated from the heap - for instance, as a result of a *cons* operation. Cells allocated to and written by *(and only by)* a task T_i^a constitute the *Heap Segment* H_i^a of the task. Only *future* activations of tasks in $A(T_i^a) \cup S(T_i^a)$ can read this cell. We have defined this set as $S_T(T_i^a)$. An error in a cell in H_i^a can, therefore, propagate only to tasks in $S_T(T_i^a)$. Reasoning along similar lines, an error in the stack segment Σ_i or in the token τ_i can propagate only to tasks in $S_T(T_i^a)$.

Definition :   We define the *domain* D_i^a of a task T_i^a as the set $H_i^a \cup \Sigma_i \cup \tau_i$.

We can therefore state,

Property 1 :   An error in D_i^a can only propagate to tasks in $S_T(T_i^a)$.

Consider the occurrence of a fault during the execution of a task, T_i^a. If due to this fault, the error is confined to D_i^a, then it can propagate only to $S_T(T_i^a)$. However, if the fault causes T_i^a to overwrite data in another write-space D_j^b, then the error could propagate to both, $S_T(T_i^a)$ and $S_T(T_j^b)$. Further, since a number of tasks in the latter set might already have been executed, recovery from the error in D_j^b may be expensive. Therefore, the domain D_i^a should be shielded from faults not "associated" with the execution of T_i^a.

Definition : We define a fault-class F_1 as the set of all faults, under the presence of which

D_i^a can be corrupted only while T_i^a is executing, and

T_i^a cannot corrupt anything but D_i^a.

In terms of this definition, our strategy for fault containment will be that of covering as large a set of hardware and software failures as possible within F_1.

24.3.2 System Design for Isolation of Heap and Stack Segments

We confine all three components - the heap segment H_i^a, the stack segment Σ_i^a, and the task token τ_i, of a domain D_i^a (corresponding to a task T_i^a) - within the same physical memory module. Thus, during the execution of a task T_i^a, the processor executing the task writes only to this MM. Further, no other processor can write to this module during the execution of T_i^a. In other words, for no pair of tasks T_i^a and T_j^b, executing concurrently at a given time, are the domains D_i^a and D_j^b in the same memory module.

Each processor can be considered as being coupled to a MM. In a distributed system such as shown in Figure 24.1a, this MM might be the processor's local memory. In a shared memory multiprocessor such as shown in Figure 24.1b, it may be any memory module. Under fault-free operation, a PE will write only to it's MM. Therefore, in the absence of faults, no two concurrent tasks write data cells in the same memory module. Of course, any task may, at any time, read data from any other memory module.

Heap space in each memory module is written in a monotone increasing order. A cell $i + 1$ in a module is written only after i has been written. Further, since only one task can be allocated heap cells from a given memory module, the heap segment H_i^a of a task T_i^a essentially consists of contiguous cells. If a PE executes tasks T_{i1}^{a1}, T_{i2}^{a2}, T_{i3}^{a3}, etc. in that order, then the corresponding memory module will have heap and stack segments as shown in Figure 24.4.

Thus, a simple counter at the memory module, serving as a *top-of-the-heap pointer*, can be used to check write accesses by the PE as illustrated in Figure 24.5. The counter is incremented at every heap segment write access to the memory module. If the processor attempts a "heap segment write" to a location other than the cell pointed to by the counter, the write is aborted and an error is signaled. Thus a task T_i^a is not allowed to destroy the contents of a heap cell not on its heap segment H_i^a. Another up/down counter is used to point to the top of the stack segment currently in use. When a PE starts using a stack segment Σ_i, it loads the counter with a pointer to the top of the segment. Subsequently, for each stack access (*push* or

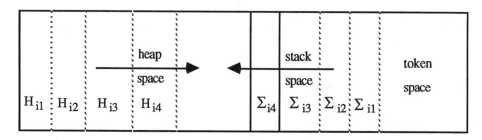

Figure 24.4: Heap, stack, and token spaces in memory modules.

pop), the pointer follows the top of the stack. When a PE presents an address for a *push* or *pop* operation, the address is compared with the contents of this counter. An attempt to write an illegal location in the stack space is flagged as an error.

24.3.3 Error Containment and Fault Coverage

The above scheme realizes the goal of isolating task domains from errors in the execution of other tasks. The consistency checks enforced by the counters are very

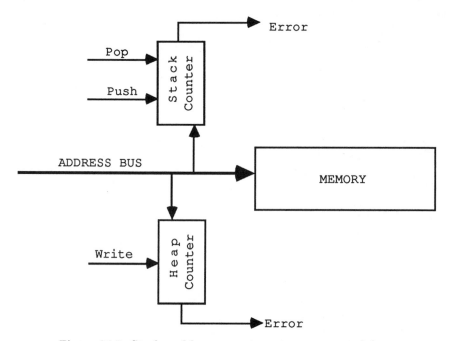

Figure 24.5: Stack and heap counters at memory modules.

strong since the counters identify the specific memory locations that can be written by the processor. Consequently, all transient and permanent addressing faults in the PE, except for those involving heap segment read accesses, will be detected in addition to any other PE malfunction leading to inconsistent memory references. Permanent interconnection network faults leading to misrouted or dropped messages will also be detected.

24.4   Secure Task Assignment

We define a task assignment protocol, called the *Dialogue* , which essentially results in a two-way "watchdog" relationship between a PE and the corresponding MM. The *Dialogue* defines messages to be used during task assignment phases and is aimed at achieving redundancy in information pertaining to the assignment of tasks. The protocol is based on the following primitive messages between a PE and a MM:

```
request_dialogue ( MM-id );
begin_dialogue ( PE-id );
suspend_dialogue ( PE-id or MM-id );
resume_dialogue ( PE-id or MM-id );
exit_dialogue ( MM-id );
respond_if_alive ( PE-id or MM-id );
```

We illustrate the protocol with the following examples.

A PE indicates to an MM that it wishes to start executing a new task by sending to the MM a `request_dialogue` message. This results in a packet < `PE-id`, `request_dialogue`, `MM-id` > to be sent to the MM. The MM responds with a `begin_dialogue` message. The PE then executes a task corresponding to a task descriptor or stack segment residing on the particular MM. When a task is complete, the PE explicitly indicates this by sending an `exit_dialogue` message to the MM.

A PE may suspend the current dialogue and request permission to join another dialogue when it wishes to place a task descriptor in the stack space in a memory module other than the one it is currently in dialogue with. Such suspension and subsequent resumption is accomplished using `suspend_dialogue` and `resume_dialogue` and is illustrated in Figure 24.6. The figure depicts two PE-MM pairs executing tasks T_{i1} and T_{i2} respectively. At some point during the execution of T_{i1}, a new task T_{i3} is to be created and assigned to PE2. PE1 suspends its current dialogue and requests a dialogue with MM2. In response, MM2 suspends its dialogue with PE2 and begins one with PE1. PE1 places a task descriptor for T_{i3} on a fresh stack segment in MM2 and exits from the dialogue. It then resumes its dialogue with

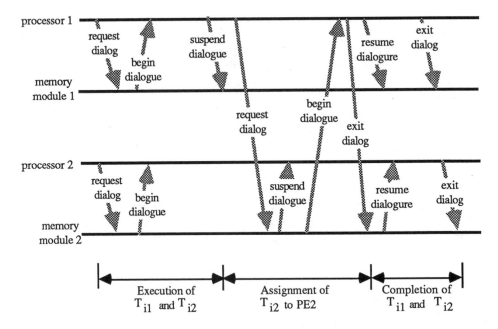

Figure 24.6: *Dialogue :* The task assignment protocol

MM1 while MM2 resumes its dialogue with PE2. The `respond_if_alive` message is a diagnostic message and is used either when an inconsistent message is received or when a response expected from a partner is not received in time.

With this protocol, redundancy in information pertaining to the assignment and execution of tasks is achieved. While a task is being executed on a PE, the corresponding MM knows the ID of the task and both expect well-defined responses from each other. An error is signaled when a partner does not meet the protocol definition due to a failure. Further, any erroneous messages received by a PE-MM pair participating in a dialogue from another PE or MM can be detected.

24.5   Error Detection by Replication

Error detection techniques described in Secs. 24.3 and 24.4 are not adequate for detection of errors such as incorrect arithmetic computation within the processor. Redundant computation using a second set of resources (processor, memory, and interconnection) and a subsequent comparison of results may be employed for detection of such errors. This section discusses how dual-redundancy may be incorporated within the applicative task framework. The approach serves not only to detect an error, but also to identify the particular task or set of tasks required to be re-executed for recovery.

The task tree is duplicated by invoking two copies of the top-level (main) function. We assume that the scheduling mechanism guarantees that the resources used by the two copies of each task are distinct. It is also assumed that the scheduling mechanism is such that the two copies of each task execute close to each other in time so as to minimize error detection latency.

A compiler inserts *Compare_results* tasks in the task tree which compare results from sets of twin tasks. Note that a comparison of results of every set of twin tasks is not required. The *Compare_results* tasks may be inserted at random points in the task tree (and one at the top). Choosing the density of "compare points" essentially involves a tradeoff between the cost of the comparisons and the error detection latency. Since tasks may return single values (atoms) or may return pointers to data structures, comparisons for tasks returning atoms will cost much less than for those returning larger data structures. Consequently, it is preferable that a large fraction of the comparison points cover tasks that return atoms rather than large data structures.

While the *Compare_results* task appears in both copies of the program, it is executed only once for each pair of twin tasks. For the twin task that completes first, the **compare** task does nothing because results from the second are not yet available. When the second completes, the parent **compare** task actually carries out a comparison of the results. Once an error is found as a result of the comparison, the domain D_i^a in which the error first appeared is to be identified. This is accomplished by comparing results of tasks along the erroneous path in the task tree.

24.6 System Memory Backups

Backups of the system memory onto a secondary storage are required for recovery after the detection of an error. While systems based on imperative programming need to synchronize and backup the entire system memory at every checkpoint, our architecture permits distributed, unsynchronized and incremental memory backup.

Data cells on heap segments are written only once during the execution of an application. At any given time, the value of the *top-of-heap* pointer indicates what locations have been written until that point in time. Thus the heap space may be backed up incrementally - only memory areas written after the previous backup have to be stored. Further, the backup process need not synchronize with the computation.

Unlike the heap space, memory space used by stack segments is written more than once. Stack segments are allocated when a task is spawned, are purged when the task completes, and may subsequently be reused for another task. However, incremental backups are still possible. A stack segment for a task may be copied out to a backup store when it spawns off child tasks and suspends itself.

The memory space allocated for task tokens is the only area which may be modified by the processors at random. The token space in each memory module, therefore, has to be backed up periodically in entirety. However, this should constitute only a small fraction of the memory space. In this case too, the backup may proceed independently of the computation.

Task Descriptors may be written out to a backup store when a task is created. Since TDs for any task can be regenerated if needed by re-executing its parent, the recovery procedures do not require a backup of all TDs. Thus, only TDs for tasks at, say, every n-th level in the task tree are stored. There is essentially a tradeoff to be made between the overhead of re-executing ancestor tasks to regenerate a given TD, and storing a large number of TDs.

It is the "write-once" characteristic and the consequent determinacy of applicative environments which leads to the inexpensive backup scheme. In contrast, imperative environments not only require a synchronization of all processors at a barrier, but also have to save the entire system memory at each checkpoint. Distributed checkpointing schemes may lead to the potentially disastrous *domino effect* .

24.7   Recovery Procedures

24.7.1   Recovery From Errors Detected in D_i^a During Execution of T_i^a

When an error due to a fault of class F_1 is detected during the execution of a task T_i^a, recovery is accomplished by the following procedure :

Recovery Procedure P1 :

If a = 1, then

(1) Kill any tasks in $S_T(T_i^a)$ that have already been spawned.

(2) Purge the stack segment Σ_i.

(3) Re-execute the parent task T_j^1 using the task descriptor, TD_j^1 to reconstruct the task descriptor TD_i^1.

(4) Retry T_i^1 using TD_i^1 and with a new heap segment $\hat{H}_i^1$ and using a new stack segment $\hat{\Sigma}_i$.

If a = 2, then steps 1 and 2 are the same. Step 3 and 4 become :

(3') Re-execute T_i^1 using the task descriptors TD_i^1 to reconstruct $\hat{\Sigma}_i$ for T_i^2.

(4') Retry T_i^2 using $\hat{\Sigma}_i$ and with a new heap segment $\hat{H}_i^2$.

Property 2 : Recovery procedure P1 preserves correctness of results.

Proof: Consider steps 1 and 2. From the information transfer relationship between tasks shown in Figure 24.3, it is clear that any results computed in $S_T(T_i^a)$ are not referred to by any tasks in $\overline{S_T(T_i^a)}$. Thus killing any tasks in this set that may already have been spawned does not affect any uncorrupted tasks in $\overline{S_T(T_i^a)}$. Of course, the tokens for the killed tasks are reset to state X and their stack segments purged.

In step 3, T_j^1, the parent of T_i^1, is re-executed in a new heap segment $\hat{H}_j^1$ to re-construct $\hat{\Sigma}_i$. T_j^1 thus has two heap segments, H_j^1 and $\hat{H}_j^1$. While the re-execution of T_i^1 will use the latter, other children of T_j^1 will continue to use the former. H_i^1 and $\hat{H}_i^1$ are identical except for pointers that point within the same heap segment. That is, a pointer p in H_i^1 to a cell within H_i^1 will not be equal to the corresponding pointer $\hat{p}$ in $\hat{H}_i^1$. All pointers in H_i^1 to cells not in H_i^1 will be identical to corresponding ones in $\hat{H}_i^1$. Further, any atoms in H_i^1 will be identical to those in $\hat{H}_i^1$. The information contained in the two heap segments is, thus, equivalent. Similarly, the new task descriptor for T_i^a is also equivalent to first.

Correctness for the case $a = 2$ (steps 3' and 4') can be established along similar lines.

24.7.2   Recovery from Errors Detected in D_i^a After Execution of T_i^a

A. Lost Data Cells

Case (1) : *The lost data cell is available in backup store.*

The contents are simply restored from the secondary backup memory.

Case (2) : *The lost data cell was created since the last backup.*

In this case, T_i^a and all tasks in $S_T(T_i^a)$ have to be re-executed. Note that since T_i^a was executed after the last backup, the set $S_T(T_i^a)$ contains only a small number of tasks. Damage, therefore, is small.

B. Lost Stack Segments

Stack segments are created when a task is created and disappear when the task is complete. Therefore, while a stack segment Σ_i exists, a task could be in states C, A1, A2, or S. One or more cells may be lost during this time. Recovery for cases when a Σ_i is lost while the task T_i^a is in one of the active states, A1 or A2 has already been considered in Section 24.7.1. Here, we consider the case when the task is in state C or S:

Case (1) : Task-ID is not lost.

In this case, Σ_i is reconstructed using the steps 3 and 3' in procedure P1 in Sec 24.7.1.

Case (2) : Task-ID is lost.

Recovery in this case may appear difficult. Since the task-ID is not known, which stack segment to reconstruct is not known. The approach taken is that of purging the stack segment and waiting for an eventual deadlock state where there are no active tasks in the system. In this state, all tasks in $A(T_i^a)$ are suspended because the parent of T_i^a is waiting for results from T_i^a. Furthermore, tasks in $S(T_i^a)$ will all be in state D (done) or X (do not exist). From this situation, the Task-ID can be inferred. Subsequently, Σ_i can be reconstructed and computation can proceed.

C. Lost Tokens

The token for a task T_i is referenced by the scheduler to determine whether the task is done and by the parent task which uses the result field of the token. In both cases, the token is examined only under the situation where the parent is suspended. A token can be reconstructed as follows. We first check the status of the children of T_i. If all children have a status X or D, then reconstruct the task descriptor for T_i and re-execute T_i to reconstruct the token. *Else* , since the children have been created and are active or suspended, the task T_i is suspended. Therefore, the status field in the token is set to S.

24.7.3 Recovery from Loss of a Memory Module or PE

The loss of a memory module implies the loss of heap segments, stack segments, and tokens for all tasks corresponding to that module. However, if a spare memory module is available and can be reconfigured into the system so as to replace the defective one, recovery can be easily accomplished. The procedure consists of the following steps:

Restore, from backup storage, the saved contents of the lost memory module. Since the restored contents represent the state at the time of the backup, some information may be lost.

For each heap segment H_i^a lost, kill all tasks in $S_T(T_i^a)$ and re-execute T_i^a. Once again, it should be noted that this set $S_T(T_i^a)$ is small because T_i^a was executed after the last memory backup for the module.

Table 24.1: Cost of recovery from each type of error.

Error Class	Type of Error	Recovery Cost
E_1(Sec.24.7.1)	Errors in D_i^a detected during execution of T_i^a	Re-execution of single task.
E_2(Sec.24.7.2A)	Lost data cells in D_i^a detected by task other than T_i^a	If available in backup then recovered from backup; else re-execution of few tasks.
E_3(Sec.24.7.2B)	Lost cells in Σ_i detected by task other than T_i^a	Re-execution of single task.
E_4(Sec.24.7.2C)	Lost token T_i detected by task other than T_i^a	Re-execution of single task in worst case.
E_5(Sec.24.7.3)	Loss of PE	Re-execution of single task. System continues to operate with performance degradation corresponding to loss of PE.
E_6(Sec.24.7.3)	Loss of memory module	Re-execution of a small number of tasks.

When a task loses a PE, the task is simply re-executed on another PE when one becomes available. The performance of the system degrades due to the loss of the PE.

A summary of recovery costs for each case discussed above is presented in Table 24.1. It is clear that even for such cases as loss of an entire memory module, the cost of recovery is limited to re-execution of a small set of tasks.

24.8 Conclusions and Future Research

This paper has provided evidence that functional languages are highly amenable to fault tolerant multiprocessor design. Three primary characteristics of functional environments are exploited in the proposed architecture :

- *Well-defined behavioral characteristics of applicative tasks,* which lead to upset exposure techniques relying on verification of the behavior, and which signal errors on detection of any illegal behavior.

- *Determinacy,* or the property that given a set of arguments, the application of the function will always produce the same result. Determinacy leads to inexpensive recovery procedures.

- *Side-effect free nature and the single assignment rule,* as a result of which we can do away with expensive recovery cache mechanisms and can design recovery procedures free from the Domino Effect. Further, the single assignment rule permits incremental machine state backups.

Issues pertaining to garbage collection and other non-functional operating system tasks have not been discussed in this paper. Future work is needed to incorporate them within the functional framework for fault tolerance. Approaches for handling data structures such as (infinite) streams, exception mechanisms such as *catch/throw* , and lazy/eager evaluation strategies, are also candidates for future research.

References

[Gabr84] R.P. Gabriel and J. McCarthy, *Queue-based Multi-processing LISP,* Proceedings, 1984 ACM Symposium on LISP and Functional Programming.

[Grit84] D. H. Grit, *Towards Fault Tolerance in a Distributed Applicative Multiprocessor,* Proceedings, International Symposium on Fault Tolerant Computing, Jun 1984, pp 272-277.

[Hals84] R.H. Halstead, Jr., *Implementation of Multilisp : LISP on a Multiprocessor,* Proceedings, 1984 ACM Symposium on LISP and Functional Programming.

[Hugh83] J. L. A. Hughes, *Error Detection and Correction Techniques for Data-Flow Systems,* Proceedings, International Symposium on Fault Tolerant Computing, June 1983, pp 318-321.

[Kell79] R. M. Keller, G. Lindstrom, and S. Patil, *A Loosely-Coupled Applicative Multi-processing System,* AFIPS Conference Proceedings, June 1979, pp 613-622.

[Kell84] R. M. Keller, F. C. H. Lin, and J. Tanaka, *Rediflow Multiprocessing,* Proceedings, COMPCON Spring 84, Feb. 1984, pp 410-417.

[Leun80] C. K. C. Leung and J. B. Dennis, *Design of a Fault-Tolerant Packet Communication Computer Architecture,* Proceedings, International Symposium on Fault Tolerant Computing, 1980, pp 328-335.

[LiKe86] F.C.H. Lin and R.M. Keller,*Distributed Recovery in Applicative Systems,* Proceedings, 1986 International Conference on Parallel Processing.

[Misu76] D. P. Misunas, *Error Detection and Recovery in a Data-flow Computer,* Proceedings, 1976 Conference on Parallel Processing, August 1976, pp 123-131.

[Rand84] B. Randell, *Fault Tolerance and System Structuring*, Proceedings, 4th Jerusalem Conference on Information Technology, 1984.

[Srin85] V.P. Srini, *A Fault Tolerant Dataflow System*, IEEE Computer, March 1985, pp 54-68.

[Vegd84] S.R. Vegdahl, *A Survey of Proposed Architectures for the Execution of Functional Languages*, IEEE Trans. on Computers, Dec. 1984, pp 1050-1071.

Chapter 25

Fault-Tolerant Multistage Interconnection Networks for Multiprocessor Systems

V. P. Kumar [1]
S.M. Reddy [2]

Abstract

In this paper we present a class of multipath multistage interconnection networks (MINs) called Augmented Shuffle-Exchange Networks. These MINs can be designed to have the degree of switch fault tolerance desired. They feature links among switches belonging to the same stage, and the number of links between adjacent stages is the same as in unique path MINs. The paths available from a source to a destination have varying lengths. Rerouting in the presence of faults or blocking can be accomplished in these MINs dynamically, without resorting to backtracking. In addition to tolerating faults in individual switches, the proposed networks make it possible to tolerate faults in groups of switches, thus facilitating on-line repair. Reliability and performance studies show that the proposed MINs achieve a significant improvement over unique path MINs and compare favorably with other multiple path MINs.

25.1   Introduction

Multistage networks have long been studied for use in telephone switching [1] and in multiprocessor systems [2-11]. Several excellent surveys on this subject are available

---

[1] AT&T Bell Laboratories, Holmdel, NJ
[2] University of Iowa, Iowa City, IA

for the interested reader [2-4]. Since the early 70's several Multistage Interconnection Networks (MINs) have been proposed to meet the communication needs in multiprocessor systems in a cost-effective manner [5-11]. Many multiprocessor systems proposed and developed, or currently under development, make use of MINs [12-18]. While these MINs have many interesting properties that make them attractive for multiprocessors, they suffer from degrading performance and reliability with increasing network size, the alleviation of which is the subject of this paper.

A typical MIN is designed for $N = m^n$ inputs and N outputs, and utilizes n stages of N/m crossbar switching elements of size $m \times m$. The switches in adjacent stages are connected by *links* in such a way that a path can be found from any input to any output (Figure 25.1). Among the many advantages MINs have are $O(N \log N)$ hardware cost as opposed to $O(N^2)$ cost of crossbar switches, $O(\log N)$ path lengths, the existence of simple and distributed routing algorithms, and the ability to provide up to N simultaneous connections. These MINs have the properties that there exists

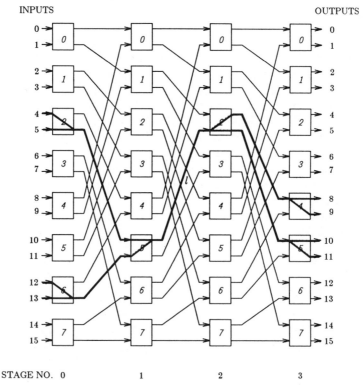

Figure 25.1: A unique path multistage interconnection network with 16 inputs and 16 outputs.

a unique path from any input to any output, and that distinct input-output paths may have common links. For instance, Figure 25.1 shows that any path from a source in {4,5,12,13} to a destination in {8,9,10,11} must pass through link *l*. This leads to two chief disadvantages. First, an input-output connection may be *blocked* by a previously established connection, even if the output required is not busy, thus causing *poor performance* in a random access environment. Second, the failure of even a single link or switch will disconnect several input-output paths, leading to a *lack of fault-tolerance*. Further, the reduction in performance due to blocking and the decrease in reliability due to lack of fault-tolerance become increasingly serious with the size of the network as the number of paths passing through a given link or a switch increases linearly with N.

In this paper we present methods of augmenting unique path MINs to create redundant paths between every input-output pair and achieve fault-tolerance and improved performance. The fault-tolerant MINs discussed in this paper, called *Augmented Shuffle-Exchange MINs* (ASENs), feature links among switches belonging to the same stage, which result in a dramatic improvement in reliability and a significant increase in performance at a reasonable cost. There are also several practical advantages such as ease of repair and maintenance that make the proposed MINs more attractive for use in large-scale systems.

The paper is organized as follows. In Section 25.2 several aspects of multiple path MINs are discussed and recent research results in this area are summarized. In Section 25.3 we give the construction methods for ASENs, and outline several possible variations. An Adaptive routing algorithm for ASENs is given in Section 25.4, which treat the presence of faults and the occurrence of blocking equivalently and attempt to avoid them. The fault-tolerant properties of ASENs realized by the routing algorithm are also highlighted in this section. In Section 25.5 we discuss the reliability of ASENs. Section 25.6 outlines the performance improvement and the paper concludes with Section 25.7.

25.2 Multiple Path MINs

The problems of poor performance and low reliability of unique path MINs have received the attention of several researchers recently. Many designs of MINs which provide multiple paths between input-output pairs have been proposed towards that end [19-39]. Multiple path MINs can have improved performance as well as improved reliability since alternate paths can be chosen to avoid busy links as well as faulty links or faulty switches. Redundancy of paths may be inherent in the definition of a MIN, or it may be the result of augmenting unique path MINs by extra stages, switches, links, etc.

Multiple path MINs have their own hardware cost, are effective to varying degrees in improving the performance and reliability of unique path MINs, and have their own peculiar properties and implementation requirements. Here we make a quantitative as well as a qualitative comparison of several multiple path MINs proposed and summarize the comparison in Table 25.1. For quantitative comparison we choose the hardware cost, degree of fault-tolerance provided, certain reliability measures (terminal reliability and mean time to failure), and circuit-switched performance measures (probability of acceptance of a connection request). Certain topological properties of multiple path MINs may impose restrictions on the manner in which they can be operated. Such properties can only be compared qualitatively. While most of the comparisons between different multiple path MINs are made in this section, the reliability and performance comparisons of those MINs for which the data are available, are left for Sections 25.5 and 25.6, respectively. The rest of this section contains a general discussion of multiple path MINs and an explanation of the column headings in Table 25.1.

The multipath MINs chosen for comparison in Table 25.1 are the following: The Extra Stage Cube (ESC) [20], the Inverse Augmented Data Manipulator (IADM) [23], the Enhanced Inverse Augmented Data Manipulator (EIADM) [24], the Gamma Network [25], the Bigamma and the Monogamma Networks [26], the INDRA Network [32], the F-Network [27], the Modified Omega Network [29], the Augmented C-Network (ACN) [31], the Chained Baseline Network (CBN) [35], the Load Sharing Banyan Network (LSB) [39], the Dynamic Redundancy Network (DR) [36], and the networks presented in this paper. The reader is directed to the references cited for a description of the above networks.

25.2.1   Hardware Complexity

Compared to unique path MINs, multiple path MINs have a higher hardware complexity in terms of one or more of the following:

1. the number of stages of switches [19-21,33],

2. the number of switches per stage [22-27,32,36], and

3. the size of the switching elements [22-31,34-39].

These three factors contribute to what will be termed the switch complexity of the MIN, which is equal to the complexity of the switching element times the number of switching elements per stage times the number of stages. The complexity of the switching element depends on its implementation. Some switching elements are implemented as full crossbars (Type I switches) but others are implemented as a multiplexer-demultiplexer combination which allows connection of any one input link to one or more output links (Type II switches). The complexity of a switching element is estimated by the number of crosspoints necessary to implement it (see

Table 25.1: A comparison of multiple path MINs. (N is the size of the network and $n = log_2 N$.)

MIN	Switch Type and Size	No. of Stages	Switch Complexity	Link Complexity Interstage	Intrastage
ESC	I; 2×2	$(n+1)^*$	$2N(n+1)+8N$	N	-
IADM	II; 3×3	$(n-1)^*$	$6Nn$	$3N$	-
EIADM	II; 5×5	$(n-1)^*$	$10Nn$	$5N$	-
Gamma	I; 3×3	$(n-1)^*$	$6N+9N(n-1)$	$3N$	-
BiGamma	I; 3×3	$(n)^*$	$6N+9Nn$	$3N$	-
MonoGamma	I; 3×3	$(n-1)^*$	$6N+9N(n-1)$	$3N$	-
INDRA	I; $R \times R$	$1+\log_R N$	$R^2 N(1+\log_R N)$	RN	-
F-Net	II; 4×4	$(n-1)^*$	$8Nn$	$4N$	-
Mod. Omega	I; $B \times B$	$\lceil \log_B N \rceil$	$BN\lceil \log_B N \rceil$	N	-
ACN	I; 4×4	$n-1$	$8N(n-1)$	$2N$	-
CBN	I; 3×3	n^*	$9Nn/2+3N$	N	$N/2$
LSB	I; 4×4	n	$4N(n-1)+2N$	N	-
DR	I; 3×3	$(n-1)^*$	$6(N+S)(n)$	$3(N+S)$	-
ASEN	I; 3×3	$(n-1)^*$	$(9N/2)\cdot(n-1)+4N$	N	$N/2$

MIN	Degree of Fault Tolerance	Reroutability	Fault-tolerance to Permutations; No. of Passes
ESC	1	Static	Yes;2 Passes
IADM	0	Dynamic	-
EIADM	1 or 2†	Dynamic	-
Gamma	0	Dynamic	-
BiGamma	1	Dynamic	-
MonoGamma	1	Dynamic	-
INDRA	$R-1$	Static	Yes; 1 Pass
F-Net	1	Dynamic	Yes; 2 Passes
Mod. Omega	$\left(\dfrac{B^{\lceil \log_B N \rceil}}{N} \right) - 1^{**}$	Static	Yes**; 2 Passes
ACN	1	Dynamic	Yes; 1 Pass
CBN	1††	Dynamic	Yes††; 2 Passes
LSB	1**	Dynamic	Yes**; 2 Passes
DR	S	Static	Yes; 1 Pass
ASEN	1	Dynamic	Yes; 2 Passes

* Plus additional stages of multiplexers and demultiplexers, at the output and the input of the MIN. They are included in the Switch Complexity column.

† If the switches in stage $n-2$ know about faults in stage n.

** Tolerance for faults in intermediate stages only.

†† Tolerance to link faults only. A switch fault in the final stage can cause network failure.

Figure 25.2). Thus a $m \times m$ switch of Type I has complexity m^2, a $m \times m$ switch of Type II has complexity $2m$. In many cases it is also necessary to use multiplexers and/or demultiplexers at the input side and/or the output side of the MIN. It can be easily seen that a $m \times 1$ multiplexer and a $1 \times m$ demultiplexer have complexity m.

Another measure of the cost of a MIN is its *link complexity*, which is specified by three components: the number of stages, the number of interstage links, and the number of intrastage links. This measure is important because the implementation of MINs is often interconnect-limited, at every level of integration. At the integrated circuit (IC) level, the size of the switch (assuming that each IC contains one switch) is usually determined by the number of pins available and not the complexity of the logic in the switch. At the printed circuit board level, the number of connectors available on the board is a critical factor in determining the number of switches that can be placed on a board. At the wafer scale integration level, if a MIN with a large number of inputs and outputs were to be laid out on a single wafer [40,41],

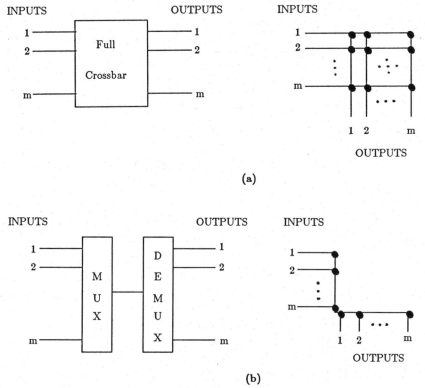

Figure 25.2: A cost model for the switches used in multistage networks: Type I switches (a), and Type II switches (b).

most of the area would be taken up by the links and not the switching elements. Of the two types of links mentioned, interstage links are more expensive than the intrastage links, as will be explained later in this paper.

In the design of multipath MINs, it is important to minimize not only the switch complexity and the link complexity, but also the number of input and output links of switching elements, while attempting to increase the fault-tolerance, reliability and performance as much as possible. The hardware complexity measures of the multipath MINs being considered are included in columns 2 through 5 of Table 25.1.

25.2.2   Selection of Paths

A key issue in multipath MINs is the manner in which rerouting or selection of alternate paths is achieved. If the topology of a multipath MIN allows rerouting to be done only at the source [3] or at some fixed points in the network, then the MIN is *statically reroutable*. If a busy link, a faulty link, or a faulty switch is encountered in setting up a path, it may be necessary to backtrack to a stage where a fork occurred and attempt to set up an alternate path from there. A multipath MIN is *dynamically reroutable* if the paths between any given input-output pair have a fork at every stage of the MIN, thus permitting rerouting decisions to be made by the switches at any stage, on the fly, as faulty links or switches are encountered or as blocking occurs. Statically reroutable MINs are inconvenient to implement since they require bidirectional paths and reverse queues for backtracking, and global fault information for possible performance enhancement by avoiding backtracking [30]. On the other hand dynamically reroutable MINs tend to have higher link complexity [22-27,31,34,35]. The type of rerouting permitted in different MINs is indicated in column 7 of Table 25.1.

25.2.3   Fault Models and Fault-Tolerance

Faults occurring in a MIN may affect either individual links, or entire switching elements. In the *link-fault model* failures are assumed to occur in links and the switching elements associated with the faulty links are assumed to be partially operational. In the *switch-fault model* failures affect switching elements which are assumed to become totally unusable or out of operation. (Since the switch fault model is the stronger of the two, and since link faults can be conservatively modeled by switch faults, we use the switch fault model for the networks presented in this

---

[3]The resources attached to the inputs of the MIN are called *sources*, and those attached to the outputs of the MIN are called *destinations*.

paper.) The robustness of a multipath MIN can be measured by its *degree of fault-tolerance,* which is equal to the number of switch faults guaranteed to be tolerated in the MIN while maintaining its ability to provide a fault-free path from every source to every destination.

A convenient way to determine the degree of fault-tolerance of a multipath MIN, and to find out whether it allows dynamic or static rerouting, is by the use of its *redundancy graph.* The redundancy graph for a source-destination pair in a MIN shows all the available paths between the source-destination pair[26,30]. There are two distinguished nodes in a redundancy graph, the source S and the destination D, and the rest of the nodes correspond to the switches along the paths between S and D. The redundancy graphs of all the source-destination pairs may be isomorphic, in which case the MIN is said to be *symmetric,* or the redundancy graph may vary from one source-destination pair to another, in which case the MIN is *asymmetric.* Let the *terminal connectivity* of a source-destination pair (S,D) be the minimum number of switch nodes that must be deleted from the redundancy graph of (S,D) to disconnect S from D. Let c be the minimum of the terminal connectivities over all the pairs (S,D) in a MIN. Then the number of faults guaranteed to be tolerated, or the degree of fault-tolerance, is equal to $c - 1$. It is also possible to determine whether or not a MIN is dynamically reroutable by examining the points at which forks occur in the redundancy graphs of the MIN. The degree of fault tolerance of the multipath MINs under consideration is indicated in column 6 of Table 25.1, and their reroutability in column 7.

For SIMD applications, it is important for multipath MINs to support Omega [6] or similar permutations, and to be able to tolerate faults in realizing those permutations. A multipath MIN may be able to realize a permutation in a single pass while tolerating certain faults, or it may incur a time overhead by requiring two or more passes to realize a permutation under faults. Whether the multipath MINs under consideration are capable of tolerating faults in realizing permutations, and, if so, the number of passes they require, is indicated in the last column of Table 25.1. This column is incomplete since we only consider the MINs which can realize Omega [6] type permutations.

The entries in Table 25.1 for ASENs will become clear later in the paper.

25.3   Augmented Shuffle-Exchange MINs

In this section we describe the construction of a class of multiple path MINs called Augmented Shuffle-Exchange MINs, which have the following properties:

1. They can be designed to have the desired degree of switch fault-tolerance.

2. They have dynamic reroutability.

3. They have the same number of interstage links as unique path MINs.

4. The size of the switches is small.

5. They have easy repairability, and hence potentially high availability.

Properties 1 through 4 are the result of providing additional links between switches of the same stage of a MIN, or by connecting switches in the same stage of two or more MINs. There are many possible ways of adding these intrastage links, but a judicious choice in placing these links guarantees tolerance of not only individual switch faults but also faults in entire groups of switches. Furthermore, the removal of the faulty group of switches from the network need not interrupt the operation of the MIN. Thus such groups of switches could be implemented as replaceable units, and repairing faults would involve replacing the unit with faulty switch(es) by a new one. Property 5 is a consequence of this.

The augmentation techniques given in this paper can be applied to a broad class of unique path networks called Delta Networks[8], which basically have the property that a *routing tag* obtained from the address of the destination can be used to perform routing in a distributed fashion. However the networks resulting from the construction procedure are particularly interesting when applied to a subclass of delta networks called Shuffle-Exchange MINs. Shuffle-exchange MINs are topologically equivalent[9] to several well-known MINs, such as the Omega Network [6], the Indirect Binary n-Cube [7], certain banyan networks[5], the generalized cube [11], the STARAN flip network [10], the baseline network [9] etc., so the constructions given in this paper can be extended to all of the above networks.

25.3.1 Preliminaries

A shuffle-exchange MIN with $N = m^n$ *inputs* and N *outputs* consists of n stages of $m \times m$ crossbar switches. N is called the *size* of the network and m the *degree* of the network. The stages are numbered from 0 to $n - 1$. Each stage consists of m^{n-1} switches, numbered from 0 to $m^{n-1} - 1$ and identified by m -ary labels $(a_0, a_1, \ldots, a_{n-2})$. The input and output links of a switch are labeled from 0 to $m - 1$. Also, an input or output link of a stage is identified by an m -ary label $(b_0, b_1, \ldots, b_{n-2}, b_{n-1})$ where $(b_0, b_1, \ldots, b_{n-2})$ is the label of the switch to which the link belongs, and b_{n-1} is the label of the link within that switch. The interconnection pattern between stages is the $m \times m^{n-1}$ shuffle permutation which can be described in terms of the m -ary labels as follows: The output link $(b_0, b_1, \ldots, b_{n-1})$ of stage i is connected to the input link $(b_1, b_2, \cdots, b_{n-1}, b_0)$ of stage $i + 1$, for $0 \le i \le n - 2$. The MIN input with the label $(a_0, a_1, \ldots, a_{n-1})$ is connected to the input link of stage 0 with the same label (and thus to the switch $(a_0, a_1, \ldots, a_{n-2})$). The MIN output labeled $(a_0, a_1, \ldots, a_{n-1})$ is connected to the output link with an identical label in stage $n - 1$. The shuffle-exchange MIN for $N = 16$ and $m = 2$ is shown in Figure 25.1.

In the following, switch i in stage j is denoted by $SW_{i,j}$.

Definition 1: A switch $SW_{i,j}$ is a *predecessor* of switch $SW_{l,j+1}$ if an output link of $SW_{i,j}$ is connected to an input link of $SW_{l,j+1}$. $SW_{l,j+1}$ is then a *successor* of $SW_{i,j}$. Further, $SW_{l,j+1}$ is the *d-successor* if the output link of $SW_{i,j}$ with the label d is connected to $SW_{l,j+1}$, and denoted as $SW_{l,j+1} = SUCC^d(SW_{i,j})$. Also, if the input link b of $SW_{l,j+1}$ is connected to the $SW_{i,j}$, then $SW_{i,j} = PRED^b(SW_{l,j+1})$.

Definition 2: Two switches $SW_{i,j}$ and $SW_{l,j}$ are said to be *first-order conjugates* of each other iff $SUCC^d(SW_{i,j}) = SUCC^d(SW_{l,j})$, for $0 \le d \le m-1$. We express this by $SW_{i,j} = CONJ^1(SW_{l,j})$ and $SW_{l,j} = CONJ^1(SW_{i,j})$.

Two switches $SW_{i,j}$ and $SW_{l,j}$ are k -th *order conjugates* iff $SUCC^d(SW_{i,j}) = CONJ^{k-1}(SUCC^d(SW_{l,j}))$, for $0 \le d \le m-1$ and $2 \le k \le n-1-j$. This is represented as $SW_{i,j} = CONJ^k(SW_{l,j})$ and $SW_{l,j} = CONJ^k(SW_{i,j})$. Two switches which are k -th order conjugates in stage i have paths leading to the same subset of switches in stage $i+k$. For first order conjugates the superscript is sometimes dropped and $CONJ^1(SW_{l,j})$ is simply written as $CONJ(SW_{l,j})$.

The following can be easily verified for the shuffle-exchange MIN.

The successors of $SW_{i,j} = (s_0, s_1, \ldots, s_{n-2})$ are of the form $(s_1, s_2, \ldots, s_{n-2}, d)$. Two first-order conjugate switches have labels which differ only in the leftmost position. Two switches $SW_{i,j} = (s_0, s_1, .., s_{n-2})$ and $SW_{l,j} = (\hat{s}_0, \hat{s}_1, \ldots, \hat{s}_{n-2})$ are k -th order conjugates if and only if $s_p = \hat{s}_p, k \le p \le n-2$. Clearly, the set of switches in a given stage can be partitioned into disjoint subsets, each subset consisting of switches which are k -th order conjugates. Since each such subset consists of m^k switches and each stage consists of m^{n-1} switches, the number of subsets is m^{n-1-k}. In particular, we are interested in the $(n-1-j)$ -th order conjugate subsets in stage j. The switches in each such subset have the property that they all have paths to the same subset of switches in the last stage and hence to the same subset of destinations. There are m^j such subsets in stage j. They are simply called *conjugate subsets* of stage j and denoted as $C_{0,j}, C_{1,j}, \ldots, C_{m^j-1,j}$. If a switch $SW_{l,j} = (s_0, s_1, \ldots, s_{n-2})$ is a member of $C_{i,j}$, then $(s_{n-1-j}, s_{n-j}, \ldots, s_{n-2})$ equals the m -ary value of i. The conjugate subsets are central to the construction procedure given next.

25.3.2   Construction Procedure

The construction procedure for Augmented Shuffle-Exchange MINs (ASENs) can be specified in such a way that it can be applied to shuffle-exchange MINs with arbitrary degree and to obtain MINs of arbitrary degree of switch-fault tolerance. However, for the sake of simplicity and clarity, the construction procedure given below is restricted to networks which have degree 2 and achieve single switch-fault tolerance. Generalizations possible are discussed in Section 3.3.

In the following, *it is assumed that each source and each destination has two separate interfaces to the network.*   This is essential for any network to be single fault-tolerant, for if a source or a destination has a single interface to the network, the failure of that interface will disconnect the source or destination from the rest of the network.

The augmentation procedure is carried out in the following steps:

1. Place a 1×2 multiplexer at each of the input links of the switches in stage 0. Connect each source $S = (s_0, s_1, \ldots, s_{n-1})$ to the switches $(s_0, s_1, \ldots, s_{n-2})$ (the *primary switch*) and $(\bar{s}_0, s_1, \ldots, s_{n-2})$ (the *secondary switch*) via the multiplexers. Notice that this step essentially provides an additional connection for each source, to the conjugate of the switch to which the source was originally connected (Figure 25.3(a)).

2. Replace each switch $(d_0, d_1, \ldots, d_{n-2})$ in stage $n-1$ with two 1×2 demultiplexers and label them $(d_0, d_1, \ldots, d_{n-2}, 0)$ and $(d_0, d_1, \ldots, d_{n-2}, 1)$. Connect the upper output link of demultiplexer $(d_0, d_1, \ldots, d_{n-2}, d_{n-1})$ to the destination $D_0 = (d_1, d_2, \ldots, d_{n-1}, 0)$ and the lower output link to the destination $D_1 = (d_1, d_2, \ldots, d_{n-1}, 1)$. This step connects each destination to two demultiplexers (Figure 25.3(b)).

3. Replace each 2×2 switch in stages 0 through $n-3$ by the 3×3 switches shown in Figure 25.3(c). The input and output links labeled 0 and 1 are to be connected exactly as before. The links *a-in* and *a-out*, called *auxiliary* input and output links, respectively, are used to make connections among switches belonging to the same stage, to form loops of switches that provide dynamic reroutability. For this purpose, it is essential that two switches which are connected by auxiliary links have paths leading to the same subset of destinations. From the description in Section 25.3.1, it follows that any arbitrary way of connecting switches within each conjugate subset $C_{i,j}$ is acceptable. However, an additional restriction that no two switches in a given loop be conjugate switches leads to some interesting properties. Two specific schemes that meet this restriction are given below:

(a) Let $SW_{i,j} = PRED^b(SW_{l,j+1})$.
Then connect $SW_{i,j}$ to $PRED^b(CONJ(SW_{l,j+1}))$ and vice versa. This implies that the switches $(s_0, s_1, \ldots, s_{n-2})$ and $(s_0, \bar{s}_1, \ldots, s_{n-2})$ are connected using the auxiliary links to form a loop. The switches which are connected by auxiliary links are called *associate* switches. Since the loops in this network contain two switches, it is called ASEN-2.

In the final stage (stage $n-1$) no conjugate switches are defined. Hence no auxiliary links can be added in stage $n-2$. This is the reason why the 2×2 switches in that stage are not replaced. The ASEN-2 for $N = 16$ is shown in Figure 25.4(a) and its redundancy graph is given in Figure 25.4(b). The demultiplexers at the output of a switch in stage $n-2$

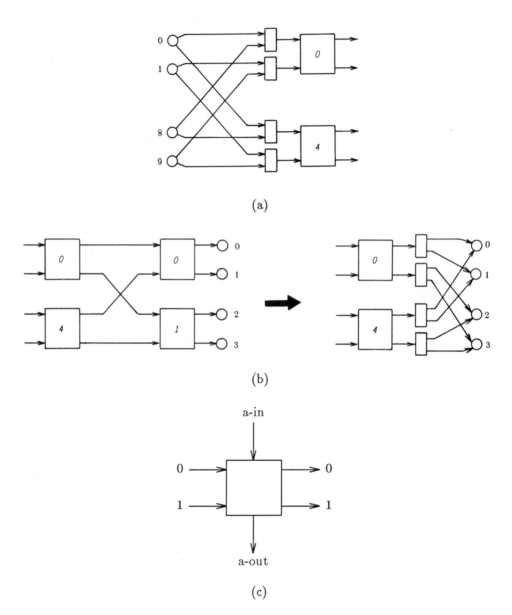

Figure 25.3: Construction of ASENs: Addition of multiplexers at the input stage (a), replacement of the switches in the last stage by 1×2 demultiplexers (b), and the 3×3 switch used to replace the original switches in stages 0 through $n - 3$ (c).

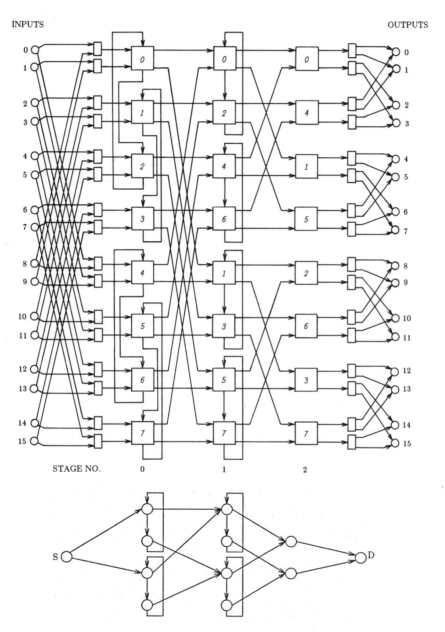

Figure 25.4: The ASEN-2 for $N = 16$ and its redundancy graph.

will usually be considered as part of that switch. Hence the redundancy graph in Figure 4(b) shows only three stages.

(b) The second scheme of connecting the auxiliary links involves partitioning each conjugate set $C_{i,j}$ into two subsets $A_{i,j}$ and $B_{i,j}$ such that no two switches belonging to a subset are first order conjugates. One way to do this is to place all the switches in $C_{i,j}$ with the most significant bit as 0 in $A_{i,j}$ and the rest in $B_{i,j}$. Then, connect the switches in $A_{i,j}$ and $B_{i,j}$, respectively, to form two loops. In stage $n-2$, each $C_{i,n-2}$ contains exactly two switches, so $A_{i,n-2}$ and $B_{i,n-2}$ contain one switch each. Hence the auxiliary links are not useful here, which explains the presence of 2×2 switches in stage $n-2$. Since the loops formed in this fashion contain the maximum number of switches subject to the constraints that the switches belong to the same conjugate set and that no two switches in a loop be first order conjugates, the resulting network is called ASEN-Max.

An ASEN-Max for $N = 16$ and its redundancy graph are given in Figure 25.5.

The following properties of ASEN-2 and ASEN-Max can be verified by inspecting Figures 25.4 and 25.5:

1. Both ASEN-2 and ASEN-Max are symmetric networks, i.e., the redundancy graph of a source-destination pair is isomorphic to that of any other source-destination pair.

2. The loops formed in a stage are such that for every loop there exists another loop which is connected to the same set of switches in the next stage. Such pairs of loops are called *conjugate loops*. As a result of the augmentation, for every switch in stage $n-2$ there exists another switch which is connected to the same subset of destinations (via the demultiplexers in stage $n-1$).

3. No two switches in a loop have the same successor switches in the next stage (i.e., no two switches in a loop are first order conjugate switches).

ASEN-2 and ASEN-Max are special cases of many possible networks that can be obtained by connecting switches in the same stage into loops, in that ASEN-2 has the minimum number of switches connected in loops in each stage, and ASEN-Max has the maximum number of switches in a loop in each stage, while satisfying property 2 given above. If satisfaction of property 2 is not required, it is possible to increase the loop size even further (up to twice the size of loops in ASEN-Max). Obviously several intermediate loop sizes are also possible. In the analytical results given in Sections 25.5 and 25.6 we will be considering ASENs with different loop sizes. An ASEN which has loop size 1 in stage $n-2$, 2 in stage $n-3$, and 4 in

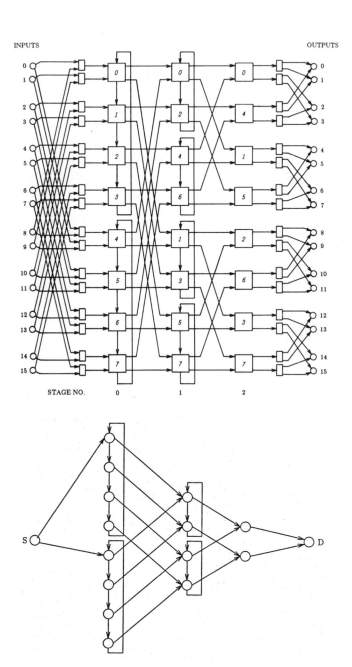

Figure 25.5: The ASEN-Max for $N = 16$ and its redundancy graph.

all the stages up to stage $n - 4$, will be denoted as ASEN-4. ASEN-8 denotes an ASEN which has a maximum loop size of 8. In general, the suffix following "ASEN" indicates the maximum loop size in the MIN.

The number of switches that can be connected in a loop, which can be left as a design parameter, could be constrained by physical design considerations. To take the best advantage of the structure of the ASENs and reduce I/O requirements of printed circuit boards (PCBs) or VLSI chips, one could place all the switches in a loop on the same board or chip. This would confine all the auxiliary links to be on the same board. If a loop consisting of L switches is placed on a PCB, then the board must have enough connectors for $2L$ input links and $2L$ output links. Thus, the number of switches one can connect in a loop may become dependent on the I/O capacity of the PCBs. Another advantage of implementing the loops as modules is manifested in the convenience of maintenance and repair of the networks. As will be shown in Section 25.4, the tolerance provided in the ASENs is not only for single switch faults, but also for faults affecting *all* the switches in a given loop. In other words, removal of a loop from the network need not disrupt the continued operation of the network. Thus, in the event that a switch becomes faulty, the loop containing the switch can be removed from the network and a replacement board inserted without interrupting the operation of the network. It may be economically attractive to have all the loops of the same size so that only one type of replacement board need be maintained. This may be an advantage of the ASEN-2 over other possible ASENs.

The augmentation of unique path MINs by adding links among switches of the same stage to create multiple paths was independently proposed by Tzeng, Yew and Zhu in [35]. Certain key differences between their networks and the ASENs are pointed out here. First the manner in which links are added in our scheme creates conjugate loops, which provide, in addition to fault-tolerance, easy repairability and maintainability. Second, the ASENs are designed to tolerate switch faults, as will be shown later, while the Chained Baseline Networks in [35] are primarily designed to tolerate link faults. The Chained Baseline Networks can tolerate single switch faults in all the stages except the final stage. Finally, the ASENs have slightly lower switch complexity since they have one stage less than the Chained Baseline Networks.

25.3.3   Generalized ASENs

The construction procedure given above could be regarded as deleting the last stage of a shuffle-exchange MIN, and then adding a multiplexer stage at the input, a demultiplexer stage at the output, and the auxiliary links to form loops which may contain a variable number of switches. But a shuffle-exchange MIN of size N and degree 2 with the last stage deleted is equivalent to two shuffle-exchange MINs of size

$N/2$. Thus there is an equivalent procedure which starts from two shuffle-exchange MINs of size $N/2$ to form an ASEN of size N. In general, it is possible to take m copies of a shuffle-exchange MIN of degree m, and connect sources and destinations to switches in each copy, and form m -tuples of conjugate loops. Recall that in a shuffle-exchange MIN of degree m, a set of first-order conjugate switches contains m elements. Thus each conjugate subset $C_{i,j}^k, 0 \leq k \leq m - 1$ (the superscript denotes the number of the subnetwork), can be partitioned into m subsets, such that no two first-order conjugate switches are contained in the same subset. These subsets can then be connected in a manner analogous to the one described for 2×2 networks to form m loops of switches which are conjugate. The switch size would be $(m + 1) \times (m + 1)$ if one auxiliary input link and one auxiliary output link is added to every switch. It is, however, possible to add $m' \leq (m - 1)$ auxiliary input and output links, increasing the switch size to $(m + m') \times (m + m')$. The additional links could be used to add *chordal links* [42] within each loop of switches. The degree of fault-tolerance in each case would be $(m - 1)$. However, if $m' < (m - 1)$ auxiliary links per switch exist, then up to m' faults can be avoided by dynamic rerouting, but tolerance of more than m' faults may necessitate backtracking.

Another interesting way to generalize ASEN-2 type networks is to consider a shuffle-exchange MIN of degree m, where m is an even number. Then, only two copies of the network, copy 0 and copy 1, can be used to form loops of size two, as follows (the superscripts denote the copy to which the switches belong): Connect the a-out link of the switch $SW_{i,j}^0 = (s_0, s_1, \ldots, s_{n-2})^0$ to the a-in link of the switch $SW_{k,j}^1 = (\hat{s}_0, s_1, \ldots, s_{n-2})^1$, where $\hat{s}_0 = s_0 + 1$, if s_0 is even and $\hat{s}_0 = s_0 - 1$, if s_0 is odd. The size of the switches in this type of network will also be $(m + 1) \times (m + 1)$.

In general, one could start from t copies of a shuffle-exchange MIN of size N and degree m, where t is a submultiple of m, and add $t' \leq (t - 1)$ auxiliary link pairs per switch and connect them in a manner similar to Step 3 in Section 25.3.2, to obtain an ASEN of size $N \cdot t$ and switch size $(m + t') \times (m + t')$. This would guarantee tolerance of up to t' faults without backtracking and up to $t - 1$ faults with backtracking.

Extension of the augmentation procedure to make it applicable for Generalized Shuffle Networks [43] is quite straight-forward and hence is not included here.

25.4   Fault Tolerant Routing

Although the constructions for many possible ASENs were given in the previous section, in the rest of the paper we will study in detail the ASENs constructed from shuffle-exchange MINs of degree 2, such as ASEN-2, ASEN-Max, ASEN-4 etc. In this section we give a routing scheme for such ASENs. The routing algorithm is adaptive and inherently fault-tolerant - it will use a preferred path when the network is fault-free, and also take corrective action, if possible, as and when faulty switches or links are encountered. To use the routing schemes given here, it is essential for sources and switching elements to be able to detect faults in the switches to which they are connected. Several techniques for detecting faults in multistage networks have been reported [44-46]. Such techniques, however, require off-line application of test inputs and/or make the test-results available only to the external resources attached to the network. Since fault information is required at the individual switching elements, as and when faults occur, these methods are not adequate for the proposed multipath MINs. A design of switching elements which enables a switching element to detect faults in an adjoining switching element, concurrently with normal operation, is given in [47]. Thus in the following we assume that a switch is aware of the status (faulty or not faulty) of the other switches it is connected to.

As pointed out in Section 25.1, multipath MINs are used for increased reliability as well as throughput (or bandwidth). The throughput is increased by choosing an alternate path when a chosen path is blocked due to the required switch output being used in servicing another request. The need to choose an alternate path also arises when certain switches or links in the MIN are faulty. In most MINs the routing algorithms can be designed such that rerouting in the presence of blocking due to busy switches/links and due to faulty switches/links can be treated equivalently. This is true with the routing algorithm presented next. However in the interest of brevity we present these algorithms in terms of rerouting in the presence of switch failures. The procedure would be able to handle rerouting due to blocking by busy switches and failed links identically. Analysis of network effectiveness (e.g., fault-tolerance, terminal reliability, throughput, etc.) would however be different and depend on the particular parameter of interest. For example, in evaluating the fault-tolerance of the MIN the point of interest is the success or failure of the routing algorithm in the presence of failed switches or links, whereas in calculating the throughput of the fault-free MINs the point of interest is how the routing algorithm succeeds in servicing a request for connection in the presence of other requests.

The routing algorithms for ASENs rely upon the digit-routable property of the delta networks [8]. Briefly, digit routing in a delta network is accomplished by the source attaching the destination label $(d_0, d_1, ..., d_{n-1})$ as the routing tag along with a request for connection. The request progresses through the stages of the delta network as the switch in stage i uses the digit d_i, from the destination tag, to route the incoming request via the output link with the label d_i. The digit-routability property of delta networks guarantees that the request reaches the correct destination. For ASEN, this routing scheme is modified to make use of the additional links available for tolerating faults and for reducing blocking along a path.

The routing scheme, being distributed in nature, is described in terms of the actions taken by the sources of the messages and by the switches in different stages.

Algorithm **ASEN_ROUTE**

1. For each source S: To request a connection with a destination, submit its label $(d_0, d_1, \ldots, d_{n-2}, d_{n-1})$ as the routing tag, to the primary switch. If the primary switch is faulty, submit the request and routing tag to the secondary switch.

2. For each switch in stage $i, 0 \leq i \leq n - 3$: For a request received on the input link 0, 1, or a-in, examine bit d_i of the destination tag. Route the request via the output link labeled d_i if the corresponding successor switch is fault-free. If the successor switch is faulty, then route the request via the a-out link to the associate switch.

3. For each switch in stage $n - 2$: For a request received on the input link labeled 0 or 1, examine the bit d_{n-2} and route via the output link labeled d_{n-2}.

4. For each demultiplexer in stage $n - 1$: If a request is received, route it by the upper output link if bit d_{n-1} is a 0, and by the lower output link if the bit is a 1.

Given the digit-routability of shuffle-exchange MINs, and the construction procedure of the previous section, it is easy to verify that this routing algorithm in fact makes a connection to the correct destination. In addition, since there are at least two choices to route at each step, except in stage $n - 2$ (where it is assumed that a fault in a demultiplexer at the output of a switch is a fault in that switch, and that the destinations are fault-free) it is clear that the routing procedure delivers a request from a source to any required destination in the presence of single switch failures. This claim is formally stated below.

Theorem 1: In the absence of any conflicting requests, Algorithm ASEN_ROUTE delivers a request generated by a source to the correct destination tolerating any single switch fault.

Quite often, the criterion applied to determine whether a MIN is faulty or not is that there be at least one fault-free path from every source to every destination. We now examine the robustness of ASEN-2, and the adaptive routing algorithm given above, with respect to this criterion. We first characterize the switch faults that ASEN-2 is guaranteed to survive, and the faults that will always cause the network to fail. It is clear that a minimum of two switch faults can disrupt all available paths between some source-destination pair in ASEN-2. For example, if both the switches that a source or destination is connected to become faulty, then that source or destination becomes disconnected from the rest of the network. However, if such *critical faults* do not occur, several switch faults can be tolerated. Consider the redundancy graph of ASEN-2. We have the following theorem characterizing certain multiple switch faults that are always tolerated in ASEN-2.

Theorem 2: If the faults occurring in the ASEN-2 are such that they affect switches in at most one loop in every pair of conjugate loops, and at most one switch in every pair of switches $SW_{i,n-2}$ and $CONJ(SW_{i,n-2})$ (a fault in a demultiplexer at the output of a switch in stage $n-2$ is considered as a fault in that switch), then there exists at least one path from every source to every destination of the network.

It may be remarked here that in any ASEN, regardless of the number of switches in a loop, if in every pair of conjugate loops at least one loop is free from faults, then the redundancy graph of every source-destination pair remains connected. The reader can convince himself of this by examining the redundancy graphs. This is the reason why ASENs lend themselves to easy repair and maintenance. When a switch is detected to be faulty, the loop of switches containing the faulty switch can be removed from the network and a replacement plugged in, without having to disrupt the continued operation of the network.

There are certain types of multiple faults in ASEN-2 that will always cause some source to be disconnected from some destination and these are characterized by the following theorem.

Theorem 3: In ASEN-2, if the switches $SW_{i,j}$ and $CONJ(SW_{i,j})$ are faulty at the same time, then some source is disconnected from some destination.

The proof of this theorem follows from the fact that in the redundancy graph of ASEN-2, in every stage there are two switches which act as the gateways to that stage. These two switches are, in fact, first-order conjugate switches. Moreover, every pair of conjugate switches occupy these gateway positions in some redundancy graph. Thus if both the switches in some conjugate pair become faulty, then some redundancy graph will be disconnected.

From Theorem 2, note that the maximum number of switch faults that can be tolerated in ASEN-2 is at least half of the total number of switches, i.e., $(N/4)(\log_2 N - 1)$. Further, by Theorem 3, at most half of the switches in ASEN-2 are permitted to be faulty, for otherwise both the switches will become faulty in some conjugate pair. Thus the maximum number of switch faults tolerated is $(N/4)(\log_2 N - 1)$.

The fault-tolerance of ASENs of larger loop sizes is bounded from below by that of ASEN-2.

Theorem 4: If under a combination of multiple switch faults, ASEN-2 remains connected, then an ASEN containing loops of larger size will also remain connected under the same combination of faults.

Once again, this can be verified from the redundancy graphs.

25.5   Reliability of the ASENs

In this section we give probabilistic measures of the effect of faults on the operation of ASENs, and examine the effectiveness of the multiple paths in improving the reliability of the MINs.

A general criterion for the evaluation of the robustness of a MIN is that every member of a subset of sources must have paths to every member of a subset of destinations. The probability that the above criterion is satisfied is called *multiterminal reliability* [26]. Two special cases of this criterion are usually considered. The first case is that of the subsets of sources and destinations containing exactly one element each. This special case leads to a measure called *two-terminal reliability*, or simply *terminal reliability*, which is the probability that a given source-destination pair has at least one fault-free path between them, given that each switch has a certain reliability (the *reliability* of a switch is the probability that it is fault-free). The terminal reliability of a MIN can be evaluated using its redundancy graph. The other special case of the multiterminal reliability criterion is that of the subsets of sources and destinations containing all the sources and all the destinations, respectively. This special case leads to the assumption that the MIN is faulty whenever all the paths are disconnected between some source-destination pair, and gives us the *reliability of the MIN*. If the reliability is time-variant one can obtain the *Mean Time To Failure (MTTF)* of the MIN, which is the expected time elapsed before some source is disconnected from some destination.

As the loop size in an ASEN increases, the number of alternate paths between a source-destination pair increases. Thus one might expect an increase in the terminal reliability with increasing loop size. However, the terminal reliability reaches a saturation after a loop size of four, with the measure being practically equal for ASEN-4 and ASEN-Max [47]. The terminal reliabilities of ASEN-2 and ASEN-4 are given in Table 25.2, assuming that the reliability p of a switch is 0.9. The terminal reliabilities of several other multiple path MINs are also given in Table 25.2, again assuming that $p = 0.9$. It must be noted here that although the switches of different MINs have different complexities, if they are implemented in VLSI, their failure probabilities will not be significantly different for the purposes of the terminal reliability evaluation. Also note that, ACN has higher terminal reliability than the ASENs, but it also has a higher hardware complexity (Table 25.1).

To evaluate the mean time to failure some simple series-parallel reliability models for ASENs can be constructed based on Theorems 2 and 3[47,48]. These models enable us to give an optimistic estimate and a pessimistic estimate for the MTTF

Table 25.2: The terminal reliabilities of some multiple path MINs.

No. of	Terminal Reliability ($p = 0.9$)						
Inputs	ASEN-2	ASEN-4	ACN	Indra ($R=2$)	Gamma†	Mod. Omega	2×2 Delta
8	0.971	0.971	0.980	0.963	0.875	0.810	0.729
16	0.953	0.960	0.970	0.946	0.859	-	0.656
32	0.935	0.949	0.961	0.925	0.844	0.802	0.590
64	0.917	0.939	0.951	0.900	0.830	-	0.531
128	0.899	0.929	0.941	0.869	0.815	0.781	0.478
256	0.882	0.919	0.932	0.835	0.801	-	0.430
512	0.866	0.909	0.923	0.798	0.787	0.751	0.387
1024	0.849	0.899	0.914	0.758	0.773	-	0.345

† Maximum

of ASEN-2, which turn out to be very close. These estimates for ASEN-2, and the MTTF figures known for other multiple path MINs, are given in Table 25.3. We have assumed that the failure rate λ is the same for the switches in all the MINs included in the table. This is an approximation. However, as mentioned before, the failure rates of switching elements implemented in VLSI increase less than linearly, perhaps logarithmically, with the size of the switching element. Hence the approximation does not result in a serious error. Also, a reasonable estimate for λ is about 10^{-6} per hour.

Table 25.3: The Mean Time To Failure ($MTTF$) of some multipath MINs.

No. of	Mean Time To Failure ($\times \lambda^{-1}$)						
Inputs	ASEN-2*	ASEN-2**	ACN	Indra ($R=2$)	F-Net	Mod. Omega	2×2 Delta
8	0.513	0.582	0.582	0.110	0.300	0.250	0.083
16	0.246	0.300	0.300	4.31×10^{-2}	0.178	-	3.125×10^{-2}
32	0.136	0.173	0.173	1.77×10^{-2}	0.110	5.38×10^{-2}	1.25×10^{-2}
64	0.081	0.106	0.106	7.47×10^{-3}	6.99×10^{-2}	-	5.21×10^{-3}
128	0.050	0.067	0.067	3.23×10^{-3}	4.52×10^{-2}	1.30×10^{-2}	2.23×10^{-3}
256	0.032	0.043	0.043	1.42×10^{-3}	2.96×10^{-2}	-	9.77×10^{-4}
512	0.021	0.028	0.028	6.36×10^{-4}	1.96×10^{-2}	2.9×10^{-3}	4.34×10^{-4}
1024	0.014	0.019	0.019	2.87×10^{-4}	1.30×10^{-2}	-	1.95×10^{-4}

*   Pessimistic estimate

**   Optimistic estimate

Table 25.4: Circuit switching performance of multipath MINs: the probability of acceptance for request generation probability 1.

Size, N	ASEN-4	ACN	Indra $(R=2)$	F-net	Mod. Omega	Delta	Crossbar
8	0.597	0.656	0.580	0.527	0.597	0.517	0.656
16	0.556	0.634	0.539	0.432	-	0.450	0.644
32	0.527	0.615	0.503	0.367	0.476	0.399	0.638
64	0.506	0.597	0.472	0.320	-	0.359	0.635
128	0.488	0.581	0.444	0.283	0.397	0.327	0.634
256	0.473	0.566	0.420	0.255	-	0.300	0.633
512	0.460	0.552	0.398	0.231	0.342	0.278	0.632
1024	0.449	0.539	0.378	0.212	-	0.259	0.632

25.6 Performance of ASENs

The improvement in the performance of ASENs in a circuit-switching environment due to the availability of multiple paths is discussed this section. The *probability of acceptance* of a request submitted by a source is usually used as a measure of the performance of a circuit-switching MIN. This is the probability that a request submitted by a source reaches the required destination without getting blocked in the network, given that each source generates a request with a certain probability and aims at the destinations with equal probability. In addition to this measure, we also consider the path length distribution in ASENs. This is important since the probability of acceptance is improved by using the additional paths which have a more length. If the increase in the path length is excessive, it could mean more latency in setting up connections. Fortunately, that is not the case.

Analysis and simulation show that the probability of acceptance in ASEN-4 is noticeably higher than in ASEN-2. But the figures for ASEN-Max are practically the same as those of ASEN-4 [47]. Even though ASEN-Max has more paths than ASEN-4, the additional paths are also longer, and the probability of successfully setting up a connection by using longer path lengths rapidly diminishes.

In Table 25.4 we have given the probability of acceptance of requests for ASEN-2, ASEN-4 and other multipath MINs, assuming that sources generate requests with probability 1. These figures have been derived either from the references cited, or by making calculations similar to the ones made for the ASENs. The figures for the crossbar switch and the unique path delta network are also included for comparison. Once again, it is noted here that ACNs have better performance but they are also more expensive.

The probability distribution of the path length of a successful connection in the ASEN-2 with $N = 2^{10}$, with the sources generating requests with probability 1, is given in Figure 25.6. As can be seen, the probability distribution is weighted towards shorter path lengths indicating that the path lengths do not increase excessively.

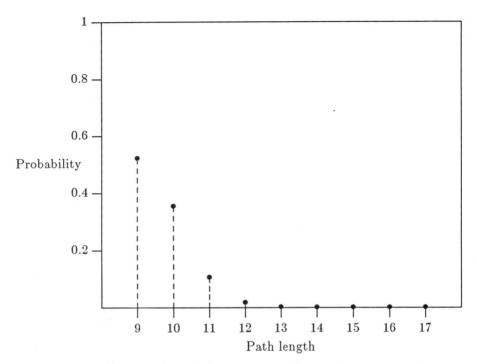

Figure 25.6: The probability distribution of the path-length of a successful connection in an ASEN-2 of size 1024, for request generation probability 1.

Thus the price paid for the increase in acceptance probability does not seem to be unacceptably high.

25.7   Conclusions

We have presented a class of fault-tolerant multistage interconnection networks, called Augmented Shuffle-Exchange MINs (ASENs), featuring links among switches belonging to the same stage. These additional links are used to form loops of switches in each stage, which result in multiple paths for fault-tolerance and improved performance. To gain further insight into the theory underlying the proposed networks, one may note that while unique path MINs such as shuffle-exchange MINs or banyan networks may be regarded as superposed binary trees, the ASENs can be regarded as superposed X-Trees [49], i.e., binary trees with links added between nodes on the same level.

Reliability and performance studies have shown that the proposed networks are quite effective. The figures show not only that the improvement over unique path

MINs is impressive, but also that ASENs compare favorably with other multipath MINs. Further analysis [50] has shown that because of the on-line repairability, the reliability of ASENs with repair of faulty loops is dramatically increased. Analytical evaluation [47] also shows that increasing loop size of ASENs has the effect of improving the terminal reliability and the circuit-switching performance up to a point, but beyond that the improvement is negligible. Further, the saturation is reached fairly quickly, at about loop size of four. Since having large number of switches in the loops may entail certain disadvantages in the physical design of these MINs, we recommend loop sizes of no more than four or eight.

In conclusion, ASENs achieve fault-tolerance, high reliability, and good performance while retaining most of the basic advantages of unique path shuffle-exchange MINs. Given the importance of high reliability and high performance in large scale multi-processor systems, ASENs appear to be good candidates for use in such systems.

References

[1] V.E. Benes, *Mathematical Theory of Connecting Networks and Telephone Traffic*, Academic Press, New York, NY, 1965.

[2] T.Y. Feng, "A Survey of Interconnection Networks," *Computer*, Vol. 14, No. 12, December 1981, pp. 12-27.

[3] G.M Masson, G.C. Gingher, Shinji Nakamura, "A Sampler of Circuit Switching Networks," *Computer* June 1979, pp. 32-48.

[4] H.J. Siegel, "Interconnection Machines for SIMD Machines," *Computer* June 1979, pp. 57-66.

[5] L.R. Goke, G.J. Lipovski, "Banyan Networks for Partitioning Multiprocessor Systems," *Proc. 1st Annual Symposium on Computer Architecture*, December 1971, pp. 21-28.

[6] D.H. Lawrie, "Access and Alignment of Data in an Array Processor," *IEEE Trans. Comp.*, Vol. C-24, December 1975, pp. 1145-1155.

[7] M.C. Pease, "The Indirect Binary n-Cube Microprocessor Array," *IEEE Trans. Comp.*, Vol. C-26, May 1977, pp. 458-473.

[8] J.H. Patel, "Performance of Processor-Memory Interconnections for Multiprocessors," *IEEE Trans. Comp.*, October 1981, pp. 771-780.

[9] C. Wu, T. Feng, "On a Class of Multistage Interconnection Networks," *IEEE Trans. Comp.*, Vol. C-29, August 1980, pp. 696-702.

[10] K.E. Batcher, "The Flip Network in STARAN," *1976 Intl. Conf. Parallel Processing,* August 1976, pp. 65-71.

[11] H.J. Siegel, R.J. McMillen, "The Multistage Cube: A Versatile Interconnection Network," *Computer,* Vol. 14, No. 12, December 1981, pp. 65-76.

[12] G.H Barnes, S.F. Lundstrom, "Design and Validation of a Connection Network for Many-Processor Multiprocessor Systems," *Computer,* December 1981, pp. 31-41.

[13] U.V. Premkumar, R. Kapur, M. Malek, G.J. Lipovski, and P. Horne, "Design and Implementation of the Banyan Interconnection Network in TRAC," *AFIPS 1980 Nat'l. Computer Conf.,* June 1980, pp. 643-653.

[14] H.J. Siegel, L.J. Siegel, F.C. Kemmerer, P.T. Mueller,Jr., H.E. Smalley, Jr., and S.D. Smith, "PASM: A Partitionable SIMD/MIMD System for Image Processing and Pattern Recognition," *IEEE Trans. Computers,* Vol. C-30, December 1981, pp. 934-947.

[15] A. Gottlieb, R. Grishman, C.P. Kruskal, K.P. McAuliffe, L. Rudolph, and M. Snir, "The NYU Ultracomputer - Designing an MIMD Shared Memory Parallel Computer," *IEEE Trans. Computers,* Vol. C-32, February 1983, pp. 175-189.

[16] D. Gajski, D. Kuck, D. Lawrie and A. Sameh, "Cedar - A Large Scale Multiprocessor," *1983 Intl. Conf. Parallel Processing,* August 1983, pp. 524-529.

[17] W. Crowther, J. Goodhue, E. Starr, R. Thomas, W. Milliken, T. Blackadar, "Performance Measurements on a 128-Node Butterfly Parallel Processor," *1985 Intl. Conf. Parallel Processing,* August 1985, pp. 531-540.

[18] G.F. Pfister, W.C. Brantley, D.A. George, S.L. Harvey, W.J. Kleinfelder, K.P. McAuliffe, E.A. Melton, V.A. Norton and J. Weiss, "The IBM Research Parallel Processor Prototype (RP3): Introduction and Architecture," *1985 Intl. Conf. Parallel Processing,* August 1985, pp. 764-771.

[19] K.M. Falavarjani and D.K. Pradhan, "A Design of Fault-Tolerant Interconnection Networks," Unpublished Memo, 1981.

[20] G.B. Adams, H.J. Siegel, "The Extra Stage Cube: A Fault-Tolerant Interconnection Network for Supersystems," *IEEE Trans. Comp.,* Vol. 31, May 1982, pp. 443-454.

[21] D.M. Dias and J.R. Jump, "Augmented and Pruned N log N Multistage Networks," *1982 Intl.Conf. Parallel Processing,* August 1985, pp. 10-11.

[22] R.J. McMillen, H.J. Siegel, "Routing Schemes for the Augmented Data Manipulator Network in an MIMD System," *IEEE Trans. Comp.,* Vol. C-31, December 1982, pp. 1202-1214.

[23] H.J. Siegel, R.J. McMillen, "Dynamic Rerouting Tag Schemes for the Augmented Data Manipulator Network," *8th Intl. Symp. on Computer Architecture,* May 1981, pp 505-516.

[24] R.J. McMillen, H.J. Siegel, "Performance and Fault-Tolerance Improvements in the Inverse Augmented Data Manipulator Network," *Proc. 9th Annual Symposium on Computer Architecture,* April 1982, pp. 63-72.

[25] D.S. Parker, C.S. Raghavendra, "The Gamma Network: A Multiprocessor Interconnection Network with Redundant Paths," *Proc. 9th Annual Symposium on Computer Architecture,* June 1982, pp. 73-80.

[26] C.S. Raghavendra, D.S. Parker, "Reliability Analysis of an Interconnection Network," *Proc. 4th International Conference on Distributed Computing Systems,* May 1984, pp. 461-471.

[27] L. Ciminiera, A. Serra, "A Fault-Tolerant Connecting Network for Multiprocessor Systems," *Proc. of the 1982 International Conference in Parallel Processing,* August 1982, pp. 113-122.

[28] K. Padmanabhan, D.H. Lawrie, "Fault-Tolerance Schemes in Shuffle-Exchange Type Interconnection Networks," *Proc. of the 1983 International Conference on Parallel Processing,* August 1983, pp. 71-75.

[29] K. Padmanabhan, D.H. Lawrie, "A Class of Redundant Path Multistage Interconnection Networks," *IEEE Trans. Comp.,* Vol. C-32, December 1983, pp. 1099-1108.

[30] K. Padmanabhan, "Fault Tolerance and Performance Improvement in Multiprocessor Interconnection Networks," Ph.D. Thesis, Dept. of Comp. Science, Univ. of Illinois, Urbana-Champaign, May 1984.

[31] S.M. Reddy, V.P. Kumar, "On Fault-Tolerant Multistage Interconnection Networks," *Proc. of the 1984 International Conference of Parallel Processing,* August 1984, pp. 155-164.

[32] C.S. Raghavendra, A. Varma, "INDRA: A Class of Interconnection Networks with Redundant Paths," *1984 Real Time Systems Symposium,* December 1984.

[33] V. Cherkassky, E. Opper, and M. Malek, "Reliability and Fault Diagnosis Analysis of Fault Tolerant Multistage Interconnection Networks," *14th Intl. Symp. Fault-Tolerant Computing,* June 1984, pp. 246-251.

[34] V.P. Kumar and S.M. Reddy, "Design and Analysis of Fault-Tolerant Multistage Interconnection Networks With Low Link Complexity," *12th Intl. Symp. Computer Architecture,* June 1985, pp.376-386.

[35] N. Tzeng, P. Yew, C. Zhu, "A Fault-Tolerant Scheme for Multistage Interconnection Networks," *12th Intl. Symp. Computer Architecture*, June 1985, pp. 368-375.

[36] M. Jeng and H.J. Siegel, "A Fault-Tolerant Multistage Interconnection Network for Multiprocessor Systems Using Dynamic Redundancy," *6th Intl. Conf. Distributed Computing Systems*, May 1986, pp. 70-77.

[37] G.B. Adams III and H.J. Siegel, "A Survey and Comparison of Fault-Tolerant Multistage Networks," *Computer*, June 1987, pp. 14-27.

[38] V.P. Kumar and S.M. Reddy, "Augmented Shuffle-Exchange Multistage Interconnection Networks," *Computer* June 1987, pp. 30-40.

[39] C.-T. A. Lea, "A Load-Sharing Banyan Network," *1985 Intl. Conf. Parallel Processing*, August 1985, pp. 317-324.

[40] D.S. Wise, "Compact Layout of Banyan/FFT Networks," *Proc. CMU Conf. VLSI Systems and Computations*, Computer Science Press (1981), pp. 186-195.

[41] M.A. Franklin, "VLSI Performance Comparison of Banyan and Crossbar Communication Networks," *IEEE Trans. Computers*, vol C-30, April 1981, pp. 283-290.

[42] B.W. Arden and H. Lee, "Analysis of Chordal Ring Network," *IEEE Trans. Computers*, Vol. C-30, April 1981, pp. 291-295.

[43] L.N. Bhuyan, D.P. Agrawal, "Design and Performance of Generalized Interconnection Networks," *IEEE Trans. Comp.*, Vol. C-32, December 1983, pp. 1081-1090.

[44] T.-y. Feng and C.-l. Wu, "Fault Diagnosis for a Class of Multistage Interconnection Networks," *IEEE Trans. Computers*, Vol C-30, October 1981, pp. 743-758.

[45] D.P. Agrawal, "Testing and Fault-Tolerance of Multistage Interconnection Networks," *Computer*, April 1982, pp. 41-53.

[46] W.K. Fuchs, J.A. Abraham, and K.-H. Huang, "Concurrent Error Detection in VLSI Interconnection Networks," *10th Intl. Symp. Computer Architecture* June 1983, pp. 309-315.

[47] V.P. Kumar, *On Highly Reliable, High Performance Multistage Interconnection Networks*, Ph.D. Thesis, University of Iowa, December 1985.

[48] K.S. Trivedi, *Probability and Statistics with Reliability, Queueing and Computer Science Applications* Prentice-Hall, Englewood Cliffs, N.J., 1982.

[49] A.M. Despain and D.A. Patterson, "X-Tree: A Tree-Structured Multiprocessor Computer Architecture," *5th Ann. Symp. Computer Architecture*, April 1978, pp. 144-151.

[50] J.T. Blake, *Comparative Analysis of Multistage Interconnection Networks*, Ph.D. Thesis, Duke University, 1987.

Chapter 26

Analyzing the Connectivity and Bandwidth of Multiprocessors with Multi-stage Interconnection Networks [1]

Israel Koren [2]
Zahava Koren [3]

Abstract

When implementing multi-processing systems consisting of a large number of processors, memory modules and interconnection switches, we must expect some of the system elements to become faulty. These faults may be the result of manufacturing defects or failures that occur while the system is already in operation. If the system is allowed to continue its operation in the presence of a few faulty elements, we need to predict the performance of the degraded system.

In this paper, we analyze the performance of multi-processor systems with a multistage interconnection network in the presence of faulty elements. We propose the use of two measures for performance, namely, bandwidth and connectivity. We then derive expressions for these measures for a non-redundant system and for a system with redundancy in its interconnection network. Finally, we compare the two systems through some numerical examples.

---

[1] This work was supported in part by NSF under contract DCR–85–09423.
[2] University of Massachusetts, Amherst, MA. On leave from the Technion, Haifa, Israel
[3] University of Massachusetts, Amherst, MA

26.1   Introduction

Recent advances in VLSI technology and development of new computer-aided design tools like silicon compilers, enable the design and implementation of multi-processing systems consisting of a very large number of processors. One important class of these multi-processing systems includes the shared-memory multi-processors where all processors can access a set of memory modules through an interconnection network. This interconnection network can be a crossbar network, a multiple bus network or a multi-stage interconnection network.

When implementing a complex multi-processor, some of its elements (like processors, memory modules or interconnection switches) are expected to become faulty. These faults may be the result of manufacturing defects or failures that occur while the system is already in operation. In many cases the faulty elements can not be immediately repaired or replaced and we would still like to use the system at a degraded rate of performance until a repair and/or replacement can take place.

An important question then is how well the gracefully degrading system performs in the presence of faults. This question may arise in different situations. In one case faulty elements can be neither repaired nor replaced. An example might be a multi-processor integrated on a small number of large area VLSI chips or even wafers where some defective elements may be present (e.g., [4]). A different situation is when the faulty elements can be replaced but the mean time to repair or replacement is considered to be too long. An example might be a real-time computing system where even a relatively short down-time period may be intolerable.

In this paper we attempt to answer the above question for one type of shared-memory multi-processors, namely, those interconnected through a multi-stage network.

Multi-stage interconnection networks were proposed as a cost-effective alternative to the expensive crossbar networks. However, these networks are inherently very sensitive to failures of any kind. They usually provide a unique path between any processor and any memory module and therefore, a single fault in any internal switch or link will render some memories unreachable from certain processors. Several schemes for introducing fault-tolerance into the architecture of these interconnection networks have been suggested in recent publications (e.g., [1], [3], [5], [6] and [8]). A survey of these and other schemes is presented in [2].

Most of the proposed schemes provide redundant paths between every source/destination pair so that all single faults and many multiple faults can be tolerated. This is achieved by augmenting an existing topology either by adding an extra stage of switches (e.g., [1]), or by augmenting the switching elements and their interconnections (e.g., [5], [6] and [8]). The multiple disjoint paths provided by these schemes may in some cases also increase the performance of the network. For example, it has been shown in [8] that the bandwidth of their proposed fault-tolerant multi-stage network is comparable even to that of a crossbar network.

In this paper we examine the performance of multi-stage multi-processors (with and without redundancy) in the presence of faults. We also propose objective functions for measuring the performance of these systems. In the next section we present the proposed performance measures and introduce several basic assumptions and notations. In Section 26.3 the non-redundant network is analyzed. A similar analysis is then repeated in Section 26.4 for a network with built-in redundancy, namely, the Extra Stage Cube network [1]. These networks are then compared through some numerical examples in Section 26.5. Final conclusions are presented in Section 26.6.

26.2 Preliminaries and Notations

Consider N processors (where $N = 2^k$) connected to N memories through a multi-stage interconnection network designed out of 2×2 switches. Our analysis can be generalized to the case where the number of processors is not necessarily a power of 2, the number of memories is different from the number of processors and finally, the network is built from $a \times b$ switches. For the sake of clarity and brevity however, we restrict our discussion here to the above mentioned simpler case.

An $N \times N$ interconnection network with no redundancy is constructed of $k = log N$ stages, each containing $N/2$ switches as illustrated in Fig. 26.1. Redundant networks have a larger number of stages and/or more switches per stage and/or use more complex switches. Non-redundant networks and some networks with internal redundancy have been previously analyzed but in most previous studies it has been assumed that faults can occur only in the interconnection network while the processors and memories are assumed to be fault-free. Clearly, this assumption is invalid in our environment and we have to consider faulty processors and faulty memories in addition to faults in the interconnection network.

Let q_r denote the probability that a processor is faulty at some given time instant t and let $p_r = 1 - q_r$ denote the probability of a fault-free processor. If $t = 0$ then q_r is the probability that manufacturing defects have occurred in the processor. If $t > 0$ then q_r is the probability that the processor either had defects at $t = 0$ or became faulty later on. Similarly, we denote by q_m (p_m) the probability of a faulty (fault-free) memory and by q_l (p_l) the probability of a faulty (fault-free) link. Our fault model for the interconnection network is the link fault model (e.g., [2]), however, we allow multiple link faults so that switch faults are covered as well.

We are interested in comparing the performance of alternative architectures for multi-stage networks in this environment where processors, memories and links can fail. The performance measure to be used should capture the capabilities of the

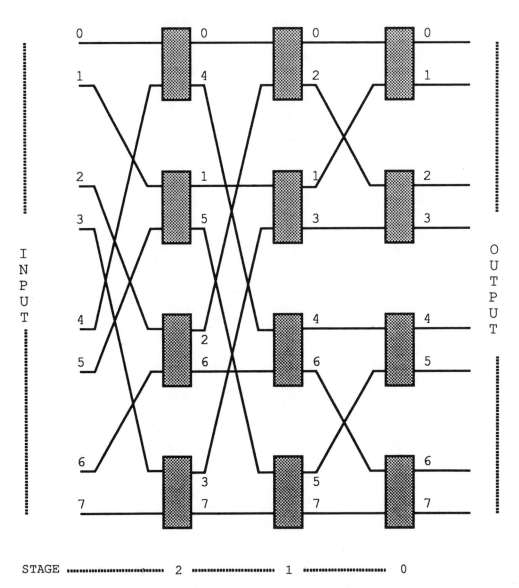

Figure 26.1: An 8 × 8 non-redundant interconnection network.

gracefully degrading multi-stage system, e.g., the number of fault-free processors and memories, the number of fault-free paths within the interconnection network, and the number of memory requests (from the processors) that can be transmitted through the interconnection network simultaneously.

A commonly used measure for the performance of an interconnection network is the *bandwidth* [7]. The bandwidth is defined as the expected number of requests for the shared memory which are accepted per cycle, given that each processor generates, with probability p_a, a request during a cycle, and that a request blocked at any stage is lost. The bandwidth measures the effect of blocking which results from the fact that in multi-stage interconnection networks paths are shared by two or more processor-memory pairs. A processor-memory connection can be blocked by a previously established connection even if the memories involved are distinct.

The bandwidth measure is concerned only with the interconnection network and is therefore, insufficient for our purposes. It is strongly affected by the amount of traffic in the network (i.e., the probability of request p_a) and hence, it provides only limited insight regarding the *connectivity* of the multi-processing system, i.e., the number of fault-free processors which are still connected to some fault-free memories. System connectivity can be the basis for measuring the processing power of the system or its computational availability. In what follows we present two measures for connectivity to be used in addition to the bandwidth measure.

One proposed measure for connectivity is C - the average number of operational processor-to-memory paths where both processor and memory are fault-free. A shortcoming of this objective function is that it provides no indication on how many *distinct* processors and memories are still connected. We propose therefore, as an additional measure for connectivity, the tuple (N_r, N_m) where N_r denotes the average number of fault-free processors which are connected to at least one fault-free memory module and N_m is similarly defined for memories. Note that this tuple does not necessarily imply that a complete fault-free $N_r \times N_m$ interconnection network exists.

The bandwidth and the two connectivity measures, namely, C and the tuple (N_r, N_m), can together adequately characterize the capabilities of a multi-stage multi-processor in the presence of faulty elements.

26.3 A Non-redundant Interconnection Network

In this section we analyze a non-redundant interconnection network and derive expressions for its bandwidth and the different connectivity measures as previously defined. An analysis of the bandwidth for a fault-free network appears in [7]. In what follows we generalize it to an environment in which faults may be present.

We adopt here the simplifying assumption that the destinations of the memory requests are independent and uniformly distributed among the N memories. Therefore, the network bandwidth can be obtained by multiplying the number of memories
N by the probability that a given memory module is non-faulty and has a request
at its input. This last probability is calculated iteratively, following a path leading
to this memory , i.e, the probability of a request on an output link of a switch is
calculated from the probability that such a request has been accepted at the input
links to the same switch.

To simplify our discussion we say that a link is in state 1 (0) if it has (has not) a
request for the memory. A faulty link is considered to be in state 0. The probability
of a request on a link is thus the probability that this link is in state 1. We assign
numbers to the $k = logN$ stages in a descending order so that stage 0 is the last
stage and its output links are connected to the memories, stage $(k-1)$ is the first
one and its inputs are connected to the processors (see Fig. 26.1). Consider a
switch in stage i and denote its outputs $X^{(i)}, Y^{(i)}$. Its input links are the outputs
of (two different) switches in stage $(i+1)$ and are denoted by $X^{(i+1)}$ and $Y^{(i+1)}$.
Based on our assumption that memory requests are uniformly distributed among
the memories, the probability that an incoming request will be routed to any output
link is the same. Hence, it is sufficient to consider only a single output link and
derive an expression for the probability that it is at state 1, i.e., $P\{X^{(i)} = 1\}$.

Since a request for a memory module can reach the output link of a switch through
any of the two input links, the state probability $P\{X^{(i)} = 1\}$ of the given output
link has to be calculated from the joint probabilities of these input links, i.e.,

$$P\{(X^{(i+1)}, Y^{(i+1)}) = (u, v)\}; \quad u, v = 0, 1.$$

This calculation is performed using transition probabilities which take into account
the status (faulty or fault-free) of the (physical) input links and the destinations of
the incoming requests. Since memory modules are assumed to be equivalent, the
incoming requests are routed to any of the two output links with probability 0.5.
Consequently, the transition probabilities between the two inputs and the output
of a switch are,

$$
\begin{aligned}
&P\{X^{(i)} = 1 \, / \, (X^{(i+1)}, Y^{(i+1)}) = (0,0)\} = 0 \\
&P\{X^{(i)} = 1 \, / \, (X^{(i+1)}, Y^{(i+1)}) = (0,1)\} = \tfrac{1}{2} \, p_l \\
&P\{X^{(i)} = 1 \, / \, (X^{(i+1)}, Y^{(i+1)}) = (1,0)\} = \tfrac{1}{2} \, p_l \\
&P\{X^{(i)} = 1 \, / \, (X^{(i+1)}, Y^{(i+1)}) = (1,1)\} = p_l - \tfrac{1}{4} \, p_l^2
\end{aligned}
\tag{3.1}
$$

Note that only input link faults are taken into account. Faults at the output links
are considered as input link faults at the next stage.

The state probability $P\{X^{(i)} = 1\}$ of the output link is given by,

$$P\{X^{(i)} = 1\} = \frac{1}{2} \, p_l \left[P\{(X^{(i+1)}, Y^{(i+1)}) = (0,1)\} + P\{(X^{(i+1)}, Y^{(i+1)}) = (1,0)\} \right]$$

$$+p_l(1 - \frac{1}{4}\, p_l) \times P\{(X^{(i+1)}, Y^{(i+1)}) = (1,1)\} \tag{3.2}$$

For the non-redundant network the inputs into each switch are independent. Hence,

$$P\{(X^{(i+1)}, Y^{(i+1)}) = (u, v)\} = P\{X^{(i+1)} = u\} \times P\{Y^{(i+1)} = v\}; \quad u, v = 0,1 \tag{3.3}$$

Using (3.3) and the following equation

$$P\{Y^{(i+1)} = 0\} = P\{X^{(i+1)} = 0\} = 1 - P\{X^{(i+1)} = 1\}$$

we obtain from (3.2) after some algebraic manipulations,

$$P\{X^{(i)} = 1\} = p_l \times P\{X^{(i+1)} = 1\} - \frac{1}{4}\, p_l^2 \times (P\{X^{(i+1)} = 1\})^2 \tag{3.4}$$

This expression is identical to the one derived in [7] if fault-free (i.e., $p_l = 1$) 2×2 switches are assumed.

This simple recursion formula enables us to calculate the successive state probabilities, starting from the processors outputs up to the memory inputs.
For the processors output links we have

$$P\{X^{(k)} = 1\} = p_a\, p_r \tag{3.5}$$

Recursively, we calculate $P\{X^{(0)} = 1\}$. To compute the bandwidth note that the memory and its input link can be faulty as well, hence,

$$BW = N \times P\{X^{(0)} = 1\} \times p_m\, p_l \tag{3.6}$$

The next part of this section is devoted to the analysis of the network connectivity, as measured by C - the average number of connected processor-memory pairs, and by N_r and N_m - the average number of processors connected to at least one memory, and of memories connected to at least one processor, respectively.

In a non-redundant interconnection network there is exactly one path between a processor and a memory and consequently, the calculation of C is straightforward. C is obtained by multiplying the number of processor-memory pairs by the probability of a fault-free path. The latter probability is,

$$p_r\, p_l^{k+1}\, p_m \tag{3.7}$$

where $(k + 1)$ is the number of links along the path. Therefore,

$$C = N^2 \times p_r\, p_l^{k+1}\, p_m \tag{3.8}$$

Let Φ_r be the probability that a given processor (say processor 0) is fault-free and is connected to at least one fault-free memory. N_r is obtained by multiplying N by Φ_r. Due to the overlapping of the paths leading to the same processor, we must utilize the inclusion and exclusion principle to calculate the probability Φ_r. Define

E_j as the event in which memory j is connected to processor 0. Φ_r can now be expressed in terms of the events E_j as the probability that at least one of the events E_j occurs,

$$\Phi_r = p(E_1 \cup E_2 \cup ... \cup E_N) \tag{3.9}$$

The inclusion and exclusion formula states that

$$p(\cup E_j) = \sum_{i=1}^{N} (-1)^{(i-1)} W(i) \tag{3.10}$$

where $W(i)$ is the sum over all $\begin{pmatrix} N \\ i \end{pmatrix}$ subsets $\{j_1, j_2, ..., j_i\}$ of size i, of the probability that all paths in the subset are operational, namely,

$$W(i) = \sum_{\{j_1, ..., j_i\}} P(E_{j_1} \cap E_{j_2} \cap ... \cap E_{j_i}) \tag{3.11}$$

For a subset of i paths to be operational, all links in the subset must be fault-free. Hence, the required probability $P(E_{j_1} \cap E_{j_2} \cap ... \cap E_{j_i})$ depends not only on i but also on the number of links that the paths in the given subset have in common, since each link must be taken into account exactly once. Let d denote the number of distinct links in the subset, then

$$P(E_{j_1} \cap E_{j_2} \cap ... \cap E_{j_i}) = p_m^i \, p_r \, p_l^d \tag{3.12}$$

and

$$W(i) = p_m^i \, p_r \sum_d S_{i,d} \, p_l^d$$

where $S_{i,d}$ is the number of subsets of size i which consist of exactly d distinct links. Note that d can be expressed as the sum $d_0 + d_1 + ... + d_k$, where d_n is the number of distinct links in the subset at level n. Also note that for any subset of size i, $d_k = 1$ and $d_0 = i$. Using combinatorial arguments which are omitted here for the sake of brevity, we obtain the following equation for $S_{i,d}$,

$$S_{i,d} = \sum_{d_0 + d_1 + ... + d_k = d} \begin{pmatrix} 2 \\ d_{k-1} \end{pmatrix} 2^{d_1 + ... + d_{k-2} + 2d_{k-1} - i} \prod_{n=1}^{k-1} \begin{pmatrix} d_n \\ d_{n-1} - d_n \end{pmatrix} \tag{3.13}$$

As an example, for the 8×8 interconnection network in Fig. 26.1 we obtain,

$$\begin{aligned}
W(1) &= (p_m \, p_l) \, p_r \, p_l \, 8p_l^2 \\
W(2) &= (p_m \, p_l)^2 \, p_r \, p_l \, (4p_l^2 + 8p_l^3 + 16p_l^4) \\
W(3) &= (p_m \, p_l)^3 \, p_r \, p_l \, (8p_l^3 + 16p_l^4 + 32p_l^5) \\
W(4) &= (p_m \, p_l)^4 \, p_r \, p_l \, (2p_l^3 + 4p_l^4 + 48p_l^5 + 16p_l^6) \\
W(5) &= (p_m \, p_l)^5 \, p_r \, p_l \, (24p_l^5 + 32p_l^6) \\
W(6) &= (p_m \, p_l)^6 \, p_r \, p_l \, (4p_l^5 + 24p_l^6) \\
W(7) &= (p_m \, p_l)^7 \, p_r \, p_l \, 8p_l^6 \\
W(8) &= (p_m \, p_l)^8 \, p_r \, p_l \, p_l^6
\end{aligned} \tag{3.14}$$

Substituting $W(1), ..., W(N)$ into (3.10) and then multiplying by N yields N_r. N_m is obtained similarly by interchanging p_r and p_m.

26.4 The Extra Stage Interconnection Network

The analysis of the non-redundant network was simplified by the independence between the two inputs to any switch, enabling the straightforward calculation of the joint probabilities in (3.2). The incorporation of redundancy into the multi-stage interconnection network, resulting in two (or more) paths connecting any given processor-memory pair, introduces dependency among the links. Equation (3.3) is no longer valid in the general case and a different analysis is required, depending on the network's topology.

As an example for an interconnection network with redundancy we analyze in this section the Extra Stage Cube Network (ESC) [1] which includes $k + 1 = log N + 1$ stages and is depicted in Fig. 26.2. The analysis of this network is further complicated by the existence of multiplexers and demultiplexers at the input and output stages, respectively. The purpose of these circuits is to avoid the disconnection of a fault-free processor (or a fault-free memory) upon the failure of a single link. When calculating the bandwidth of the network we have therefore, to distinguish between the first and last stages on one hand and the internal stages on the other hand. All internal stages will be analyzed in one way while the first stage (stage k) and the last stage (stage 0) require a different treatment.

To calculate the state probability $P\{X^{(i)} = 1\}$ for an internal stage we must utilize the joint probabilities $P\{(X^{(i+1)}, Y^{(i+1)}) = (u, v)\}; u, v = 0, 1$. Since the input links $X^{(i+1)}$ and $Y^{(i+1)}$ are dependent (there is at least one processor which may send its memory requests through either one of them), their joint probability has to be calculated from the joint probabilities of the four links at level $(i + 2)$ through which all incoming requests pass. For example, to calculate the joint probability of output links 0 and 1 of stage 1 in Fig. 26.2 we need the joint probabilities of the output links 0, 1, 2 and 3 of stage 2. These probabilities might in turn, require the knowledge of the joint probabilities of eight links (and so on for larger ESC networks) making the analysis mathematically intractable.

However, each processor connected to the ESC has only two alternative paths to any given memory and therefore, no more than two links out of every four leading to two switches are dependent at any stage. For example, output links 0 and 1 of stage 2 in Fig. 26.2 are dependent since processor 0 (and 1) can send requests to memory 0 through either one of them. Similarly, links 2 and 3 are dependent. The pair 0,1 is however, independent of the pair 2,3.

In general, for the internal stages in the ESC network, we have

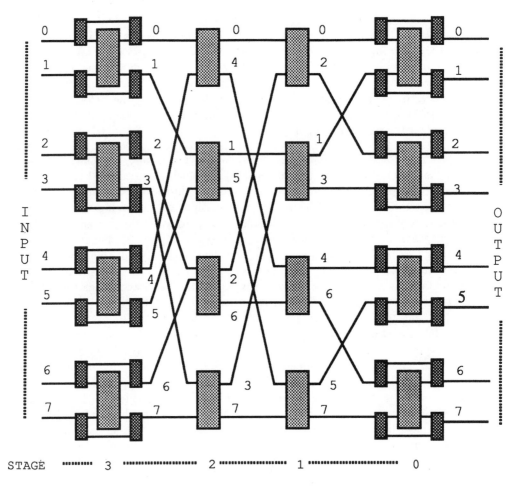

Figure 26.2: An 8 × 8 Extra Stage Cube network (ESC).

$$P\{(X^{(i)}, Y^{(i)}) = (u, v)\}$$

$$= \sum_{(s_0,\dots,s_3)=(0000)}^{(s_0,\dots,s_3)=(1111)} P\{(X^{(i+1)}, Y^{(i+1)}, Z^{(i+1)}, W^{(i+1)}) = (s_0, s_1, s_2, s_3)\}$$

$$\times P\{(X^{(i)}, Y^{(i)}) = (u, v) \,/\, (X^{(i+1)}, Y^{(i+1)}, Z^{(i+1)}, W^{(i+1)}) = (s_0, s_1, s_2, s_3)\}$$

$$= \sum_{(s_0,\dots,s_3)=(0000)}^{(s_0,\dots,s_3)=(1111)} P\{(X^{(i+1)}, Y^{(i+1)}) = (s_0, s_1)\} \times P\{(Z^{(i+1)}, W^{(i+1)}) = (s_2, s_3)\}$$

$$\times P\{X^{(i)} = u/(X^{(i+1)}, Z^{(i+1)}) = (s_0, s_2)\} \times P\{Y^{(i)} = v/(Y^{(i+1)}, W^{(i+1)}) = (s_1, s_3)\}$$

$$(4.1)$$

$$u, v = 0, 1$$

Consequently, only joint probabilities of two links are required and these can be calculated recursively beginning at the last stage (stage 0) down to the first stage (stage k). Once we reach the first stage, the independence between the two processors enables us to calculate the joint probabilities of their states, similarly to (3.3). However, the existence of multiplexers makes the calculation of the joint probabilities of stage k slightly more complicated than in the non-redundant network. By observing the ESC network we see that the joint probabilities for the two outputs of any switch in the first stage are:

$$P\{(X^{(k)}, Y^{(k)}) = (0,0)\} = (P\{X^{(k+1)} = 0\})^2$$
$$\qquad + q_l^3 \left[2\, P\{X^{(k+1)} = 0\} \times P\{X^{(k+1)} = 1\} + q_l (P\{X^{(k+1)} = 1\})^2 \right]$$
$$P\{(X^{(k)}, Y^{(k)}) = (0,1)\} = (1 - q_l^3)\, P\{X^{(k+1)} = 1\} \times P\{X^{(k+1)} = 0\}$$
$$\qquad + (2 p_l q_l^3 + p_l^2 q_l^2) P\{(X^{(k)}, Y^{(k)}) = (1,0)\} = P\{(X^{(k)}, Y^{(k)}) = (0,1)\}$$
$$P\{(X^{(k)}, Y^{(k)}) = (1,1)\} = p_l^2 (2 - p_l)^2 \, (P\{X^{(k+1)} = 1\})^2$$

$$(4.2)$$

where $P\{X^{(k+1)} = 1\}$ is given by (3.5) and $P\{X^{(k+1)} = 0\} = 1 - P\{X^{(k+1)} = 1\}$.

For the last stage (stage 0) which includes demultiplexers we use the following transition probabilities:

$$P\{X^{(0)} = 1 \,/\, (X^{(1)}, Y^{(1)}) = (0,0)\} = 0$$
$$P\{X^{(0)} = 1 \,/\, (X^{(1)}, Y^{(1)}) = (0,1)\} = \tfrac{1}{2}\, p_l$$
$$P\{X^{(0)} = 1 \,/\, (X^{(1)}, Y^{(1)}) = (1,0)\} = \tfrac{1}{2}\, (1 - q_l^2)$$
$$P\{X^{(0)} = 1 \,/\, (X^{(1)}, Y^{(1)}) = (1,1)\} = \tfrac{5}{4} p_l - \tfrac{1}{2}\, p_l^2$$

$$(4.3)$$

The bandwidth is then calculated from

$$BW = N \times P\{X^{(0)} = 1\} \times p_m \, p_l \qquad (4.4)$$

In what follows we calculate the two measures for network connectivity. The measure C can be expressed as the product of the number of processor-memory pairs (i.e., N^2) times the probability that at least one fault-free path between a given processor-memory pair exists. Each processor-memory pair in the ESC network is connected by two disjoint paths (except for both ends), hence

$$P\{At \ least \ one \ path \ is \ fault-free\} = P\{First \ path \ is \ fault-free\}$$

$$+P\{Second \ path \ is \ fault-free\} - P\{Both \ paths \ are \ fault-free\} \qquad (4.5)$$

$$= p_r(1 - q_i^2)p_l^k(1 - q_i^2)p_m + p_r p_l^{k+2}p_m - p_r p_l^{2k+2}p_m = p_r p_m p_l^{k+2}(5 - 4p_l + p_l^2 - p_l^k)$$

Multiplying by N^2 yields C.

N_r and N_m are calculated following the same steps as in Section 26.3. Define Φ_r as the probability that a given processor (say processor 0) is connected to at least one memory, and E_j as the event in which the j-th path emanating from processor 0 is fault-free. Recalling that each processor is connected to all memories through $2N$ paths, equations (3.9) (3.10) become,

$$\Phi_r = P\{E_1 \cup E_2 \cup ... \cup E_{2N}\} = \sum_{i=1}^{2N}(-1)^{i-1}W(i) \qquad (4.6)$$

where $W(i)$ is defined in (3.11). To obtain the probability that a given subset of paths $\{j_1, ..., j_i\}$ is fault-free, note that in the ESC network each path has $(k+2)$ links, hence the number of distinct links in a subset of paths can be expressed as $d_0 + d_1 + ... + d_{k+1}$ where $d_{k+1} = 1$ and d_0 is the number of memories that the paths in the subset lead to. Given d and d_o,

$$P(E_{j_1} \cap ... \cap E_{j_i}) = p_r \, p_m^{d_o} \, p_l^{d-d_0-1} \, (1 - q_i^2)^{d_0+1} \qquad (4.7)$$

This equation differs from (3.12) because of the existence of multiplexers and demultiplexers in the ESC network. Denote by S_{i,d,d_0} the number of subsets of size i which consist of exactly d distinct links, out of which d_0 are at level 0, then

$$W(i) = p_r \sum_{d,d_0} S_{i,d,d_0} \, p_m^{d_o} \, p_l^{d-d_0-1} \, (1 - q_i^2)^{d_0+1} \qquad (4.8)$$

Using combinatorial arguments which are omitted here, we obtain

$$S_{i,d,d_0} = 2^{d-d_0-i-1} \cdot \sum_{d_1+...+d_k=d-d_0-1} \prod_{n=2}^{k+1} \binom{d_n}{d_{n-1}-d_n} \times weight(d_0/d_1, d_2) \qquad (4.9)$$

where

$$weight(d_0/d_1, d_2) = \sum_{a=0}^{d_1-d_2} \sum_{b=0}^{2d_2-d_1} \binom{d_1 - d_2}{a} \binom{2d_2 - d_1}{b}$$

$$\times \binom{d_1 - d_2 - a}{a + b - d_1 - d_2 - i} \binom{a}{2d_2 - b - d_0} 2^{2d_1-3a-b-i} \qquad (4.10)$$

Substituting $W(1), ..., W(2N)$ into (4.6) and then multiplying by N yields N_r. N_m is obtained similarly by interchanging p_r and p_m.

26.5 Numerical Results

In this section we present some numerical comparisons between the two previously analyzed networks. The bandwidth of the two systems of size 16×16 has been calculated as a function of p_a (the probability of request), for three different sets of values of the probabilities p_l, p_r and p_m. The results are depicted in Fig. 26.3. A very important conclusion that can be drawn from this comparison is that both networks have exactly the same bandwidth if there are no faulty elements (case (i) in Fig. 26.3). The additional path between every pair of processor-memory that the ESC network provides, does not increase its bandwidth over that of a non-redundant network, due to the sharing of the redundant paths by other processor-memory pairs.

The situation is different when faulty elements are present in the system (cases (ii) and (iii) in Fig. 26.3). The ESC network shows a smaller reduction in bandwidth in the presence of faulty elements. Here, the redundant paths in the ESC network

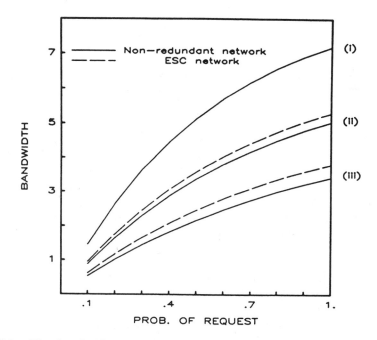

Figure 26.3: The bandwidth of two 16×16 networks as a function of p_a for (i) $p_l = p_r = p_m = 1$, (ii) $p_l = 0.95$, $p_r = 0.85$, $p_m = 0.9$ and (iii) $p_l = 0.9$, $p_r = 0.75$, $p_m = 0.8$.

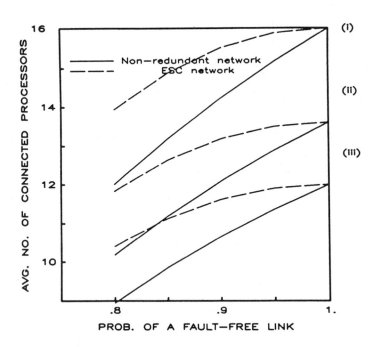

Figure 26.4: The average number of connected fault-free processors N_r for two 16×16 networks as a function of p_l for (i) $p_r = p_m = 1$, (ii) $p_r = 0.85$, $p_m = 0.9$ and (iii) $p_r = 0.75$, $p_m = 0.8$.

reduce the effect of faulty links on the bandwidth. And, as is evident from Fig. 26.3, the advantage of the ESC network over the non-redundant one increases as the probability p_l of a fault-free link decreases.

Fig. 26.4 depicts the average number of fault-free connected processors N_r as a function of p_l for the two systems (for three cases similar to those in Fig. 26.3). This figure shows that the ESC network is less sensitive to link faults than the non-redundant one when N_r is employed as a measure for system connectivity. We have also tried to separate the effect of the extra stage from the effect of the additional multiplexers and demultiplexers. When the latter were removed, the ESC network showed no advantage over the non-redundant one and both systems produced the same N_r curves.

The other measure for system connectivity, i.e., C - the average number of fault-free connected processor-memory pairs is depicted in Fig. 26.5. As in Fig. 26.4, this connectivity measure is shown as a function of p_l for the same three cases. Here again we can see the advantage of the network with redundancy over the non-redundant one. Combining Figures 26.4 and 26.5 we can conclude that not only is the number of connected processors N_r larger in the ESC network than in the non-redundant system, but in addition, each fault-free processor in the ESC network is connected (on the average) to a larger number of fault-free memories.

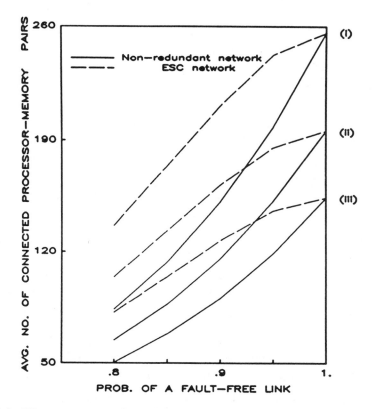

Figure 26.5: The average number of connected fault-free processor-memory pairs for two 16×16 networks as a function of p_l for (i) $p_r = p_m = 1$, (ii) $p_r = 0.85$, $p_m = 0.9$ and (iii) $p_r = 0.75$, $p_m = 0.8$.

26.6 Conclusions

The performance of two multi-processor systems with a multi-stage interconnection network in the presence of faulty elements has been analyzed in this paper. The first is a non-redundant network and the second is the Extra Stage Cube network which was selected as an example for a network with redundancy. The bandwidth and connectivity of the multi-processing system have been suggested as measures and used to analyze and compare the performance of these two systems.

The approach to performance analysis which was introduced in this paper, can be applied to other schemes for incorporating redundancy into multi-stage networks. Such an analysis will allow a more accurate comparison of the performance of these architectures in the presence of faulty elements. It can also suggest ways to develop new fault-tolerant architectures.

References

[1] G. B. Adams and H. J. Siegel, "The Extra Stage Cube: A Fault-Tolerant Interconnection Network for Supersystems," *IEEE Trans. on Computers,* Vol. C-31, May 1982, pp. 443-454.

[2] G. B. Adams, D. P. Agrawal and H. J. Siegel, "A Survey and Comparison of Fault-Tolerant Multistage Interconnection Networks," *Computer,* Vol. 20, June 1987, pp. 14-27.

[3] M. Jeng and H. J. Siegel, "A Fault-Tolerant Multistage Interconnection Network for Multiprocessor Systems using Dynamic Redundancy," *Proc. of the 1986 Symp. on Distributed Computing Systems,* pp. 70-77.

[4] I. Koren and D.K. Pradhan, "Modeling the Effect of Redundancy on Yield and Performance of VLSI Systems," *IEEE Trans. on Computers,* Vol. C-36, March 1987, pp.344-355.

[5] V. P. Kumar and S. M. Reddy, "Augmented Shuffle-Exchange Multistage Interconnection Networks," *Computer,* Vol. 20, June 1987, pp. 30-40.

[6] K. Padmanabhan and D. H. Lawrie, "A Class of Redundant Path Multistage Interconnection Networks," *IEEE Trans. on Computers,* Vol. C-32, Dec. 1983, pp. 1099-1108.

[7] J. H. Patel, "Performance of Processor-Memory Interconnection for Multiprocessors," *IEEE Trans. on Computers,* Vol. C-30, Oct. 1981, pp. 771-780.

[8] N. Tzeng, P. Yew and C. Zhu, "A Fault-Tolerant Scheme for Multistage Interconnection Networks," *Proc. of the 12-th Annual Symp. on Comp. Arch.,* June 1985, pp. 368-375.

Chapter 27

Partially Augmented Data Manipulator Networks: Minimal Designs and Fault Tolerance [1]

Darwen Rau [2]
Jose A. B. Fortes [2]

Abstract

Augmented data manipulator networks are multistage interconnection networks which implement at each stage interconnection functions present in the single stage network known as PM2I network or barrel shifter. These multistage networks include the ADM (Augmented Data Manipulator) and IADM (Inverse Augmented Data Manipulator) networks, which have been extensively studied and proposed for use in multiprocessor systems. This paper derives new partially augmented networks based on the solution to the shortest path problem in the PM2I network. The new networks include: the HADM (Half Augmented Data Manipulator) and HIADM (Half Inverse Augmented Data Manipulator) networks which have half the number of stages of the ADM and IADM networks, the MADM (Minimum Augmented Data Manipulator) and the MIADM (Minimum Inverse Augmented Data Manipulator) networks which have the minimum link complexity required for one-to-one connections in a network of size N with $\log_4 N$ stages of uniform switches,

[1] This research was supported in part by the National Science Foundation under Grant DC1-8419745 and by the Innovative Science and Technology Office of the Strategic Defense Initiative Organization and was administered through the Office of Naval Research under contract No. 00014-85-k-0588.
[2] Purdue University, West Lafayette, IN

and the Extra Stage MADM and MIADM networks which are fault-tolerant versions of the MADM and MIADM networks that can tolerate at least three switch failures. The derivations of these networks are presented and their properties and advantages over other designs are analyzed.

27.1 Introduction

Multistage interconnection networks are often designed by implementing at each stage interconnection functions characteristic of a single-stage network. This paper proposes new multistage networks which offer advantages over previously known designs based on the PM2I network [Sie77]. The new networks are derived from the solution to the shortest path problem in the PM2I network. Further analysis leads to the derivation of designs with minimal link complexity and fault-tolerance.

The plus-minus 2^i *(PM2I)* network [Sie77] is a single-stage network defined by the PM2I interconnection functions:

$$PM2I_{+i}(S) = (S + 2^i) \bmod N, \ 0 \leq i \leq n - 1,$$

$$PM2I_{-i}(S) = (S - 2^i) \bmod N, \ 0 \leq i \leq n - 1$$

where $N = 2^n$ corresponds to the number of network nodes and $S, 0 \leq S \leq N - 1$, denotes a node address. Thus, in the PM2I network there exist links from a node S to nodes $PM2I_{+i}(S), \ 0 \leq i \leq n - 1$, as well as links to nodes $PM2I_{-i}(S)$, $0 \leq i \leq n - 1$. These links are referred to as the $+2^i$ *links* and -2^i *links* , respectively. A PM2I network of $N = 8$ nodes is illustrated in Figure 27.1.

The class of data manipulator networks, introduced in [Fen74], are constructed based on the PM2I functions. It includes, among others, the Augmented Data Manipulator (ADM) network [SiS78], the IADM network [McS82] and the Gamma network [PaR82][PaR84]. The IADM network and the ADM network differ only in that the input side of one of them corresponds to the output side of the other and vice versa. The Gamma and the IADM networks are topologically equivalent; however, they use switches of different types. Each 3×3 crossbar switch used in the Gamma network can connect simultaneously all three inputs to all three outputs whereas each switch used in the IADM network can connect only one of its three inputs to one or more of its three outputs.

The *ADM* network is composed of $n = \log N$ stages labeled from 0 to $n - 1$ from the output side to the input side. Each stage consists of $3N$ connection links and N switches. The switches are labeled from 0 to $N - 1$ from the top to the bottom. An extra column of switches is appended at the end of the last stage and is referred to as stage n. Each switch j of stage $i + 1$ has three output links to switches $(j - 2^i) \bmod N, \ j$ and $(j + 2^i) \bmod N$ of stage i. The link joining j of stage $i + 1$ and j of stage i is called a *straight link* , the link joining $(j - 2^i) \bmod N$ of stage $i + 1$

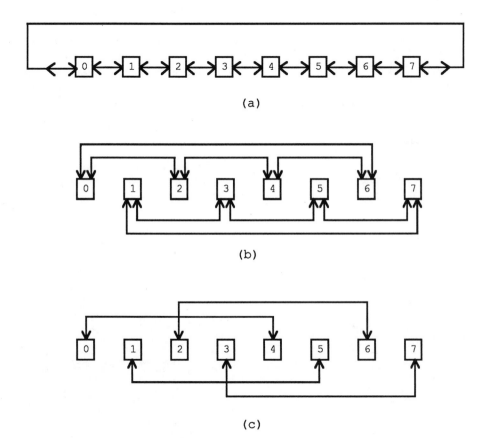

(a)

(b)

(c)

Figure 27.1: The $PM2I$ network for $N = 8$. A: $PM2I_0$ network. B: $PM2I_1$ network. C: $PM2I_2$ network.

and j of stage i is a *plus* $(+2^i)$ *link* [McS82], and the link joining $(j + 2^i)$ mod N of stage $i + 1$ and j of stage i is a *minus* (-2^i) *link* . Each switch selects one of its input links and connects it to one or more output links. Figure 27.2 illustrates an ADM network of size $N = 8$.

Because the only difference between the ADM and IADM networks is that their input and output sides are reversed, the stages of the IADM network are labeled from 0 to $n - 1$ from the input side to the output side. Each switch j of stage i in the IADM network is connected to switches $(j - 2^i)$ mod N, j and $(j + 2^i)$ mod N of stage $i + 1$. A plus link in the IADM network from switch j of stage i is connected to switch $j + 2^i$ of stage $i + 1$ is the same link as the minus link in the ADM network from switch $j + 2^i$ of stage $i + 1$ to switch j of stage i. Similar relationship applies to a minus link in the IADM network and a plus link in the ADM network. Due to the reversal of the input and output sides of the ADM and IADM network, stage i of the ADM network corresponds to the switches of stage i of the IADM network and the links of stage $i - 1$ of the IADM network.

The results of this paper are based on the study of shortest path problem in the PM2I network. The solution to the shortest path problem for the PM2I network is derived from an algorithm [PaR82] that generates routing tags for the Gamma

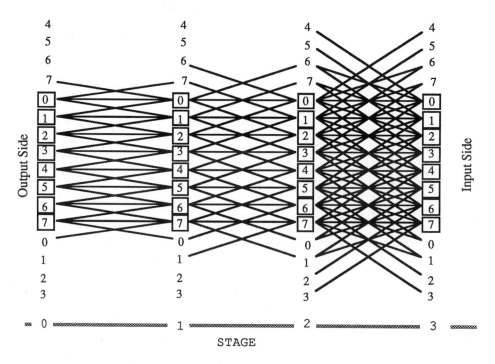

Figure 27.2: The ADM network for $N = 8$.

network. Because the IADM and Gamma network are topologically equivalent and the ADM and IADM networks differ only in their input and output sides, the results in this paper apply to all of these networks. However, the main interest of this paper is the study of the ADM network and the discussions are centered on the properties of the ADM network.

Given a string of n digits, $t = t_0 t_1 \cdots t_{n-1}$, the notation $t_{p/q}$ denotes the digits of t starting at t_p and ending at t_q. Throughout this paper, j and $j + a$ (where a is some constant) represent labels of switches. Also modulo N arithmetic is assumed, e.g. $j + a$ implies $(j + a) \bmod N$. The notation j^i is used to indicate that a switch j belongs to stage i and (j^{i+1}, j^i) is used to represent a link joining j^{i+1} and j^i. A sequence of switches of contiguous stages $(j^{i+k}, j^{i+k-1}, \ldots, j^i)$ is used to represent a path from j^{i+k} to j^i.

Section 27.2 of the paper considers the formulation and solution of the shortest path problem for the PM2I network. In Section 27.3 these results are used to derive new networks that require less hardware complexity and transmission delay than other known augmented data manipulator networks. These new networks are called *partially augmented data manipulator networks* . Details of routing schemes for these networks are also discussed in Section 27.3. Fault-tolerant topologies are proposed in Section 27.4 by adding an extra stage to these networks, with the result that four disjoint paths exist between any source and any destination in the networks. Section 27.5 concludes the paper.

27.2 Shortest Path Problem in the PM2I Network

Given a source node S and a destination node D in the PM2I network, the shortest path problem is to find a path from S to D which contains a minimal number of links. When circuit switching is used for communication between nodes, delays are identical for any link and transmission delay is directly proportional to the number of links on a path. Thus, the shortest path is also the one for which transmission delay is minimum.

Given a source node S and destination node D in the PM2I network, define *distance* Δ to be $(D - S) \bmod N$; thus the range of Δ is $0 \le \Delta \le (N - 1)$. Routing from a source S to a destination D in the PM2I network can be characterized by the *combination tag* $t_{0/2n-1} = t_0 t_1 \cdots t_{2n-1}$ such that

$$\Delta = (\sum_{i=0}^{n-1} t_i 2^i + \sum_{i=n}^{2n-1} t_i(-2^{i-n})) \bmod 2^n \tag{1}$$

where Δ is the distance from the source S to the destination D and t_i 's are non-negative integers. A positive value of t_i indicates that link $+2^i$, for $0 \le i \le n-1$, or link -2^{i-n}, for $n \le i \le 2n - 1$, is used in the routing path whereas $t_i = 0$ indicates that the link is not used. A combination tag, as suggested by its name, specifies a combination of PM2I links that can be used to cover the distance between the source and the destination. However, the combination tag $t_{0/n-1}$ does not specify the sequence in which the links are used. Several distinct paths can be derived from a combination tag and all these paths contains the same number of links. Since the combination tag depends only on the distance Δ, it is often identified as a *combination tag of distance* Δ. A shortest path is specified by a combination tag for which the number of links $\sum_{i=0}^{2n-1} t_i$ is minimum and the problem of finding such a tag - called *minimum weight combination tag* - can be stated as follows:

Problem (P) Find $t^* = t^*{}_{0/n-1}$ such that

$$H_\Delta = \min \sum_{i=0}^{2n-1} t_i = \sum_{i=0}^{2n-1} t_i^*$$

subject to

$$\Delta = (\sum_{i=0}^{n-1} t_i 2^i + \sum_{i=n}^{2n-1} t_i(-2^{i-n})) \bmod 2^n$$

$$0 \le t_i \text{ for } 0 \le i \le 2n - 1$$

$$0 \le \Delta \le 2^n - 1$$

A feasible solution to this problem corresponds to a combination tag, and an optimal solution to it corresponds to a minimum weight combination tag. For convenience of discussion, the terms (i) a feasible solution and a combination tag, and (ii) an optimal solution and a minimum weight combination tag are used interchangeably.

The next two lemmas reduce the size of the set of feasible solutions.

Lemma 2.1 If t^* is the optimal solution to (P), then $t_i \in \{0,1\}$, $0 \le i \le 2n - 1$.

Proof: The proof is by contradiction. Assume that the optimal solution t^* contains $t_k^* \ge 2$, for some k, $0 \le k \le n - 2$ or $n \le k \le 2n - 2$. Then there exist alternate paths that, compared with paths defined by t^*, reduce traversal through link $+2^k$ (or -2^{k-n}) twice and increase its traversal through link $+2^{k+1}$ (or -2^{k+1-n}) once; i.e.

$$t_k^* 2^k + t_{k+1}^* 2^{k+1} = (t_k^* - 2)2^k + (t_{k+1}^* + 1)2^{k+1}$$

Comparing with the total delay of the path defined by t^*, the total delay of the alternate paths is reduced by one, which is contradictory to the hypothesis that t^* minimizes the routing delay. If $k = n - 1$ (or $2n - 1$) such that $t_{n-1}^* \ge 2$ (or $t_{2n-1}^* \ge 2$), a carry is generated in the highest order digit and t_{n-1}^* is discounted by two, denoted $t'_{n-1} = t_{n-1}^* - 2$. The carry vanishes due to (mod2^n) operation and the total delay of the alternate paths is reduced by two; again a contradiction results.   □

Lemma 2.2 If t^* is the optimal solution to (P), then $t_i^* \cdot t_{n+i}^* = 0$, $0 \leq i \leq n - 1$; i.e. the shortest path between any source and any destination in the PM2I network cannot contain both link $+2^i$ and link -2^i for any i, $0 \leq i \leq n - 1$.

Proof: The proof is by contradiction. Suppose the opposite is true. From Lemma 2.1, a digit of the tag representing a shortest path, can only be 0 or 1; by assumption of having both $+2^i$ and -2^i links on the routing path, $t_i^* = t_{n+i}^* = 1$. The effects of $+2^i$ and -2^i cancel each other. Thus the values for t_i^* and t_{n+i}^* can be substituted by 0 and still satisfy the equality constraint in (P) (also equation (1)). The routing delay is thus reduced by two. A contradiction results. $\square$

From Lemma 2.2, either t_i^* or t_{n+i}^* is zero, $0 \leq i \leq n - 1$, so that the two sums $\sum_{i=0}^{n-1} t_i 2^i$ and $\sum_{i=n}^{2n-1} t_i (-2^{i-n})$ in equation (1) can be combined to form $\sum_{i=0}^{n-1} t_i 2^i$, with the extension of the values for t_i to include negative integers. The result in Lemma 2.1 confines the values for each t_i of a tag representing a shortest path to be 0 and 1. Together with the necessary extension to include negative integers, the possible values for t_i of an optimal solution are -1, 0 or 1. Thus, the problem of finding a minimum weight combination tag can be reformulated as follows:

Problem $(\bar{P})$ Find $t^* = t^*_{0/n-1}$ such that

$$H_\Delta = \min \sum_{i=0}^{n-1} |t_i| = \sum_{i=0}^{n-1} |t_i^*|$$

subject to

$$\Delta = \left(\sum_{i=0}^{n-1} t_i 2^i \right) \bmod 2^n$$

$$t_i \in \{-1, 0, 1\} \text{ for } 0 \leq i \leq n - 1$$

$$0 \leq \Delta \leq 2^n - 1$$

A branch-and-bound approach is used to find the optimal solution for $(\bar{P})$, which is also a minimum weight combination tag. This approach is based on an algorithm proposed in [PaR82] that can find all signed-digit representations for the distance between any source and any destination in the Gamma network. Each signed-digit representation corresponds to a routing tag for the source/destination pair. Moreover, since the IADM network and the Gamma network are topologically equivalent, the routing tags generated by the algorithm are also valid routing tags for the IADM network. The Gamma network is constructed based on the PM2I functions and a routing tag uniquely specifies a path in it. In particular, each stage is composed of 2^n switches, and at each stage i, $0 \leq i \leq n - 1$, each switch is connected to three output links $+2^i$, -2^i and straight link, and only one of them is on the routing path; in addition, the path in the Gamma network traverses a distance of $((D - S) \bmod 2^n)$ from S to D. These correspond to the constraints in

$(\bar{P})$: $\Delta = (\sum_{i=0}^{n-1} t_i 2^i) \bmod 2^n$, $t_i \in \{-1, 0, 1\}$ and $-(2^n - 1) \leq \Delta \leq 2^n - 1$. Thus a routing tag that specifies a path from S to D in the Gamma network is also a feasible solution to $(\bar{P})$. Note that $t_i = 0$ indicates that a straight link is used at stage i for routing in the Gamma network.

A routing tag for the Gamma network can be converted to a combination tag for the PM2I network: if the ith bit of the routing tag is 1, $t_i = 1$, if it is $\bar{1}, t_{i+n} = 1$ (hereafter the signed-digit representation $\bar{1}$ [Avi61] is used to represent -1), and if it is 0, $t_i = t_{i+n} = 0$. A combination tag satisfying conditions (a) $t_i \in \{-1,0,1\}$ and (b) $t_i \cdot t_{i+1} = 0$ can also be converted to a routing tag for the Gamma network: if $t_i = 1$, the ith bit of the routing tag is 1, if $t_{i+n} = 1$, the ith bit of the routing tag is $\bar{1}$, and if $t_i = t_{i+n} = 0$, the ith bit of the routing tag is 0. The optimal solution to $(\bar{P})$ certainly satisfies conditions (a) and (b) and is also a minimum weight tag. Because we are only interested in the shortest path (which can be characterized by a minimum weight combination tag) in the PM2I network, given the one-to-one correspondence between a minimum weight routing tag and a minimum weight combination tag, they are used interchangeably. The algorithm in [PaR82] is stated as follows.

Algorithm ALL-TAGS($\Delta, t_{0/n-1}$)
$\Delta_0 = \Delta$
for $i = 0$ to $n - 1$ do
    if Δ_i is even then $t_i = 0$, $\Delta_{i+1} = \Delta_{\frac{i}{2}}$
$$\text{else} \quad \left\{ \begin{array}{l} t_i = 1, \Delta_{i+1} = \frac{\Delta_{i-1}}{2} \\ t_i = \bar{1}, \Delta_{i+1} = \frac{\Delta_{i+1}}{2} \end{array} \right\}$$
    endif
enddo

In the algorithm, t_i is uniquely determined ($= 0$) if Δ_i is even whereas freedom exists in choosing the value for t_i (1 or $\bar{1}$) if Δ_i is odd. An example is shown in Table 27.1 that generates all tags for routing from $S^0 = 1$ to $D^2 = 4$ in the IADM network of size $N = 8$. In this case, $\Delta = 3 = -5 \bmod 8$.

Table 27.1:

$[\Delta_0]$	t_0	$[\Delta_1]$	t_1	$[\Delta_2]$	t_2	$[\Delta]$
[3]	1	[1]	1	[0]	0	$(= +3)$
[3]	1	[1]	$\bar{1}$	[1]	1	$(= +3)$
[3]	1	[1]	$\bar{1}$	[1]	$\bar{1}$	$(= -5)$
[3]	$\bar{1}$	[2]	0	[1]	1	$(= +3)$
[3]	$\bar{1}$	[2]	0	[1]	$\bar{1}$	$(= -5)$

As mentioned previously the focus of this paper is the ADM network, the tags generated by algorithm ALL-TAGS can also be used in the ADM network because the ADM and IADM network differ only in that the input side of one of them corresponds to the output side of the other and vice versa. In the IADM network routing is from a switch of the lowest order stage to the highest order stage while routing in the ADM network is just the opposite. Therefore, the lowest order digit of the tag is first examined by a switch for routing in the IADM network and the highest order digit is first examined for routing in the ADM network. At stage i, $0 \leq i \leq n - 1$, of both networks, if t_i is 0, straight link is used for routing; if it is 1, link $+2^i$ is used; if it is $\bar{1}$, link -2^i is used. In particular, routing from S^0 to D^n in the IADM network is equivalent to routing from D^n to S^0 in the ADM network. Let $t_{0/n-1}$ be the tag for the routing from S^0 to D^n in the IADM network, it can be readily verified that tag $t^c_{0/n-1}$, where $t^c_i = -t_i$, $0 \leq i \leq n - 1$ can be used for routing from D^n to S^0 in the ADM network. The two tags $t_{0/n-1}$ and $t^c_{0/n-1}$ represent the same path, with different interpretations in the ADM and IADM networks. For example, the tags 110, $1\bar{1}1$, $1\bar{1}\bar{1}$, $\bar{1}01$ and $\bar{1}0\bar{1}$ for the IADM network in the above table can be converted to $\bar{1}\bar{1}0$, $\bar{1}1\bar{1}$, $\bar{1}11$, $10\bar{1}$ and 101 for the ADM network, respectively. Figure 27.3 illustrates the routing from $D^2 = 4$ to $S^0 = 1$ in the ADM network using these tags.

The possibility of having two values, 1 and $\bar{1}$, for t_i, if Δ_i is odd can be used to find the optimal solution to $(\bar{P})$. It is shown below how to choose the value for t_i so that t_{i+1} can be pre-determined as desired.

Lemma 2.3 In the process of generating tags in algorithm ALL-TAGS, if Δ_i is odd, it is always possible to make $t_{i+1} = 0$ by properly choosing the value for t_i.

Proof: Since $\frac{\Delta_i+1}{2}$ and $\frac{\Delta_i-1}{2}$ differ exactly by one, one of them is even and the other is odd. Suppose that, without loss of generality, $\frac{\Delta_i+1}{2}$ is even. Then t_i can be chosen to be -1 so that $\Delta_{i+1} = \frac{\Delta_i+1}{2}$, which makes $t_{i+1} = 0$. $\qquad\square$

For example, one of the paths illustrated in Figure 27.3 is represented by a tag $t_{0/2} = 10\bar{1}$ of distance $\Delta = \Delta_0 = 3$; in this case t_0 is chosen to be 1 so that $t_1 = 0$.

Theorem 2.4 There exists an optimal solution t^* to $(\bar{P})$ which has no adjacent nonzero digits; i.e., $t^*_{i+1} \cdot t^*_i = 0$ for $0 \leq i \leq n - 2$. If $t^*_{n-1} = 0$ then t^* is the unique optimal solution with no adjacent nonzero digits; otherwise, there exists another optimal solution t^c with no adjacent nonzero digits, where $t^c_i = t^*_i$, $0 \leq i \leq n - 2$, and $t^c_{n-1} = -t^*_{n-1}$.

Proof: The proof consists of three parts. Part (i) finds a minimum weight tag, part (ii) proves the uniqueness of the minimum weight tag when $t^*_{n-1} = 0$, and part (iii) finds another minimum weight tag if $t^*_{n-1} \neq 0$.

(i) An algorithm which results from modifying algorithm ALL-TAGS is first given to construct a minimum weight tag; it is followed by a proof of its optimality.

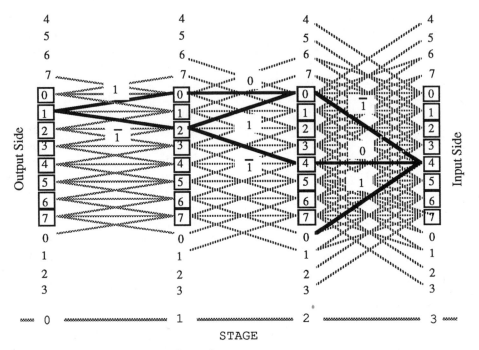

Figure 27.3: Rerouting from 4^3 to 1^0 in the ADM network for $N = 8$. The solid lines are the links on the routing paths and the dotted lines are other links of the ADM network. Labels on the links are the digits of the routing tags.

Algorithm SHORTEST-PATH $(\Delta, t^*{}_{0/n-1})$

$\Delta_0 = \Delta$

for $i = 0$ to $n - 1$ do

    if Δ_i is even then $t_i^* = 0$, $\Delta_{i+1} = \frac{\Delta_i}{2}$

    else if $\frac{\Delta_i - 1}{2}$ is even then $t_i^* = 1$, $\Delta_{i+1} = \frac{\Delta_i - 1}{2}$

        else $t_i^* = \bar{1}$, $\Delta_{i+1} = \frac{\Delta_i + 1}{2}$

        endif

    endif

enddo

Since the set of tags generated by algorithm SHORTEST-PATH is a subset of those generated by algorithm ALL-TAGS, algorithm SHORTEST-PATH correctly generates a tag of distance Δ. It remains to show that the tag has minimum weight. The strategy used in the algorithm SHORTEST-PATH is to generate a zero digit whenever possible (for Δ_i is odd, t_i is chosen to be such that Δ_{i+1} is even, which makes $t_{i+1} = 0$). To see why this is a good strategy, let i be the smallest index such that Δ_i is odd, and let t^* and t be the solutions found by applying this strategy and

and by not complying with this strategy, respectively. Assume that, without loss of generality, $\frac{\Delta_i - 1}{2}$ is even. It is shown that there are four possible cases and the terminating conditions for each case can be continued by applying the discussion for one of the four cases recursively.

Case 1

Table 27.2 below illustrates the discussion for case 1 based on this assumption. Since $\frac{\Delta_i - 1}{2}$ ($= \Delta_{i+1}$ for t^*, denoted $\Delta_{i+1}(t^*)$) is assumed to be even, $t^*_{i+1} = 0$. Because $\Delta_{i+1}(t) = \frac{\Delta_i + 1}{2}$ is odd, there are two possible values, 1 and $\bar{1}$, for t_{i+1}. If $t_{i+1} = 1$, $\Delta_{i+2}(t) = \Delta_{i+2}(t^*)$. The discussion for case 1 terminates here.

Table 27.2:

case 1	$[\Delta_i]$	t_i	$[\Delta_{i+1}]$	t_{i+1}	$[\Delta_{i+2}]$
$t*$	$[\Delta_i]$	1	$[\frac{\Delta_i - 1}{2}]$	0	$[\frac{\Delta_i - 1}{4}]$
t	$[\Delta_i]$	$\bar{1}$	$[\frac{\Delta_i + 1}{2}]$	1	$[\frac{\Delta_i - 1}{4}]$

Case 2

The alternative is that $t_{i+1} = \bar{1}$, which is illustrated in cases 2, 3 and 4 in the tables below. In case 2, $\frac{\Delta_i - 1}{4}$ is assumed to be even. Case 2, illustrated in Table 27.3 terminates here with $\Delta_{i+2}(t^*)$ being even and $\Delta_{i+2}(t)$ being odd.

Case 3

Table 27.3:

case 2	$[\Delta_i]$	t_i	$[\Delta_{i+1}]$	t_{i+1}	$[\Delta_{i+2}]$
$t*$	$[\Delta_i]$	1	$[\frac{\Delta_i - 1}{2}]$	0	$[\frac{\Delta_i - 1}{4}]$
t	$[\Delta_i]$	$\bar{1}$	$[\frac{\Delta_i + 1}{2}]$	$\bar{1}$	$[\frac{\Delta_i + 3}{4}]$

In cases 3 and 4, $\frac{\Delta_i - 1}{4}$ is assumed to be odd. Case 3 is illustrated in the Table 27.4 with the assumption that $\frac{\Delta_i + 3}{8}$ is even. Since $\Delta_{i+2}(t^*) = \frac{\Delta_i - 1}{4}$ is odd, $t^*_{i+2} = \bar{1}$ or 1, and $t_{i+2} = 0$. Since $\frac{\Delta_i + 3}{8}$ is even, the algorithm chooses $t^*_{i+2} = \bar{1}$, and $\Delta_{i+3}(t) = \Delta_{i+3}(t^*)$. The discussion for case 3 terminates here.

Table 27.4:

case 3	$[\Delta_i]$	t_i	$[\Delta_{i+1}]$	t_{i+1}	$[\Delta_{i+2}]$	t_{i+2}	$[\Delta_{i+3}]$
$t*$	$[\Delta_i]$	1	$[\frac{\Delta_i - 1}{2}]$	0	$[\frac{\Delta_i - 1}{4}]$	$\bar{1}$	$[\frac{\Delta_i + 3}{8}]$
t	$[\Delta_i]$	$\bar{1}$	$[\frac{\Delta_i + 1}{2}]$	$\bar{1}$	$[\frac{\Delta_i + 3}{4}]$	0	$[\frac{\Delta_i + 3}{8}]$

Case 4

Case 4 for which $\frac{\Delta_i-5}{8}$ is even is illustrated in Table 27.5. If $\frac{\Delta_i-5}{8}$ is even, the algorithm chooses $t^*_{i+2} = 1$. In this case, $\Delta_{i+3}(t^*) = \frac{\Delta_i-5}{8}$ and $\Delta_{i+3}(t) = \frac{\Delta_i+3}{8}$. The discussion for case 4 terminates here.

Table 27.5:

case 4	$[\Delta_i]$	t_i	$[\Delta_{i+1}]$	t_{i+1}	$[\Delta_{i+2}]$	t_{i+2}	$[\Delta_{i+3}]$
$t*$	$[\Delta_i]$	1	$[\frac{\Delta_i-1}{2}]$	0	$[\frac{\Delta_i-1}{4}]$	1	$[\frac{\Delta_i-5}{8}]$
t	$[\Delta_i]$	$\bar{1}$	$[\frac{\Delta_i+1}{2}]$	$\bar{1}$	$[\frac{\Delta_i+3}{4}]$	0	$[\frac{\Delta_i+3}{8}]$

To conclude, cases 1 and 3 have the terminating conditions that $\Delta_{i+2}(t^*) = \Delta_{i+2}(t)$ and $\Delta_{i+3}(t^*) = \Delta_{i+3}(t)$, respectively. The discussion for $\Delta_i(t^*) = \Delta_i(t)$, which is the condition where the discussion for all cases begin, can be applied again to these terminating conditions. In cases 2 and 4, the terminating conditions are that $\Delta_{i+2}(t^*)$ is even and $\Delta_{i+2}(t)$ is odd, and $\Delta_{i+3}(t^*)$ is even and $\Delta_{i+3}(t)$ is odd, respectively. The discussions done for each case for iteration $i + 1$ when $\Delta_{i+1}(t^*)$ is even and $\Delta_{i+1}(t)$ is odd can be applied again to them. Let $|t_{i/j}|$ denote the number of nonzero bits of $t_{i/j}$. In case 1, $|t^*_{i/i+1}| = 1$ and $|t_{i/i+1}| = 2$; in case 2, $|t^*_{i/i+2}| = 2$ and $|t_{i/i+2}| = 2$; in case 3, $|t^*_{i/i+1}| = 1$ and $|t_{i/i+1}| = 2$; in case 4, $|t^*_{i/i+2}| = 2$ and $|t_{i/i+2}| = 2$. Thus all possible cases are exhausted and no t yields a tag of smaller weight than t^*.

(ii) Next the proof of uniqueness for the tag generated by algorithm SHORTEST-PATH is shown; the proof is by contradiction. Suppose there exists another tag $t_{0/n-1}$ that also has no adjacent nonzero digits. Let i be the lowest index such that $t_i \neq t^*_i$; thus $t_{0/i-1} = t^*_{0/i-1}$ so that $\Delta_i(t) = \Delta_i(t^*)$. There are three possible cases, (a), (b) and (c), for $t_i \neq t^*_i$. (a) $t_i = \bar{1}$ and $t^*_i = 1$ (or vice versa); then $t_{i+1} = 0 = t^*_{i+1}$, since $t_i \cdot t_{i+1} = 0$ and $t^*_i \cdot t^*_{i+1} = 0$. But this is impossible because $\Delta_i(t)(= \Delta_i(t^*))$ is odd so that only either $t_{i+1} = 0$ or $t^*_{i+1} = 0$ (Lemma 2.3). A contradiction results. (b) $t_i = 1$ and $t^*_i = 0$ (or vice versa). Then $\Delta_i(t)$ is odd and $\Delta_i(t^*)$ is even. But this is impossible because $\Delta_i(t) = \Delta_i(t^*)$. A contradiction results. (c) $t_i = \bar{1}$ and $t^*_i = 0$ (or vice versa). The discussion is exactly the same as case (b).

(iii) Existence of the other optimal solution is shown for $t^*_{n-1} = 1$. The case that $t^*_{n-1} = \bar{1}$ can be treated analogously. If $t^*_{n-1} = 1$, then $\Delta = (\sum_{i=0}^{n-2} t^*_i 2^i + 2^{n-1})$ $\bmod 2^n = (\sum_{i=0}^{n-2} t^*_i 2^i + 2^{n-1} - 2^n) \bmod 2^n = (\sum_{i=0}^{n-2} t^*_i 2^i - 2^{n-1}) \bmod 2^n$; so t^*_{n-1} can also be $\bar{1}$ and the rest of digits remain unchanged. $\square$

Actually the proof of Theorem 2.4 has a much stronger implication regarding optimality of a tag than just verifying existence of a minimum weight tag that has no adjacent nonzero digits. It is stated as Corollary 2.5.

Corollary 2.5 A feasible solution to $(\bar{P})$ is optimal if it has no adjacent nonzero digits.

Proof: From the process of generating each digit in algorithm SHORTEST-PATH, the feasible solutions with no adjacent nonzero digits are either unique or different only at t_{n-1} (1 or $\bar{1}$). There exists an optimal solution to $(\bar{P})$ that has no adjacent nonzero digits. So the feasible solution with no adjacent nonzero digits must be also an optimal solution. □

Corollary 2.5 only guarantees optimality of a tag that has no adjacent nonzero digits; a tag with adjacent nonzero digits may as well be a minimum weight tag. For instance, for $n = 4$, $\Delta = -6$, the tag of distance Δ can be $t_{0/3} = 0\bar{1}\bar{1}0$ or $t_{0/3} = 010\bar{1}$; both tags have a minimum weight of two.

Corollary 2.6 [3] The maximum number of links on the shortest path in the PM2I network from any source to any destination is $\lceil n/2 \rceil$, i.e.

$$\max_{0 \le \Delta \le (N-1)} H_\Delta = \lceil n/2 \rceil$$

Proof: From Theorem 2.4, there exists a minimum weight tag with no adjacent nonzero digits for every distance Δ. The maximum number of nonzero digits of such a minimum weight tag is $\lceil n/2 \rceil$; i.e. the tag consists of alternating 1 and 0 digits. □

Algorithm SHORTEST-PATH is capable of finding a minimum weight routing tag for the ADM network, which can be converted to a combination tag for the PM2I network, and also deduces that the number of hops is bounded above by $\lceil n/2 \rceil$. This knowledge can be further used to investigate properties of the ADM network.

27.3 Construction of Half Augmented Data Manipulator Networks

Corollary 2.6 indicates that the shortest path between any two nodes in PM2I network uses at most $\lceil n/2 \rceil$ links, which implies that $\lceil n/2 \rceil$ is the least number of stages needed in a multistage network based on PM2I functions, where any source can be connected to any destination in one pass. Furthermore, from Theorem 2.4 it is possible to infer how such a network can be constructed. For convenience of discussion, assume n to be even hereafter. The path in the ADM network defined by the routing tag that has no adjacent nonzero digits includes only one of the links -2^{2k+1}, -2^{2k}, $+2^{2k}$, $+2^{2k+1}$ and straight link, for every stage k, $0 \le k \le (n/2) - 1$. This implies that the links of two adjacent stages $2k$ and $2k+1$ in the ADM network can be coalesced into one stage and thus the total number of stages is reduced to

---

[3] An equivalent result is reported in [HwB84]. We were unable to identify the original reference which first reported this result.

$n/2$. The network is called a *Half ADM (HADM) network* . The HADM network consists of $n/2$ stages ordered from 0 to $(n/2)$ -1 from the output side to the input side. An extra column of switches is appended in the input side and is referred to as stage $n/2$. A source is a switch at stage $n/2$ and a destination is a switch at stage 0. Switch j of stage $k+1$ has five output links to switches of stage k : $(j + 2^{2k+1})$, $(j + 2^{2k})$, j, $(j - 2^{2k})$ and $(j - 2^{2k+1})$. An HADM network of size $N = 16$ is shown in Figure 27.4.

The tag generated by algorithm SHORTEST-PATH can be used as a routing tag in the HADM network. Close examination of the topology of the HADM network reveals that there exists latitude in using tags other than the ones with no adjacent nonzero digits to control routing in the HADM network; i.e. two adjacent digits of a routing tag can be both nonzero. Since, for a given source/destination pair, only one of the links -2^{2k+1}, -2^{2k}, $+2^{2k}$, $+2^{2k+1}$ and straight link is used for routing in the HADM network, as long as the tag satisfies the constraint that $t_{2k} \cdot t_{2k+1} = 0$ for $0 \leq k \leq (n/2) - 1$, it is a valid routing tag in the HADM network. There are five possible combinations for such a pair of digits $t_{2k}t_{2k+1}$: $\bar{1}0$, 10, 00, 01 and $0\bar{1}$. If $t_{2k}t_{2k+1} = \bar{1}0$, link -2^{2k} is used; if $t_{2k}t_{2k+1} = 10$, link $+2^{2k}$ is used; if $t_{2k}t_{2k+1} = 00$, straight link is used; if $t_{2k}t_{2k+1} = 01$, link $+2^{2k+1}$ is used; if $t_{2k}t_{2k+1} = 0\bar{1}$, link -2^{2k+1} is used. The routing tags representing the same distance Δ in the HADM network are called the *equivalent routing tags* . The multitude of equivalent routing tags suggests that there may exist multiple paths for some source/destination pairs. If a routing tag has no equivalent routing tags, it is unique, and only one routing path exists for the source/destination pair.

Recall that algorithm SHORTEST-PATH always generates a zero digit whenever possible. If Δ_i is even, t_i is uniquely confined to be 0; if Δ_i is odd (for which t_i can be 1 or $\bar{1}$), then it chooses the value for t_i such that $t_{i+1} = 0$.

This constraint can be relaxed for generating equivalent routing tags for the HADM network. For $t_{2k} = 0$ and Δ_{2k+1} odd, two subsets of equivalent tags can be generated by choosing 1 for t_{2k+1} for one of them and by choosing $\bar{1}$ for t_{2k+1} for the other. That is, if $t_{2k}t_{2k+1} = 0\,\bar{1}$ or 01, both 1 or $\bar{1}$ can be considered for t_{2k+1} to form equivalent routing tags, since it's always possible to make t_{2k+3} zero by properly choosing a value for t_{2k+2} (Lemma 2.3) and satisfy the constraint $t_{2k} \cdot t_{2k+1} = 0$ and $t_{2k+2} \cdot t_{2k+3} = 0$. For example, there are two paths from $S = 3^2$ to $D = 13^0$ in an HADM network of size $N = 16$, which are specified by the tags $t_{0/3} = 0\bar{1}\bar{1}0$ ($\Delta = -6$) and $t_{0/3} = 0101$ ($\Delta = 10$), respectively. In this example, t_0t_1 can be $0\bar{1}$ or 01; particularly the tag $t_{0/3} = 0\bar{1}\bar{1}0$ is obtained by choosing $t_2 = \bar{1}$ so that $t_3 = 0$.

Similar to the relationship between the ADM and IADM networks, the *Half IADM (HIADM) network* has the same topology as the HADM network with the input and output sides exchanged. A tag $t_{0/n-1}$ for routing from $S^{n/2}$ to D^0 in the HADM network can be conveniently converted to $t^c{}_{0/n-1}$, where $t^c{}_i = -t_i$, $0 \leq i \leq n-1$, for routing from D^0 to $S^{n/2}$ in the Half IADM network. Note that tag $t^c{}_{0/n-1}$ also satisfies the constraint $t^c{}_{2k} \cdot t^c{}_{2k+1} = 0$, $0 \leq k \leq (n/2) - 1$.

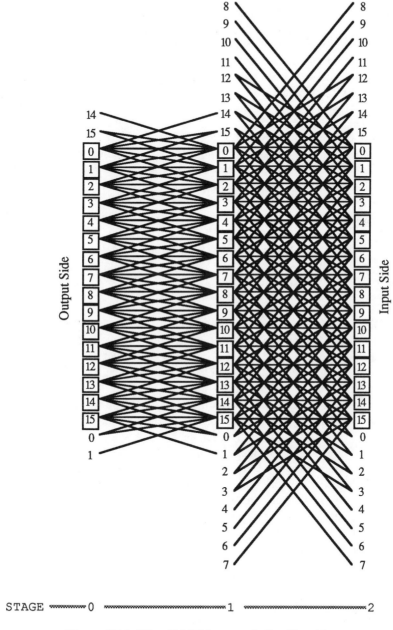

Figure 27.4: The $HADM$ network for $N = 16$.

It was shown that for some source and some destination in the ADM network, there exists only a path between them; the same is true in the HADM network. For example, routing for a distance $\Delta = 0$ in an HADM network of size $N = 16$ has a unique tag $t_{0/3} = 0000$, which represents a path consisting of all straight links. Thus the HADM network is not fault-tolerant. It is interesting to attempt further reduction of the network complexity while maintaining the connection between any source and any destination. It is shown in Theorem 3.1 that actually only four output links for each switch would suffice to provide connection for any source/destination pair in the HADM network.

Consider a quad-tree that consists of $\log_4 N$ levels and N leaves. Clearly the out-degree of four for each node in the quad-tree is the smallest out-degree such that the root can reach any leaf; if any node except a leaf has an out-degree less than four, some leaves can not be reached by the root. Similarly, for a network of size N that consists of $\log_4 N$ stages of uniform switches, at least four output links for each switch are needed so that any source can communicate with any destination. Such a network has the minimum number of output links for each switch required for one-to-one connections and is called a *Minimum ADM (MADM) Network* . It consists of $n/2$ stages of 4×4 switches. Each switch of stage $k+1$, $0 \le k \le (n/2)-1$, is connected to four output links: straight link, $+2^{2k}$, -2^{2k} and $+2^{2k+1}$. Figure 27.5 illustrates a MADM network of size $N = 16$.

The MADM and HADM networks differ only in that each switch of stage k in the MADM network is connected to only one of the $+2^{2k+1}$ and -2^{2k+1} links while each switch of stage k in the HADM network is connected to both links. So only a subset of routing tags for the HADM network are valid routing tags for the MADM network. In addition to the constraint that $t_{2k} \cdot t_{2k+1} = 0$ for $0 \le k \le (n/2) - 1$, which a routing tag for the HADM network must satisfy, a valid tag for the MADM network must also satisfy the second constraint that, for Δ_{2k+1} odd, t_{2k+1} must be 1 if link $+2^{2k+1}$ is used and t_{2k+1} must be $\bar{1}$ if link -2^{2k+1} is used. The second constraint does not specify which of links $+2^{2k+1}$ and -2^{2k+1} is used at stage k; each stage can choose freely a plus or minus link. As a result, there are as many as $2^{n/2}$ types of MADM network; they differ in their choice of link $+2^{2k+1}$ or -2^{2k+1} at some stage k. The algorithm MADM-TAGS below demonstrates an example of generating routing tags for a particular type of MADM network that contains $+2^{2k+1}$ link at every stage k, $0 \le k \le (n/2) - 1$. For convenience of discussion, this network is referred to as the MADM network.

Algorithm MADM-TAGS $(\Delta, t_{0/n-1})$
$\Delta_0 = \Delta$
for $i = 0$ to $n - 1$ do
    if Δ_i is even then $t_i = 0$, $\Delta_{i+1} = \frac{\Delta_i}{2}$
    else if i is even then if $\frac{\Delta_i - 1}{2}$ is even then $t_i = 1$, $\Delta_{i+1} = \frac{\Delta_i - 1}{2}$
                else $t_i = \bar{1}$, $\Delta_{i+1} = \frac{\Delta_i + 1}{2}$
                endif
            endif
        endif
    enddo

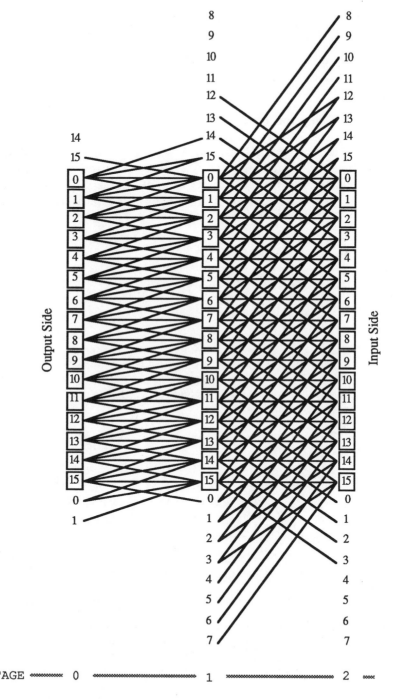

Figure 27.5: The MADM network for $N = 16$.

The difference between the processes of generating tags for the HADM network and for the MADM network is that, for Δ_{2k+1} odd, t_{2k+1} can be 1 or $\bar{1}$ for the HADM network while t_{2k+1} can only be 1 for generating routing tags for the MADM network. So each digit is uniquely determined in algorithm MADM-TAGS. This indicates that there exists a unique tag for each distinct Δ, which corresponds to a unique path for each source/destination pair in the MADM network.

Since there are only four output links for each switch in the MADM network, two bits per stage suffice to represent the choice of one of the four output links of a switch to send data. A total of n bits are needed to implement the signed-digit representations for routing tags. Let $r_{0/n-1}$ be such a routing tag, in which a digit can be represented by a bit. Each switch at stage k in the MADM network examines bits $r_{2k}r_{2k+1}$ to determine the output link via which data are routed. One possible implementation is shown below.

$$r_{2k}r_{2k+1} = \begin{cases} 11 \rightarrow +2^{2k+1} \\ 00 \rightarrow \text{straight} \\ 01 \rightarrow +2^{2k} \\ 10 \rightarrow -2^{2k} & \text{for } 0 \leq k \leq (n/2) - 1 \end{cases}$$

where $\rightarrow$ means "en route".

However, for the generation of tags in algorithm MADM-TAGS, two bits may be needed to represent a digit of the routing tag and thus a total of $2n$ bits are needed. Once the computation is done, the tag can be converted to $r_{0/n-1}$ for actual routing, which requires only n bits per tag.

Theorem 3.1 There exists a unique path between any source and any destination in the MADM network.

Proof: It is shown that a routing tag $t_{0/n-1}$ for the HADM network that contains $t_{2k}t_{2k+1} = 0\bar{1}$ can be recoded to become $t'_{0/n-1}$ such that $t'_{2k}t'_{2k+1} = 01$ and $t'_{2j} \cdot t'_{2j+1} = 0$, for $0 \leq j \leq (n/2)-1$. Case (i) If $t_{2k+2} = 0$ such that $t_{2k/2k+2} = 0\bar{1}0$, then $t'_{2k/2k+3} = 01\bar{1}t'_{2k+3}$ or $011t'_{2k+3}$. Since Δ_{2k+2} is odd ($t'_{2k+2} = 1$ or $\bar{1}$), from Lemma 2.3, either $011t'_{2k+3}$ or $01\bar{1}t'_{2k+3}$ has $t'_{2k+3} = 0$. Case (ii) If $t_{2k+2} \neq 0$ then t_{2k+3} must be equal to 0, because a tag for HADM network must satisfy the constraint $t_{2k+2} \cdot t_{2k+3} = 0$. If $t_{2k/2k+3} = 0\bar{1}10$, $t'_{2k/2k+3} = 0100$, and if $t_{2k/2k+3} = 0\bar{1}\bar{1}0$ then $t'_{2k/2k+3} = 010\bar{1}$. The discussion for recoding $t'_{2k+2}t'_{2k+3} = 0\bar{1}$ is analogous to that for recoding $t_{2k}t_{2k+1} = 0\bar{1}$. Next uniqueness of the routing path is shown. Since the out-degree of every switch in the MADM network is four and there are $n/2 = (\log N)/2 = \log_4 N$ stages, each source switch and all switches connected to it form a quad-tree. The source switch is the root and the switches connected to it are the nodes in the quad-tree, with the switches of stage 0 as the leaves. There exists a unique path from a root to a leaf in the quad-tree and thus also a unique path from a source to a destination in the MADM network. ☐

The topology of the *Minimum Inverse ADM* (*MIADM*) network is the same as the MADM network, with the input and output sides reversed, much like the relationship between the ADM and IADM networks and between the HADM and HIADM networks. Especially the routing tag conversion technique used for the HADM and HIADM networks can be readily applied and the proposed routing scheme for the HADM network can also be used in the HIADM network.

27.4 The Extra Stage MADM Network

Complexity of the MADM network is minimum in the sense that, given the constraint of network size N and $\log_4 N$ stages of uniform switches, it can provide communication for any source/destination pair in the network by using minimum number of interstage links per stage. However, this kind of topology has a drawback that it does not provide fault-tolerance; a switch failure would prevent some source/destination pairs from communicating to each other. The lack of fault-tolerance suggests the use of augmentation techniques [AdS81] to improve fault-tolerance for the MADM network. First an important observation for routing in the MADM network is made.

Theorem 4.1 In the MADM network, the paths from a source S, to destinations D, ($D + (N/4)$), ($D - (N/2)$) and ($D - (N/4)$) are all disjoint.

Proof: The proof shows only that the two paths from S to D and from S to $(D - (N/4))$ are disjoint; the other cases can be treated similarly. The proof consists of two parts: (A) given the tag $t_{0/n-1}$ for routing from S to D, a tag $t'_{0/n-1}$ for routing from S to $(D - (N/4))$ can be derived from it and they differ only in digits $n - 2$ and $n - 1$, and (B) proof of disjointness of the two paths based on the results in (A).

(A) Since $t_{0/n-1}$ is the routing tag from S to D, $(D - S) = (\sum_{i=0}^{n-2} t_i 2^i + t_{n-1} 2^{n-1}) \bmod 2^n$. So $(D - S - (N/4)) = (\sum_{i=0}^{n-3} t_i 2^i + (t_{n-2} - 1)2^{n-2} + t_{n-1} 2^{n-1}) \bmod 2^n$. There are three possible values, 1, $\bar{1}$ and 0, for t_{n-2}, which are discussed in cases (i), (ii) and (iii), respectively, as follows. (i) If $t_{n-2} = 1$, $(D - S - (N/4)) = (\sum_{i=0}^{n-3} t_i 2^i + 0 \cdot 2^{n-2} + t_{n-1} 2^{n-1}) \bmod 2^n$. That is, $t'_{0/n-1} = t_{0/n-3} 0 t_{n-1}$. (ii) If $t_{n-2} = \bar{1}$, t_{n-1} must be 0 because $t_{n-2} \cdot t_{n-1} = 0$ and $(D - S - (N/4)) = (\sum_{i=0}^{n-3} t_i 2^i + 0 \cdot 2^{n-2} + (t_{n-1} - 1)2^{n-1}) \bmod 2^n$. Then $t'_{0/n-1} = (\sum_{i=0}^{n-3} t_i 2^i + 0 \cdot 2^{n-2} - 2^{n-1}) \bmod 2^n = (\sum_{i=0}^{n-3} t_i 2^i + 0 \cdot 2^{n-2} - 2^{n-1} + 2^n) \bmod 2^n = (\sum_{i=0}^{n-3} t_i 2^i + 0 \cdot 2^{n-2} + 2^{n-1}) \bmod 2^n$. So $t'_{0/n-1} = t_{0/n-3} 01$. (iii) If $t_{n-2} = 0$, $(D - S - (N/4)) = (\sum_{i=0}^{n-3} t_i 2^i - 2^{n-2} + t_{n-1} 2^{n-1}) \bmod 2^n$. There are two possible values, 0 and 1, for t_{n-1}, which are discussed in cases (a) and (b), respectively. t_{n-1} can not be $\bar{1}$ because it is assumed that link $+2^{n-1}$ is used at stage $n/2$ in the MADM network. (a) If $t_{n-1} = 0$, $t'_{0/n-1} = t_{0/n-3} \bar{1} 0$. (b) If $t_{n-1} = 1$, $(D - S - (N/4)) = $

$(\sum_{i=0}^{n-3} t_i 2^i - 2^{n-2} + 2^{n-1}) \bmod 2^n = (\sum_{i=0}^{n-3} t_i 2^i + 2^{n-2} + 0 \cdot 2^{n-1}) \bmod 2^n$ so that $t'_{0/n-1} = t_{0/n-3} 10$.

(B) From (A), it is seen that the two routing tags for the two paths from S to D and from S to ($D - (N/4)$) differ only in digits $n - 2$ and $n - 1$; i.e. $t_i = t'_i$ for $0 \le i \le n - 3$. The two tags are the unique tags for routing from S to D and from S to ($D - (N/4)$), respectively. Let F and F' be the two switches at stage $(n/2) - 1$ on the paths from S to D and from S to ($D-(N/4)$), respectively. Since $t_i = t'_i$, for $0 \le i \le n - 3$, $\sum_{i=0}^{n-3} t_i 2^i = \sum_{i=0}^{n-3} t'_i 2^i$ (i.e. the distances that the two paths traverse from stage $(n/2) - 1$ to 0 are the same), and the distance between the two destinations D and $(D - (N/4))$ is ($(N/4) \bmod N$); hence the distance between F and F' must be also ($(N/4) \bmod N$), denoted $|F - F'| = (N/4) \bmod N$. The intermediary switches at stage k, $0 \le k \le (n/2) - 2$, on the two paths are $F + \delta_k$ and $F' + \delta_k$, respectively, where $\delta_k = \sum_{i=k}^{(n/2)-2}(t_{2i} 2^{2i} + t_{2i+1} 2^{2i+1})$. But ($F + \delta_k) \ne (F' + \delta_k) \bmod 2^n$ because $|F - F'| = (N/4) \bmod N$. That is, the two paths never share a common intermediary switch and thus are disjoint.   ⬜

The identification of disjoint paths from a source to different destinations in Theorem 4.1 can be used to improve fault-tolerance for the MADM network. The technique is to add an extra stage to the MADM network. The extra stage can be placed in the output side of the MADM network such that each switch D at the extra stage is connected to four switches at the first stage of the MADM network: D, $(D + (N/4))$, $(D - (N/2))$ and $(D - (N/4))$. Data can be sent from source S to any of the four switches and then to the destination via the extra stage. Thus there exist four disjoint paths from any source to any destination in the extra stage network. Such a network with an extra stage in the output side of the MADM network is called an *extra stage MADM network*. An extra stage MADM network consists of $(n/2)+1$ stages labeled from 0 to $n/2$ from the output side to the input side, with an additional column of switches in the input side referred to as stage $(n/2) + 1$. The extra stage in the extra stage MADM network consists of the switches of stage 0 and the input links of the switches. The topology of the extra stage MADM network from stage 1 to $(n/2)+1$ is the same as that of the MADM network from stage 0 to $n/2$. The extra stage MADM network is three-fault-tolerant because of the existence of four disjoint paths for every source/destination pair and thus can withstand at least three switch failures (except the input and output switches). Since each destination in the MADM network has four input links, which are connected to four switches in the preceding stage, at most three-fault-tolerance is possible. By appending an extra stage to the MADM network, the optimal fault tolerance is achieved (except the input and output switches).

Since the four output links of a switch at the extra stage are straight link and links -2^{n-2} ($= -N/4$), $+2^{n-2}$ ($= N/4$) and -2^{n-1} ($= -N/2$), the extra stage has the same connection patterns as stage $n/2$ of the MADM network. The routing tags for the MADM network can be used directly for routing from stage $(n/2)+1$ to stage 1 in the extra stage MADM network. Using the tags of distances

$\Delta = (D-S)$, $\Delta = (D-S-(N/4))$, $\Delta = (D-S-(N/2))$, and $\Delta = (D-S+(N/4))$, respectively, a source S (a switch at stage $(n/2)+1$) in the extra stage MADM network can send data to any of the four switches D, $(D+(N/4))$, $(D-(N/4))$ and $(D-(N/2))$ at stage 1, and then reaches the destination D at stage 0. The routing from D^1 to D^0 is controlled by tag bits 00, from $(D+(N/4))^1$ to D^0, by $\bar{1}0$, from $(D-(N/2))^1$ to D^0, by 01, and from $(D-(N/4))^1$ to D^0, by 10. So in the extra stage $MADM$ network, $n+2$ bits are needed to represent a routing tag. Note that since the four tags of distances $\Delta = (D-S)$, $\Delta = (D-S-(N/4))$, $\Delta = (D-S-(N/2))$, and $\Delta = (D-S+(N/4))$ differ only in digits $n-2$ and $n-1$, once one of them is computed, the other can be readily computed by recoding the last two digits. The proof of Theorem 4.1 demonstrates the example of recoding the tag of distance $\Delta = (D-S)$ to a tag of distance $\Delta = (D-S-(N/4))$. Table 27.6 below summarizes the recoding of digits $t_{n-2}t_{n-1}$ of a tag into the other three tags that are of distance $+N/4$, $-N/4$ and $-N/2$ from it.

Figure 27.6 illustrates an extra stage MADM network of size $N = 16$. It is also shown the four disjoint paths from $S = 3^3$ to $D = 12^0$. They are represented by the tags of distances $\Delta = 9$, $\Delta = 5$, $\Delta = 1$ and $\Delta = -3$, which are $(t_{0/3} =)$ $1001, 1010, 1000$ and $10\bar{1}0$, respectively. The routing paths are ($3^3, 11^2, 12^1, 12^0$), ($3^3, 7^2, 8^1, 12^0$), ($3^3, 3^2, 4^1, 12^0$) and ($3^3, 15^2, 0^1, 12^0$), respectively. Routing from 12^1 to 12^0 is controlled by tag bits 00, from 8^1 to 12^0, by 10, from 4^1 to 12^0, by 01, and from 0^1 to 12^0, by $\bar{1}0$.

It can be similarly shown that an extra stage can also be appended in the input side of the MADM network such that a switch S at the extra stage is connected to four switches at stage $n/2$ of the MADM network: S, $(S+1)$, $(S-1)$ and $(S+2)$. Four disjoint paths result from addition of such an extra stage to the MADM network. In this type of extra stage network, the extra stage consists of the switches of stage $(n/2)+1$ and the output links of the switches, and stage $n/2$ to stage 0 has the same topology as the MADM network. The extra stage appended in the input side has the same connection patterns as stage 1 of the MADM network. A source S at the extra stage can send data to any of the four switches at stage $n/2 : S$, $(S+1)$, $(S-1)$ and $(S+2)$ that are directly connected to it and uses tags of distances $\Delta = (D-S)$, $\Delta = (D+1-S)$, $\Delta = (D-1-S)$ and $\Delta = (D-2-S)$, respectively, to send data to the destinations D at stage 0. The routing from $S^{(n/2)+1}$ to $S^{n/2}$ is controlled by tag bits 00, from $S^{(n/2)+1}$ to $(S+1)^{n/2}$, by 10, from $S^{(n/2)+1}$ to $(S-1)^{n/2}$, by $\bar{1}0$, and from $S^{(n/2)+1}$ to $(S+2)^{n/2}$, by 01.

Table 27.6:

	$+N/4$	$-N/4$	$-N/2$
00	10	$\bar{1}0$	01
01	$\bar{1}0$	10	00
10	01	00	$\bar{1}0$
$\bar{1}0$	00	01	10

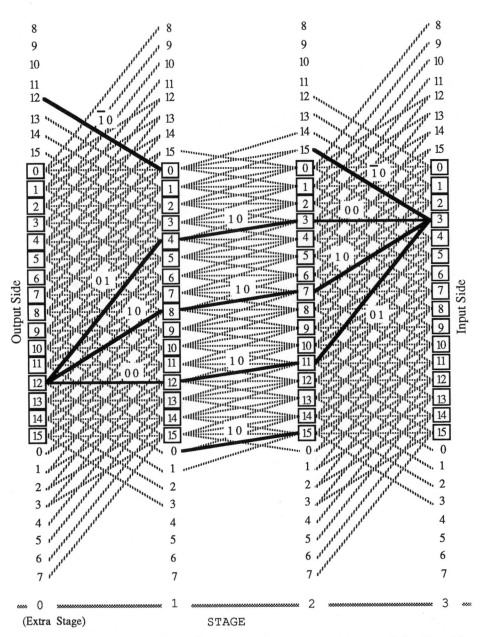

Figure 27.6: Routing from 3^3 to 12^0 in the extra stage MADM network for $N = 16$. The solid lines are links on the routing paths and the dotted lines are links of the extra stage MADM networks. Labels on the links are digits of the routing tags.

Apparently adding an extra stage to the input side of an MADM network is equivalent to adding the extra stage to the output side of the MIADM network and vice versa. Thus all discussions associated with the relationship between the MADM and MIADM networks can be applied to the extra stage networks as well.

27.5 Conclusion

This paper addresses the problem of designing multistage networks which are based on the implementation of PM2I functions at each stage. This type of multistage networks is referred to as augmented data manipulator networks and includes the well known ADM and IADM networks. Since the designs proposed in this paper use fewer stages and links than the ADM and IADM networks, they are referred to as partially augmented data manipulator networks. The HADM and HIADM networks derived in this paper have the least number of stages required in multistage networks (based on PM2I functions) where any source can be connected to any destination in one pass. The MADM and MIADM networks also have the least number of stages and, in addition, have the minimum number of links per switch required for one-to-one connections. The extra stage MADM and MIADM networks contain one more stage than the MADM and MIADM networks, respectively, and are fault-tolerant versions of the MADM network capable of tolerating at least three switch faults.

References

[AdS81] G. B. Adams III and H. J. Siegel, "The Extra Stage Cube: A Fault-Tolerant Interconnection Network for Supersystems," *IEEE Trans. Computers*, Vol. C-30, No. 5, pp. 443-454, May 1981.

[Avi61] A. Avizienis, "Signed-Digit Number Representations for Fast Parallel Arithmetic," *IRE Trans. Electronic Computers*, pp. 389-400, Sept. 1961,

[Fen74] T-Y Feng, "Data Manipulating Functions in Parallel Processors and Their Implementations," *IEEE Trans. Computers*, Vol. C-23, No. 3, pp. 309-318, Mar. 1974.

[HwB84] K. Hwang and F. A. Briggs, *Computer Architecture and Parallel Processing*, McGraw-Hill Book Company, NY, pp. 345, 1984.

[McS82] R. J. McMillen and H. J. Siegel, "Routing Schemes for the Augmented Data Manipulator Network in an MIMD System," *IEEE Trans. Computers*, Vol. C-31, No. 12, pp. 1202-1214, Dec. 1982.

[PaR82] D. S. Parker and C. S. Raghavendra, "The Gamma Network: A Multiprocessor Interconnection Network with Redundant Paths," *9th Annu. Symp. on Computer Architecture*, pp. 73-80, Apr. 1982.

[PaR84] D. S. Parker and C. S. Raghavendra, "The Gamma Network," *IEEE Trans. Computers*, Vol. C-33, No. 4, pp. 367-373, Apr. 1984.

[Sie77] H. J. Siegel, "Analysis Techniques for SIMD Machine Interconnection Networks and the Effects of Processor Address Masks," *IEEE Trans. Computer*, Vol. C-26, No. 2, pp. 153-161, Feb. 1977.

[SiS78] H. J. Siegel and S. D. Smith, "Study of Multistage SIMD Interconnection Networks," *5th Annual Symp. on Computer Architecture*, pp. 223-229, Apr. 1978.

Chapter 28

The Design of Inherently Fault-Tolerant Systems [1]

Lee A. Belfore, II [2]
Barry W. Johnson [2]
James H. Aylor [2]

28.1   Introduction

With very large scale integrated (VLSI) circuits having as many as 10 million circuit components on a silicon chip and the potential for an order of magnitude more when considering wafer scale integration (WSI), two questions arise. First, how can all these components be organized so that a useful function is performed? For a device such as a random access memory, the organization is relatively obvious. However, if processing elements are considered, the organization is significantly complicated and may change, depending upon the application. Second, how can all these circuit components work together reliably? One approach, known as fault avoidance, is to manufacture the parts using very rigorous and expensive procedures and to take into consideration all possible environmental influences when designing, fabricating, and packaging the device. A second approach is to employ fault-tolerant design techniques.

This paper attempts to address both questions through discussion, analysis and the introduction of a concept referred to as inherent fault tolerance. Simply stated, an inherently fault-tolerant system has fault-tolerant properties that are inseparable from its functional capabilities. Inherent fault tolerance is present in a number of different systems. Examples include the brain, holograms, and certain proven redundancy techniques such as triple modular redundancy.

[1] This work was supported by the Virginia Center for Innovative Technology under the contract number 5-30971
[2] University of Virginia, Charlottesville, VA

A psychologist many years ago conducted studies on the brains of rats by training the rats to run a maze [12]. After training the rats, he removed parts of their cerebral cortex and found that the rat's ability to navigate the maze was related to the quantity of the cortex removed. There was no apparent relation between maze running skill and the specific region of the cortex removed; the ability to navigate was apparently distributed uniformly throughout the brain. The researcher was attempting to remove the ability to run the maze under the assumption that the skill was localized within the brain. However, the memory was distributed throughout the brain in such a way that if part of the brain was damaged, the remaining undamaged memory would still function.

Holograms recreate a three dimensional wavefront of the object or information which is contained within them [13]. If a hologram is cut into smaller pieces, the information is recreated of the entire hologram, even though only a small piece of the original hologram is used. The resolution of the original image is degraded, however, in some relation to the amount of the hologram which has been removed.

Triple modular redundancy (TMR) is a fault-tolerant design technique where a module is triplicated and the outputs of each are voted upon to determine the result. If only the input and the output are observed (and fault-free), the module exhibits inherent fault tolerance in that a single failure within the TMR module can be tolerated. The TMR system either functions in a fault-free or failed functional capacity; there is no degraded functional state. If TMR is used with the technique of flux summing, inherent fault-tolerant properties can be observed [1]. If the failures are in such a way that the output of the failed module(s) can be compensated for, the flux summer will average the results of each module and in effect distribute the effect of the fault(s).

We will look at the neural network as an example of an inherently fault tolerant system. The purpose of this paper is to present our results in this direction and to provide an initial interpretation of these results.

The remainder of this paper is organized into five sections. Section 28.2 provides background material necessary for further developments in the paper, including an overview of the important literature. Section 28.3 describes, in detail, the neural network studied through simulations while section 28.4 summarizes the simulations performed. Section 28.5 discusses the properties of inherently fault-tolerant systems and section 28.6 gives directions toward future work.

28.2　Background

The concept of inherent fault tolerance is related to a number of different areas of research and technology; the most obvious is fault tolerance. Additional background material on neural networks and their inherent fault tolerance capabilities will be presented.

28.2.1   Fault-Tolerant System Terminology

This section provides a brief overview of the important terminology in the fault tolerance field. Fault-tolerant systems have the ability of continuing to function in the presence of a failure of one or more of its subsystems [1]. Several key terms in the fault tolerance field include failure, fault, error and malfunction. The meaning of each term is best illustrated via the four-universe model which consists of the physical, logical, informational, and external universes [2]. Failures occur in the physical universe. A failure is when something physically breaks. The logical universe is where faults occur. A fault is the incorrect state of the system as a result of a failure. The informational universe is where a discrepancy in the function occurs as a result of the fault. The external universe is where malfunctions occur, which is, the system is not performing its designed function due to some error in the system.

The effect of fault tolerance in a system can be measured in many ways. Each measure can be considered a design goal because their requirements on system fault-tolerant performance are very different. For example, *reliability* is the probability that the system will be functioning over some period of time. *Availability* is the probability that the system will be functioning at some specific time. *Safety* is the probability that the system will either operate correctly or fail in a safe way. In other words, the system knows that it is failed and can force itself into an innocuous state. *Maintainability* is the probability that a failed system will become operational within some period of time [1]. *Performability* is the probability that a system will have at least some level of performance [3].

28.2.2   Neural Networks

A neural network is some interconnection of biological neurons. The importance of neural networks is seen through examples of some neural systems from nature. The human brain, a very complex neural network, has the ability to efficiently solve problems in pattern recognition (face recognition), optimization (the quickest way to get from point A to B in rush hour), and control systems (any motor activity). Another important capacity is the ability to learn and apply new skills. Also, the human neural system has managed to incorporate a large amount of parallelism into its implementation. An example is the vision system where many sensors, rods, and cones in the eye are monitored, and processing of the input is done by many neurons working synergistically in parallel. The neural networks function with an asynchronous parallelism that digital systems typically do not possess.

Two methods can be used to emulate neural networks. First, an electrical model of a neural network can be constructed out of electrical components which maintain some of the properties of the biological neural network [4-7]. The electrical model can be further emulated through simulation. Second, optical neural systems have been developed that work in similar fashion [8].

In the electrical model, each neuron is constructed from three structures; these structures are the summing network, the delay network, and the nonlinear amplifier. Summing networks take the outputs from various neurons and combine them by a weighted average of those neurons. The weight for each input is modeled by a conductance (inverse of resistance). The term "weights" will be synonymous with connection or conductance throughout this paper. The smaller the conductance, the smaller the weight assigned to that input. In the most abstract sense, all neurons can be seen to connect to all other neurons, with a 0 conductance corresponding to no connection. The delay network is modeled by a parallel resistor-capacitor network that simulates the temporal properties of the neuron This provides a delay for electrical currents from the summing network to dissipate. A nonlinear amplifier is driven by the output of the summing network. The output of the nonlinear amplifier is a voltage between 0 and 1. The nonlinearity is designed so that neuronal outputs of the network will converge to values near 0 or 1. The electrical model is shown in Figure 28.1.

The neurons can be inverting in one of two senses. First, the output can be the negative of the non-inverting output. Second, the output can be simply 1.0 - (the non-inverting output) since for a properly chosen amplifier transfer function, the difference between these two representations is a constant. The shape of the transfer function will not be changed. The transfer function of the nonlinear amplifier used in our simulations is described by

$$f(x) = x^p/(x^p - (1-x)^p)$$

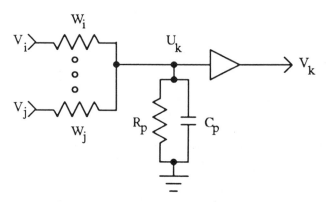

Figure 28.1: The electrical neural model.

Where the parameter p adjusts the steepness of the non-linearity. Note that $0 \leq f(x) \leq 1$ for $0 \leq x \leq 1$.

One important aspect of neural networks is that they appear to function well in the presense of failures [5,9,10]. With much of today's design emphasis based on making larger and more reliable systems, approaches involving methodologies that automatically incorporate degrees of fault tolerance are important. One goal of fault-tolerant design methodologies is to build reliable systems out of potentially faulty components and subsystems. A conventional fault-tolerant design technique for digital systems is approached by designing the system with fault tolerance in mind. Various ad hoc techniques are applied that have worked reasonably well in the past. For example, coding schemes can improve a system's reliability by masking out both temporary and permanent faults. One researcher looking at a biological neural system found that 10 of 650 connections in a highly organized neural structure were incorrect, yet the system was able to function acceptably [11].

28.3   Neural-like Network Example

In this research, the travelling salesman problem has been implemented and investigated using neural-like networks [7]. The travelling salesman problem was chosen for a number of reasons. First, a neural implementation is known and has been studied in some detail. Second, it is a computationally difficult problem from the NP-complete class of problems [14] which is important because it demonstrates that difficult problems of current interest can be solved using neural-like networks. Third, the neural implementation performs the computations collectively, which is a feature of many neural implementations.

The travelling salesman problem is, given a list of cities and all intercity distances, to find the shortest circuit or tour such that each city is visited once. A map of the six-city travelling salesman problem used in this work is shown in Figure 28.2. The cities for this example were chosen at random and placed on a unit plane. This problem size was chosen because it is large enough that its behavior may not be obvious, yet small enough that simulations could be run in a reasonable amount of time. As for size, there are 36 neurons and 720 interconnections in this circuit.

One shortcoming of using the neural implementation of Hopfield and Tank [7] is that it does not appear to work well for larger problem sizes. No relationship exists between problem size and certain critical network parameters. Indeed, Hopfield and Tank had difficulties with a thirty city problem size or 900 neurons in total. Biological neural systems, however, appear to be organized in general into units of twenty to 100 neurons [10] and possibly not more than 1000 neurons [17].

Figure 28.3 shows a simplified schematic of a six city implementation of the travelling salesman problem. A two dimensional array of neurons is set up to represent

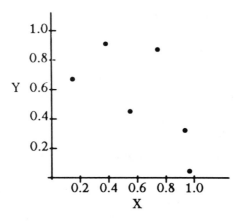

Figure 28.2: Six city map.

the problem with each neuron depicted in Figure 28.3 by a circle. The size of each dimension is the size of the problem with the row dimension corresponding to the city and the column dimension corresponding to the position within the tour that a city occupies. A valid tour corresponds to one neuron converging to 1 in each column and row and the rest being 0. If a neuron is near the 1 state it inhibits other neurons in the same row and column from being in the 1 state which guarantees that a valid state or tour will result. Neural interconnections are represented by weights. These weights can be divided into two classes where the first are those neurons guaranteeing valid solutions and the second programming the particular

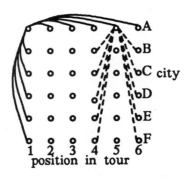

Figure 28.3: Simplified network schematic.

problem instance. The weights guaranteeing valid states are shown for a single neuron with the solid lines in Figure 28.3. The intercity distance information that corresponds to the particular problem instance, is programmed between columns. Longer distances between cities are programmed as larger inhibitions between the neurons corresponding to the state allowing these cities to be consecutive in the tour. The distance weights are shown for a single neuron with the dashed lines in the figure.

28.4 Simulation Results on Neuron Circuit Models

Two classes of simulations were performed on the neural network. The first one is a rather brute force approach of running a large number of simulations on a given problem and compiling the simulated results into a histogram. Because of the symmetry of the neural network, each simulation needs some perturbation injected into it at the start. The perturbation is with respect to the initial conditions or output voltage of the neurons. Our approach was to make this perturbation random and to look at the results from a statistical perspective. Statements on the behavior of the circuits can be observed from the shape a histogram produced from a number of simulations.

The second approach is a parametric analysis of fault behavior with respect to initial starting conditions of the network. A two dimensional plane is set up where the axes correspond to the relative initial conditions of two of the neurons required to solve the problem. The remaining neurons are set to the same initial condition. The resulting solution is plotted at the coordinates corresponding to the relative initial conditions. A shade is assigned to each solution with better solutions being assigned lighter shades. The patterns of the plotted results allow behavior to be observed.

28.4.1 Faults Studied

Three basic types of faults were investigated. The first is the random perturbation that causes movement off an unstable equilibrium point for the neural system simulation. This allows the system to converge to a legal solution. The second is stuck faults which have the characteristic that some circuit parameter is stuck in some state. The third is open circuit in the connections between neurons. While these could be considered stuck faults because the open connections remain open, they are treated differently because of the number of connections in the network. It is possible that the open connections could be the most significant failure mode for larger systems. For the travelling salesman problem, the number of neurons increase as the square of the problem size while the number of interconnections increase to the fourth power of the problem size.

The random perturbation is defined as follows. The initial condition for the network is 0.5 $\pm\delta$ for the output of each neuron where δ is the random perturbation. The control case with respect to the random perturbation is when no faults are in the circuit.

A set of simulations is run for each class of faults and magnitude of perturbation. Each simulation from this set consists of some appropriate element chosen at random to be faulted. All simulations in this set have the same magnitude of perturbation.

The following are classes of stuck faults that are simulated. A stuck-at-1 fault is when the output of a neuron is always 1. A stuck-at-0 fault is when the output of a neuron is always 0. A stuck-at-high impedance fault is when the output of a neuron does not drive the network to which it is connected.

Open connections were chosen at random at the beginning of each individual simulation. 1, 2, 5, 10, 15 and 20 open connections in the network were simulated. No distinction is made between connections that enforce valid solutions and those that program the particular problem instance.

28.4.2   First Approach to the Fault Analysis of Neural Like Networks

Description of Approach 1

The first approach was to run a large set of simulations each for a specified number of simulation cycles. Each simulation had a nominal starting point perturbed by a pseudo-random coin flipping. The results of the simulations were compiled into a histogram of the possible legal ending states. The legal ending states are ordered from best to worst. The best state is that tour that has the shortest, or optimal, tour. The worst state is that tour that has the longest possible tour. By viewing the histogram, properties of the circuit can be deduced. The simulations varied in two ways. The first is the magnitude of the perturbation, and the second are the faults injected into the circuit.

Simulation Set-up

Each simulation of the problem requires some sort of perturbation in order to converge to a usable result. By using this property, behavior can be observed by using different random starting perturbations. The intent is that the actual behavior could be observed by running a large number of simulations and looking at the results of this group of simulations as a whole. The number of simulations chosen was 500 because it was felt that this was large enough to provide meaningful results and because this number of simulations could be performed in a reasonable amount of time. Simulations were run for 500 simulation cycles. Again, 500 was chosen because it was deemed to be a large enough period of time to produce reasonable results. It is also a guard against cases which may converge to either an illegal tour or not converge at all in the period allotted for simulation. This issue is discussed in below .

Analysis of Histogram

The histogram of the simulation results can be used to approximate the probability density function of the expected results. Any of the 60 possible tours can be a solution. The histogram counts how many of each of these tours resulted from the 500 simulations. The histogram shows only those simulations that converged. See *Study of Nonconvergence* below for further discussions. The tours are plotted with respect to the length of the tour on one axis and with respect to the number of tours that resulted on the other. Initial observations of the histograms indicated that under some circumstances an exponential curve would best approximate the behavior of the histogram. A least mean square metric was used in some cases to find the parameters of this exponential. The circumstances under which this exponential parametric analysis is useful appear to be when the initial perturbation is relatively large. Another parametric approach is to compute the average tour from the histogram. Better performance is indicated by a shorter average tour. All results are presented in terms of the weighted average.

Discussion of Results for Approach 1

The analysis of the results is considered the average length of the tours for each set of simulations. In total, 28 sets of simulations were run. They correspond to two complete groups of 10 sets of simulations with the same perturbation and various fault scenarios. The additional 4 sets of simulations investigate the size of the perturbation on a fault-free simulation. The simulations are summarized in Table 28.1. Entries where a number is present correspond to that set of simulations being run. The number is extracted from the simulation data and is simply the average tour length. Example histogram results are shown in Figures 28.4a-c.

Performance versus Significance of Fault

For the most part, the greater the significance of a fault, the worse the performance of the network. The degree of badness can be described in one of two ways. In some cases, the histogram looks very much like an exponential distribution. A single parameter can be used to describe this distribution. Fitting the histogram to this distribution, one reduces the results of the large number of simulations to a

(a)

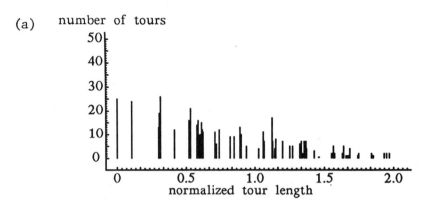

(b)

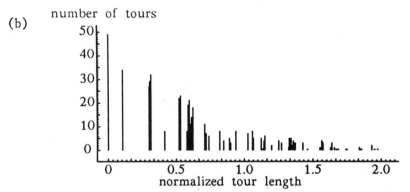

(c)

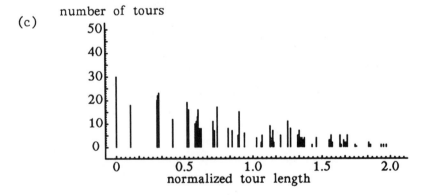

Figure 28.4: Example histogram results from simulations.

Table 28.1: Simulation Results, Average Tour Length

δ	fault free	SA 0	SA 1	SA z	number of open connections					
					1	2	5	10	15	20
10^{-1}	3.45	3.53	3.44	3.53						
10^{-2}	3.27	3.39	3.33	3.42	3.31	3.32	3.33	3.38	3.43	3.47
10^{-3}	2.95	3.10	2.97	3.10	2.96	2.98	3.04	3.14	3.20	3.29
10^{-4}	2.94									
10^{-5}	2.92									
10^{-6}	2.91									

single parameter. Better performance is indicated by a larger slope (implying larger area) in the vicinity of the better tours. Flatter histograms would have a smaller parameter and thus be considered worse performance. Because not all histograms appeared to be exponential distributions this method was not used here. The second way to measure the performance of the network would be to consider the weighted average of the histogram. The advantage of this method is that it is independent of the shape of the histogram, but it is not as descriptive.

In general, better results are obtained with a smaller δ This is illustrated in Figure 28.5. One interesting observation is that as the perturbation becomes increasingly smaller the improvement is less significant. Fault-free networks perform better than faulty networks with the same δ. This is reflected in the table of the weighted averages.

The stuck-at-0 fault always resulted in worse performance than the stuck-at-1 fault. The stuck-at-0 fault is a more significant fault by considering the following. First, a neuron with a state 0 implies that all other neurons in that row and column are 1. A neuron stuck-at-0 reduces the flexibility of the search given the initial

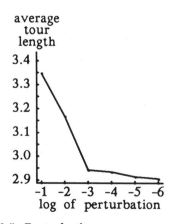

Figure 28.5: Perturbation vs average solution.

perturbation. However, when a neuron is 1, this implies that one neuron in the row and the column is 0 and the rest are 1. Much more search flexibility remains when a neuron is stuck-at-1.

The neural system searches out some solution which is a local energy minimum of the system and perhaps a global minimum. These stuck faults constrain the search and may not allow the system enough freedom to find a better result. A conflicting argument could be that since the stuck faults can be incorporated into a legal solution, the search process will be shortened in time and the results will be improved. This conflicting argument is not supported by the data generated. The one piece of information not considered in this argument is the initial random perturbation. The stuck fault could be detrimental since the random perturbation is an initial direction in the search for a solution. If the search is impeded by a stuck fault, the solution found may not be as good as in the fault-free case.

The high impedance faults were looked at because they model a possible hardware failure. The high impedance fault performs markedly worse than all other fault configurations with the exception of 20 open connections. An explanation for this is that ordinarily, a neuron will drive all networks to which it is connected. A high impedance fault will electrically connect all these networks. The effect is more wide ranging than the stuck faults because the electrical dynamics are drastically changed. If a hardware implementation is attempted, this type of fault should be minimized or eliminated at all costs.

One interesting aspect of open connection faults is that information about the problem to be solved is lost and solution is obtained in the presence of imperfect information. As the number of open connections is increased, the amount of information about the problem decreases. The neural system appears to find the best solution given the information supplied. Looking at $\delta = 0.01$ and $\delta = 0.001$, as the number of open connections increase, the average solution obtained progressively gets worse as is illustrated in Figure 28.6. For reference, the average tour length if all tours were equally likely to result is 3.80. The shortest tour is 2.69 and the longest tour is 4.65.

Study of Nonconvergence

Table 28.2 presents the percentage of simulations from each set of simulations that did not converge to an acceptable state. The acceptable states are those states that do not result in an encoding corresponding to one of the sixty possible tours.

The neural implementation of a circuit "solving" the travelling salesman problem corresponds to a heuristic method. Heuristic methods give a good but not necessarily the best solution with reasonable resource usage. Indeed, heuristic methods may not give a solution even though one may exist. One example is a heuristic algo-

Table 28.2: Percentage of simulations not converging

δ	fault free	SA 0	SA 1	SA z	number of open connections					
					1	2	5	10	15	20
10^{-1}	2.6	3.8	4.8	36.4						
10^{-2}	4.2	8.6	4.8	33.2	4.6	5.4	5.0	8.2	11.0	13.8
10^{-3}	0.8	5.8	5.2	36.6	3.4	3.6	6.4	12.4	12.0	13.0
10^{-4}	0.2									
10^{-5}	3.0									
10^{-6}	17.4									

rithm that performs switchbox routing [16]. In some cases, this algorithm will not find a solution even though one exists. With some manual intervention, however, this algorithm will find a solution. A couple of heuristic techniques were employed in the neural solution travelling salesman problem. The first is the randomization through the perturbation at the start of each simulation. Some starting points may not provide solutions but a high enough probability of such random starting points from the set of all possible starting points will yield good solutions. The second is the termination of the simulation after 500 simulation cycles. This disqualifies those simulations that either would require possibly much more than 500 cycles to converge or those that would never converge. If it is allowable for a certain fraction of runs not to result in usable solutions, arrangements have to be made to run enough simulations to collect an acceptable number of solutions.

Studying Table 2, one observes faulty simulations will converge less frequently than fault free simulations. Also, as the number of faults increase, the frequency of convergence decreases. The fraction of simulations not converging is an important issue. Depending on the application, the results of Table 2 may or may not be acceptable.

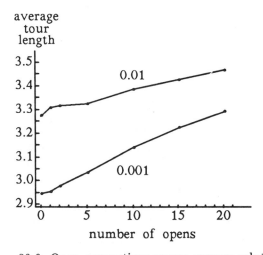

Figure 28.6: Open connections versus average solution.

Three possible events can be happening. These events were not investigated but are stated as follows. First, the time required to converge may increase due to the fault. The fault is either constraining the search process in the case of stuck faults or removing information about the problem as in the case of removal of weight connections. Second, the network may be in some sort of state that it never converges. Instabilities caused by the introduction of faults may result in the network never converging to a stable state. Third, the network may converge to a state not corresponding to a solution encoding for a tour.

Considering the perturbation, as the magnitude of the perturbation gets very small, the rate of convergence decreases. Simulations were run for a perturbation of $10^{-7}\delta$ and none of the simulations converged after 500 simulation cycles. The limit was raised to 1000 simulation cycles and a second set of 500 simulations were run. All 500 converged to a usable solution. In some, but not all cases, the nonconvergence can be considered a time penalty. Some fraction of those tours that did not converge in 500 cycles will converge after some longer period of time.

28.4.3   Second Approach to the Fault Analysis of Neural Like Networks

The second approach was to look at specific initial conditions with respect to the tours to which the circuit converged. Neurons were chosen two at a time and the tours to which they converged was plotted in a two dimensional plane as a function of the initial conditions of these neurons. All other neurons are set to the same nominal value. Both good and faulty simulations were run, and convergence patterns between the two were compared.

By constructing images from simulations under different, deterministic, starting positions, behavior of the neural-like networks in the presense of faults was observed. The scales on each of the axes is logarithmic so that small- and large-scale effects can be observed. Neurons that are plotted against each other were chosen because of the relationships between them.

Five neurons were chosen such that two will have the following relationships between them. The relationship between neurons is they are coupled or connected. The coupling is determined by connections between the neurons. There are four different classes of coupling including: (1) row inhibition, (2) column inhibition, (3) distance inhibition, and (4) indirectly coupled. Case (4) where neurons indirectly coupled means that no direct connection exists between the neurons being examined.

The simulations run included (A) fault-free, (B) stuck-at-1, (C) stuck-at-0, and (D) open connection. Simulations will be identified by a number followed by a letter, the number corresponds to the neuron choice and the letter corresponds to the fault being studied.

Four specific simulation results are presented. They correspond to 1A, 1B, and 1D. The plots of the solutions found are shown in figures 28.7 - 28.9. The initial conditions of the networks can be reproduced by inserting the axes scale into the following equation

$$SGN(n) * \frac{2^{(ABS(n)-10)}}{10}$$

where $SGN(n)$ is the sign of n and $ABS(n)$ is the absolute value of n.

Looking at the figures, some interesting patterns can be seen. First, around the border where the perturbation is relatively large, the patterns are very similar. The effect of the perturbation is more significant than the effect of the fault. Near the center of the plots, more significant changes occur. Looking at the stuck-at-1 fault, the entire central region converges to a single solution. For the open connection case, the central region changes a great deal with no central region effects found as in the stuck-at-1 case. The region that is connected to the point -5,-8 seems to pervade the central regions of the fault-free and open connection simulations.

28.5 Properties of Inherently Fault-Tolerant Systems

Only one property is firmly established through the simulations performed on neural-like networks. This property is that the effect of the fault is distributed

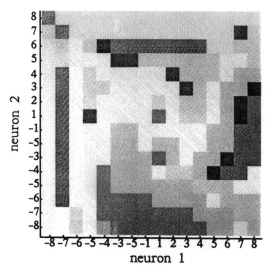

Figure 28.7: Fault-free, row inhibition.

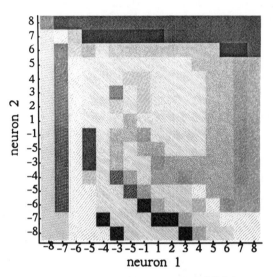

Figure 28.8: Stuck-at-1, row inhibition.

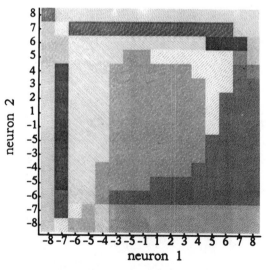

Figure 28.9: Open connection, row inhibition.

throughout the system. Since the input to each electrical neuron is a summing network from other neurons, the input to the nonlinear amplifier can be considered the average of all inputs to it. In this case, the performance of the system is degraded in relation to the extent of the damage to the system. Holograms will lose detailed resolution of images as a result of cutting portions away from the original. ¿From the initial investigations of neural networks, they appear to average the result of faults throughout the system. This is supported by the degraded simulation results. Since the initial starting point of the simulation is an unstable equilibrium point plus some perturbation, the effect of a fault is to modify the slope locally to this point. Ideally, this will show shifting of the converging states which is apparent from examining Figures 28.7 and 28.9.

28.6   Future Efforts and Conclusions

Possible future efforts can go in many different directions. Some directions are in the more general area of inherent fault tolerance. Others are looking specifically at neural-like networks.

Since only a single property has been established, it would be appropriate to consider what other properties could be established. When this process reaches some level of completeness, these properties may be able to be applied to a design or the design process an example property in the neural network is that information propagates in one direction only. The simulation results from the stuck-at-high impedance case weakly illustrate this property. Another test could be to simulate a fault of shorting the input to the output of the nonlinear amplifier and run the simulations. A biological neuron has information transmission in one direction only [15]. No possible failure could cause information transfer in the opposite direction because the function of the neuron prevents this occurrence. Holograms also pass information in one direction. A beam of light goes through the hologram in one direction only.

Two postulates are necessary to establish inherent fault tolerance. First, inherently fault-tolerant systems can be built up from faulty components. Second, a system built up entirely from inherently fault-tolerant subsystems will be inherently fault-tolerant. A goal could be to establish the validity of these statements through proof.

Any hardware implementation implementation is going to have to deal with circuit component tolerances. The question is how much the component tolerances affect the quality of solutions. One goal of this research-direction could be to relate circuit component tolerances to the random perturbation.

Only one problem instance was used in all the simulations performed. For completeness, more problem instances should be considered before any serious conclusions are drawn. In particular, the fault-tolerant behavior of systems containing many more neurons would be interesting.

Similarly, only one problem implementation was studied for the simulation effort. For completeness, other problems should be considered before any serious conclusions can be drawn.

It may be possible to optimize circuit parameters to minimize the effect of faults. If this is possible, basically, the same hardware can achieve different levels of fault tolerance given the circuit parameters.

Acknowledgements

This work was funded under the Virginia Center for Innovative Technology. We would also like to acknowledge Dr. William B. (Chip) Levy for our introduction to neural networks.

References

[1] Barry W. Johnson, *The Design and Analysis of Fault Tolerant Digital Systems*, manuscript in preparation, Jan. 1987.

[2] A. Aviziensis, "The Four-Universe Information System Model for the Study of Fault Tolerance", *Proceedings of the 12th Annual Symposium on Fault-Tolerant Computing*, Santa Monica, California, June 1982, pp. 6-13.

[3] Jean-Claude Laprie, "Dependable Computing and Fault Tolerance: Concepts and Terminology", *Proceedings of the 15th Annual Symposium on Fault-Tolerant Computing*, June 1985, pp. 2-11.

[4] J.J. Hopfield, "Neurons with graded response have collective computational properties like those of two-state neurons," *Proc. Natl. Acad. Sci. USA*, Vol. 79, April 1982, pp. 2554-2558.

[5] J.J. Hopfield, "Neural networks and physical systems with emergent collective computational abilities", *Proc. Natl. Acad. Sci. USA*, Vol. 81, May 1984, pp. 3088-3092.

[6] J.J. Hopfield, D.W. Tank, "Computing with neural circuits: a model," *Science*, vol. 233, no 4764, l986, pp. 625-633.

[7] J.J. Hopfield and D. W. Tank, " 'Neural' Computation of Decisions in Optimization Problems," *Biological Cybernetics*, vol. 52, 1985, pp. 141-152.

[8] Yaser S. Abu-Mostafa and Demetri Psaltis, "Optical Neural Computers," *Scientific American*, March, 1987, pp. 88-95.

[9] L.D. Jackel, R.E. Howard, H.P. Graf, B. Straughn, J.S. Denker, "Artificial neural networks for computing," *J. Vac. Sci. Technol. B.*, vol. 4, No 1, Jan/Feb 1986, pp. 61-63.

[10] E. Harth and N. S. Lewis, "The Escape of *Tritonia:* Dynamics of a Neuromuscular Control Mechanism," *J. Theor. Biol.*, 1975, vol. 55, pp. 201-228.

[11] G. A. Horridge, and I. A. Meinertzhagen, "The accuracy of the patterns of connexions of the first- and second-order neurons of the visual system of *Calliphora*," *Proc. Roy. Soc. Lond*, B, vol. 175, 1970, pp. 69-82.

[12] Douglas R. Hofstadter, *Godel, Escher, Bach: an Eternal Golden Braid*, Vintage Books, New York, 1979, p.342. The reference originally cited in this book was from Steven Rose, *The Conscious Brain*, Vintage Books, New York, 1976. Rose included a summary of the work by Karl Lashley whose work was done in the 1920's.

[13] Jay O'Rear, *Physics*, Macmillan Publishing Co., Inc., New York, 1979, pp. 496-500.

[14] Michael R. Garey and David S. Johnson, *Computers and Intractability: A Guide to the Theory of NP-Completeness*, W. H. Freeman and Company, San Francisco, 1979.

[15] William Hughes, *Aspects of Biophysics*, John Wiley & Sons, New York, 1979.

[16] Gordon T. Hamachi and John. K. Ousterhout, "A Switchbox Router with Obstacle Avoidance," *ACM/IEEE 21st Design Automation Conference Proceedings*, Albuquerque (June 1984), pp. 173-179.

[17] Theodore Holmes Bullock, "Reassessment of Neural Connectivity and Its Specification," *Information Processing in the Nervous System*, H. M. Pinsker and W. D. Willis editors, 1980, pp. 199-220.

Chapter 29

Fault-Tolerant LU-Decomposition in a Two-Dimensional Systolic Array [1]

J.H. Kim [2]
and S.M. Reddy [3]

Abstract

In this paper, we present a new systolic algorithm to solve LU-decomposition effeciently in a 2-D systolic array. The LU-decomposition is the main step to solve a system of linear equations, a problem encountered in many scientific and industrial applications. Continuing growth of interest in systolic arrays poses new problems in ensuring the reliability of computations performed by such systems. A concurrent error detection scheme proposed earlier is applied to the new systolic LU-decomposition algorithm. The proposed concurrent error detection scheme requires small hardware overhead and no time overhead, since it utilizes the inherent idle cycles in the cells of the array.

[1] The research reported has been supported in part by SDIO/IST contract No. N00014-87-K-0419 managed by U.S. Office of Naval Research.
[2] University of Iowa, Present address: Center for Advanced Computer Studies, University of Southern Louisiana.
[3] University of Iowa, Iowa City, IA.

29.1   Introduction

Systolic arrays consisting of identical or nearly identical computational cells with synchronous data flow are considered to be preferred architectures for executing linear algebraic operations. The continuing growth of interest in systolic arrays poses new problems in ensuring the reliability of computations performed by such systems [5] - [8]. Since each processor in such a system contributes to the computation process, the failure of even a single processor in the system could cause the results computed by such a system to be unacceptable. It is therefore desirable that systolic arrays be designed to tolerate physical faults in the system and still produce correct results, or at least be able to identify a faulty result when it occurs.

With the present possibility of reduced voltage levels for VLSI devices and the subsequent reduction in noise margins [4], susceptibility of dense VLSI circuits to transient faults also increases. Therefore, techniques are needed to detect faults concurrently with normal operation, especially in real-time applications.

In this paper, we present a new systolic algorithm to solve LU-decomposition efficiently in a 2-D systolic array. The LU-decomposition is the main step to solve a system of linear equations that are encountered in many scientific and industrial applications [12]. A concurrent error detection scheme proposed earlier [11] is applied to the new systolic LU-decomposition algorithm to detect an error concurrently with normal operations. The concurrent error detection scheme requires small hardware overhead and no time overhead, since it utilizes the inherent idle cycles in the cells of the array. The remainder of this paper is organized as follows. Preliminary concepts are developed in Section 2. Section 3 describes a new systolic algorithm to solve LU-decomposition. Fault model and concurrent error detection scheme are given in Section 4. Concluding remarks are given in Section 5.

29.2   Preliminaries

The problem of factoring a symmetric positive-definite matrix A into lower and upper triangular matrices L and U is called LU-decomposition. Figure 29.1 illustrates LU-decomposition of a 4 × 4 full matrix. Once the L and U factors are known, it is relatively easy to invert a matrix A or solve the linear system Ax = b. The triangular matrices $L = (l_{ij})$ and $U = (u_{ij})$ are computed according to the following recurrences [1]:

$$a_{ij}^{(1)} = a_{ij},$$

$$a_{ij}^{(k+1)} = a_{ij}^{(k)} + l_{ik}(-u_{kj}),$$

$$l_{ik} = \begin{cases} 0 \\ 1 \\ a_{ik}^{(k)} * u_{kk}^{-1} \end{cases}$$

$$
\begin{pmatrix} a_{11} & a_{12} & a_{13} & a_{14} \\ a_{21} & a_{22} & a_{23} & a_{24} \\ a_{31} & a_{32} & a_{33} & a_{34} \\ a_{41} & a_{42} & a_{43} & a_{44} \end{pmatrix} = \begin{pmatrix} 1 & & & \mathbf{0} \\ l_{21} & 1 & & \\ l_{31} & l_{32} & 1 & \\ l_{41} & l_{42} & l_{43} & 1 \end{pmatrix} \times \begin{pmatrix} u_{11} & u_{12} & u_{13} & u_{14} \\ & u_{22} & u_{23} & u_{24} \\ & & u_{33} & u_{34} \\ \mathbf{0} & & & u_{44} \end{pmatrix}
$$

Figure 29.1: LU-decomposition of a matrix A

$$
u_{kj} = \begin{cases} 0 \\ a_{kj}^{(k)} \end{cases}
$$

where $a_{ij}^{(k+1)}$ means the value of a_{ij} after k iterations.

The evaluation of these recurrences can be pipelined on a hexagonal systolic array[1]. This pipelined computation is shown in Figure 29.2 for the LU-decomposition problem given in Figure 29.1. The hexagonal systolic array in Figure 29.2 is constructed as follows : The PEs below the upper boundaries are the inner product step processors given in Figure 29.3. The processor at the top is a special processor. It computes the reciprocal of its input and passes the result southwest and also passes the same input northward unchanged. The other PEs on the upper boundaries are again inner product step processors, but their orientation is changed : the ones in the upper left boundary are rotated 120 degrees clockwise; the ones in the upper right boundary are rotated 120 degrees counterclockwise. Each PE only operates every third time step, and the hexagonal array outputs at one output every three units of time. For detailed explanation about this algorithm see [1]. If A is an n×n matrix, the nxn hexagonal array can compute the L and U matrices in $O(4n)$ time units. Since in a hexagonal array every PE except the boundary PEs is connected to 6 neighbouring PEs, a drawback of the hexagonal array is that reconfiguration to avoid faulty cells is difficult. For example, for the case where there is a single faulty PE in an nxn rectagular array and an nxn hexagonal array and both arrays are reconfigured employing the same reconfiguration scheme proposed in [2], the largest size of rectangular array is (n-1)×(n-1) while that of the reconfigured hexagonal array is (n-2)×(n-2) [3].

29.3 The Proposed Systolic Algorithm

In this section, we describe a new LU-decomposition algorithm using a proposed 2-D systolic array which lends itself to easy reconfiguration. The proposed 2-D array is constructed by using rows of linear bidirectional systolic arrays studied earlier [10]. For this reason we call it a 2-D bidirectional systolic array. Figure 29.4 depicts a 2-D bidirectional systolic array for the LU-decomposition of matrix A given in

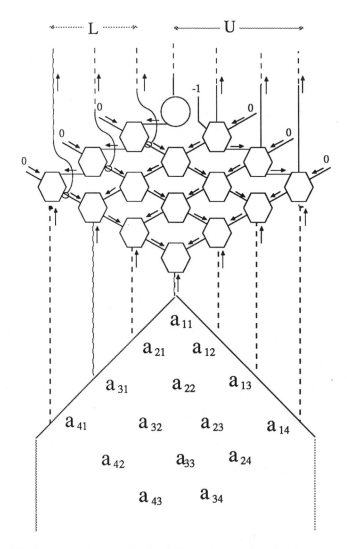

Figure 29.2: The hexagonal array for the LU-decomposition in Figure 29.1. Adapted from [9].

$$a_{out}$$

$$l_{in} \qquad u_{in}$$

$$u_{out} \quad a_{in} \quad l_{out}$$

$$a_{out} = a_{in} + l_{in} * u_{in}$$
$$l_{out} = l_{in}$$
$$u_{out} = u_{in}$$

Figure 29.3: The function of a PE in 29.2.

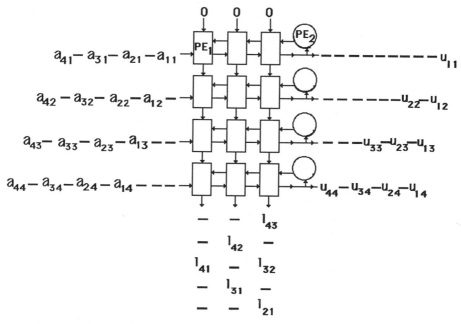

Figure 29.4: A new systolic LU-decomposition algorithm.

Figure 29.1. Each PE is numbered according to a matrix ordering: the PE on row i, column j is designated as PE[i,j]. Two kinds of PEs, called PE_1 and PE_2, are used. A-data as well as U-data has a tag bit. If a_{ij} datum is an element of a main diagonal of matrix A (i.e. i=j in a_{ij}), the tag bit of A-data is set to 1, otherwise the tag bit of A-data is set to 0. When PE_2 receives the a_{ij} datum with a tag bit set to 1 (0), it produces u_{ij} data with the tag bit set to 1 (0). The functions of two kinds of PEs are shown in Figure 29.5. When PE_1 receives U-data whose tag bit is 0, PE_1 functions as follows:

$$a_{out} = a_{in} + l_{in} * u_{in}, \quad l_{out} = l_{in}, \quad u_{out} = u_{in}$$

When PE_1 receives U-data whose tag bit is 1, PE_1 produces new L-data as follows:

$$l_{out} = a_{in} * u_{in}, \quad a_{out} = a_{in}, \quad u_{out} = u_{in}$$

When PE_2 receives A-data whose tag bit is 0, PE_2 produces U-data, $u_{out} = -a_{in}$. When PE_2 receives A-data whose tag bit is 1, PE_2 produces U-data, $u_{out} = a_{in}^{-1}$. For a given nxn matrix A, the array needs to use n rows of cells, with n cells in each row.

Figure 29.6 shows snapshots of the 2-D bidirectional array at various stages of the computation.

At t=4, u_{11} which is $a_{11}^{(1)}$ is output at the 1st row, and PE[1,4] receives A-data $a_{11}^{(1)}$ with tag=1 and produces U-data u_{11}^{-1} with tag=1.

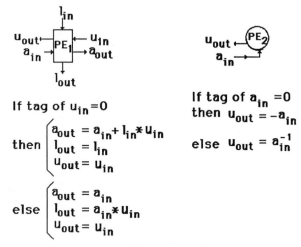

Figure 29.5: The function of two kinds of cells.

At t=5, PE[1,3] receives U-data u_{11}^{-1} with tag=1 and A-data $a_{21}^{(1)}$. Since the tag of U-data is 1, PE[1,3] produces l_{21} as follows: $l_{out} = a_{21}^{(1)} * u_{11}^{-1} = l_{21}$. At the same time, u_{12} which is $a_{12}^{(1)}$ is output at the 2nd row, and PE[2,4] receives A-data $a_{12}^{(1)}$ with tag=0 and produces U-data $-u_{12}$ with tag=0.

At t=6, the following three operations occur simultaneously :

PE[1,2] receives U-data u_{11}^{-1} with tag=1 and A-data, $a_{31}^{(1)}$. Since the tag of U-data is 1, PE[1,2] produces l_{31} as follows: $l_{out} = a_{31}^{(1)} * u_{11}^{-1} = l_{31}$.

PE[2,3] receives U-data $-u_{12}$ with tag=0, A-data $a_{22}^{(1)}$ and L-data l_{21} which was output at PE[1,3]. Since the tag of U-data is 0, PE[2,3] updates A-data as follows:
$a_{out} = a_{22}^{(2)} = a_{22}^{(1)} + l_{21} * (-u_{12})$

u_{13} which is $a_{13}^{(1)}$ is output at the 3rd row, and PE[3,4] receives A-data $a_{13}^{(1)}$ with tag=0 and produces U-data $-u_{13}$ with tag=0.

At t=7, the following five operations occur simultaneously :

PE[1,1] receives U-data u_{11}^{-1} with tag=1 and A-data, $a_{41}^{(1)}$. Since the tag of U-data is 1, PE[1,1] produces l_{41} as follows: $l_{out} = a_{41}^{(1)} * u_{11}^{-1} = l_{41}$,

PE[2,2] receives U-data $-u_{12}$ with tag=0, A-data $a_{32}^{(1)}$ and L-data l_{31} which was output at PE[1,2]. Since the tag of U-data is 0, PE[2,2] updates A-data as follows:
$a_{out} = a_{32}^{(2)} = a_{32}^{(1)} + l_{31} * (-u_{12})$

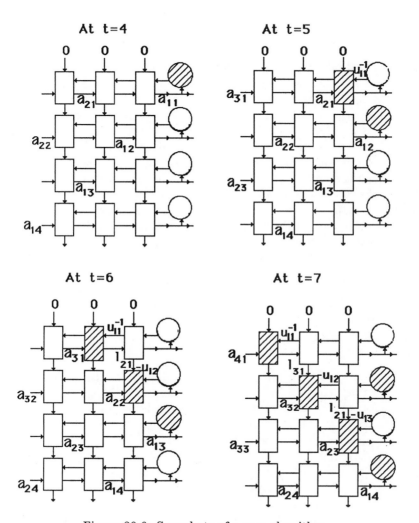

Figure 29.6: Snapshots of a new algorithm.

u_{22} which is $a_{22}^{(2)}$ is output at the 2nd row, and PE[2,4] receives A-data $a_{22}^{(2)}$ with tag=1 and produces U-data u_{22}^{-1} with tag=1.

PE[3,3] receives U-data $-u_{13}$ with tag=0, A-data $a_{23}^{(1)}$ and L-data l_{21}. Since the tag of U-data is 0, PE[3,3] updates A-data as follows: $a_{out} = a_{23}^{(2)} = a_{23}^{(1)} + l_{21} * (-u_{13})$

u_{14} which is $a_{14}^{(1)}$ is output at the 4th row, and PE[4,4] receives A-data $a_{14}^{(1)}$ with tag=0 and produces U-data $-u_{14}$ with tag=0.

It can be noted that A-data $a_{kj}^{(1)}$ for k≤j, entering the array is transformed into $a_{kj}^{(k)}$ when a_{kj} reaches the last column. The $l_{(i+j)i}$ is output at PE[n-j+1,i] and u_{ik} is output at the kth row of the nxn 2-D array. To make each l_{ij} meet each u_{jk}, consecutive l_{ij} of L-data stream and consecutive u_{jk} of the U-data stream should be separated by two time units. Therefore, a PE becomes idle at every other cycle, and every other PE in the array is idle at any given time. This array outputs at the rate of one output every two units of time. Therefore, the throughput of the proposed algorithm is 150% of the previous design which uses as many cells and outputs at the rate of one output every three units of time [1]. Additionally, since every PE except the boundary PE is connected to 6 neighboring PEs in a hexagonal array and 4 neighboring PEs in a 2-D bidirectional array, the reconfigurability of a hexagonal array is lower than that of a 2-D bidirectional array.

29.4   Concurrent Error Detection

29.4.1   Fault Model

An appropriate way to deal with failures in VLSI circuits is at the functional level. Therefore, we assume that faults will make their effect felt at the level of a small part of a large network in terms of altered values of the outputs. The functional level chosen is a cell of the array and is thought to include the output links emanating from it. In this paper we assume that at most one PE_1 is faulty in every three consecutive PE_1s in a row and at most one PE_2 is faulty in every three consecutive PE_2s in a column of the array.

29.4.2   The Proposed Method

The concurrent error detection scheme proposed earlier and called Comparison with Concurrent Redundant Computation (CCRC) [11] is the one we have adopted. CCRC is based on the observation that there is inherent spatial redundancy in the array which could be exploited to perform a concurrent redundant computation. We could launch two computations in a way that they are performed on different regions

Table 29.1: Fault location table.

Comparison results		
$M_1[i,j]$	$M_1[i,j+1]$	
match	match	Cells are fault-free
match	mismatch	PE$[i,j+1]$ is faulty
mismatch	match	PE$[i,j-1]$ is faulty
mismatch	mismatch	PE$[i,j]$ is faulty

of the array. Then, at the time instant when the computational wavefront of the required computation reaches a faulty cell, the shadow (redundant) computation reaches a fault-free cell and this computation would be confined to a fault-free region of the array, and thus, a comparison of the corresponding results would lead to detection of the fault. CCRC proposed earlier can only detect faults but cannot locate faults. By adding additional logic, a single faulty cell among three consecutive cells can be located from the outcome of two consecutive comparisons under permanent fault assumption according to Table 29.1. For duplicating any computation, the two cells involved must receive the same inputs. Figure 29.7 shows the implementation which achieves duplication of inputs in two cells involved. For concurrent error detection only, one additional PE_2 is not needed since both PE_2s in the (2k+1)th row and the (2k+2)th row can always engage in duplicate computations. MUX1 and MUX2 are driven by a modulo 2 counter which is required to be set to 1 whenever a primary input and u_{ij} input from PE_2 are applied to the

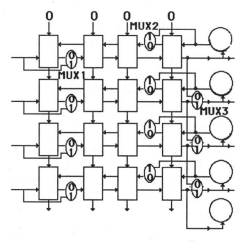

Figure 29.7: Application of CCRC to a new algorithm.

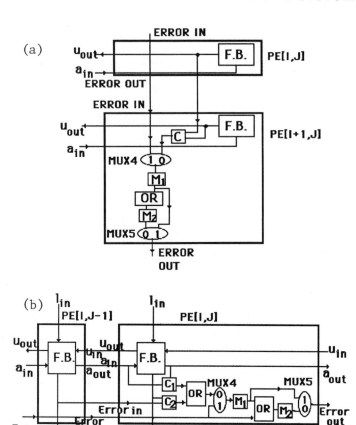

Figure 29.8: Error detection and location logic of PE_1 and PE_2.

array, respectively. When MUX1 is set to 1, a primary input is selected at PEs in the 2nd column. This results in the initiation of required computation at PEs in the 2nd column and its redundant computation concurrently at PEs in the 1st column, whenever primary inputs are applied. MUX3 is driven by a modulo 2 counter which is required to be set to 1 whenever A-data comes from the PE located in the same row.

Figure 29.8 illustrates a two cell boundary of the augmented array. PE[i,j] shows the additional logic required in cells for error detection and location, which is active only in cells engaged in the required computation, and is, therefore, not shown in PE[i,j-1] (cf. Figure 29.8(a)) for the sake of clarity. For concurrent error detection only, we do not need MUX4, MUX5, memory element M_2 and one OR gate which is used to scan-out the individual error indication of each cell. These are needed if fault location is desired. C_1 and C_2 are comparators. The output of the OR gate of two comparators is clocked into the memory element M_1. M_1 output is ORed with the error signal from the previous cell. The output of the OR gate is clocked into the memory element M_2, and so the error indication moves synchronously with the computation wavefront. During normal operations, MUX4 and MUX5 are set to 0. When the error signal reaches the outside of the array, the array stops normal operations and individual error indication stored at the memory element M_1 is brought out from every cell with MUX4 and MUX5 set to 1. A single permanent-faulty cell among three consecutive cells can be located according to Table 1. Note that cells in the 1st column do not need the error detection logic. Assuming that the error detection logic is fault-free we state the following theorem on fault coverage of CCRC.

Theorem 1 : If we assume that at most one PE_1 is faulty in every three consecutive PE_1s in a row and at most one PE_2 is faulty in every three consecutive PE_2s in a column of the array, CCRC would detect faults and locate permanent-faulty cells.

The hardware overhead of CCRC is one column of PE_1s, one additional PE_2, 3 multiplexers, links required for error detection and propagation, and the error detection logic required in every cell except the 1st column, as shown in Figure 29.8. Since redundant computations are performed using naturally occuring idle states, there is no time overhead.

29.5   Concluding Remarks

In this paper, we presented a new systolic algorithm to solve LU-decomposition effeciently in a proposed 2-D bidirectional systolic array which lends itself to easy reconfiguration. The LU-decomposition is the main step to solve a system of linear equations encountered in many scientific and industrial applications. Rao proposed another LU-decomposition systolic algorithm which requires half as many cells as our proposed algorithm [13]. But, in his algorithm the output of each element of an upper triangular matrix U, has to be stored in each cell and is not available to the boundary cells until the completion of entire operation. Concurrent error detection scheme proposed earlier is applied to the new systolic LU-decomposition algorithm to detect faults and locate permanent faulty cells. The concurrent error detection scheme requires small hardware overhead and no time overhead, since it utilizes the inherent idle cycles in each processimg element of the array.

References

[1] C.Mead and L. Conway, *Introduction to VLSI systems* , chapter 8, Addison-Wesley, 1980.

[2] I. Koren, "A Reconfigurable and Fault-Tolerant VLSI Multiprocessor Array," *The 8th Symposium on Computer Architecture* , IEEE & ACM, May 1981, pp. 425-442.

[3] D.Gordon, I.Koren, and G.M.Silberman, "Restructuring hexagonal arrays of processors in the presence of faults," to appear in *J. VLSI Comput. Syst.*

[4] D.F.Barbe, "VHSIC Systems and Technology," *IEEE Computer* , Feb. 1981, pp.13-22.

[5] Y.H.Choi, et al, "Fault Diagnosis of reconfigurable systolic arrays," *Int'l Conf. Computer Design 84*, pp. 451-455.

[6] K.H.Huang and J.A.Abraham, "Low cost schemes for fault tolerant matrix operation with processor arrays", *FTCS-12* , June 1982.

[7] M.Sami and R.Stefanelli, "Reconfiguration Architectures for VLSI Processing Array," *Proc. of the IEEE* , May 1986, pp.712-722.

[8] J.H.Kim and S.M.Reddy, "A Fault-Tolerant Systolic Array Design Using TMR Method, *Int'l Conf. Computer Design 85*, pp. 769-773.

[9] C. Mead and L. Conway, *Introduction to VLSI Systems*, Addison-Wesley Pub. Co., Reading MA (1980), p. 282.

[10] H.T.Kung, "Let's design algorithms for VLSI systems," *Proc. Caltech. Conf. on VLSI*, Jan. 1979, pp. 65-90.

[11] R.K.Gulati and S.M.Reddy, "Concurrent Error Detection in VLSI Array Structures," *Int'l Conf.Computer Design 86* , pp.488-491.

[12] S.J.Leon, "Linear Algebra with Application," Macmillan Publishing Co., NY, 1980.

[13] S.K Rao, private communication, Oct., 1987.

PART V
System Issues

Introduction

The papers included in this Section, loosely entitled System Issues, generally address concurrent computation from the perspective of an overall system or a VLSI implementation. Here, various issues considered in earlier sections of this book must be integrated into a functional implementation. Most of the papers concern special purpose applications. Others seek more general purpose parallel computing environments. Although a vast number of alternative concurrent computing and parallel processing systems have been built or proposed, the implementations described here illustrate the general issues encountered in realizing parallel computing environments and describe some new directions being pursued.

Reeves provides an overview of parallel programming environments, addressing general topics, highly parallel SIMD and MIMD architectures, communications, programming environments and languages. This broad range of topics is connected by emphasizing highly parallel scientific computers. Such scientific computation applications are a major motivation for several of the computing systems which have been designed. Experience in this emerging area will strongly impact general purpose, parallel computing systems.

Reeves' discussion of general purpose parallel computing systems is followed by Kung and Huang's discussion of systolic computing systems, the other major application which has been implemented. **Kung and Huang** consider not only a classical signal processing function (Kalman filtering) but also neural computation algorithms. This merges onto a common system organization two distinct classes of algorithms, qualitatively seeming to be at opposite extremes of the spectrum of concurrent computing.

Image processing has been a premier application for concurrent computation, spawning several special purpose pixel-based processor arrays. **Shu and Nash** discuss image processing from the perspective of a complete image processing environment (The Image Understanding Architecture being developed by Hughes Research Labs and the University of Massachusetts), with several hierarchical levels of computation resources. Such multi-layer computation systems are likely to evolve, using very high performance concurrent computation units at the lowest level of the architecture and more general purpose, programmable processing units toward the top of the hierarchy. The communications environment at the lowest level (bit serial processing arrays) is emphasized.

VLSI models of computation provide a formal basis for optimizing the design of algorithms implemented on VLSI IC's. **Roychowdhury et al.** provide an example of the application of such models to the efficient VLSI design of the Viterbi algorithm. By considering the relationship between processor interconnections and state transistion diagrams representing the convolutional encoders, the encoding function can be mapped to a shuffle-exchange network. This in turn can be mapped into layouts achieving lower bounds on chip area.

Cameron et al. describe a highly parallel supercomputing environment (the IC* environment) based on the IC* model of computation developed at Bell Communications Research. The IC* model's inherent fine grained parallelism illustrates the advantages achieved in incrementally specifying a system's design and behavior. Drawing on this fine grained level of description, a broad range of software tools can be developed in support of system design and programming.

Holsztynski and Raghavan review a distributed, SIMD-type system they are implementing from the perspective of the macro controller, memory management and use of stacks rather than randomly accessed variables. **Krishnaswamy et al.** consider the physical design and control of high level shared memory for a MIMD-style architecture. Their *Linda Machine* combines special VLSI IC's, a programming abstraction and a communication environment based on a logically shared, associative memory.

Chapter 30

Programming Environments for Highly Parallel Scientific Computers

Anthony P. Reeves [1]

30.1 Introduction

Very high speed computing resources are important for a number of scientific applications. In recent years highly parallel computer architectures have been developed
which offer an alternative to the conventional pipelined vector approach to supercomputers. An attractive feature of the highly parallel systems is the potential for
further speedup by simply adding additional processor resources. One of the main
disadvantages of the parallel approach is that the conventional Fortran framework
cannot be used and new programming techniques must be mastered.

Two main highly parallel architectural types, for which there have been a number
of commercial realizations, are the processor array, a SIMD organization consisting
of a single control unit and a very large number of slave processing elements, and
multiprocessor computers which have a MIMD organization and which have recently
been typified by hypercube systems.

A key issue for the development of the highly parallel systems is the development
of appropriate programming environments. For the SIMD systems, high level languages have been developed which are well matched to the architecture. These languages have primitives similar to matrix algebra operations which are directly implemented

---

[1] Univeristy of Illinois at Urbana-Champaign, Urbana,IL.

on a processor array. Typical of these languages is Parallel Pascal [1] which has been implemented on NASA's Massively Parallel Processor (MPP) which has 16384 bit-serial PE's [2].

Appropriate programming environments are much more difficult to specify for MIMD systems since a wide range of different programming paradigms are possible. On a hypercube system, different applications require different operating environments for optimal performance. A typical limited environment, which is available with most hypercube systems, is a conventional language for programming each node, such as Fortran or C, with some message passing extensions. This is too low a level for most applications since the programmer must essentially develop a multiprocessor operating environment in addition to the application program. For a large number of current applications, a simple SIMD-like environment, such as that provided by Parallel Pascal, has been shown to be suitable. This is especially true for systems with vector processor nodes, such as the FPS T-Series [3], since the ratio of processing speed to interprocessor bandwidth is greater for these systems than for other hypercube systems.

For the future, new programming environments must be developed which can take full advantage of MIMD resources for single user scientific applications. Such systems should have provisions for graceful fault tolerance, dynamic load balancing, dynamic algorithm selection, and automatic task decomposition and allocation.

30.2   Issues in Highly Parallel Programming

Highly parallel systems offer a mechanism for increasing processing speed over conventional computer systems beyond that possible by technological improvements alone. Unfortunately, efficient programs for these architectures cannot be developed in conventional programming languages and are more difficult to develop than conventional programs. This has been a major impediment to their acceptance by the general high speed processing community. This situation may be changed as better programming tools and environments are developed and second generation systems offer better performance improvements over conventional systems.

30.2.1   Efficiency and Convenience

An important factor in a highly parallel programming environment is *efficiency* ; since additional hardware is being replicated in a system in order to increase speed. There is a tradeoff between efficiency and *convenience* which frequently occurs in programming languages. For example, languages such as APL, Setl, and Lisp are frequently more convenient for the programmer than more conventional languages such as Fortran, C, or Pascal. However, on most computation intensive applications, one of the latter languages is used since efficiency becomes essential.

Must an efficient language be inconvenient? Perhaps ultimately this may not be the case; however, with the current state of the art in compiler construction, some inconvenience is necessary. In an ideal environment the programmer would be free to specify a problem in the notation of his or her choice without regard to the underlying processing environment. However for conventional systems, the programmer may have to be knowledgeable about numerical techniques, storage limitations and an inconvenient programming language in order to create an effective program.

For vector supercomputers there is the further art of *vectorization* to be mastered. The programmer must know much about the architecture and particular programming techniques of the specific machine in order to a achieve possibly an order of magnitude speed improvement over a simple vectorizing compiler.

In highly parallel computer architectures a new dimension of parallelism (and complexity) is introduced and programmers who wish to take advantage of these new capabilities must become familiar with yet more *machine parameters* . SIMD systems can usually be categorized with a small number of additional parameters such as: degree of parallelism, interconnection topology and and interprocessor communication speed. For the bit-serial SIMD systems the arithmetic speeds for various data types will also be significantly different from conventional systems. For MIMD systems several more parameters are needed such as interprocessor latency, and task switching time. Furthermore, with MIMD systems we are faced with the interaction of a number of asynchronous processes distributed on a number of independent processors. In the conventional programming environment, such activities are usually controlled by the operating system which is transparent to the programmer. However, for highly parallel MIMD systems there are many different programming paradigms of which current operating systems only implement a small subset. Therefore, only a small number of the potential applications can be easily programmed. A programmer may be faced with developing many parts of an *operating system* in order to efficiently utilize a highly parallel MIMD system.

30.2.2 Problem Classes

Problems for MIMD systems may be categorized into three major classes *trivial*, *SIMD*, and *complex* . Trivial problems are those which exhibit a high degree of parallelism without interactions. For example consider the task of processing 1000 independent data sets; the simple solution is to run each data set on a different processor. With this technique up to 1000 processors can be efficiently utilized without significantly more overhead than for a single conventional processor.

SIMD problems are those for which efficient solutions on a SIMD parallel processor are known. In many cases, an effective technique which has been developed for

SIMD systems can be easily ported to a MIMD environment. One scheme to do this is to implement a SIMD language on the MIMD system.

The complex class consists of problems which, although having the potential for a high degree of parallelism, present some significant amount of irregular interactions between processing units. The challenge is to balance the parallel system resources of storage, communications, and computation to achieve the most optimal performance. Frequently a large number of problem decomposition strategies are possible; the effectiveness of many solutions depend upon both machine parameters and the data set being processed.

30.2.3   Fault Tolerance

Fault tolerance is becoming an increasingly important aspect of parallel processing systems for scientific applications. As technology allows the degree of parallelism of such systems to increase, the probability of a faulty processor and the cost associated with it also increases. Fault tolerance is relatively simple to deal with for the trivial class of problems and also for many complex problems since the topological relationship between processors and the exact number of processors is not critical. For the SIMD class, however, the situation is generally more difficult. For SIMD systems the usual solution is to be able to switch in extra hardware when a fault occurs. This is very costly; for example, the MPP has 3% additional hardware to be able to deal with a single fault out of 16384 processors. In general, it cannot continue with two or more processor faults. We have developed techniques for task reassignment which permit MIMD systems with a large number of faulty processors to efficiently continue processing SIMD class problems [4]. However, this technique will only work if a programming strategy is used which does not specifically allocate data processors.

30.2.4   Load balancing and new algorithms

Load balancing is frequently a critical task for highly parallel distributed systems, especially for the complex class of problems. In conventional systems load balancing in a multiprocessing domain is typically so complex that simple heuristics are used by the operating system which produce acceptable approximate solutions. In highly parallel MIMD systems the situation is even more complex since the cost of moving tasks between processors must be considered and also the number of processes is usually much larger.

Another opportunity with MIMD systems is the possibility of developing algorithms which are sensitive to their spatial location in the processing structure. In conventional algorithms only temporal relationships are usually considered i.e. all operations occur in some predefined sequence. However, in distributed systems there is a cost associated with accessing spatially remote data and the possibility of doing additional local processing when held up by remote resources is interesting to consider. Such new approaches to algorithms will require new programming language features.

30.3 Highly Parallel Architectures

In highly parallel systems we are interested in utilizing the resources of hundreds or thousands of processors. Such systems can take full advantage of the high functional complexity made available with VLSI technology and also with full wafer technology as it becomes available.

The classical memory processor bottleneck of conventional and pipelined processors is avoided by using VLSI processors which are matched in speed to current high density memory technology and by providing each processor with its own fast memory. The new problem which this creates is the need to communicate information between the different processors.

We are interested here in highly parallel systems by which we mean architectures which are not fundamentally limited in the degree of parallelism. A key concept in designing algorithms for such systems is to make them independent of the number of processors used. In this way higher speed for the algorithm may be achieved by simply adding more processors. This is in contrast to serial processors and "vector" processors which offer a limited amount of parallelism for existing conventional programs but which are not able to take advantage of any additional hardware parallelism.

There are two main computation models of highly parallel architectures; these are frequently termed SIMD and MIMD. In a SIMD system a single instruction stream broadcasts instructions to a number of slave processing elements (PE's). In the MIMD case a number of independent processors work together on a single task. Most commercial SIMD systems have between 10^3 and 10^5 PE's connected with a mesh topology while most commercial MIMD systems have between 10 and 10^3 processors with a hypercube interconnection scheme dominating the highly parallel systems.

30.3.1 Highly Parallel SIMD Computers

The concept of SIMD systems is to use a large number of ALU's to simultaneously process a number of data elements. Usually, these systems consist of a set of Processing Elements (PE's) each of which contains an ALU and some local memory. There is a single "host" program control unit which broadcasts the same instruction to all PE's. The advantage of such systems is that very large numbers of PE's (many thousands) may be efficiently used for suitable applications. The main disadvantage is that users must very carefully map their problems onto the system. Conventional serial languages such as Fortran are out of the question. Usually an array based high level language is used.

A design strategy for high hardware efficiency is to use a very large number of simple PE's in combination with one complex, fast instruction unit which does not waste any PE cycles. For this type of architecture to be most effective there should be at least as many data elements to be processed as there are PE's and the algorithms should be well structured for the near neighbor interconnection structure. In practice, most systems of this type have bit-serial PE architectures; that is, the basic word size used in each PE is only one bit wide. A number of such PE's can be fabricated on a single chip. The advantages of the bit-serial approach over more conventional multi-bit words include more effective use of hardware, more optimal use of storage and the possibility of adjusting the processing time (possibly dynamically) to fit the actual precision of the data; i.e. short data words are processed faster than longer data words. In a multi-bit organization, processing is restricted to multiples of the machine "word" size which must be larger than the data precision. A precursor for these advantages is that the two conditions listed above (i.e., degree of parallelism and appropriate structure) are met. When this is true, then the processor array is a very efficient computer architecture with much more of the hardware dedicated to processing the data than with more conventional organizations.

The Massively Parallel Processor consists of 16384 bit-serial Processing Elements (PE's) connected in a 128 × 128 mesh [2,5]. That is, each PE is connected to its 4 adjacent neighbors in a planar matrix. The MPP offers a simple basic model for analysis since it involves just mesh interconnections and bit-serial PE's. The minimal architecture of the MPP is of particular interest to study, since any architecture modifications to improve performance would result in a more complex PE or a more dense interconnection strategy.

30.3.2   The MPP Processing Element

The MPP processing element is shown in Fig. 30.1. All data paths are one bit wide and there are 8 PE's on a single CMOS chip with the local memory on external memory chips. Except for the shift register, the design is essentially a minimal architecture of this type. The single bit full adder is used for arithmetic operations and the Boolean processor, which implements all 16 possible two input logical functions, is used for all other operations. The NN select unit is the interface to the interprocessor network and is used to select a value from one of the four adjacent PE's in the mesh.

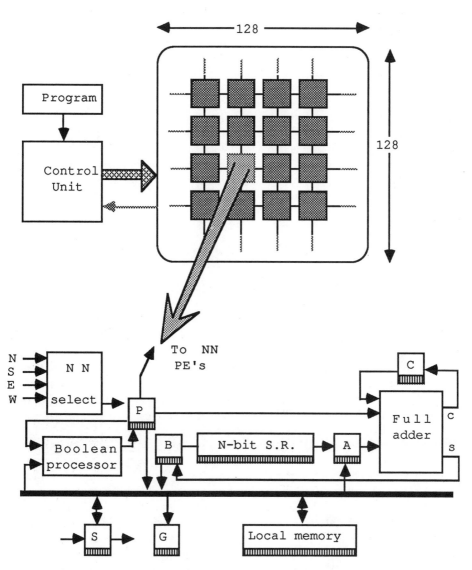

Figure 30.1: The MPP processing element.

The S register is used for I/O. A bitplane is slid into the S registers independent of the PE processing operation and it is then loaded into the local memory by cycle stealing one cycle. The G register is used in masked operations. When masking is enabled only PE's in which the G register is set perform any operations; the remainder are idle. The masked operation is a very common control feature in SIMD designs. Not shown in Fig. 30.1. is an OR bus output from the PE. All these outputs are connected (ORed) together so that the control unit can determine if any bits are set in a bitplane in a single instruction. On the MPP, the local memory has 1024 words (bits) and is implemented with bipolar chips which have a 35 ns access time.

The main novel feature of the MPP PE architecture is the reconfigurable shift register. It may be configured under program control to have a length from 2 to 30 bits. Improved performance is achieved by keeping operands circulating in the shift register which greatly reduces the number of local memory accesses and instructions. It speeds up integer multiplication by a factor of two and also has an important effect on floating-point performance.

30.3.3 Highly Parallel MIMD Computers

There are two main architecture models for MIMD systems: shared memory systems and distributed systems. In the shared memory architecture a number of independent processors have access to a single logical address space implemented by means of a number of memory modules and a processor-to-memory interconnection network. In the distributed approach each processor has its own local memory. Successful shared memory systems have been developed for small numbers of processors; however, these systems are difficult to extend to high degrees of parallelism since the memory latency and memory contention increase with the number of processors in the system.

With the distributed MIMD organization, a processor has simple fast access to its own memory but to access the memory of another processor it must communicate through an I/O channel to that processor. Typically this is done by a processor interrupt which involves a large amount of overhead for both processors. Consequently, the data processing operations and communication operations are usually separated and block data transfers are used as much as possible between processors to offset the overhead of performing a transfer. While the programming environment of a loosely coupled system is much less convenient than for a shared memory system, the potential advantage is that a very large number of processors may be used. It is only in the last few years that commercial systems with a distributed MIMD organization for scientific applications have been developed.

A typical node of a hypercube system is shown in Fig. 30.2. The heart of the node is a conventional microprocessor which performs all general control functions

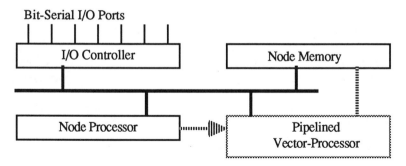

Figure 30.2: A hypercube node processor.

and some arithmetic computations. Each node has a number of I/O ports which connect it to other nodes. Typically, a node in a D-dimensional hypercube system has D+1 ports; the extra port is for I/O to the outside world. With an I/O port on every processor, the potential for very high I/O bandwidth exists. With the FPS T-series hypercube system and optionally with the Intel system, a VLSI pipelined floating-point processor is used for vector arithmetic.

The hypercube MIMD system is characterized by the a highly flexible hardware which is capable of supporting many different software strategies coupled with some of the least developed programming environments. In order to control an ensemble of independent processors, an operating system is required. However, the operating system is to support only a single user (in some cases multiple users is also a possibility). A major feature of any system is the technique used for message passing since this dramatically impacts the performance of the system and the development of algorithms.

The first generation hypercubes have been built with off-the-shelf components. A node consists of a conventional microprocessor with a standard VLSI communication chip used for each of the I/O ports. Typically a node occupies a single board.

A technologically innovative hypercube system has been developed by NCUBE [6]; the system environment it supports is representative of most other hypercube systems. Each processor node on the NCUBE system consists of a specially designed VLSI chip, which is claimed to have a computation performance similar to a VAX 11/780, and six memory chips which provide 128 Kbytes of memory. This minimal memory size will be expanded in the future with the availability of larger memory chips. The special 32-bit processor has 11 bit-serial bidirectional data channels with DMA support. The floating point performance is about 0.5 MFLOPS for scalar register operations. A single processor board contains 64 interconnected processors; i.e., a 6 cube. Most other hypercube systems have in the order of one processor on a board. An NCUBE system may contain 16 processor boards, for a total of 1024 processor nodes, in a single cabinet; i.e. a 10 cube.

Each node runs a simplified UNIX-like operating system and will support C and Fortran programs with message passing primitives. The bandwidth between adjacent nodes is about 1 Mbyte/second with a start-up latency of $300\mu s$. A store-and-forward policy is used for transferring messages between nonadjacent nodes which means that message buffers are required at each node for in-transit messages.

While the integer performance of current hypercube systems is adequate for some applications, the floating point arithmetic is much too slow for most scientific applications. For these applications each node should be enhanced with a floating point accelerator.

The FPS T-series is also a technologically innovative design which is significantly different from most other hypercube systems [3]. The node for the FPS T-Series consists of a control processor and a vector pipeline for floating point arithmetic. The control processor is a Transputer which is a high-speed 32-bit microprocessor with 4 bit-serial data channels that are under direct program control. Multiplexers are used to give each node 16 channels. The processor and memory are contained on a single board and a cabinet holds 16 node processor boards. A very large system of this type could, in theory, consist of 4096 nodes housed in 256 cabinets; however, current systems are used in experimental environments and consist of one or two cabinets. Each processor has a pipelined vector processor which has a peak processing rate in the order of 12 MFLOPS for 64-bit numbers. There is 1 Mbyte of memory on each node which could be made larger on future systems.

Hypercube systems are in a very early stage of development still. The current systems based on off-the-shelf components are good research vehicles on which to base future designs. Highly improved interprocessor communication speeds may be anticipated as recently developed architectures, which have been designed for hypercubes, are realized in VLSI components.

30.3.4 Hypercube Operating Systems

The basic programming tools provided with current hypercube systems are very similar. The usual programming environment is a conventional high level language such as Fortran or C with a library of message passing subroutines which allow data to be conveyed between different nodes. Programming environments for hypercube systems are currently a very active research area. There are a number of important issues in the method chosen for message passing and process management. Some of the main issues are as follows:

Process Management. Nodes are frequently based on conventional microprocessors; therefore they are capable of running a number of processes concurrently under an operating system. In fact, this is necessary if the I/O is done with DMA. However, it is not clear how to manage a number of processes in a single application environment. Interesting nonconventional scheduling strategies are possible. For example, since task switching takes time, one strategy is to only swap tasks when the running task is blocked. A different novel strategy being explored at JPL for real-time applications is the time-warp concept. With this scheme all messages are time stamped and the process to be scheduled is selected by the waiting message with the oldest time stamp. Yet another school of thought is that multiple application processes are unnecessary and better performance is to be obtained with a single process scheme.

Message Addressing. A second issue is the way in which processes address messages to other processes. A the lowest level a process may send a message to a node where any process on that node may receive it. At the next level a message address may specify a process within a node. The highest level is for a process to address a message to a specific process; the target process in this case could be moved to a different node. The successive levels described above involve increasing overhead costs; they also directly impact the types of algorithms which can be implemented.

Message Protocols. Perhaps one of the most controversial issues of hypercube design is the protocol used for message transfer. The central issue is how to most effectively utilize the bandwidth of the hypercube interconnection network; the most appropriate protocol depends on both the available hardware facilities and the properties of the messages such as length distribution, frequency, and coherence of the message paths. The message properties are highly application dependent. A first issue is should data transfers be DMA or processor polling; polling may require less overhead for synchronized processes and DMA may be more efficient for unsynchronized long messages. Second, how should a message be routed through a sequence of nodes. Possibilities include: (a) only near neighbor data transfers are permitted, (b) messages are transferred between adjacent nodes and then automatically routed towards their destination, this scheme requires a scheduling policy at each node to select between messages which want to use the same I/O port, (c) messages may create virtual circuits between source and destination nodes; all intermediate links are reserved until the complete message has been transferred.

Another issue is whether synchronous or asynchronous messages are supported. In the synchronous case a receiving process must acknowledge that it has received a message before a sending process is permitted to continue execution; in the asynchronous case the sending process continues execution without waiting for an acknowledgement. One problem, especially with asynchronous message passing, is that message buffers are required at each node for transferring messages and for pending messages. Since messages can be of a variable size it is very difficult to allocate the appropriate amount of each nodes memory for message buffers; most current systems have limited buffer pools and cannot recover if the messages exceed the allocated space. One approach to this problem is "packetizing" the messages into fixed length packets; this greatly simplifies the message buffer management; an extension of this concept is to combine packetizing with the virtual switch concept such that short messages use the strict virtual circuit while long messages can be interrupted to permit other traffic to flow. The more complex protocols have very attractive bandwidth properties at the cost of additional message processing overhead. Different applications (and algorithms) are suited to different communication protocols. It is quite possible that the optimal solution for some problems may involve the use of several protocols.

Hardware Support for Communication. Each of the communication protocols mentioned above could be made more efficient if a special hardware I/O processor was designed to implement them. Most current systems use simple DMA I/O ports since these are familiar from conventional processor systems. Greatly improved I/O hardware may be anticipated in future designs when the interprocess communication properties of hypercube systems are better understood.

30.4   Programming Environments

Highly parallel computer systems cannot be effectively programmed with conventional serial languages such as Fortran. For SIMD systems only a limited set of programming strategies are possible; a typical language is outlined in the following section. In contrast, for hypercube systems a wide range of programming strategies are possible and only a few have been tentatively explored.

30.4.1   SIMD Languages

High level programming languages for mesh connected SIMD computers are well developed; they usually have operations similar to matrix algebra primitives since entire arrays are manipulated with each machine instruction. The features of a typical high level language for processor arrays, called Parallel Pascal, are outlined below.

Parallel Pascal [1] is designed for the convenient and efficient programming of parallel computers and is an upward compatible extended version of the standard Pascal programming language. It is the first high level programming language to be implemented on the MPP. Parallel Pascal was designed with the MPP as the initial target architecture; however, it is also suitable for a large range of other parallel processors.

Parallel Pascal includes parallel expressions and a mechanism for processor array allocation. In addition, there are three fundamental classes of operations on array data which are frequently implemented as primitives on array computers but which are not available in conventional programming languages, these are: data reduction, data permutation and data broadcast. These operations have been included as primitives in Parallel Pascal. Mechanisms for the selection of subarrays and for selective operations on a subset of elements are also important language features.

Parallel Expressions. In Parallel Pascal all conventional expressions are extended to array data types. In a parallel expression all operations must have conformable array arguments. A scalar is considered to be conformable to any type compatible array and is conceptually converted to a conformable array with all elements having the scalar value.

Parallel data declaration. In many highly parallel computers including the MPP there are at least two different primary memory systems; one in the host and one in the processor array. Parallel Pascal provides the reserved word *parallel* to allow programmers to specify the memory in which an array should reside.

Reduction Functions. Array reduction operations are achieved with a set of standard functions in Parallel Pascal. The numeric reduction functions maximum, minimum, sum and product and the Boolean reduction functions any and all are implemented.

Permutation Functions. One of the most important features of a parallel programming language is the facility to specify parallel array data permutations. In Parallel Pascal three such operations are available as primitive standard functions: *shift, rotate* , and *transpose* . The shift and rotate primitives are found in many parallel hardware architectures and also, in many algorithms. The shift function shifts data by the amount specified for each dimension and shifts zeros (null elements) in at the edges of the array. Elements shifted out of the array are discarded. The rotate function is similar to the shift function except that data shifted out of the array is inserted at the opposite edge so that no data is lost.

While transpose is not a simple function to implement with many parallel architectures, a significant number of matrix algorithms involve this function; therefore, it has been made available as a primitive function in Parallel Pascal.

Distribution Functions. The distribution of scalars to arrays is done implicitly in parallel expressions. To distribute an array to a larger number of dimensions the *expand* standard function is available. This function increases the rank of an array by one by repeating the contents of the array along a new dimension. The first parameter of expand specifies the array to be expanded, the second parameter specifies the number of the new dimension and the last parameter specifies the range of the new dimension.

This function is used to maintain a higher degree of parallelism in a parallel statement which may result in a clearer expression of the operation and a more direct parallel implementation. In a conventional serial environment such a function would simply waste space.

Sub-Array Selection. Selection of a portion of an array by selecting either a single index value or all index values for each dimension is frequently used in many parallel algorithms; e.g., to select the ith row of a matrix which is a vector. In Parallel Pascal all index values can be specified by eliding the index value for that dimension.

Conditional Execution. An important feature of any parallel programming language is the ability to have an operation operate on a subset of the elements of an array. In Parallel Pascal a *where - do - otherwise* programming construct is available which is similar to the conventional *if - then - else* statement except that the control expression results in a Boolean array rather than a Boolean scalar. All parallel statements enclosed by the where statement must have results which are the same size as the controlling array. Only result elements which correspond to true elements in the controlling array will be modified. Unlike the if statement, both clauses of the where statement are always executed.

30.4.2   SIMD Machine Parameters

Programming a SIMD computer with a language such as Parallel Pascal is comparable in convenience to programming with a conventional language. In some ways it is simpler since the array data structure is manipulated in a much more direct manner with more powerful operators than with serial languages. The difficulty in using SIMD languages is in matching the data structures to the dimensions of the parallel hardware and specifying problems in terms of data permutations that can be efficiently implemented.

Additional information to assist the programmer is needed in the form of the processor array size and the cost of specific data permutations. For example, consider the essential parameters for the MPP. First, the conventional system parameters of arithmetic speed and storage size need to be known. The peak arithmetic performance of the MPP is in the order of 400 million floating point operations per second (MFLOPS) for 32-bit data and 3000 million operations per second (MOPS) for 8-bit integer data. On the current system, the primary data memory has a very limited size of 2 Mbytes. There is a secondary solid state memory system which can be as large as 64 Mbytes. Data is transferred between primary and secondary memory systems in 2kbyte ($128 \times 128 \times$ 1-bit) blocks: a block transfer requires about $12.8\mu s$.

Second, the programmer needs to know the topology and speed of the processor interconnection network. The MPP consists of 16384 PE's in a 128×128 mesh connected array. The only permutation function which is directly implemented by the MPP is the near neighbor rotate (or shift). The direction of the rotation may be in any of the four cardinal directions. The rotation utilizes the toroidal end around edge connections of the mesh. The *shift* function is similar except that the mesh is not toroidally connected and zeroes are shifted into elements at the edge of the array; therefore, the shift function is not a permutation function in the strict sense. The speed of the near neighbor shift or rotate is 100ns for each bit (i.e., 16384 bits for the whole array) to be transferred. This information is not in a high enough level for a programmer to easily develop efficient algorithms. What is needed is a characterization of the speed of the network for important macro permutations for multi-bit data types. To do this we define the *transfer ratio* for a given permutation and data type as the relative time taken to implement the permutation for that data type over the time taken to perform an elemental arithmetic operation on that data type. An elemental arithmetic operation time is the average of the time for an add and a multiplication operation.

Transfer times for some important data manipulations on the MPP are given in Figure 30.3. A more detailed analysis of data mappings on the MPP is given in [7]. Two data sizes are considered, the 512 × 512 array is processed as a sequence of 16 128 × 128 blocks on the MPP. The elemental arithmetic times are therefore 16 times greater than for the 128 × 128 array. Three representative base data types are considered: single-bit *Boolean* data, 8-bit *integer* data and 32-bit floating-point (*real*) data. Estimated elemental operation times data types are 200ns, $5\mu s$, and $40\mu s$ for the data types Boolean, integer, and real respectively. It should be remembered that elemental operations also include many other functions such as transcendental functions since these can be computed in times comparable to a multiplication on a bit-serial architecture.

Several interesting observations may be made directly from the transfer ratios in Figure 30.3. Data shifts of arbitrary size are not very expensive especially for integer and real data types. Row broadcast of multi-bit data is faster than might

Data Manipulation	Array Size					
	128 x 128			512 x 512		
	Boolean	integer	real	Boolean	integer	real
DATA SHIFT a) 1 element b) worst case	1.0 33	0.32 10.2	0.16 5.2	2.0 33	0.5 10	0.24 5.2
BROADCAST a) Global b) Row (or column)	2 68	0.64 3.2	0.32 0.52	0.59 18.2	0.20 0.92	0.09 0.19
SHUFFLE	640	90	42	640	90	42
TRANSPOSE FLIP	840 190	110 43	44 21	840 190	110 43	44 21
SORT	19000	1100	280	14000	790	210
SWAP	256	20	10	256	20	10

Figure 30.3: Transfer Ratios for Different Data Manipulations and Array Sizes

be expected (due to the MPP shift register). The shuffle permutation is not very fast. Implementing the FFT with shuffle operations will be very inefficient; however if regular butterfly operations are implemented with multi-element shifts then an efficient FFT implementation is possible. The figures for the swap operation express the cost of exchanging an array with the secondary memory in a more useful manner than the raw transfer speeds.

A table of transfer times, such as that shown in Figure 30.3 but with many more permutations listed, is an important aide for programmers who are developing algorithms for a given system. It is also useful for comparing different systems.

30.4.3   MIMD Languages

The usual programming environment on current hypercube systems consisting of a conventional language plus message passing subroutines is at much too low a level for a user programmer. Typically the user writes a single program which runs on every processor. A unique processor identifier is used to determine where to send messages. While the environment has a superficially familiar appearance, the organization of programs is much more complex due to the message passing between processes. In effect, the user is now expected to program many functions which are conventionally done by the operating system.

For a large number of structured scientific applications a SIMD style of operation is used. In this scheme each node runs a single user process and is responsible for processing a contiguous block of a data array. A library of subroutines supports the distributed array data structures in a similar way to a SIMD language. For example, subroutine procedures may be available to perform total array transpose, matrix multiply or FFT operations. The message passing is hidden from the user who only has to specify the dimensions of the array to be processed and call the appropriate function. This environment is similar to that of the early array processors; while it is reasonably simple to use existing library functions, it is very difficult for the user to specify new functions.

An alternative to the library approach is to implement a SIMD language such as Parallel Pascal. This provides the user with a convenient method of developing new functions while still hidding the message passing details. A major advantage of this approach is that the user programs the whole problem at a high level rather than attempting to do the subtask decomposition and develop a program which works on one block of the problem.

Parallel Pascal has been ported to the FPS T-series [8]. In this implementation, scalar variables are replicated in all processors and parallel arrays are distributed between the processors. Experience with this system has led to the consideration of several language enhancements for MIMD operation; these were: vector indexing, local permutations, sparse matrix packing and multiple execution.

Vector indexing would permit a parallel array to be used as the indices in a vector; i.e., to use a vector for a table-look-up operation. This is very simple to implement on the transputer but it cannot be directly implemented on the MPP due to the global address scheme.

Local permutations would allow the specification of permutations which do not involve interprocessor transfers since, in general these are much faster. Unfortunately, this would imply that the assignment of data to processors is known by the the programmer which limits the ability of the system to reassign data. Task reassignment might be desirable to achieve fault tolerance or to achieve load balancing in the multiple execution mode outlined below. One effective strategy for sorting on a MIMD system is to use quicksort on the nodes and bitonic sort between the nodes. Without a feature like local permutations it is not possible, in Parallel Pascal, to program the independent node quicksorts.

Sparse matrix packing would take advantage of the scatter and gather capabilities of the transputer to support a compressed vector format for very sparse data. Arithmetic operations would be much faster on this compressed form but costly conversion to the distributed form would probably be necessary for data permutation operations.

Multiple execution would permit various parts of the array to be processed with different algorithms simultaneously. For example, solutions to PDE's often require different operations for the boundary elements. A modified form of the where statement is being considered which is based more on conditional evaluation rather than conditional assignment. However, it will probably have more restrictions with respect to data permutations than the current where statement.

For applications which do not fit the SIMD constraints a special purpose operating system is possible. With this approach the user programs a problem by means of a sequence of task oriented system primitives; different sets of primitives may be implemented for different applications. The user only specifies the computation process and the format of the input and output data. The system performs all subtask decomposition, allocation and scheduling to best fit the available resources. While this is the most difficult approach to implement, it offers the greatest promise for the future.

With an appropriate high level software environment, the system, rather than the programmer, is able determine which processes and which data elements will be allocated to which nodes. This flexibility has two important consequences. First, the system can implement fault tolerance strategies. If a node fails then the system can allocate its processes to other nodes. Second, dynamic load balancing techniques are possible; if some nodes have completed the work allocated to them then they may be reassigned to take some of the load of any remaining busy nodes.

Matrix size 128 x 128			
Data Mapping	FPS 16-Nodes	FPS 256-Nodes	MPP
(MFLOPS)	46	124	400
Near Neighbor Shift	0.63	0.60	0.16
Global Broadcast	0.30	3.5	0.32
Row/Col Broadcast	1.2	4.8	0.32
Matrix size 512 x 512			
Data Mapping	FPS 16-Nodes	FPS 256-Nodes	MPP
(MFLOPS)	62	740	400
Near Neighbor Shift	0.19	0.63	0.24
Global Broadcast	0.020	1.1	0.09
Row/Col Broadcast	1.2	5.0	0.19

Figure 30.4: FPS T-Series Hypercube Permutation Performance

MIMD Machine Parameters

MIMD systems require more parameters to characterize their behavior than SIMD systems. For the complex class of problems it is very difficult to predict system performance and more research needs to be done in this area. For the SIMD class of problems, where there is only one process in each processor and communications are loosely synchronized, good predictions can be made. Transfer ratios may be determined in a similar way to SIMD systems and system comparisons can be made. Such a comparison is shown in Figure 30.4 where selected transfer ratios for the FPS hypercube system and the MPP are given.

Results in Figure 30.4 were based on parameters which were measured on an FPS system [9]. These parameters were: pipeline setup delay $100\mu s$, peak node performance 4 MFLOPS, transfer latency $19\mu s$, and internode transfer speed 600 kbytes/s. The system was considered to be statically configured as a mesh due to the large reconfiguration time.

Contiguous blocks of data are allocated to each processor and near neighbor operations only require transferring one edge of these blocks between processors. From the table we can see that, in most cases the transfer ratios for the given data manipulations are quite reasonable. The row/column broadcast cost is higher than an elemental operation in some cases. This could be reduced if the FPS system could make use of the other hypercube links. The performance of the 256-node FPS system on a 128 × 128 array is not very good due to the small amount of data (64 elements) on each node; the pipeline setup time dominates the processing cost.

30.5 Conclusion

Highly parallel computers have the potential for very high speed scientific applications; however, current programming environments are at a lower convenience than for conventional computers. Two important highly parallel computer architectures have been considered here which are at radically different stages of development. SIMD systems have been in use for a number of years now. Programming environments are developed for such systems that are at a similar level to conventional programming environments. However, these systems are not suitable for all highly parallel applications and new skills need to be acquired to effectively program them.

Highly parallel MIMD systems are at a much earlier stage of development; furthermore they are much more flexible than SIMD systems and a large amount of research still needs to be done to determine the best programming strategies for such systems. They can emulate a SIMD system but they will not be as effective for algorithms which are ideal for the SIMD system. A critical component of a hypercube computer is the software system. Current programming environments are very difficult to use; however, for specific tasks, effective systems have been demonstrated. More research in the software area is necessary for hypercube systems to come close to realizing their full potential.

References

[1] A. P. Reeves, Parallel Pascal: An Extended Pascal for Parallel Computers, Journal of Parallel and Distributed Computing, vol. 1, 1984, 64-80.

[2] K. E. Batcher, Design of a Massively Parallel Processor, IEEE Transactions on Computers, vol. C-29, no. 9, September 1981, 836-840.

[3] J. L. Gustafson, S. Hawkinson, and K. Scott, The Architecture of a Homogeneous Vector Supercomputer, Proceedings of the 1986 International Conference on Parallel Processing, August 1986, 649-652.

[4] M. U. Uyar and A. P. Reeves, Fault Reconfiguration for the Near Neighbor Problem in a Distributed MIMD Environment, Fifth International Conference on Distributed Computing, May 1985, 372-379.

[5] A. P. Reeves, The Massively Parallel Processor: A Highly Parallel Scientific Computer, Data Analysis in Astronomy II, ed. V. Di Gesu, Plenum Press, 1986, 239-252.

[6] J. P. Hayes, T. N. Mudge, Q. F. Stout, S. Colley, and J. Palmer, Architecture of a Hypercube Supercomputer, Proceedings of the 1986 International Conference on Parallel Processing, August 1986, 653-660.

[7] A. P. Reeves and C. H. Moura, Data Manipulations on the Massively Parallel Processor, Proceedings of the Nineteenth Hawaii International Conference on System Sciences, January 1986, 222-229.

[8] A. P. Reeves and D. Bergmark, Parallel Pascal and the FPS Hypercube Supercomputer, Proceedings of the 1987 International Conference on Parallel Processing, August 1987.

[9] D. Bergmark, J. M. Francioni, B. K. Helminen, and D. A. Poplawski, On the Performance of the FPS T-Series Hypercube, Second Conference on Hypercube Multiprocessors, September 1986.

Chapter 31

Systolic Designs for State Space Models: Kalman Filtering and Neural Networks [1]

S-Y Kung [2]

J. N. Huang [2]

31.1 Introduction

In this paper, a systematic mapping methodology is introduced for deriving systolic and wavefront arrays from regular computational algorithms [10]. It consists of three stages of mapping design: *(data) dependence graph (DG) design, signal flow graph (SFG) design,* and *array processor design*. This methodology allows systolic design with many desirable properties, such as local communication and fastest pipelining rates, etc. Based on this methodology, we shall develop systolic array designs for two important applications of adaptive state-space models. One is for the Kalman filtering algorithm which is popular in many digital signal processing applications. The other one is the Hopfield model for artificial neural networks (ANN), which has recently received increasing attention from AI and parallel processing research community.

---

[1] This paper was also presented at the 26th IEEE Conf. on Decision and Control, Los Angeles, CA, Dec. 9-11, 1987 and appeared in Proc. 26th IEEE Conf. Decision and Control, pp. 1461-1467, 1987. ©1987 IEEE.

[2] This research was supported in part by the National Science Foundation under Grant ECS-82-13358, by the Semiconductor Research Corporation under USC SRC program, and by the Innovative Science and Technology Office of the Strategic Defense Initiative Organization and was administered through the Office of Naval Research under Contract No. N00014-85-K-0469 and N00014-85-K-0599.

[3] Princeton University, Princeton, NJ

The Systolic Kalman (SK) filter designs based on a triangular array (triarray) configuration is presented. In order to facilitate the systolic design, the original algorithm for the Kalman filter estimation is reformulated in a new least squares formulation. The design has advantages in both numerical accuracy and computational efficiency. For the case of white additive noise, the SK-W filter design uses approximately $n^2/2$ processors and provides a speed-up of $n^2/2$, with a nearly 100% utilization rate. (Here n is the order of the state-space model.) For the case of colored additive noise, the proposed SK-C filter design also offers comparable speed-up performance.

The Hopfield and Tank model for neural networks can be programmed to perform associative retrieval or as computational networks for optimization problems. Based on this model, a locally interconnected systolic architecture for artificial neural networks is proposed. Some advantages of this architecture over the analog neural circuit and optical neural networks are: high pipelining rate, high precision, capability of learning, fast convergence speed, and global optimal solution searching.

31.2   Mapping Algorithms onto VLSI Array Processors

The major emphasis of VLSI system design is to reduce the overall interconnection complexity and keeping the overall architecture highly regular, parallel, pipelined. It stresses the importance of local communication in the array processor. *Systolic* and *wavefront* arrays are a new class of pipelined array architectures very suitable for VLSI implementation, because they feature the important properties of modularity, regularity, local interconnection, and a high degree of pipelining.

The main concern in algorithm-oriented array processor design is: *given an algorithm, how is an array processor systematically derived?* The ultimate design should begin with a powerful algorithmic notation to express the recurrence and parallelism associated with the description of the space-time activities. Next, this description will be converted into a VLSI hardware description, or into executable array processor machine codes. In deriving systolic/wavefront arrays for a given algorithm, there are three major stages [10]:

31.2.1   Deriving Dependence Graph from Algorithms

A (data) dependence graph (DG) is a directed graph which specifies the data dependencies of an algorithm. In a DG, *nodes* represent computations, and *arcs* specify the data dependencies between computations. In our notation, with respect to a dependence arc, the terminating node depends on the initiating node. For regular and recursive algorithms, the DGs will also be regular and can be represented by a grid model; therefore, the nodes can be specified by simple indices, such as (i, j, k).

Design of locally linked DG is a critical step leading to the design of pipeline array processors, such as systolic or wavefront arrays.

31.2.2 Mapping the DG onto Signal Flow Graph Arrays

Signal Flow Graphs (SFGs) The abstract array resulting from a projection and associated linear schedule can be represented by a *signal flow graph (SFG)* [11]. In an SFG each node represents a computation element, and the arcs represent the communication. Communication and computation is assumed to be instantaneous. Time is explicitly modeled by *delays* on the arcs. In representing the projection of a DG, the SFG nodes correspond to the projection of the DG nodes, and the SFG arcs correspond to the projection of the DG arcs. The delays on an SFG arc are determined by the number of schedule planes separating the endpoints of the DG arc that it originated from.

Projection and Scheduling For regular DGs, we can use *projections* as the mapping procedure to derive the array processor structure. We define a *projection vector $\vec{d}$* (cf. Figure 31.1(a)), such that index points that lie on the same line in the projection direction are mapped to the same processor. The order in which these nodes are executed is determined by a *linear schedule*. The linear schedule can be visualized as a series of parallel planes or lines (cf. Figure 31.1(b)). All points on the same plane are computed at the same time. A linear schedule can be specified by a *schedule vector $\vec{s}$*, defined as the vector perpendicular to the schedule planes and pointing in the direction of increasing time[3].

31.2.3 Transforming the SFG to Systolic/Wavefront Arrays

The SFG obtained from DG projection can be mapped to a systolic array or a wavefront array. To convert an SFG array into a systolic array, a cut-set systolization (retiming) procedure can be adopted [10]. Similarly, a simple procedure can be used to convert the SFG array into a wavefront array [12]. In this paper, we only concentrate on the design of systolic arrays. However, it should be understood that after an SFG is obtained, it is straightforward to convert it to a wavefront array.

---

[3]Note that the schedule planes cannot be parallel to the projection direction, otherwise multiple points in the plane will be projected to a single PE, and they cannot be computed at the same time. This means that the projection and schedule directions are constrained to be non-orthogonal. In addition to this constraint, the linear schedule has to meet the data dependence constraints in the DG, that is, if point A depends on the value of point B, then we cannot schedule A before B.

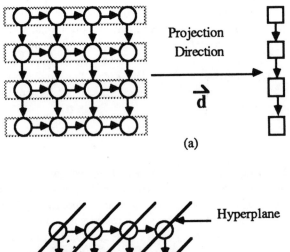

(a)

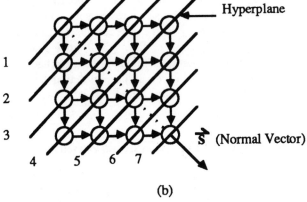

(b)

Figure 31.1: Illustrations of (a) a linear projection with projection vector $\vec{d}$; (b) a linear schedule $\vec{s}$ and its hyperplanes.

31.3 Systolic Designs for Kalman Filtering

The Kalman filter is an optimal linear minimum variance predictor (or estimator) originally introduced in [8]. By using a state space model, it facilitates the prediction of the n-dimensional state vector recursively given each new m-dimensional measurement vector in a discrete time-varying dynamic system;

$$
\begin{aligned}
\mathbf{x}(k+1) &= \mathbf{F}(k)\mathbf{x}(k) + \mathbf{w}(k+1) \\
\mathbf{y}(k) &= \mathbf{C}(k)\mathbf{x}(k) + \mathbf{v}(k)
\end{aligned}
\tag{31.1}
$$

where $\mathbf{F}(k)$ and $\mathbf{C}(k)$ are coefficient matrices with dimension $n \times n$ and $m \times n$; $\mathbf{x}(k)$ and $\mathbf{w}(k+1)$ are the n-dimensional state vector and system noise vector respectively; $\mathbf{y}(k)$ and $\mathbf{v}(k)$ are the m-dimensional measurement vector and measurement noise vector respectively. The noise vectors are zero mean, independent processes with known covariance matrices $\mathbf{R_W}(k+1)$ and $\mathbf{R_V}(k)$ respectively. The noise $\mathbf{w}$ is also assumed to be uncorrelated with $\mathbf{v}$ (i.e., $E[\mathbf{w}(i)\mathbf{v}^T(j)] = \mathbf{0}$, for all i, j).

Least squares system formulation for Kalman filtering was proposed in [18], which is an *expanded matrix representation* of the state space iteration in Eq. 31.1. Based on this formulation, a trapezoidal systolic array is proposed [17] for both the measurement and time updates. Along this line, this paper presents a triangular array (triarray) configuration which is much more efficient than the designs previously proposed in terms of speed-up factor and processor utilization efficiency.

31.3.1 Whitening and Least Squares Formulation

When noises are colored, the noise covariance matrices, $\mathbf{R_W}(k+1)$ and $\mathbf{R_V}(k)$, are not identity matrices. In order to use the least squares formulation, a first step is to whiten the system and measurement noises. The covariance matrices of the two noise vectors can be expressed as: $\mathbf{R_W^{-1}}(k+1) = \mathbf{W}(k+1)^T\,\mathbf{W}(k+1)$, $\mathbf{R_V^{-1}}(k) = \mathbf{V}(k)^T\,\mathbf{V}(k)$, where $\mathbf{W}(k+1)$ and $\mathbf{V}(k)$ are *upper triangular matrices* and can be obtained by the *reverse Cholesky decomposition* of $\mathbf{R_W}(k+1)$ and $\mathbf{R_V}(k)$.

By applying premultiplication of the whitening operators $\mathbf{W}(k+1)$ and $\mathbf{V}(k)$ to Eq. 31.1 and grouping together the consecutive state vectors (up to stage k) and measurement vectors (up to stage $k-1$), two accumulated vectors $\mathbf{X}(k)$ and $\mathbf{Y}(k)$ can be formed, and a least squares formulation can be obtained [18] [17]

$$
\tilde{\mathbf{U}}(k) = \tilde{\mathbf{A}}(k)\mathbf{X}(k) + \tilde{\mathbf{Y}}(k)
\tag{31.2}
$$

where $\mathbf{X}(k) = [\mathbf{x}^T(1)\ \mathbf{x}^T(2)\ \ldots\ \mathbf{x}^T(k)]^T$, $\tilde{\mathbf{U}}(k) = [\tilde{\mathbf{w}}^T(1)\ \tilde{\mathbf{v}}^T(1)\ \tilde{\mathbf{w}}^T(2)\ \tilde{\mathbf{v}}^T(2)\ \ldots\ \tilde{\mathbf{v}}^T(k-1)\ \tilde{\mathbf{w}}^T(k)]^T$, $\tilde{\mathbf{Y}}(k) = [\mathbf{0}\ \tilde{\mathbf{y}}^T(1)\ \mathbf{0}\ \tilde{\mathbf{y}}^T(2)\ \ldots\ \tilde{\mathbf{y}}^T(k-1)\ \mathbf{0}]^T$, and

$$
\tilde{\mathbf{A}}(k) \;=\;
\begin{bmatrix}
\mathbf{W}(1) & & & & & \\
\tilde{\mathbf{C}}(1) & & & & \mathbf{0} & \\
\tilde{\mathbf{F}}(1) & \mathbf{W}(2) & & & & \\
 & \tilde{\mathbf{C}}(2) & & & & \\
 & \tilde{\mathbf{F}}(2) & \mathbf{W}(3) & & & \\
 & & & \ddots & & \\
 & & & & \mathbf{W}(k-1) & \\
 & \mathbf{0} & & & \tilde{\mathbf{C}}(k-1) & \\
 & & & & \tilde{\mathbf{F}}(k-1) & \mathbf{W}(k)
\end{bmatrix}
$$

where we assumed $\mathbf{x}(0) = 0$ and therefore $\tilde{\mathbf{w}}(1) = \mathbf{W}(1)\mathbf{x}(1)$. Also $\tilde{\mathbf{F}}(k) = -\mathbf{W}(k+1)\mathbf{F}(k)$, $\tilde{\mathbf{C}}(k) = \mathbf{V}(k)\mathbf{C}(k)$, $\tilde{\mathbf{y}}(k) = -\mathbf{V}(k)\mathbf{y}(k)$, $\tilde{\mathbf{w}}(k+1) = \mathbf{W}(k+1)\mathbf{w}(k+1)$, $\tilde{\mathbf{v}}(k) = -\mathbf{V}(k)\mathbf{v}(k)$, therefore, the covariance matrices $\mathbf{R}_{\tilde{\mathbf{w}}}(k+1)$ and $\mathbf{R}_{\tilde{\mathbf{v}}}(k)$ become identity matrices. The above operations are termed *whitening* for $\mathbf{F}(k)$, $\mathbf{C}(k)$, and $\mathbf{y}(k)$.

31.3.2   Recursive Least Squares Updating

Since the noise vector $\tilde{\mathbf{U}}(k)$ in Eq. 31.2 has an identity covariance matrix with zero mean, the best predictor, $\hat{\mathbf{x}}(k)$, given $\{\mathbf{y}(1)\cdots\mathbf{y}(k-1)\}$, can now be formulated as a least squares estimation problem and solved by the QR decomposition method. Applying an orthogonal tranformation matrix $\mathbf{Q}$, with a dimension of $[(k-1)m + kn] \times [(k-1)m + kn]$ at stage k, to both sides of Eq. 31.2, yields:

$$
\mathbf{Q}\tilde{\mathbf{U}}(k) = \mathbf{Q}\tilde{\mathbf{A}}(k)\mathbf{X}(k) + \mathbf{Q}\tilde{\mathbf{Y}}(k) \tag{31.3}
$$

where

$$
[\,\mathbf{Q}\tilde{\mathbf{A}}(k)\mid\mathbf{Q}\tilde{\mathbf{Y}}(k)\,] =
\left[
\begin{array}{ccccc|c}
\mathbf{R}_{11} & \mathbf{R}_{12} & & & & \mathbf{b}_1' \\
 & \mathbf{R}_{22} & \mathbf{R}_{23} & & \mathbf{0} & \mathbf{b}_2' \\
 & & & \ddots & & \vdots \\
 & \mathbf{0} & & \mathbf{R}_{k-1,k-1} & \mathbf{R}_{k-1,k} & \mathbf{b}_{k-1}' \\
 & & & & \mathbf{R}(k) & \mathbf{b}_k \\
\mathbf{0} & \mathbf{0} & \mathbf{0} & \cdots & \mathbf{0} & \mathbf{r}_1 \\
\vdots & \vdots & \vdots & \ddots & \vdots & \vdots \\
\mathbf{0} & \mathbf{0} & \mathbf{0} & \cdots & \mathbf{0} & \mathbf{r}_{k-1}
\end{array}
\right]
$$

where $\mathbf{R}_{ii}$ and $\mathbf{R}(k)$ are upper triangular. The optimal predictor $\hat{\mathbf{x}}(k)$ depends only on the vector $\mathbf{b}_k$:

$$\hat{\mathbf{x}}(k) = -\mathbf{R}^{-1}(k)\mathbf{b}_k \tag{31.4}$$

Since $\mathbf{R}(k)$ is an upper triangular matrix, Eq. 31.4 can be solved by back substitution.

At the next recursion, with the new measurement $\mathbf{y}(k)$, the updated system equation for estimating $\hat{\mathbf{x}}(k+1)$, is given by a modified matrix-vector form:

$$\hat{\mathbf{U}}(k+1) = \hat{\mathbf{A}}(k+1)\mathbf{X}(k+1) + \hat{\mathbf{Y}}(k+1) \tag{31.5}$$

where

$$\hat{\mathbf{U}}(k+1) = \begin{bmatrix} \mathbf{Q}_1\tilde{\mathbf{U}}(k) \\ - - - \\ \tilde{\mathbf{v}}(k) \\ \tilde{\mathbf{w}}(k+1) \end{bmatrix}$$

and

$[\hat{\mathbf{A}}(k+1)|\hat{\mathbf{Y}}(k+1)] =$

$$\begin{bmatrix}
\mathbf{R}_{11} & \mathbf{R}_{12} & & & & & | & \mathbf{b}'_1 \\
& \mathbf{R}_{22} & \mathbf{R}_{23} & & 0 & & | & \mathbf{b}'_2 \\
& & \ddots & & & & | & \vdots \\
& 0 & & \mathbf{R}_{k-1,k-1} & \mathbf{R}_{k-1,k} & & | & \mathbf{b}'_{k-1} \\
& & & & \mathbf{R}(k) & & | & \mathbf{b}_k \\
0 & 0 & \cdots & 0 & 0 & 0 & | & \mathbf{r}_1 \\
\vdots & \vdots & \ddots & \vdots & \vdots & \vdots & | & \vdots \\
0 & 0 & \cdots & 0 & 0 & 0 & | & \mathbf{r}_{k-1} \\
0 & 0 & \cdots & 0 & \tilde{\mathbf{C}}(k) & 0 & | & \tilde{\mathbf{y}}(k) \\
0 & 0 & \cdots & 0 & \tilde{\mathbf{F}}(k) & \mathbf{W}(k+1) & | & 0
\end{bmatrix}$$

To compute $\hat{\mathbf{x}}(k+1)$ by QR decomposition on $\hat{\mathbf{A}}(k+1)$ we need only be concerned with the a $(2n+m) \times (2n+1)$ matrix as opposed to the large $\hat{\mathbf{A}}(k+1)$ matrix. This leads to the following computation which is required for each iteration:

$$\mathbf{Q}_1 \begin{bmatrix} \mathbf{R}(k) & \mathbf{0} & | & \mathbf{b} \\ \tilde{\mathbf{C}}(k) & \mathbf{0} & | & \tilde{\mathbf{y}}(k) \\ \tilde{\mathbf{F}}(k) & \mathbf{W}(k+1) & | & \mathbf{0} \end{bmatrix} = \begin{bmatrix} \mathbf{R}_{k,k} & \mathbf{R}_{k,k+1} & | & \mathbf{b}'_k \\ \mathbf{0} & \mathbf{R}(k+1) & | & \mathbf{b}_{k+1} \\ \mathbf{0} & \mathbf{0} & | & \mathbf{r}_k \end{bmatrix} \quad (31.6)$$

where $\mathbf{R}_{k,k}$ and $\mathbf{R}(k+1)$ are $n \times n$ upper triangular matrices, and $\mathbf{r}_k$ is a residual vector.

31.3.3   Mapping Kalman Filtering Algorithms to Triarrays

Givens Transformation for QR Triangularization   A commonly used orthogonal transformation for QR triangularization is the Givens transformation [3]. The Givens transformation is a numerically stable orthogonal operator that performs a plane rotation to the $N \times M$ matrix $\mathbf{A}$. The purpose of these rotations is to nullify the subdiagonal elements of matrix $\mathbf{A}$ (and finally reduce it to upper triangular form in least squares solution).

In order to nullify the (l, k) element of the matrix $\mathbf{A}$ using the data elements in k-th row of the same matrix, an $N \times N$ orthogonal matrix $\mathbf{Q}_l^{(k)}$ can be introduced, which agrees with the identity matrix everywhere except in some 2-by-2 principal submatrix;

$$\mathbf{Q}_l^{(k)} = \begin{array}{c} \\ \\ \\ \\ \\ \\ \\ \\ \\ \\ \end{array} \begin{bmatrix} 1 & \cdots & 0 & \cdots & 0 & \cdots & 0 & \cdots & 0 \\ \vdots & \ddots & \vdots & \cdots & \vdots & \cdots & \vdots & \cdots & 0 \\ 0 & & \cos\theta & & 0 & & \sin\theta & & 0 \\ & & & \ddots & & \vdots & & & \vdots \\ 0 & & 0 & & 1 & & 0 & & 0 \\ \vdots & & & & \vdots & \ddots & & & \vdots \\ 0 & & -\sin\theta & & 0 & & \cos\theta & & 0 \\ \vdots & \vdots & & & \vdots & & & \ddots & \vdots \\ 0 & \cdots & 0 & & \cdots & & 0 & \cdots & 1 \end{bmatrix} \begin{array}{c} \\ \\ k \\ \\ \\ \\ l \\ \\ \\ \end{array}$$

where

$$\theta = \tan^{-1} [\, a_{l,k} / a_{k,k} \,] \quad (31.7)$$

We shall term the above operation of creating the angle parameters as Givens Generation (GG). Premultiplication by $\mathbf{Q}_l^{(k)}$ amounts to a rotation of θ degrees to the matrix $\mathbf{A}$:

$$\mathbf{A}' = \mathbf{Q}_l^{(k)}\mathbf{A}$$

where $\mathbf{A}' = \{a'_{ij}\}$

$$
\begin{aligned}
a'_{k,j} &= a_{k,j}\cos\theta + a_{l,j}\sin\theta \\
a'_{l,j} &= -a_{k,j}\sin\theta + a_{l,j}\cos\theta
\end{aligned}
\tag{31.8}
$$

for all $j = 1, \ldots, M$. In this paper, we shall call the operations in Eq. 31.8 as Givens Rotations (GR).

31.3.4 Mapping QR/Postrotation Algorithms to Triarray

With reference to Eq. 31.6, the situation arises in using Givens transformation for matrix triangularization involves the triangularization of an $(M + N) \times (M + N)$ matrix $[\mathbf{A} \mid \mathbf{B}]$ using an $(M + N) \times (M + N)$ orthogonal matrix:

$$
\mathbf{Q}[\mathbf{A} \mid \mathbf{B}] = \mathbf{Q}\left[\begin{array}{cc} \mathbf{R}' & 0 \\ \mathbf{F} & \mathbf{W} \end{array}\right] = \left[\begin{array}{cc} \mathbf{R}'' & \mathbf{Z} \\ 0 & \mathbf{W}' \end{array}\right]
$$

where $\mathbf{R}'$ and $\mathbf{W}$ are $M \times M$, and $N \times N$ upper triangular matrices, respectively. The $N \times M$ full matrix $\mathbf{F}$ is to be nullified by Givens transformation.

By taking advantage of the special matrix structure of $[\mathbf{A} \mid \mathbf{B}]$, it is possible that the triangular structure of $\mathbf{W}$ remains intact throughout the triangularization procedure. In order to achieve this goal, some processing ordering must be adopted when performing nullification operations on $\mathbf{F}$. This can be illustrated by showing how the substeps progress and displaying their snapshots in Figure 31.2. We note that the $\mathbf{F}$ matrix is nullified in a *bottom up* order, as shown in Figure 31.2, hence allowing $\mathbf{W}$ to remain upper triangular throughout entire process. For convenience, we will divide this triangularization procedure into two stages. The process of nullifying $\mathbf{F}$ is called *QR nullification* stage, while the modification of $\mathbf{W}$ is called *postrotation* stage.

Array Processor Design for QR Nullification

Let's first discuss the QR nullification stage. It involves the triangularization of an $(M + N) \times M$ matrix $\mathbf{A}$ using an $(M + N) \times (M + N)$ orthogonal matrix $\mathbf{Q}$:

$$[\,\mathbf{A}\,|\,\mathbf{B}\,] \;=\; \left[\begin{array}{c|c} \mathbf{R}' & \mathbf{0} \\ \hline \mathbf{F} & \mathbf{W} \end{array}\right] \equiv$$

$$\left[\begin{array}{ccc|ccc}
x & x & x & & & \\
  & x & x & & & \\
  &   & x & & & \\
\hline
x & x & x & x & x & x \\
x & x & x &   & x & x \\
x & x & x &   &   & x
\end{array}\right]
\Rightarrow
\left[\begin{array}{ccc|ccc}
g & g & g & & & g \\
  & x & x & & & \\
  &   & x & & & \\
\hline
x & x & x & x & x & x \\
x & x & x &   & x & x \\
\mathbf{0} & g & g &   &   & g
\end{array}\right]
\Rightarrow$$

$$\left[\begin{array}{ccc|ccc}
g & g & g &   & g & g \\
  & x & x &   &   &   \\
  &   & x &   &   &   \\
\hline
x & x & x & x & x & x \\
\mathbf{0} & g & g &   & g & g \\
0 & x & x &   &   & x
\end{array}\right]
\Rightarrow
\left[\begin{array}{ccc|ccc}
g & g & g & g & g & g \\
  & x & x &   &   &   \\
  &   & x &   &   &   \\
\hline
\mathbf{0} & g & g & g & g & g \\
0 & x & x &   & x & x \\
0 & x & x &   &   & x
\end{array}\right]
\Rightarrow$$

$$\left[\begin{array}{ccc|ccc}
x & x & x & x & x & x \\
  & g & g &   &   & g \\
  &   & x &   &   &   \\
\hline
0 & x & x & x & x & x \\
0 & x & x &   & x & x \\
0 & \mathbf{0} & g &   &   & g
\end{array}\right]
\Rightarrow
\left[\begin{array}{ccc|ccc}
x & x & x & x & x & x \\
  & g & g &   & g & g \\
  &   & x &   &   &   \\
\hline
0 & x & x & x & x & x \\
0 & \mathbf{0} & g &   & g & g \\
0 & 0 & x &   &   & x
\end{array}\right]
\Rightarrow$$

$$\left[\begin{array}{ccc|ccc}
x & x & x & x & x & x \\
  & g & g & g & g & g \\
  &   & x &   &   &   \\
\hline
0 & \mathbf{0} & g & g & g & g \\
0 & 0 & x &   & x & x \\
0 & 0 & x &   &   & x
\end{array}\right]
\Rightarrow
\left[\begin{array}{ccc|ccc}
x & x & x & x & x & x \\
  & x & x & x & x & x \\
  &   & g &   &   & g \\
\hline
0 & 0 & x & x & x & x \\
0 & 0 & x &   & x & x \\
0 & 0 & \mathbf{0} &   &   & g
\end{array}\right]
\Rightarrow$$

$$\left[\begin{array}{ccc|ccc}
x & x & x & x & x & x \\
  & x & x & x & x & x \\
  &   & g &   & g & g \\
\hline
0 & 0 & x & x & x & x \\
0 & 0 & \mathbf{0} &   & g & g \\
0 & 0 & 0 &   &   & x
\end{array}\right]
\Rightarrow
\left[\begin{array}{ccc|ccc}
x & x & x & x & x & x \\
  & x & x & x & x & x \\
  &   & g & g & g & g \\
\hline
0 & 0 & \mathbf{0} & g & g & g \\
0 & 0 & 0 &   & x & x \\
0 & 0 & 0 &   &   & x
\end{array}\right]$$

Figure 31.2: The progress of the nullification of the $\mathbf{F}$ matrix, while retaining its triangular structure of $\mathbf{W}$, where the blank elements are also regarded as 0. The g elements indicate the active row elements involved in that substep of the Givens transformation, and a $\mathbf{0}$ indicates where nullification is taking place.

$$QA = Q \begin{bmatrix} \mathbf{R}' \\ \mathbf{F} \end{bmatrix} = \begin{bmatrix} \mathbf{R}'' \\ 0 \end{bmatrix}$$

where $\mathbf{R}'$ is an $M \times M$ upper triangular matrix, and $\mathbf{F}$ is an $N \times M$ full matrix to be nullified by the Givens transformation.

With reference to Figure 31.2, the $\mathbf{F}$ matrix is originally set to be equal to $\mathbf{F}^{(M)}$, and will be converted into an $N \times (M - 1)$ matrix $\mathbf{F}^{(M-1)}$ after the first column of $\mathbf{F}^{(M)}$ is nullified in the first iteration. After k iterations, k columns of $\mathbf{F}$ will be nullified and an $N \times (M - k)$ matrix $\mathbf{F}^{(M-k)}$ is thus generated.

DG Design for QR Nullification Algorithm We shall develop a DG for the above QR nullification algorithm. We note that, the same type of DG may also be applied to other algorithms which will be encountered in a later section. In performing the GG or GR operations, there are two input elements and two output elements. Correspondingly, there are two inputs and two outputs for each node in the DG design. Four variables are introduced in each node (i, j, k), with the two input elements denoted as $x(i, j, k)$ and $y(i, j, k)$ and the two output elements as $x'(i, j, k)$ and $y'(i, j, k)$. With reference to Figure 31.2, the single assignment form for the above QR nullification on an $(M + N) \times M$ matrix $\mathbf{A}$ is:

For k from 1 to M

For i from N to 1

For j from k to M

$$x(i, j, k) \quad \longleftarrow \quad \begin{cases} r'_{ij} & \text{if } i = N \\ x'(i - 1, j, k) & \text{if } i < N \end{cases}$$

$$y(i, j, k) \quad \longleftarrow \quad \begin{cases} f_{ij} & \text{if } k = 1 \\ y'(i, j, k - 1) & \text{if } k > 1 \end{cases}$$

$$\theta(i, j, k) \quad \longleftarrow \quad \begin{cases} \tan^{-1} y(i, j, k)/x(i, j, k) & \text{if } j = k \\ \theta(i, j - 1, k) & \text{if } j > k \end{cases}$$

$$x'(i, j, k) \quad \longleftarrow \quad x(i, j, k) \cos\theta(i, j, k) + y(i, j, k) \sin\theta(i, j, k)$$

$$y'(i, j, k) \quad \longleftarrow \quad -x(i, j, k) \sin\theta(i, j, k) + y(i, j, k) \cos\theta(i, j, k) \qquad (31.9)$$

where $r'_{i,j}$ and f_{ij} are the (i,j) elements of the input matrices $\mathbf{R}'$ and $\mathbf{F}$, separately. The final output data are $r''_{ij} = x'(i,j,i)$.

The DG of the above single assignment program has a form as shown in Figure 31.3(a). The dependence arcs in each horizontal DG plane are given in Figure 31.3(b), while the vertical dependence arcs between k-planes, i.e., from $(i, j, k-1)$ to (i, j, k), are not explicitly shown.

Triarray for QR Nullification Algorithm To derive an SFG, let us project the DG in Figure 31.3(a) in the $\vec{d} = [-1\ 0\ 0]$ direction with the schedule plane orthogonal to this direction, i.e., $\vec{s} = [-1\ 0\ 0]$. As a result, a triangular SFG array can be obtained as shown in Figure 31.3(c) [13]. The corresponding systolic triangular array (triarray) is shown in Figure 31.3(d). Compared with all other array configurations obtained from different directions of projection, this QR triarray has a fastest pipelining rate and fewer PEs [7], [16]. In this QR triarray, the diagonal PEs perform the GG operations given in Eq. 31.7 and the off-diagonal PEs perform the GR operations given in Eq. 31.8.

Array Processor Design for Postrotation Algorithm

With reference to Figure 31.2, the postrotation stage involves the modification of an $(M + N) \times N$ matrix $\mathbf{B}$ using an $(M + N) \times (M + N)$ orthogonal matrix:

$$\mathbf{QB} = \mathbf{Q}\begin{bmatrix} \mathbf{0} \\ \mathbf{W} \end{bmatrix} = \begin{bmatrix} \mathbf{Z} \\ \mathbf{W}' \end{bmatrix}$$

There exists a duality relationship between the QR nullification of $\mathbf{A} = [\mathbf{R}'^T\ \mathbf{F}^T]^T$, and the postrotation of $\mathbf{B} = [\mathbf{0}^T\ \mathbf{W}^T]^T$. In QR nullification of $\mathbf{A}$, the upper triangular structure of $\mathbf{R}'$ is retained all over the nullification procedure, and *nonzero* elements of $\mathbf{F}$ matrix are gradually decreasing (i.e., the width or height of $\mathbf{F}$ matrix is shrinking). Similarly, the upper triangular structure of $\mathbf{W}$ matrix is also retained all over the postrotation algorithm, while the *nonzero* elements of $\mathbf{Z}$ matrix are increasing (i.e., the width or height of $\mathbf{Z}$ is growing). This suggests the use of the triarray for the postrotation operation just like the QR nullification on $\mathbf{A}$ (see Figure 31.4(b)).

31.3.5   Systolic Kalman Filter – White Noise Case

For most applications, the system and measurement noises are assumed to be *white*, i.e., $\mathbf{R_W}(k+1) = diag[\sigma_{w1}, \sigma_{w2}, \ldots, \sigma_{wn}]$, $\mathbf{R_V}(k) = diag[\sigma_{v1}, \sigma_{v2}, \ldots, \sigma_{vm}]$. So

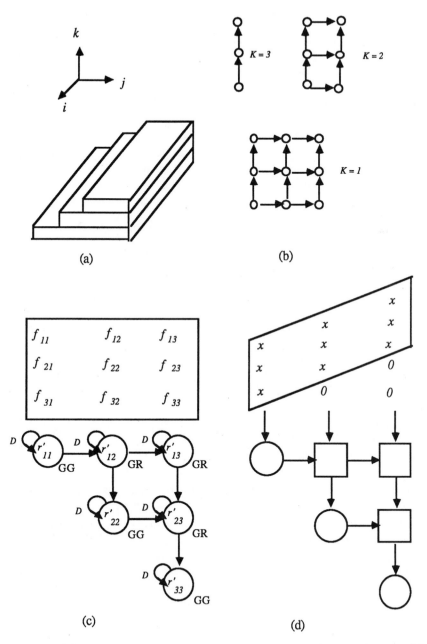

Figure 31.3: (a) DG for QR algorithms in the $N = 3$ and $M = 3$ case. (b) The dependence arcs in each horizontal DG plane. (c) Triangular SFG array for the DG. (d) The corresponding systolic QR triarray.

it is useful to design a systolic array (SK-W filter) for the white noise case. Based on this assumption, we have

$$\begin{aligned}
\mathbf{W}(k+1) &= diag[\sigma_{w1}^{-1/2}, \sigma_{w2}^{-1/2}, \ldots, \sigma_{wn}^{-1/2}] \\
\mathbf{V}(k) &= diag[\sigma_{v1}^{-1/2}, \sigma_{v2}^{-1/2}, \ldots, \sigma_{vm}^{-1/2}]
\end{aligned}$$

Therefore, the whitening operators $\mathbf{W}(k+1)$ and $\mathbf{V}(k)$ requires square-root and division operations. And $\tilde{\mathbf{F}}(k)$, $\tilde{\mathbf{C}}(k)$, $\tilde{\mathbf{y}}(k)$ can be obtained using simple scaling operations. According to Eq. 31.6, the overall Kalman filtering procedure now becomes basically a recursive QR triangularization process. This consists of the following two-phase operations:

1. Nullification of $[\tilde{\mathbf{C}}^T(k) \ \tilde{\mathbf{F}}^T(k)]^T$ by QR nullification using $\mathbf{R}(k)$.

2. Postrotation of $\mathbf{W}(k+1)$ using the rotation angle information obtained from the nullification process of $\tilde{\mathbf{F}}(k)$.

With a very careful schedule design, the two-phase operations for the recursive updating procedure can be performed by a single $n \times n$ QR triarray as discussed next.

Triarray for SK-W Filter Design

Operation on $\tilde{\mathbf{C}}(k)$ The new arriving matrix $\tilde{\mathbf{C}}(k)$ can be nullified by rotating with the resident triangular matrix $\mathbf{R}(k)$ (see Figure 31.4(a)).

Operation on $\tilde{\mathbf{F}}(k)$ As shown in Figure 31.4(a), the nullification on $\tilde{\mathbf{F}}(k)$ will continue right after (m time units) the operation on $\tilde{\mathbf{C}}(k)$ in a similar manner in the triarray. In this operation, a diagonal PE performs Givens generation (GG) and then sends the angle parameters right-ward to the remaining PEs in the same row, where the Givens rotation (GR) are performed. Most importantly, after $n+m$ time units, the parameters of the rotation angles $\{\theta_{ij}\}$, (i.e., $\{\cos\theta_{ij}, \sin\theta_{ij}\}$), start to emerge from the right side of the triarray and are stored in a data buffer to be used for the purpose of the second phase processing.

Postrotation Operation of $\mathbf{W}(k+1)$ We will prove in a moment that, right after the nullification of $\tilde{\mathbf{F}}(k)$ is completed at $t = 3n + m$, the matrix $\mathbf{W}(k+1)$ is already loaded into the triarray and the rotation parameters of $\{\theta_{ij}\}$ also become available for use. It can be shown that the postrotation on $\mathbf{W}(k+1)$ can also be performed in a triarray (see Figure 31.4(b)). Note that the triarray should now provide upward and rightward communication channels, as opposed to the downward and rightward ones provided in the first phase processing.

Rotation Parameters The parameters of the rotation angles, $\{\theta_{ij}\}$, emerge from the right side of the triarray in a skewed data pattern. The data along any anti-diagonal line will be sent to the same diagonal PE in the triarray. A natural design is to use a *data buffer* to store these parameters, with the top row of the data buffer directly linked to the diagonal PEs of the triarray. At $t = 3n + m$, with all the parameters of $\{\theta_{ij}\}$ in place, they are sent row by row to the triarray to perform the postrotation on $\mathbf{W}(k + 1)$.

Loading $\mathbf{W}(k + 1)$ As shown in Figure 31.5, $\mathbf{W}(k + 1)$ is loaded into the triar-ray immediately after $\tilde{\mathbf{F}}(k)$. Note that there is an extra triangle of zeros inserted between $\tilde{\mathbf{F}}(k)$ and $\mathbf{W}(k + 1)$ blocks, so that a perfect synchrony can be achieved among $\tilde{\mathbf{W}}(k + 1)$ (from the left, the $\{\theta_{ij}\}$ parameters (from the left) and $\mathbf{Z}(k)$ data (from the bottom) (see Figure 31.5).

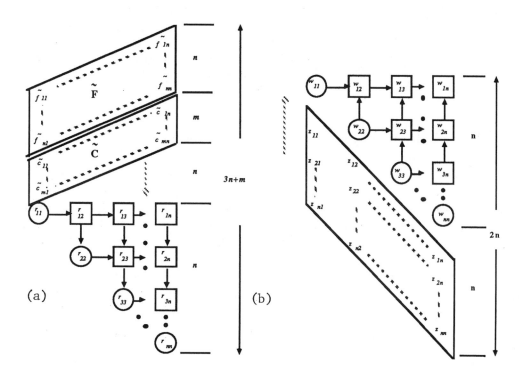

Figure 31.4: The two steps of one recursion of Kalman filter updating. (a) Using $\mathbf{R}(k)$ to nullify the $\tilde{\mathbf{C}}(k)$ and $\tilde{\mathbf{F}}(k)$ matrices. (b) Rotating $\mathbf{W}(k + 1)$ and $\mathbf{Z}(k)$.

Other Considerations in SK-W **Filter Design**

Whitening Scaling Procedure One additional linear array of length n can be used for the scaling operations encountered in the whitening stage of SK-W filter design. In this linear array, a fast square-rooter should be provided for transforming the $\{\sigma_{wi}\}$ and $\{\sigma_{vi}\}$ into $\{\sigma_{wi}^{-1/2}\}$ and $\{\sigma_{vi}^{-1/2}\}$. The diagonal elements of $\mathbf{W}(k+1)$ and $\mathbf{V}(k)$ are pipelined through this array from the left side, so that proper scaling operations can be performed to obtain the whitened matrices $\tilde{\mathbf{F}}(k)$, $\tilde{\mathbf{C}}(k)$, and $\tilde{\mathbf{y}}(k)$.

Solution of $\hat{\mathbf{x}}(k+1)$ The same scaling linear array can also be used for obtaining the best prediction of the state vector $\hat{\mathbf{x}}(k+1)$ by solving Eq. 31.4. This is easily done using the back substitution method on this linear array.

Processing Time for Recursive Updating Procedure By using a QR tri-array and a linear scaling array for this recursive updating process, the processing time is contributed from two-phase operations. The first phase takes $3n + m$ time units and the second phase takes $2n$ time units to complete their respective operations. However, immediately after n time units of the second phase processing, PE(1,1) becomes free and can be used to start processing the new recursion (for the $(k + 1)$st stage). Thus the processing time for each recursion is only $4n + m$ (instead of $5n + m$). The overall data arrangements for the triarray are displayed in Figure 31.5. The time lag for the $\{\theta_{ij}\}$ parameters reflects the time for the triarray and buffer processing.

31.3.6   Systolic Kalman Filter – Colored Noise Case

In some applications, where the additive noises are *colored*, then the systolic Kalman filter design is more involved. Before the SK-W filter can be used, a whitening array design should be provided. The output of this whitening array will be the input of the SK-W filter. The overall systolic Kalman filter design for both stages is termed SK-C filter design. The whitening stage, which consists of the following two operations.

1. Perform the reverse Cholesky decomposition (RCD) of $\mathbf{R}_\mathbf{W}(k+1)$ and $\mathbf{R}_\mathbf{V}(k)$ to obtain the whitening operators $\mathbf{W}(k+1)$ and $\mathbf{V}(k)$.

2. Given the whitening operators, perform the whitening operations on $\mathbf{F}(k)$, $\mathbf{C}(k)$, and $\mathbf{y}(k)$.

An efficient triarray design for both operations in the whitening stage and also the overall configuration of the SK-C design have been proposed [14].

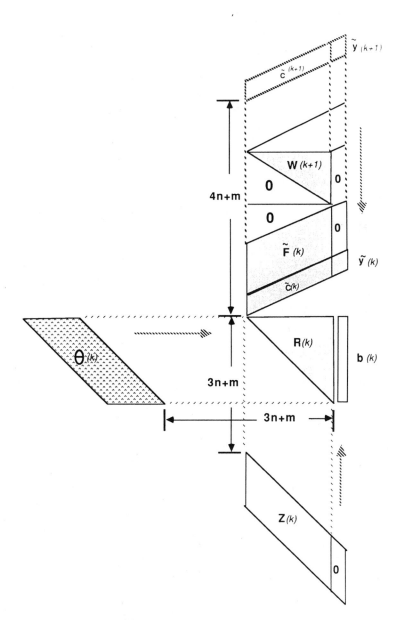

Figure 31.5: The overall data arrangements for one recursive updating in the Kalman filter algorithm.

31.4   Systolic Design for Neural Networks

Neuroscientists have revealed that the massive parallel processing power in the human brain lies in the global and dense interconnections among a large number of
identical logic elements or *neurons*. These neurons are connected to each other with
variable strengths by a network of *synapses*. The original discrete-state Hopfield
model [4], and the continuous-state Hopfield-Tank model (including the proposed
analog neural circuits) [5], [6] have recently become popular in the realization of
artificial neural networks (ANNs). They can be programmed to perform computational networks for associative retrieval or for optimization problems. In order
to take advantages the potential of VLSI design, a locally interconnected systolic
architecture for an ANN is proposed below.

31.4.1   Hopfield and Tank Model

In order to imitate the continuous input-output relationship of real neurons, and
also to simulate the integrative time delay due to the capacitance of real neuron,
the continuous-state model proposed by Hopfield and Tank can be approximated
by the following dynamic equations [5], [6], [19]:

$$U_i(k) - U_i(k-1) \;=\; \sum_{j}^{N} T_{ij} V_j(k) + I_i \tag{31.10}$$

$$V_i(k+1) \;=\; g[U_i(k)] \tag{31.11}$$

where $g[x]$ is a nonlinear function, e.g.,

$$g[x] = (1/2)[1 + tanh(x/x_0)] \tag{31.12}$$

which approaches a unit step function as x_0 tends to zero. The right hand side of Eq.
31.10 can be considered as the new excitation source, which effects "modification"
of the states as shown in the left hand side.

For both models, an energy function is defined as [4], [5]

$$E = -(1/2) \sum_{i=1}^{N} \sum_{j=1}^{N} T_{ij} V_i V_j - \sum_{i=1}^{N} I_i V_i \tag{31.13}$$

Hopfield has shown that, if $T_{ij} = T_{ji}$, then neurons in the continuous-state model
always change their states in such a manner that the energy function is reduced [5]:

The Hopfield model can be formulated as a consecutive matrix-vector multiplication problem with some prespecified thresholding operations. For example, a matrix-form expression of Eq. 31.10 and Eq. 31.11 can be written as

$$\begin{aligned} \mathbf{u}(k) &= \mathbf{T}\mathbf{v}(k) + \mathbf{i} + \mathbf{u}(k-1) \\ \mathbf{v}(k+1) &= G[\mathbf{u}(k)] \end{aligned} \qquad (31.14)$$

where $G[\mathbf{x}]$ function specifies the nonlinear thresholding of each element of the vector $\mathbf{x}$, and the vectors and matrices used are given as:

$$\mathbf{u} = [U_1, U_2, \cdots, U_N]^T$$

$$\mathbf{v} = [V_1, V_2, \cdots, V_N]^T$$

$$\mathbf{i} = [I_1, I_2, \cdots, I_N]^T$$

$$\mathbf{T} = \begin{bmatrix} T_{11} & T_{12} & \cdots & T_{1N} \\ T_{21} & T_{22} & \cdots & T_{2N} \\ \vdots & \vdots & \ddots & \vdots \\ T_{N1} & T_{N2} & \cdots & T_{NN} \end{bmatrix} \qquad (31.15)$$

Example: Solving Image Restoration Problem   Artificial neural networks have been successfully applied to low level vision processing [9]. Here, an image restoration example is given to illustrate how to map applicational problems onto ANNs. Consider an observed degraded image vector $\mathbf{g}$, which can be mathematically expressed as

$$\mathbf{g} = \mathbf{H}\mathbf{f} + \mathbf{n} \qquad (31.16)$$

where $\mathbf{H}$ is a known blurring matrix. The statistical properties of the noise vector $\mathbf{n}$ are also assumed known. Image restoration is the scheme whereby an image vector $\mathbf{f}$ is restored from the linear blurring degradation mechanism and the additive noise. In many cases, however a priori information about the image properties (e.g., smoothness, intensity distribution) is known. In order that the estimated solution also reflects this information, a modified least squares formulation should be adopted:

$$\min \ (\mathbf{g} - \mathbf{H}\hat{\mathbf{f}})^T(\mathbf{g} - \mathbf{H}\hat{\mathbf{f}}) + \gamma(\mathbf{W}\hat{\mathbf{f}})^T(\mathbf{W}\hat{\mathbf{f}}) \qquad (31.17)$$

where the $\mathbf{W}$ matrix represents the intensity weighting for the overall smoothness measure of the image, and γ is a proper regularization parameter.

To formulate the regularized image restoration problem in terms of ANN, the key step is to derive an energy function so that the lowest energy state (the most stable state of the network) would correspond to the best restored image. Once the energy function is determined, the synaptic strengths and input can be immediately derived. Let each pixel of image f_i correspond to the neuron state V_i, then the derived energy function is equal to the expression given in Eq. 31.17. By comparing Eq. 31.13 and Eq. 31.17, the corresponding $\mathbf{T}$ matrix and $\mathbf{i}$ vector are found to be

$$
\begin{aligned}
\mathbf{T} &= -2(\mathbf{H}^T\mathbf{H} + \gamma\mathbf{W}^T\mathbf{W}) \\
\mathbf{i} &= 2\mathbf{H}^T\mathbf{g}
\end{aligned}
\tag{31.18}
$$

Once $\mathbf{T}$ and $\mathbf{i}$ are determined, the ANN can be programmed accordingly to solve the image restoration problem.

31.4.2   Systolic Design of ANN via Cascade DG

The DG for consecutive matrix-vector multiplication formulation of the Hopfield and Tank model in Eq. 31.14 is derived next. It is possible to rearrange the data ordering of the T_{ij} elements, so that the input direction $V_i(k)$ becomes parallel to the output direction of $V_i(k+1)$. Such a modified DG is depicted in Figure 31.6(a). In this DG, for $i = 1, 2, \ldots, N$, the i-th column of the T_{ij} data array is circularly shifted-up by $i-1$ positions. This DG is not totally localized due to the presence of the global spiral communication arcs. However, the input direction (from the top) and the output direction (from the bottom) are parallel. The advantage is that when many such DGs are cascaded top-down, the inputs and outputs data can be matched perfectly.

Ring Systolic Design   For the top-down cascaded DG, the projection can be taken along the vertical direction, which will result in a ring systolic array architecture as shown in Figure 31.6(b). In the ANN implementation each PE, say the i-th PE, is treated as a neuron, and the synaptic strengths $(T_{i1}, T_{i2}, \ldots, T_{iN})$ are stored in it. At the k-th iteration, the operation of the PE is as follows:

1. Each of the neuron outputs $(V_1, V_2, \ldots, V_N)$ is cycling through the ring array, and will pass through the i-th PE once during the N clock cycles.

2. When V_j passes through the i-th PE, it is multiplied with T_{ij}, and the result is added to the sum of $U_i(k-1)$ and I_i (according to the Eq. 31.10).

3. After all N clock cycles, the computation for $U_i(k)$ is completed, and it is ready for the thresholding operation.

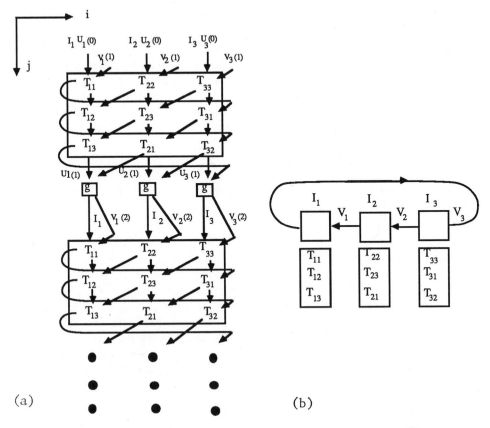

Figure 31.6: (a) DG for consecutive matrix-vector multiplication formulation of Hopfield model. (b) Ring systolic array for consecutive matrix-vector multiplication formulation of Hopfield model.

4. After the thresholding operations, the PE sends the thresholded neuron output $V_i(k+1)$ to the left-side neighbor PE.

The above procedures is repeated until a convergence is reached. For implementing a large number of neurons, the problem of the long wrap-around line can be solved by a special 2-D arrangement scheme [10].

Design for ANN with Locally Correlated Synaptic Strengths In many important applications, with human neural net included, it is practical to assume that synaptic strengths are the strongest among locallly neighboring neurons. In fact, the strengths gradually diminish as the distances between the neurons increase. This means that in practice the synaptic matrix **T** can be approximated by a banded matrix. The previously proposed ANN systolic design can be effectively adapted to exploit the banded (or circularly banded) structure of the synaptic matrix. Let M denote the *effective correlation distance*, i.e. the range within which synaptic strengths are not negligible. This implies that the the bandwidth of the synaptic matrix is M. Then the DG design in Figure 31.6 can be modified to the DG as shown in Figure 31.7. (Without loss of generality, we assume both N and M to be odd numbers.) Note that the same ring systolic array can be used to process the modified DG, with a saving of processing time by a factor of N/M, which is very significant for most applications. For example, if $N = 10^4$ and $M = 10$, it would imply a speed-up of $O(10^3)$.

Advantages of Systolic ANN   Compared to the analog RC neural circuit and the optical neural networks [2], [15], the proposed digital implementation of ANN can achieve higher precision in computation. There are other advantages uniquely pertaining to the proposed array. Firstly, the pipelining period $\alpha = 1$, which implies 100% utilization efficiency during the iteration process. Secondly, only M synaptic strengths of T_{ij} are stored in each PE. It takes only M time units to complete one iteration of computations. Finally, this design permits easy modification of the synaptic strengths, making possible the incorporation of the "learning" capability into ANN.

31.4.3   Incorporating Simulated Annealing into ANNs

Using this proposed systolic architecture, the gain parameters can be easily updated during the iterations. It is observed that if the gain control parameter of the sigmoid input-output relation in the neuron can be dynamically changed, then a faster convergence and better performance can be obtained [6].

In many optimization problems, the minimization of the given energy function suffices to obtain the correct solution. However, the minimization process in the continuous-state ANN will sometimes converge only to a local minimum instead of the global minimum. In order to overcome this difficulty, the idea of simulated annealing can be adopted and incorporated into the ANN model [1].

By appropriate substitution of the gain control parameter x_0 in Eq. 31.12 of the continuous model with the temperature control parameter T, then the thresholding operation in Eq. 31.11 becomes

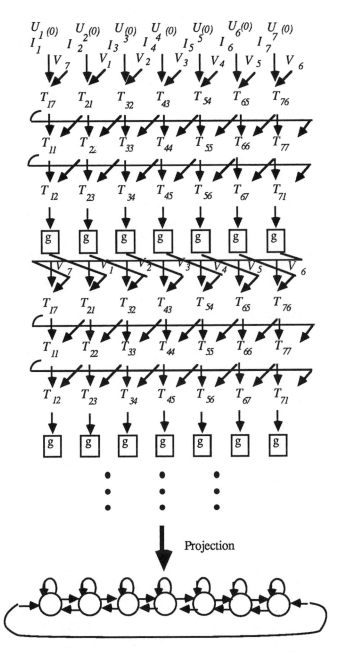

Figure 31.7: The DG for the band-matrix ANN processing with $N = 7$ and $M = 3$. As compared with the DG in Figure 31.7, this DG has the $\{T_{ij}\}$ circularly shifted downward by $(M - 1)/2$ rows, i.e., one row in this case. (Without affecting the computation, all rows of the *zero* $\{T_{ij}\}$ nodes are removed from the modified DG.) Correspondingly, the input data $\{V_i(k)\}$ are also circularly shifted rightward by $(M - 1)/2$ columns via semi-global wires. So are the intermediate output data $\{V_i(k + 1)\}$.

$$
\begin{aligned}
V_i(k+1) &= g[U_i(k)] \\
&= (1/2)[1 + tanh(U_i(k)/T)] \\
&= \frac{1}{1 + e^{-2U_i(k)/T}}
\end{aligned}
\tag{31.19}
$$

where T represents the temperature control parameter in the annealing process.

If we further confine the neuron response to be discrete valued (0 and 1 only) with a stochastic decision mechanism (i.e., $V_i(k+1)$ can be set "1" with probability given in Eq. 31.19), this modified version of the Hopfield model is essentially the same form as the Boltzmann machine [1], [10]. The deterministic thresholding operation in the continuous-state Hopfield model is now replaced by the on-off (discrete-state) stochastic decision mechanism, allowing the network to gradually realize the global optimum. The proposed systolic ANN can be easily adapted to the simulated annealing version, which involves the dynamic decrease of the gain control parameter in the thresholding operation, and stochastic decision mechanism.

31.5   Concluding Remarks

In this paper, we have demonstrated an top-down integrated array architecture design for the implementation of the Kalman filtering and artificial neural networks.

For Kalman filtering, a very efficient systolic processor is proposed, which requires only $4n + m$ time units for processing one recursion.

For ANN, a ring systolic design is proposed, which offers properties of high pipelining rate, high precision, capability of learning, fast convergence speed, and global optimal solution searching.

References

[1] D. H. Ackley, G. E. Hinton, and T. J. Sejnowski. A learning algorithm for Boltzmann machines. *Cognitive Science*, Vol. 9: pp. 147 – 169, 1985.

[2] N. H. Farhat, D. Psaltis, A. Prata and E. Paek. Optical implementation of the Hopfield model. *Applied Optics*, Vol. 24: pp. 1469– 1475, May 1985.

[3] G.H. Golub and C. F. Van Loan. *Matrix Computations*. Johns Hopkins University Press, 1983.

[4] J. J. Hopfield. Neural network and physical systems with emergent collective computational abilities. In *Proc. Natl.. Acad. Sci. USA*, Vol. 79, pp. 2554-2558, 1982.

[5] J. J. Hopfield. Neurons with graded response have collective computational properties like those of two-state neurons. In *Proc. Natl.. Acad. Sci. USA*,Vol. 81, pp. 3088–3092, 1984.

[6] J. J. Hopfield and D. W. Tank. Neural computation of decision in optimization problems. *Biological Cybernetics*,Vol. 52, pp. 141-152, 1985.

[7] H. T. Kung and W. M. Gentleman. Matrix triangularization by systolic arrays. *Proc. SPIE, Real Time Signal Processing*, 1983.

[8] R. E. Kalman. A new approach to linear filtering and prediction problems. *J. Basic Engineering*, 82: pp 35–45, 1960.

[9] C. Koch, J. Marroquin and A. Yuille. Analog "neuronal" networks in early vision. *Proc. of National Academy Science*, Vol. 83: pp. 4263–4267, 1986.

[10] S. Y. Kung. *VLSI Array Processors.* Prentice Hall Inc. N. J., 1987.

[11] S.Y. Kung. On supercomputing with systolic/wavefront array processors. *Proceedings of the IEEE*, 72: pp. 867–884, July 1984.

[12] S. Y. Kung, S. C. Lo, and P. S. Lewis. Timing analysis and optimization of VLSI data flow arrays. In *Proc. IEEE ICPP'86*, pp. 600–607, August 1986.

[13] S. Y. Kung, J. N. Hwang, and S. C. Lo. Mapping digital signal processing algorithms onto VLSI systolic/wavefront arrays. In *Proc. 12th Annual Asilomar Conf. on Signals, Systems and Computers*, pp. 6–12, November 1986.

[14] S. Y. Kung and J. N. Hwang. Systolic array designs for Kalman filtering. Submitted to *IEEE Trans. on Acoustics, Speech, and Signal Processing*, 1987.

[15] H. Mada. Architecture for optical computing using holographic associative memories. *Applied Optics*, Vol. 24: 1985.

[16] J.G. McWhirter. Recursive least-squares minimization using a systolic array. In *SPIE, In Proc. Real Time Signal Processing VI*, pp 105–110, SPIE, 1983.

[17] M. J. Chen and K. Yao. On realization and implementation of Kalman filtering by systolic array. In *Proc. 21st Conf. on Inf. Science and Systems*, The John Hopkins University, pp. 375-380, 1987.

[18] C. C. Paige and M. A. Saunders. Least squares estimation of discrete linear dynamic systems using orthogonal transformation. *SIAM J. Numer. Anal.*, 14: pp. 180–193, 1977.

[19] M. Takeda and J.W. Goodman. Neural networks for computation: number representations and programming complexity. *Applied Optics*, Vol. 25: pp. 3033–3046, September 1986.

Chapter 32

The Gated Interconnection Network for Dynamic Programming [1]

David B. Shu [2]
Greg Nash [2]

32.1 Introduction

The SIMD character of fine grain cellular architectures presents serious limitations for all but the first levels of image processing performed. Typically, an image is processed to the point where segmentation has occurred and it is now necessary for region dependent analysis. This might begin with simple data dependent operations such as region labeling followed by extraction of primitive symbolic information (area, perimeter, minimum bounding rectangle) and the gathering of other statistics (moments, histograms). Later this processing would be followed by higher level symbolic operations intended to extract features that ultimately would lead to object recognition. In such cases it is necessary to process each region separately in some way. If there were numerous regions in an image, then a purely SIMD architecture would be efficient only to the extent that the region being processed occupies a significant fraction of the image.

---

[1] Supported in part by DARPA Grant DACA76-86-C-0015

[2] Hughes Research Laboratories, Malibu, CA

Simultaneous data dependent processing of element groups for a variety of transform and symbolic calculations represents another class of computations that are important. For example, in the Hough transform it would be convenient to group pixels along lines associated with certain directions. Also, many simple symbolic operations such as boolean matrix operations, finding the minimum spanning tree, or minimum cost path of a graph can be performed very efficiently in a bit-serial fashion if appropriate groupings are possible.

In order to perform local region based computations efficiently it is necessary to provide for a communication capability above and beyond that of the mesh connected array associated with an $n \times n$ array of bit-serial PEs. We describe here a Gated Connection Network (GCN) that provides a high degree of flexibility and programmability in communication networks. The GCN is a communication structure which is one component of the Image Understanding Architecture (IUA) being developed jointly at Hughes Research Labs and University of Massachusetts. This architecture is being designed specifically for vision related processing that depends heavily upon techniques from the domain of artificial intelligence (AI) to classify objects [1],[2].

The IUA is a hierarchical, heterogeneous architecture, consisting of three different types of processing elements. For high level, coarse grained operations, there is an array of microprocessors available. For medium grained, numerically intensive calculations, there is an array of arithmetic oriented processors. Finally, for the processor per-pixel and low level symbolic operations there is an array of fine grained bit-serial type processors. These arrays are organized in a hierarchical fashion in that a processor at any level has associated with it an array of elements beneath it. Since the types of computations associated with image understanding are typically performed in a hierarchical way with variable granularity in going from raw imagery to higher levels of abstractions, the basic IUA represents a very efficient match to the problem. It also provides an associative processing capability in the bit-serial array. Associative processing represents a significant departure from other signal/image processing techniques in that it provides a way of processing data without data shuffling. For a 512×512 per-pixel processor system, the use of associative techniques provides an effective bandwidth of approximately 10^{11} bytes/sec. The associative capabilities have been designed in such a way that they can be either local, using the GCN, or global. Both are necessary to efficiently process an image. For example, it is often important to obtain global statistics (e.g. histograms) on an image during the initial stages of computation.

At the present time only a prototype 1/64th slice of the system is being constructed. The prototype IUA slice consists of 64 custom designed CMOS bit-serial PE/glue chips, 64 commercially available digital signal processing chips, and a host processor (68020 based). Each of these processor types is associated with one of the three levels described above. The bit-serial PEs (64 per chip) each have associated with them some registers and memory. One register bit (activity bit, A) can be set in a data dependent way to turn off the PE.

32.2 Gated Connection Network

A simplified version of the GCN network, which is part of the associative, bit serial level of the IUA, consists of a mesh array of four simple transmission gates per bit-serial PE as shown in Figure 32.1a. (Note that in the figure there is also a conventional mesh interconnection network which is not shown.) A PE output to this array is via the x-register of the PE and the input to a PE is via a "some/none" line. Communication between PEs is determined by the switch setting on the transmission gates. These gate settings are determined by a four bit register which resides in the memory space of each bit-serial PE. In Figure 32.1a, if PE 1 wants to send a message to PE 2, then the appropriate gate connections are made ($E(1)$ and $W(2)$ set to pass signals, all others set to block signals) and the two PEs now have an electrical connection between them.

The GCN is precharged before use. Communication occurs by having the source PE place a "1" or a "0" into its x-register. A "1" will pull down the precharged network and a "0" will leave the network unchanged. Multiple source PEs can also send wired-or messages to the same PE. This can be useful when it is desirable to let only the minimum value of a multi-bit word pass in the GCN. We will show later how this is used in finding the minimum cost path of a graph, since one of the objectives for a given node is to receive multiple messages from all the possible sources and select the smallest value.

The procedure to find minimums using some/none response is well known [3]. Here, a simple local minimum example is given which is adapted to the GCN network. Consider five gate connected PEs having stored values of "2, 3, 4, 5, and 6," respectively, as shown in Figure32.1b. We want to find the minimum "2" of these values and store it in some result field of each PE. The procedure is as follows:

```
1) for I:=2 down to 0
2) begin
3)    x-reg:= -M(V+I)        ; x-reg of active PEs (i.e.,A-reg=1)
                             ; are updated.
4)    if some then           ; A-reg of active PEs will be
          A-reg:=x-reg        ; updated only if some/none signal
                             ; on GCN net indicates ''some''.
5)       M(RSL+I):= -some/none!
                             ; All PEs will store the some/none
                             ; result in its result field.
6) end
```

where M represents a memory value, the PE activity is controlled by the activity bit A, V is the starting bit position of the value field and RSL is the starting bit position of result field. An operation with a trailing exclamation point (!) will

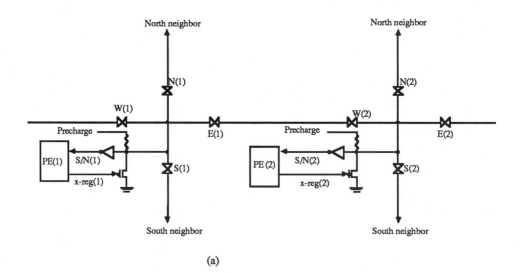

(a)

Steps	Index I	PE data binary →	2 010	3 011	4 100	5 101	6 110
0	Start	X	1	1	1	1	1
		A	1	1	1	1	1
		M(RBL+I)	0	0	0	0	0
1	2	X	1	1	0	0	0
		A	1	1	0	0	0
		M(RBL+I)	0	0	0	0	0
2	1	X	0	0	-	-	-
		A	1	1	-	-	-
		M(RBL+I)	1	1	1	1	1
3	0	X	1	0	-	-	-
		A	1	0	-	-	-
		M(RBL+I)	0	0	0	0	0

(b)

Figure 32.1: Gated connections for PE(1) and PE(2).

ignore the value of the activity bit (i.e. A-reg) and take place on all PEs.

Figure32.1b shows the contents of x-reg, A-reg, and the result field of each PE at each step in the loop. Initially, the x-reg and A-reg are set to 1. At step 1(i.e., I=2), the most significant bit (MSB) of the value is inverted and loaded into the x-reg. This is shown in Step 1 in Figure 32.1b where the x-reg of PEs 2 and 3 have the value "1" (since their MSB=0), whereas the other PEs (4, 5 and 6) have their x-reg set to "0." In such manner, it is possible to eliminate PEs 4, 5 and 6 as potential candidates for the minimum value.

Then, a some/none test is made to determine if some or none of the x-regs in the GCN network PEs have been set to the value "1." The contents of the A-reg will be changed or updated to reflect the contents of their associated x-reg only if the some/none test is positive. Since in Step 1 the test was positive, the A-regs are set to "1" for those PEs with MSB=0. In Step2, the "shrinking" process continues to the second significant bit for those PEs that are active. The x-regs of PEs 2 and 3 are again set to the inverse of the second significant bit and thus equal to zero. A some/none test is conducted again. In this case, none of the active PEs have their x-reg at a 1 level. Consequently, their A-regs remain the same. Statement 4 of the code prevents turning off of all the PEs if all x-regs are zero. In addition, the result field for the second significant bit of all of the PEs is set to 1 (the inverse of the some/none test result). This is correct since the ultimate objective is to set the binary value 010 as the result field. The some/none result of other PEs in the array will not affect this group since they are electrically disconnected.

In loop step I=0, line 3, the x-reg are set to the inverse of the least significant bit (LSB). In this case, the x-reg of PE2 is set to "1" since its LSB is a "0" whereas the x-reg of PE3 equals "0" since the LSB of PE3 equals 1. Again, a some/none test is conducted to narrow down PE 2 as the only active PE with A-reg=1 at the end. At the same time, the LSB of the result field is set to the inverse of the some/none test, i.e., "0." Thus, all connected PEs in this group will have their result fields set to the minimum binary value "010".

32.3 Minimum Cost Path

Dynamic programming is a technique for solving optimization problems that has found use in a variety of application areas. We consider here a generic form of the problem, finding the minimum cost path in a graph and showing how it can be efficiently solved on the GCN in a time proportional to the number of links in the minimum cost path from Q to all other nodes.

The dynamic programming example we will describe is that of finding the minimum cost path (MCP) on a graph. We assume a weighted, undirected graph $G = [V, E]$, $|V| \leq n$, represented by the $n \times n$ adjacency matrix, W of weights. Elements $w(i,j)$

in W, $1 \leq i, j \leq n$ denotes the weight of the undirected edge(i, j). If there is no edge between vertices i and j, then $w(i, j) = $ maxint(infinite). We assume that the edge weights are distinct, since ties in W can be broken by considering the weight of an edge to be the lexicographically ordered tuple, and resolving it by using a "find first" capability of the GCN. Given two vertices P, Q of G, the problem is then to find a path from Q to P along which the sum of the weights (SOW) is minimum.

Two types of gate settings are required, one for finding the local minimum for the SOW of "column" edges in the "LinkCost" matrix described later and one for finding local minimum for those of "row" edges. These settings are:

<N, E, W, S,> = < 1, 0, 0, 1> ; column GCN,

and

<N, E, W, S> = < 0, 1, 1, 0> ; row GCN.

The procedure described below consists of a modification of Hillis' algorithm on the GCN architecture [4]. The adjacency matrix is mapped into the 512×512 bit-serial array with edge (i, j) assigned to $PE(i, j)$. In other words, PEs in row i are associated with vertex i, so that i is called the row vertex index. Similarly, PEs in column j are associated with vertex j, so that j is called the column vertex index. Each iteration in computing the shortest path consists of each vertex sending to each of its neighbors the sum of weights (SOW) from the source (vertex Q) to itself plus the weight of the connecting edge along which the message is sent; we call this new sum the "LinkCost." The row vertices and column vertices alternate as the transmitters/receivers of the message. In order to terminate the process, we consider a given vertex as an active transmitter only when the SOW, computed at current iteration, from the source Q is less than the SOW obtained from the previous iteration. When there are no active transmitters in the whole array, the algorithm stops.

We denote the number of active transmitters at the kth iteration by $m(k)$ and let the set $T(k) = \{T1, T2, ..., Tm(k)\}$, be the set of vertex indices of active transmitters. At the kth iteration, if the row vertices are the transmitters, then all the edges (i, J) of row vertices, where i is in $T(k)$, will share the same vertical(column) GCN to send the LinkCost message to column vertex J; however, using the many to one message passing technique discussed previously, only the message sent by edge (I, J) with the minimum LinkCost will be received by vertex J. Therefore, finding the minimum SOW sent by all the edges incident on the vertex J is performed on the fly during the message passing operation, which is very efficient. This minimum SOW is received by all the edges (i, J), $1 \leq i \leq n$, associated with the column vertex J. Four variables are maintained by each edge (i, j) cell: the SOW, LinkCost, ActiveFlag and Pointer to the previous vertex along the MCP. In addition, there are two message buffers called MinBuf and PtrBuf. The number of cycles is based on 32 bits integer additions and floating point comparisons(find minimum). The initial values of these variables are

SOW, LinkCost, MinBuf := maxint

and

ActiveFlag := False ; Pointer := PtrBuf := NIL.

Also, for source vertex Q , we set edge(Q, Q) variables as follows:

```
begin
    LinkCost := 0;
    ActiveFlag := True;
    Pointer := Q;
end.
```

The above algorithm can best be understood in connection with a simplified example, such as the graph associated with the adjacency matrix in Figure 32.2. Given two vertices 2 and 5 of G, the problem is to find a path from vertex 2 to vertex 5 along which the sum of the weights(SOW) is minimum. After initialization, the single edge $PE(5,5)$ will be activated with its variables LinkCost and Pointer set to 0 and 5, respectively. It implies that vertex 5 can reach to itself at a cost of zero and the pointer to the previous vertex along the MCP is to itself.

Row \ Col	V1	V2	V3	V4	V5	V6	V7	V8	V9	V10
Vertex 1	0	2	5	1	2	3	8	3	5	2
2	2	0	4	9	1	2	1	8	2	7
3	5	4	0	1	6	9	10	9	2	5
4	1	9	1	0	2	5	3	6	1	1
5	2	11	6	2	0	3	3	9	4	8
6	3	2	9	5	3	0	8	6	1	8
7	8	1	10	3	3	8	0	6	2	9
8	3	8	9	6	9	6	6	0	2	5
9	5	2	2	1	4	1	2	2	0	8
10	2	7	5	1.	8	8	9	5	8	0

Figure 32.2: Adjacency Matrix of Graph G and Weights

32.3.1　First Iteration

Column Vertices are Transmitters

Step 0 **(a)** A single edge(5,5) is the active transmitter; row vertex 5 is the only receiver of the minimum LinkCost of 0. The resulting MinBuf fields of every edge are shown in Figure 32.3(a). This step spreads the linkcost of 0 to every edge in the row vertex 5.

(b) The PtrBuf, pointing to previous vertex in the MCP, of every edge in row vertex 5 is set to point to vertex 5 as shown in Figure 32.3(b).

Step 1 Since SOW(5,5) was preset to ∞, there are no changes for SOW of every edge, as can be seen in Figure 32.3(c). In this case, Step 1 is redundant.

Step 2 The SOW of every edge in row vertex 5 is initialized to zero, and row vertex 5 is activated as transmittersas shown in Figure 32.3(d). The Pointer fields of the active edges are updated as shown in Figure 32.3(e).

Step 3 Row vertex 5 sends to each of its neighbors the SOW from the source (vertex 5) to itself(i.e. SOW=0) plus the weight of the connecting edge (LinkCost) along which the message is sent, as is shown in Figure 32.3(f). For example, the LinkCost of edge cell(5,1) is 2, which implies that row vertex 5 sends a message to column vertex 1 and telling him at the current iteration, the cost to reach him along the MCP is 2. Similarly, vertex 5 is sending message to vertex 2 via edge cell(5,2) with LinkCost of 11.

Row Vertices are Transmitters

Step 4 **(a)** All edges $(5, j)$, $1 \le j \le 10$, associated with row vertex 5 are active transmitters and all column vertices are now the receivers of the corresponding minimum of LinkCost (MOL). The resulting MinBuf fields of every edge are shown in Figure 32.4(a).This step spreads the MOL to every edge in the column vertices.

(b) The PtrBuf, pointing to previous vertices in the MCP, of every edge are set to point to vertex 5 as shown in Figure 32.4(b) since all the MOL messages originate from row vertex 5

Step 5 Because only one diagonal edge(5,5) has SOW changed to zero, there are no changes in the SOW for every other column vertex, as can be seen in Figure 32.4(c).

Step 6 Comparing the MinBuf field and the SOW as shown in Figure 32.4(a) and (c), all the column vertices are activated as the transmitters except vertex 5 as shown in Figure 32.4(d). This is expected since vertex 5 is the original

Row \ Col	V1	V2	V3	V4	V5	V6	V7	V8	V9	V10
Vertex 1	∞	∞	∞	∞	∞	∞	∞	∞	∞	∞
2	∞	∞	∞	∞	∞	∞	∞	∞	∞	∞
3	∞	∞	∞	∞	∞	∞	∞	∞	∞	∞
4	∞	∞	∞	∞	∞	∞	∞	∞	∞	∞
5	0	0	0	0	0	0	0	0	0	0
6	∞	∞	∞	∞	∞	∞	∞	∞	∞	∞
7	∞	∞	∞	∞	∞	∞	∞	∞	∞	∞
8	∞	∞	∞	∞	∞	∞	∞	∞	∞	∞
9	∞	∞	∞	∞	∞	∞	∞	∞	∞	∞
10	∞	∞	∞	∞	∞	∞	∞	∞	∞	∞

(a) Minibuf - Step (0) (a)

Row \ Col	V1	V2	V3	V4	V5	V6	V7	V8	V9	V10
Vertex 1	NIL	NIL	NIL	NIL	NIL	NIL	NIL	NIL	NIL	NIL
2	NIL	NIL	NIL	NIL	NIL	NIL	NIL	NIL	NIL	NIL
3	NIL	NIL	NIL	NIL	NIL	NIL	NIL	NIL	NIL	NIL
4	NIL	NIL	NIL	NIL	NIL	NIL	NIL	NIL	NIL	NIL
5	5	5	5	5	5	5	5	5	5	5
6	NIL	NIL	NIL	NIL	NIL	NIL	NIL	NIL	NIL	NIL
7	NIL	NIL	NIL	NIL	NIL	NIL	NIL	NIL	NIL	NIL
8	NIL	NIL	NIL	NIL	NIL	NIL	NIL	NIL	NIL	NIL
9	NIL	NIL	NIL	NIL	NIL	NIL	NIL	NIL	NIL	NIL
10	NIL	NIL	NIL	NIL	NIL	NIL	NIL	NIL	NIL	NIL

(b) PtrBuf - Step (0) (b)

Row \ Col	V1	V2	V3	V4	V5	V6	V7	V8	V9	V10
Vertex 1	∞	∞	∞	∞	∞	∞	∞	∞	∞	∞
2	∞	∞	∞	∞	∞	∞	∞	∞	∞	∞
3	∞	∞	∞	∞	∞	∞	∞	∞	∞	∞
4	∞	∞	∞	∞	∞	∞	∞	∞	∞	∞
5	∞	∞	∞	∞	∞	∞	∞	∞	∞	∞
6	∞	∞	∞	∞	∞	∞	∞	∞	∞	∞
7	∞	∞	∞	∞	∞	∞	∞	∞	∞	∞
8	∞	∞	∞	∞	∞	∞	∞	∞	∞	∞
9	∞	∞	∞	∞	∞	∞	∞	∞	∞	∞
10	∞	∞	∞	∞	∞	∞	∞	∞	∞	∞

(c) SOW - Step (1)

Row \ Col	V1	V2	V3	V4	V5	V6	V7	V8	V9	V10
Vertex 1	F	F	F	F	F	F	F	F	F	F
2	F	F	F	F	F	F	F	F	F	F
3	F	F	F	F	F	F	F	F	F	F
4	F	F	F	F	F	F	F	F	F	F
5	T	T	T	T	T	T	T	T	T	T
6	F	F	F	F	F	F	F	F	F	F
7	F	F	F	F	F	F	F	F	F	F
8	F	F	F	F	F	F	F	F	F	F
9	F	F	F	F	F	F	F	F	F	F
10	F	F	F	F	F	F	F	F	F	F

(d) Active Flag - Step (2) (a)

Row \ Col	V1	V2	V3	V4	V5	V6	V7	V8	V9	V10
Vertex 1	NIL	NIL	NIL	NIL	NIL	NIL	NIL	NIL	NIL	NIL
2	NIL	NIL	NIL	NIL	NIL	NIL	NIL	NIL	NIL	NIL
3	NIL	NIL	NIL	NIL	NIL	NIL	NIL	NIL	NIL	NIL
4	NIL	NIL	NIL	NIL	NIL	NIL	NIL	NIL	NIL	NIL
5	5	5	5	5	5	5	5	5	5	5
6	NIL	NIL	NIL	NIL	NIL	NIL	NIL	NIL	NIL	NIL
7	NIL	NIL	NIL	NIL	NIL	NIL	NIL	NIL	NIL	NIL
8	NIL	NIL	NIL	NIL	NIL	NIL	NIL	NIL	NIL	NIL
9	NIL	NIL	NIL	NIL	NIL	NIL	NIL	NIL	NIL	NIL
10	NIL	NIL	NIL	NIL	NIL	NIL	NIL	NIL	NIL	NIL

(e) Pointer - Step (2) (a)

Row \ Col	V1	V2	V3	V4	V5	V6	V7	V8	V9	V10
Vertex 1	∞	∞	∞	∞	∞	∞	∞	∞	∞	∞
2	∞	∞	∞	∞	∞	∞	∞	∞	∞	∞
3	∞	∞	∞	∞	∞	∞	∞	∞	∞	∞
4	∞	∞	∞	∞	∞	∞	∞	∞	∞	∞
5	2	11	6	2	0	3	3	9	4	8
6	∞	∞	∞	∞	∞	∞	∞	∞	∞	∞
7	∞	∞	∞	∞	∞	∞	∞	∞	∞	∞
8	∞	∞	∞	∞	∞	∞	∞	∞	∞	∞
9	∞	∞	∞	∞	∞	∞	∞	∞	∞	∞
10	∞	∞	∞	∞	∞	∞	∞	∞	∞	∞

(f) linkcost - Step (3)

Figure 32.3: First iteration - Steps (0) through (3)

source and should be prevented from sending the same message over again. Figure 32.4(e) updates the Pointer fields of the active edges.

Step 7 All column vertices except vertex 5 send to each of their neighbors the LinkCost (e.g., the SOW from the source(vertex5) to itself plus the weight of the connecting edge along which the message is sent) as is shown in Figure 32.4(f). For example, the LinkCost of edge cell(2,1) is 4, it implies that column vertex 1 is sending message to row vertex 2 and telling it at the current iteration, the cost to reach him via vertex 1 of the MCP is 4 (i.e. the cost of $v5 \rightarrow v1 \rightarrow v2$ is $W(5,1) + W(1,2) = 2 + 2$)). Similarly, vertex 6 also sends a message to row vertex 2 via edge PE(2,6) and telling it at the current

Row \ Col	V1	V2	V3	V4	V5	V6	V7	V8	V9	V10
Vertex 1	2	11	6	2	0	3	3	9	4	8
2	2	11	6	2	0	3	3	9	4	8
3	2	11	6	2	0	3	3	9	4	8
4	2	11	6	2	0	3	3	9	4	8
5	2	11	6	2	0	3	3	9	4	8
6	2	11	6	2	0	3	3	9	4	8
7	2	11	6	2	0	3	3	9	4	8
8	2	11	6	2	0	3	3	9	4	8
9	2	11	6	2	0	3	3	9	4	8
10	2	11	6	2	0	3	3	9	4	8

(a) Minibuf - Step (4) (a)

Row \ Col	V1	V2	V3	V4	V5	V6	V7	V8	V9	V10
Vertex 1	5	5	5	5	5	5	5	5	5	5
2	5	5	5	5	5	5	5	5	5	5
3	5	5	5	5	5	5	5	5	5	5
4	5	5	5	5	5	5	5	5	5	5
5	5	5	5	5	5	5	5	5	5	5
6	5	5	5	5	5	5	5	5	5	5
7	5	5	5	5	5	5	5	5	5	5
8	5	5	5	5	5	5	5	5	5	5
9	5	5	5	5	5	5	5	5	5	5
10	5	5	5	5	5	5	5	5	5	5

(b) PtrBuf - Step (4) (b)

Row \ Col	V1	V2	V3	V4	V5	V6	V7	V8	V9	V10
Vertex 1	∞	∞	∞	∞	0	∞	∞	∞	∞	∞
2	∞	∞	∞	∞	0	∞	∞	∞	∞	∞
3	∞	∞	∞	∞	0	∞	∞	∞	∞	∞
4	∞	∞	∞	∞	0	∞	∞	∞	∞	∞
5	∞	∞	∞	∞	0	∞	∞	∞	∞	∞
6	∞	∞	∞	∞	0	∞	∞	∞	∞	∞
7	∞	∞	∞	∞	0	∞	∞	∞	∞	∞
8	∞	∞	∞	∞	0	∞	∞	∞	∞	∞
9	∞	∞	∞	∞	0	∞	∞	∞	∞	∞
10	∞	∞	∞	∞	0	∞	∞	∞	∞	∞

(c) SOW - Step (5)

Row \ Col	V1	V2	V3	V4	V5	V6	V7	V8	V9	V10
Vertex 1	T	T	T	T	F	T	T	T	T	T
2	T	T	T	T	F	T	T	T	T	T
3	T	T	T	T	F	T	T	T	T	T
4	T	T	T	T	F	T	T	T	T	T
5	T	T	T	T	F	T	T	T	T	T
6	T	T	T	T	F	T	T	T	T	T
7	T	T	T	T	F	T	T	T	T	T
8	T	T	T	T	F	T	T	T	T	T
9	T	T	T	T	F	T	T	T	T	T
10	T	T	T	T	F	T	T	T	T	T

(d) Active Flag - Step (6) (a)

Row \ Col	V1	V2	V3	V4	V5	V6	V7	V8	V9	V10
Vertex 1	5	5	5	5	5	5	5	5	5	5
2	5	5	5	5	5	5	5	5	5	5
3	5	5	5	5	5	5	5	5	5	5
4	5	5	5	5	5	5	5	5	5	5
5	5	5	5	5	5	5	5	5	5	5
6	5	5	5	5	5	5	5	5	5	5
7	5	5	5	5	5	5	5	5	5	5
8	5	5	5	5	5	5	5	5	5	5
9	5	5	5	5	5	5	5	5	5	5
10	5	5	5	5	5	5	5	5	5	5

(e) Pointer - Step (6) (a)

Row \ Col	V1	V2	V3	V4	V5	V6	V7	V8	V9	V10
Vertex 1	2	13	11	3	✕	6	11	12	9	10
2	4	11	10	11	✕	5	4	17	6	15
3	7	15	6	3	✕	12	13	18	6	13
4	3	20	7	2	✕	8	6	15	5	9
5	4	22	12	4	✕	6	6	18	8	16
6	5	13	15	7	✕	3	11	15	5	16
7	10	12	16	5	✕	11	3	15	6	17
8	5	19	15	8	✕	9	9	9	6	13
9	7	13	8	3	✕	4	5	11	4	16
10	4	18	11	3	✕	11	12	14	12	8

(f) linkcost - Step (7)

Figure 32.4: First iteration - Steps (4) through (7)

iteration, the cost to reach him via vertex 6 of the MCP is 5 (i.e. the cost of $v5 \rightarrow v6 \rightarrow v2$ is $W(5,6) + W(6,2) = 3 + 2$). The contention of multiple messages trying to reach row vertex 2 at the same time will be resolved in Step 0 of the next iteration.

32.3.2   Second Iteration

Step 0 (a) As shown in Figure 32.4(f), all column edges in the unshaded area are active transmitters and all row vertices are receivers. For example, using the many to one message passing technique discussed earlier, row vertex 2 will receive the minimum of LinkCost of 4 (first from left) sent by column vertex 1, among all the messages sent by edges in the second row. Similarly, row vertex 3 receives the MOL of 3 sent by column vertex 4. The concept of dynamic programming is applied here. It states that for a given MCP from Q to P, and for any vertex M along the path, the remaining subpath from M to Q is itself a MCP. In the above example, row vertex 3 has many paths to reach vertex 5 (i.e. Q) at varying cost as indicated by each number shown in row 3 of Figure 32.4(f); however, if vertex 3 is part of the MCP from vertex 2 to 5, then the path from vertex

Row \ Col	V1	V2	V3	V4	V5	V6	V7	V8	V9	V10
Vertex 1	2	2	2	2	2	2	2	2	2	2
2	4	4	4	4	4	4	4	4	4	4
3	3	3	3	3	3	3	3	3	3	3
4	2	2	2	2	2	2	2	2	2	2
5	4	4	4	4	4	4	4	4	4	4
6	3	3	3	3	3	3	3	3	3	3
7	3	3	3	3	3	3	3	3	3	3
8	5	5	5	5	5	5	5	5	5	5
9	3	3	3	3	3	3	3	3	3	3
10	3	3	3	3	3	3	3	3	3	3

(a) Minibuf - Step (0) (a)

Row \ Col	V1	V2	V3	V4	V5	V6	V7	V8	V9	V10
Vertex 1	1	1	1	1	1	1	1	1	1	1
2	1	1	1	1	1	1	1	1	1	1
3	4	4	4	4	4	4	4	4	4	4
4	4	4	4	4	4	4	4	4	4	4
5	1	1	1	1	1	1	1	1	1	1
6	6	6	6	6	6	6	6	6	6	6
7	7	7	7	7	7	7	7	7	7	7
8	1	1	1	1	1	1	1	1	1	1
9	4	4	4	4	4	4	4	4	4	4
10	4	4	4	4	4	4	4	4	4	4

(b) PtrBuf - Step (0) (b)

Row \ Col	V1	V2	V3	V4	V5	V6	V7	V8	V9	V10
Vertex 1	2	2	2	2	2	2	2	2	2	2
2	11	11	11	11	11	11	11	11	11	11
3	6	6	6	6	6	6	6	6	6	6
4	2	2	2	2	2	2	2	2	2	2
5	0	0	0	0	0	0	0	0	0	0
6	3	3	3	3	3	3	3	3	3	3
7	3	3	3	3	3	3	3	3	3	3
8	9	9	9	9	9	9	9	9	9	9
9	4	4	4	4	4	4	4	4	4	4
10	8	8	8	8	8	8	8	8	8	8

(c) SOW - Step (1)

Row \ Col	V1	V2	V3	V4	V5	V6	V7	V8	V9	V10
Vertex 1	F	F	F	F	F	F	F	F	F	F
2	T	T	T	T	T	T	T	T	T	T
3	T	T	T	T	T	T	T	T	T	T
4	F	F	F	F	F	F	F	F	F	F
5	F	F	F	F	F	F	F	F	F	F
6	F	F	F	F	F	F	F	F	F	F
7	F	F	F	F	F	F	F	F	F	F
8	T	T	T	T	T	T	T	T	T	T
9	T	T	T	T	T	T	T	T	T	T
10	T	T	T	T	T	T	T	T	T	T

(d) Active Flag - Step (2) (a)

Row \ Col	V1	V2	V3	V4	V5	V6	V7	V8	V9	V10
Vertex 1	5	5	5	5	5	5	5	5	5	5
2	1	1	1	1	1	1	1	1	1	1
3	4	4	4	4	4	4	4	4	4	4
4	5	5	5	5	5	5	5	5	5	5
5	5	5	5	5	5	5	5	5	5	5
6	5	5	5	5	5	5	5	5	5	5
7	5	5	5	5	5	5	5	5	5	5
8	1	1	1	1	1	1	1	1	1	1
9	4	4	4	4	4	4	4	4	4	4
10	4	4	4	4	4	4	4	4	4	4

(e) Pointer - Step (2) (a)

Row \ Col	V1	V2	V3	V4	V5	V6	V7	V8	V9	V10
Vertex 1										
2	6	4	8	13	15	6	5	12	6	11
3	8	7	3	4	9	12	13	12	5	8
4										
5										
6										
7										
8	8	13	14	11	14	11	11	5	7	10
9	8	5	5	4	7	4	5	5	3	11
10	5	10	8	4	11	11	12	8	11	3

(f) linkcost - Step (3)

Figure 32.5: Second iteration - Steps (0) through (3)

3 to 5 must be itself a MCP, which must be originated from column vertex 4 at this iteration as mentioned above (i.e. the cost of $v5 \rightarrow v4 \rightarrow v3$ is $W(5,4) + W(4,3) = 2 + 1 = 3$, and is the minimum of all possible $v5 \rightarrow \cdots \rightarrow v3$). The resulting MinBuf fields of every edge are shown in Figure 32.5(a). It shows the cost of sub-MCP of each row vertex if it is hypothesized as part of the ultimate MCP. If these costs may turn out to be higher than those of previous iteration, then the corresponding vertices will be deactivated by Step 2.

(b) The PtrBuf, pointing to previous vertex in the sub-MCP, of every edge is shown in Figure 32.5(b). For example, row vertex 3 contains 4 indicating the sub-MCP is originated from column vertex 4 as mentioned above.

Row \ Col	V1	V2	V3	V4	V5	V6	V7	V8	V9	V10
Vertex 1	5	4	3	4	7	4	5	5	3	3
2	5	4	3	4	7	4	5	5	3	3
3	5	4	3	4	7	4	5	5	3	3
4	5	4	3	4	7	4	5	5	3	3
5	5	4	3	4	7	4	5	5	3	3
6	5	4	3	4	7	4	5	5	3	3
7	5	4	3	4	7	4	5	5	3	3
8	5	4	3	4	7	4	5	5	3	3
9	5	4	3	4	7	4	5	5	3	3
10	5	4	3	4	7	4	5	5	3	3

(a) Minibuf - Step (4) (a)

Row \ Col	V1	V2	V3	V4	V5	V6	V7	V8	V9	V10
Vertex 1	10	2	3	3	9	9	2	8	9	10
2	10	2	3	3	9	9	2	8	9	10
3	10	2	3	3	9	9	2	8	9	10
4	10	2	3	3	9	9	2	8	9	10
5	10	2	3	3	9	9	2	8	9	10
6	10	2	3	3	9	9	2	8	9	10
7	10	2	3	3	9	9	2	8	9	10
8	10	2	3	3	9	9	2	8	9	10
9	10	2	3	3	9	9	2	8	9	10
10	10	2	3	3	9	9	2	8	9	10

(b) PtrBuf - Step (4) (b)

Row \ Col	V1	V2	V3	V4	V5	V6	V7	V8	V9	V10
Vertex 1	2	4	3	2	0	3	3	5	3	3
2	2	4	3	2	0	3	3	5	3	3
3	2	4	3	2	0	3	3	5	3	3
4	2	4	3	2	0	3	3	5	3	3
5	2	4	3	2	0	3	3	5	3	3
6	2	4	3	2	0	3	3	5	3	3
7	2	4	3	2	0	3	3	5	3	3
8	2	4	3	2	0	3	3	5	3	3
9	2	4	3	2	0	3	3	5	3	3
10	2	4	3	2	0	3	3	5	3	3

(c) SOW - Step (5)

Row \ Col	V1	V2	V3	V4	V5	V6	V7	V8	V9	V10
Vertex 1	F	F	F	F	F	F	F	F	F	F
2	F	F	F	F	F	F	F	F	F	F
3	F	F	F	F	F	F	F	F	F	F
4	F	F	F	F	F	F	F	F	F	F
5	F	F	F	F	F	F	F	F	F	F
6	F	F	F	F	F	F	F	F	F	F
7	F	F	F	F	F	F	F	F	F	F
8	F	F	F	F	F	F	F	F	F	F
9	F	F	F	F	F	F	F	F	F	F
10	F	F	F	F	F	F	F	F	F	F

(d) Active Flag - Step (6) (a)

Row \ Col	V1	V2	V3	V4	V5	V6	V7	V8	V9	V10
Vertex 1	5	5	5	5	5	5	5	5	5	5
2	1	1	1	1	1	1	1	1	1	1
3	4	4	4	4	4	4	4	4	4	4
4	5	5	5	5	5	5	5	5	5	5
5	5	5	5	5	5	5	5	5	5	5
6	5	5	5	5	5	5	5	5	5	5
7	5	5	5	5	5	5	5	5	5	5
8	1	1	1	1	1	1	1	1	1	1
9	4	4	4	4	4	4	4	4	4	4
10	4	4	4	4	4	4	4	4	4	4

(e) Pointer - Step (6) (a)

(f) linkcost - Step () ()

Figure 32.6: Second iteration - Steps (4) through (7)

Step 1 Since all diagonal edges (j, j), $j \neq 5$, have SOW changed as the result of previous iteration. The SOW seen in Figure 32.5(c) is a transposed version of MinBuf shown in Figure 32.4(a) for those active vertices at the previous iteration.

Step 2 Comparing new MinBuf field and the old SOW as shown in Figure 32.5(a) and 32.5(c), row vertices 2, 3, 8, 9 and 10 remain active as the transmitters as shown in Figure 32.5(d). Their new SOW is smaller than the old one. The old one used the direct link between itself and vertex 5 as can be seen via Pointer field in Figure 32.4(e), which point to vertex 5. The new one is through an indirect path via Pointer shown in Figure 32.5(e). For example, the Pointer of row vertex 2 being 1 indicates the MCP from vertex 2 to 5 is via vertex 1, not through direct link edge(5,2). These new SOW will be propagated through the graph at next iteration.

Step 3 Row vertices 2, 3, 8, 9 and 10 will send to each of its neighbors the LinkCost as shown in Figure 32.5(f). For example, the LinkCost of edge cell(10,1) is 5, implying that row vertex 10 is sending a message to column vertex 1 and telling it at the current iteration, the cost to reach him via vertex 10 of the MCP is 5(i.e. the cost of $v5 \rightarrow v4 \rightarrow v10 \rightarrow v1$ is $W(5,4) + W(4,10) + W(10,1) = 2 + 1 + 2$). At each iteration, one more indirect link is added to the sub-MCP of active vertices.

Step 4 (a) As shown in Figure 32.5(f), all row edges in the unshaded area are active transmitters. All column vertices are receivers. For example, column vertex 1 will receive the minimum of LinkCost of 5 sent by row vertex 10, among all the messages sent by all the active edges in the first column. Similarly, column vertex 4 receives the MOL of 4 sent by row vertex 3. The resulting MinBuf fields of every column vertex are shown in Figure 32.6(a).

(b) The PtrBuf, pointing to previous vertex in the sub-MCP, of every edge is shown in Figure 32.6(b).

Step 5 The SOW seen in Figure 32.6(c) is a transposed version of MinBuf shown in Figure 32.5(a) for those active vertices at the previous iteration.

Step 6 (a) Comparing the new MinBuf field and the old SOW as shown in Figure 32.6(a) and 32.6(c), all column vertices are now deactivated as shown in Figure 32.6(d). Their new SOW are larger than the old one.

(b) Stop the algorithm since all ActiveFlags are False.

The SOW in Figure 32.6(c) contains the sum of weights of MCPs for all the vertices to reach vertex 5. Also, the Pointer in Figure 32.6(e) contains all the threads of the MCPs, which can be used to trace the MCP. For our particular problem of finding MCP from vertex 2 to 5, the cost is 4 as shown in column 2 of Figure 32.6(c). The

MCP started at vertex 2. From row 2 of Pointer field in Figure 32.6(e) we retrieve the next vertex (1), and finally from row 1, vertex 5, the final goal.

The complexity of each iteration of the algorithm is $[C1 + C2(logn)]$, where $C1$ and $C2$ are constants. The factor of $logn$ is based on the need at each iteration to send pointer addresses along the minimum cost path; however, there is a significant fixed cost, $C1$, associated this operation so that for practical purposes,i.e., $n < 1000$, each iteration is constant in n. For 32 bit operations this translates to approximately 219 bit level instruction cycles for $n = 1000$. At a bit serial clock rate of 10MHz, this corresponds to 22 μsec/iteration.

32.4   Summary

We have described a flexible communication capability that permits more effective usage of traditional mesh connected SIMD arrays of the bit-serial type. As an example of its usage we have described an efficient mapping of a dynamic programming problem. However, we have looked at a variety of problems [2], such as connected component labeling, Hough transform, minimum spanning tree, graph matching and neural computations, and have shown that these also can make very effective used of this structure.

References

[1] Charles Weems, Edward Riseman Allen Hansen, David Shu and Greg Nash, *The Image Understanding Architecture,* to be published

[2] David Shu, Greg Nash and Charles Weems, *The Image Understanding Architecture and Applications,* in Future Trends in Machine Vision, edited by Jorge Sanz, Springer Verlag, 1988.

[3] C. C. Weems, *Image Processing on a Content Addressable Array Parallel Processor,* Univ. Mass. Ph.D Thesis, 1984.

[4] D. Hillis, The Connection Machine, MIT Press, 1985, pp. 10-13.

Chapter 33

Decoding of Rate k/n Convolutional Codes in VLSI [1]

V. P. Roychowdhury [2]

P. G. Gulak [2]

A. Montalvo [2]

T. Kailath [2]

Abstract

A systematic procedure for an efficient VLSI implementation of the Viterbi algorithm for decoding convolutional codes is presented. This implementation is based on a network of simple processors, each performing an add-compare-select and branch-generation operation, that reside on a single die and are connected to execute the Viterbi algorithm in a highly parallel way. The chip area of such implementations will depend on the processor interconnections, which in turn depend on the state transition diagram of the convolutional encoder (or dual encoder). It is shown that for all rate $1/n$ convolutional codes generated by feed-forward FIR encoders the encoder state transition diagram, which is described by a de Bruijn graph, can be mapped by a simple equivalence relation to a well-known interconnection scheme in parallel processing referred to as the shuffle-exchange network, for which layout techniques that achieve a proven lower bound on implementation area in a VLSI medium have been established. These results are then extended

[1] This work was supported in part by the National Science Foundation under Grant DCI-84-21315-A1, the U. S. Army Research Office under Contract DAAL03-86-K-0045, the SDIO/IST, managed by the Army Research Office under contract DAAL03-87-k-0033 and Rockwell International Contract INT 6G3052.

[2] Stanford University, Stanford, CA.

to rate $1/n$ codes generated by (IIR) encoders containing feedback. Finally, in the case of general (feed-forward and feedback) rate k/n convolutional encoders it is shown that the state transition diagram of either the encoder or the dual encoder can be always mapped to the Cartesian product of de Bruijn graphs, and therefore of shuffle-exchange graphs; the point is that optimum VLSI layouts for the Cartesian product are easier to obtain and much less complicated than any direct VLSI layouts for the original state transition diagram.

33.1 Introduction

Optimal (maximum likelihood) decoding of convolutional codes can be accomplished by a Viterbi algorithm based on the state transition diagram of the encoder. The Viterbi algorithm (VA) [Vit67], [VO79], [For73b] is in essence a technique for estimating the state sequence of a finite state Markov process observed in memoryless noise. Central to the VA is the concept of a trellis diagram, which is a graphical representation of the state diagram drawn as a function of discrete time. There are q^v nodes at each time step in the trellis diagram, where q is the alphabet size and v is the total number of memory elements in the encoder (often referred to as the constraint length). The VA can be thought of as a dynamic programming solution to the problem of finding the shortest path in this trellis diagram [Omu69]. The essence of the algorithm is a relatively simple procedure of add, compare, and select operations that must be applied to each of the q^v nodes belonging to the same time step in the trellis diagram. The sole reason that the VA is computationally demanding is that the number of operations during each symbol interval T (defined as the time interval between two consecutive output symbols of the receiving channel) grows exponentially with the constraint length of the encoder. It has been traditionally implemented on a single processing element driven by a control unit (*e.g.* a microprocessor) using a direct sequential algorithm. This approach requires $O(q^{v+1})$ operations and $O(q^v)$ random accesses to the processor's memory during each symbol interval T. The throughput rate of such an implementation may not be acceptable in applications where high data rates are required.

The only alternative to increasing the throughput rate is to perform the operations in parallel, thus increasing the hardware complexity. Fortunately, the advent of VLSI (Very Large Scale Integration) technology has opened up opportunities for realizing the parallelism inherent in computationally intensive algorithms. The increase in hardware complexity may now be tolerable as the VLSI technology is capable of realizing chips with the hundreds of thousands of transistors required to realize the VA for constraint lengths of commercial interest. This observation has led to several attempts at providing parallel implementations of the VA in VLSI (see *e.g.,* [CK84], [GS86], [CY86], [Gul84]). Gulak and Shwedyk [GS86], [Gul84] described a fully parallel implementation of the VA based on a set of q^v processors connected according to a *shuffle-exchange* network, for which area efficient VLSI layouts are known in the literature [KLLM81], [Lei81] (there is a restriction in [GS86] that the encoder be a rate $1/n$ feed-forward (FIR) circuit). It should be pointed out that shuffle-exchange networks are functionally equivalent to a whole family of other popular networks such as hypercubes, cube-connected cycles, butterflies, omega networks *etc.,* [PV81], [Ull81] and thus the VA (for rate $1/n$ FIR encoders) can be efficiently implemented on any of these architectures. We should remark that among these architectures, the shuffle-exchange networks have the least VLSI area for the same number of nodes.

Another family of parallel implementations was presented by Chang and Yao [CY86]. They interpreted the VA (both for rate $1/n$ and rate k/n feed-forward convolutional encoders) as a sequence of matrix-vector multiplications (where the usual $+$ operation is replaced by the *min* operation and the usual multiplication operation is replaced by addition) and then implemented the VA using systolic architectures already developed in the literature for matrix-vector multiplication. The implementation uses $O(q^v)$ processors. However, the symbol interval T is at best $O(q^v/v)$. In fact the gain in speed over that of the sequential processor can be shown to be at best qv. Hence, though an exponential number of processors is used, the gain in throughput rate is at best logarithmic. This is in sharp contrast to the fully parallel (shuffle-exchange) implementations of Gulak and Shwedyk [GS86], [Gul84] where the gain in speed is directly proportional to the number of processors used. Of course, a price is paid for speed in terms of the area required to layout the architectures in VLSI. In fact, if N is the total number of processors used, then the area required by the fully parallel version is $O(N^2/\log^2 N)$ whereas the area required by the systolic implementation is $O(N)$. In [GK], we have studied other non-fully-parallel architectures that may often be preferred to linear systolic architecture (especially a so called 'cascade' architecture); however, in the rest of this paper we shall confine ourselves to fully parallel implementations. One reason is that despite the results mentioned above, several questions remain unanswered for fully parallel implementations. First, efficient VLSI implementations are not known for the k/n convolutional codes with or without feedback. Even for the rate $1/n$ case, VLSI implementations are known only for the feed-forward encoders. Moreover, the issue of optimality of the VLSI implementations (*e.g.,* those appear in [GS86], [Gul84]) have not been addressed.

In this paper we provide answers to several of these open questions. First we prove the important fact (see Section 33.2.1) that for efficient and fully parallel implementation of the VA, the processors must always be connected according to the state transition diagram of the encoder. Thus, for efficient implementation in VLSI, it will be desirable to devise layouts for encoder state transition diagrams for which both the area and the average length of interconnecting wires is minimum (for faster communication and less cost). We show that the state diagrams of rate $1/n$ feed-forward encoders, which are known as de Bruijn graphs, can be efficiently laid out in VLSI by modifying existing layouts for shuffle-exchange networks. We then show that the state transition diagram of a rate $1/n$ encoder *with feedback* is either a de Bruijn graph or a subgraph of such a graph, when the encoder realization is in a certain (controller) canonical form. Thus, for *any* rate $1/n$ encoder the VA can be efficiently implemented in VLSI on a set of processors connected according to a de Bruijn or equivalently a shuffle-exchange network. For rate k/n feed-forward encoders with certain 'obvious' realization (to be defined in section 33.3), we show that the state diagrams can be represented as Cartesian product of k, possibly distinct, de Bruijn graphs. Minimum area VLSI layouts for the product graphs are presented using a recursive layout technique that uses the optimal layout strategy for shuffle-exchange networks. For the general case of rate k/n encoders (*i.e.* with and without feedback) it is shown that the dual encoder always has the feed-forward structure for which the state diagram can be efficiently represented as a product graph. In such cases it is recommended that one should apply the VA on the trellis diagram corresponding to the dual encoder, since the state diagram of the original encoder may be arbitrarily complex and efficient VLSI implementation may not be possible. Thus in general, the VA for rate k/n encoders can be always implemented in parallel on a set of processors connected according to a Cartesian product of de Bruijn (or shuffle-exchange) graphs.

The rest of the paper is organized as follows. In section 33.2 we briefly discuss the Viterbi algorithm and introduce the relevant terminologies. We also present an equivalence between de Bruijn graphs and shuffle-exchange graphs that enables us to use the existing layouts for shuffle-exchange networks for efficiently laying out de Bruijn networks. Techniques to generalize such results so as to provide efficient implementation when the encoder has feedback are also outlined. Section 33.3 deals with rate k/n encoders. First it is shown that the state diagram for feed-forward rate k/n encoders can be represented as a Cartesian product of smaller de Bruijn graphs. Optimal layouts for such product graphs are presented and a comparison with the direct implementation is made. It is shown that the product graph representation saves an exponential factor in silicon area for only a constant factor loss in the throughput rate. Space limitations prevent us from providing rigorous proofs for several theoretically intriguing results, nonetheless they are presented in this section for completeness (the reader is referred to [RGMK87] for more details). Finally, section 4 contains some concluding remarks.

33.2   VLSI Architectures for Decoding Rate 1/n Convolutional Codes

A rate $1/n$ convolutional encoder is a finite-state linear sequential circuit operating on GF(q) that has one input and n outputs (see [For70], [For73a]). The number of memory elements (or shift registers) in the circuit is called the constraint length of the encoder. For example Fig. 33.1 shows a rate $1/2$ convolutional encoder with constraint length 2.

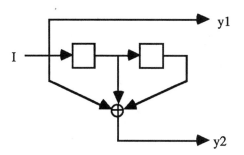

Figure 33.1: A rate 1/2 convolution encoder.

33.2.1 The Viterbi Algorithm

The VA [Vit67] was originally invented for decoding convolutionally encoded data, though since then, it has been applied to other types of problems (*e.g.*, the trellis-coded modulation schemes used in high-speed data modems). The basic theory behind the VA is readily available in the literature and for a good survey the reader is referred to Forney [For73b]. In this section we shall introduce the basic concepts that are required for describing its implementation in VLSI.

The VA can be thought of (see [Omu69]) as a dynamic programming solution to the problem of estimating the state sequence of a finite-state Markov process observed in memoryless noise. Central to the VA is the concept of the *trellis diagram*, which is a graphical representation of the state diagram of the encoder drawn as a function of discrete time. Each time step corresponds to a single symbol (or baud) interval T, and corresponds to one stage of the trellis. The number of stages in the trellis diagram is equal to the length of the input data sequence. The number of nodes (or states) at each stage of the trellis is q^v where q is the cardinality of the input alphabet set and v is the constraint length of the encoder. Each node at every stage of the trellis diagram represents one possible state of the encoder. There is an edge between the node S_i^t (*i.e.*, the node representing state S_i at stage t) to the node S_j^{t+1} if and only if there is a directed edge from state S_i to S_j in the state transition diagram of the encoder. Each such edge (*i.e.*, $S_i^t \rightarrow S_j^{t+1}$) is assigned a weight, λ_{ij}^{t+1} (called the *branch-metric*), and is a measure of the 'unlikelihood' that the channel output at time $t+1$ is caused by the state transition $S_i \rightarrow S_j$ in the encoder. Fig. 33.2 gives the state transition and trellis diagrams for the rate 1/2 encoder of Fig. 33.1.

The VA can now be described as follows. Two quantities, namely, the path metric and the survivor sequence are associated with each state of the trellis diagram. The path metric P_j^t of the state S_j at time t is the weighted length of the shortest weighted path (the weights on the edges in the trellis diagram being the branch-

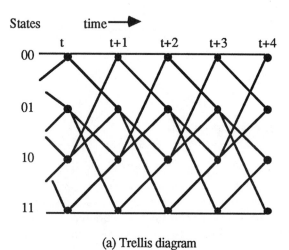

(a) Trellis diagram

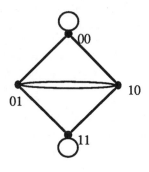

(b) State transition diagram

Figure 33.2: The trellis diagram and the state diagram of the convolution encoder in Figure 33.1.

metrics) between the starting node S_0^0 to the node S_j^t in the trellis diagram. Similarly, the survivor sequence Q_j^t for state S_j at time t is the state sequence associated with the shortest weighted path in the trellis diagram between the initial node S_0^0 and the node S_j^t. Once every baud interval, the path metrics are updated as follows:

$$P_j^{t+1} = \min(P_i^t + \lambda_{ij}^{t+1}) \quad \forall\, i \text{ such that } S_i \to S_j \tag{33.1}$$

where $S_i \to S_j$ implies that there is a valid state transition from state S_i to S_j and $P_0^0 = 0$. The old survivor sequence of the winning ancestor, is augmented with the symbol corresponding to the transition to state S_j to form the new survivor sequence for the state S_j. After sufficiently long time L, (see *e.g.*, [VO79] where the issue of how large L should be for sufficiently low probability of error is addressed) the survivor sequence of the state with the minimum path metric is chosen to be the estimate for the state sequence of the encoder; one can then complete the decoding procedure by determining the input sequence corresponding to the estimated state sequence.

It is evident that a fully parallel implementation of the VA can be realized by assigning a single processor for every node in the trellis diagram. However, for a message of length L, such a realization will have Lq^v processors (and Lq^{v+1} interconnections); and, moreover, because only q^v processors are active at any time step, such a realization is very inefficient. A somewhat more practical strategy is to use q^v processors connected according to the state transition diagram of the encoder. The processor M_j, at step t, contains the path metric P_j^t and the survivor sequence Q_j^t for state S_j. At step $t+1$ it receives the channel output y_{t+1} and the path metric P_i^t from each of its predecessor state S_i and also computes λ_{ij} for each state transition $S_i \to S_j$. It then computes P_j^{t+1}, as given by (33.1), and updates the survivor sequence. Since each state has q predecessors, the total time taken at each step (and thus the symbol interval T) is $O(q)$. Hence, a gain of $O(q^v)$ is achieved over a sequential processor (T is $O(q^{v+1})$ for sequential implementation) by using q^v processors in parallel. The major focus of the rest of the paper is to provide efficient VLSI implementations for this architecture. An important measure of efficiency in VLSI technology is the silicon area required to layout an architecture. Apart from the obvious increase in cost due to a larger area, the increased area compounds cost by drastically reducing the yield and by possibly increasing the energy required for communicating information in the system. Thus, to be able to choose the most efficient implementation, among a host of possible choices, one needs a model that can quantify the notion of 'area' of a VLSI layout. One such popular model, proposed by Thompson [Tho80], is described next.

33.2.2 VLSI Grid model

The VLSI grid model proposed by Thompson [Tho80] is simple and assumes that the chip consists of a grid of vertical and horizontal tracks, spaced apart by some unit interval. Processors are viewed as points on the grid and are located only at the intersections of grid tracks. Wires are routed through the tracks in order to connect pairs of processors. Wires can intersect only at right angles and overlapping is not allowed. From a computational point of view this model is attractive since the layout area is calculated easily as the product of the numbers of vertical and horizontal tracks in the layout. We should comment that a major motivation behind this model is the observation that the total area in VLSI is often dominated by the wiring rather than the processors; that is why the processors are assumed as just points.

33.2.3   de Bruijn and Shuffle-Exchange Graphs

The state at time t of an encoder can be represented by a v-tuple

$$S^t =< x_{v-1}\, x_{v-2}\, \cdots x_0 >$$

where x_i is the content of the ith memory element. For a feed-forward encoder the next state S^{t+1} will then be given by the tuple

$$S^{t+1} =< x_{v-2}\, x_{v-3}\, \cdots x_0\, a > \quad \forall\, a \in \{0, \cdots, q-1\}$$

where q is the alphabet size. For $q = 2$, the state transition graph generated by the above transitions is known as the de Bruijn graph or Goods diagram of order v (see [Gol82]). For the rest of the section we shall restrict ourselves to the binary alphabet (*i.e.*, $q = 2$) case, primarily because the results are simpler to state and can be easily extended for arbitrary alphabet size q (see *e.g.*, [Gul84], [RGMK87]). In our interpretation, an edge of a de Bruijn graph is called a *shuffle edge* if it is of the form $< x_{v-1}\, x_{v-2}\, \cdots x_0 > \rightarrow < x_{v-2}\, x_{v-3}\, \cdots x_0\, x_{v-1} >$ and it is called a *shuffle-exchange* edge (for reasons to be explained later) if it is of the form $< x_{v-1}\, x_{v-2}\, \cdots x_0 > \rightarrow < x_{v-2}\, x_{v-3}\, \cdots x_0\, \overline{x_{v-1}} >$. An 8 node de Bruijn graph representing the state diagram of encoder with constrain length 3 is shown in Fig. 33.3.

de Bruijn graphs are closely related to shuffle-exchange graphs, which are well-known in parallel processing (see *e.g.*, [Sto71], [Ull81]). A *shuffle-exchange* graph of

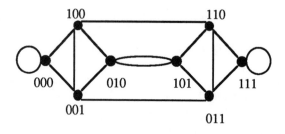

Figure 33.3: An 8 node de Bruijn graph.

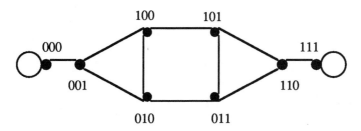

Figure 33.4: An 8 node shuffle-exchange graph.

order v has 2^v nodes, each labeled by a v tuple $< x_{v-1}\, x_{v-2}\, \cdots x_0 >$. Every node $< x_{v-1}\, x_{v-2}\, \cdots x_0 >$ is connected by a *shuffle* edge to the node $< x_{v-2}\, \cdots x_0\, x_{v-1} >$ and by an *exchange* edge to the node $< x_{v-1}\, x_{v-2}\, \cdots \overline{x_0} >$. A cycle of shuffle edges is known as a necklace. An 8 node shuffle-exchange graph is shown in Fig. 33.4.

Several authors (see *e.g.*, [Lei81], [KLLM81]) have presented good layout techniques for shuffle-exchange networks. These techniques cleave the network into necklaces and then appropriately insert the exchange edges. Thompson [Tho80] showed that the minimum VLSI layout area for a N-node shuffle-exchange graph is $\Omega(N^2/\log^2 N)$ and then Kleitman *et al.* [KLLM81] devised optimal layout techniques that achieve the lower bound. We shall show next that these optimal layouts of shuffle-exchange networks can be easily modified to yield optimal layouts for de Bruijn networks.

33.2.4 Optimal Layouts for de Bruijn Networks

The layouts for de Bruijn graphs can be easily motivated by redrawing the graph so that it looks exactly like a shuffle exchange graph of the same order. The shuffle edges can be drawn just as they are in de Bruijn graphs; however, a shuffle-exchange edge is redrawn by first routing it along a shuffle edge (*i.e.*, along $< x_{v-1}\, x_{v-2}\, \cdots x_0 > \rightarrow < x_{v-2}\, x_{v-3}\, \cdots x_0\, x_{v-1} >$) and then along an exchange edge (*i.e.*, along $< x_{v-2}\, x_{v-3}\, \cdots x_0\, x_{v-1} > \rightarrow < x_{v-2}\, x_{v-3}\, \cdots x_0\, \overline{x_{v-1}} >$); (Remark: this also justifies the term 'shuffle-exchange' edge). The resulting graph looks exactly like a shuffle-exchange graph, except that it has two parallel edges for every edge in the latter graph (see Fig. 33.5). This simple procedure can be used to modify any layout of a shuffle-exchange graph to a layout of a de Bruijn graph; the resulting layout will have two wires for every wire in the original layout. Hence, the length and the width of the new layout will be at most twice the original values. Thus, the area of the de Bruijn graph layout is at most 4 times the area of the corresponding shuffle-exchange network. A rigorous proof can be found in [RGK87], [RGMK87].

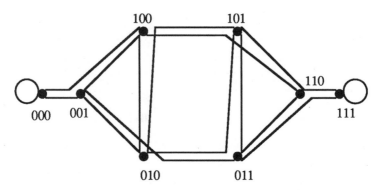

Figure 33.5: An 8 node de Bruijn graph redrawn as a shuffle-exchange graph.

Thus, we can obtain $O(N^2/\log^2 N)$ area layouts for de Bruijn networks by modifying the optimal layouts for shuffle-exchange networks. Since, the minimum area for de Bruijn graphs is $\Omega(N^2/\log^2 N)$ ($\Omega(q^2 N^2/\log^2 N)$ for arbitrary alphabet size q, see [RGK87]), the layouts obtained by modifying the optimal layouts for shuffle-exchange graphs are also optimal for de Bruijn graphs. [There is an erroneous claim in Samatham and Pradhan [SP84] that the area for de Bruijn graphs is $\Omega(N^2/\log N)$] .

33.2.5   Architectures for decoding rate $1/n$ codes with feedback

Decoding for encoders with feedback can also be done on a set of processors connected according to a de Bruijn graph of appropriate order; hence, the efficient layout strategies discussed before can be reapplied. First we shall define a canonical structure for encoders with feedback. Let x_i^t denote the content of the ith memory module (or shift-register) at time t. Then, a rate $1/n$ encoder is said to be in the controller canonical form if the memory elements can be ordered such that $x_i^{t+1} = x_{i-1}^t \ \forall \ i = 1, \cdots, \ v-1$ and $x_0^{t+1} = f(u, x_0, x_1, \cdots, v_{v-1})$, where $f(.)$ is some function defined on GF(q) (see Fig. 33.6).

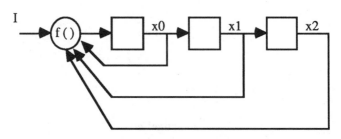

Figure 33.6: A rate 1/n encoder realized in the controller canonical form.

Theorem 1 *The state diagram of a controller canonical form is either a de Bruijn graph or a subgraph of such a graph.*

Proof: We can label a state as a v-tuple $< x_{v-1}\, x_{v-2} \cdots x_0 >$. If an edge is in the state diagram, G, of the encoder then it is of the form $< x_{v-1}\, x_{v-2} \cdots x_0 > \rightarrow < x_{v-2}\, x_{v-3} \cdots x_0\, f(.) >$. However, $f(.)$ is either 0 or 1 (assuming $q = 2$), which means that the edge is also in the de Bruijn graph of order v. Hence, G is contained in the de Bruijn graph of order v. $\qquad\Box$

Since, convolutional encoders are linear systems in GF(q), it can be shown that (see [Kai80], [For70]) any rate $1/n$ encoder can be realized in the controller canonical form and hence can be implemented on a set of q^v processors connected according to a de Bruijn graph.

33.3 VLSI Architectures for Decoding Rate k/n Convolutional Codes

A rate k/n convolutional encoder is a finite-state linear sequential circuit operating on $GF(q)$ that has k inputs and n outputs. We shall first discuss decoders for feed-forward rate k/n codes and then extend the results to codes with feedback.

A feed-forward encoder can be always realized in an obvious manner as shown in Fig. 33.7 (see [For70]). In the obvious realization, each input has a separate sequence of shift-registers associated with it (*i.e.*, the input I_i has a shift-register bank of length v_i). The outputs are given as linear functions of the internal states and the inputs. The total state of the encoder at time t can then be given by

$$S^t = < s_1^t, s_2^t, \cdots, s_k^t >$$

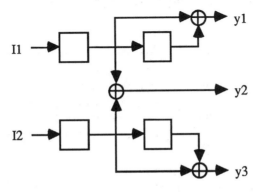

Figure 33.7: A rate 2/3 encoder realized in the obvious manner.

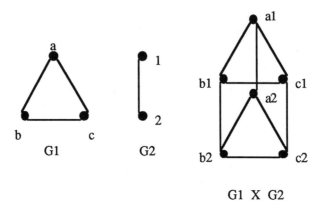

G1 X G2

Figure 33.8: An example of Cartesian product of two graphs.

where s_i^t is the state of the shift-register sequence associated with the input I_i and can itself be represented by the tuple $s_i^t = <x_{v_i-1}\ x_{v_i-2}\cdots x_0>$. The next state of the feed-forward encoder is then given by

$$S^{t+1} = <s_1^{t+1}, s_2^{t+1}, \cdots, s_k^{t+1}>$$

where s_i^{t+1} is the next state corresponding to the state s_i^t and can be represented by the tuple $s_i^{t+1} = <x_{v_i-2}\ x_{v_i-3}\ \cdots x_0\ a>$ ∀ $a\ \in\ \{0,\cdots,\ q-1\}$. One can now draw the state transition diagram for the encoder as given by the above transitions (see Fig. 33.8). It has q^{v+k} nodes, (where $v = \sum_1^k v_i$) and each node has degree $2q^k$. We know that the decoder can be always realized efficiently on a set processors connected according to the state transition diagram of the encoder (see Section 33.2.1). However, efficient layout techniques for such state transition diagrams are not readily apparent.

33.3.1   Product Graph Representation of The State Transition Diagram

However, there is an alternative representation of the state transition diagram that is easy to layout and requires considerably less area. In the new representation, a state transition $S^t \rightarrow S^{t+1}$ is done in k steps, where in the ith step only the state s_i^t changes to s_i^{t+1} and the rest of the states remain unchanged. The resulting representation of the state transition diagram can be formulated as a *Cartesian product* of the state state transition diagrams of the individual shift-register banks, *i.e.,*

$$G = G_k \times G_{k-1} \times \cdots \times G_1 \tag{33.2}$$

where G_i is the state transition diagram of the ith shift-register bank (hence, a de Bruijn graph of order v_i) and the Cartesian product of two graphs is defined as follows (see Figure 33.8 for an example of Cartesian product graphs):

Definition 1 *The Cartesian product of two graphs $G_1 = (V_1, E_1)$ and $G_2 = (V_2, E_2)$ is the graph $G = (V, E)$ where $V = V_1 \times V_2$ and an edge $< (u_1, u_2), (v_1, v_2) > \in E$ if and only if either $u_1 = v_1$ and $< u_2, v_2 > \in E_2$ or $u_2 = v_2$ and $< u_1, v_1 > \in E_2$.*

The reason that the modified state transition diagram is given by (33.2) follows directly from the above definition; a rigorous proof is given in [RGMK87]. We should note here that in the product graph representation one has to make k transitions in order to obtain a valid state transition of the encoder; the degree of every node in this representation is $2qk$ and the total number of edges is kq^{v+1}.

Efficient Layouts for Product Graphs

A convenient way of interpreting $G = G_1 \times G_2$ is that G can be obtained by replacing every node of G_1 by a copy of G_2 and then interconnecting these macro-nodes according to the interconnection pattern of G_1. This suggests a recursive VLSI layout procedure: 1) layout G_2 optimally (since every individual graph is de Bruijn, we can apply the results of previous section), 2) layout copies of G_2 and connect them in the same way as the nodes of G_1 are connected in an optimal layout of G_1. The procedure can easily continue for any $k > 2$. This intuitive recursive layout technique turns out to be also asymptotically optimal and achieves the lower bound $\Omega(q^2 N^2 / v_{max}^2)$ derived in [RGMK87] (v_{max} is the length of the longest shift-register bank, and N is the total number of nodes in the graph). The lower bound on the VLSI area for the state transition diagram is shown to be $\Omega(q^{2k} N^2 / v_{max}^2)$ in [RGMK87]. Thus, the product graph representation saves a factor of q^k in the silicon area. Since one has to make k transitions in the product graph to make a valid state transition, it can be shown that the symbol interval increases by at most a constant factor; however, the reduction in area outweighs the loss in speed. Furthermore, it can be shown that the product graph representation requires only q input and q output ports for every processor, whereas the direct implementation requires q^k such ports. If the processors are required to have a fixed number of ports, then one can show that the direct realization of the state transition diagram is no longer even faster.

33.3.2 Architectures for decoding rate k/n codes with feedback

The state transition graph will not be nicely structured for an encoder with arbitrary structure. However, for every encoder one can define a dual encoder (see [For73a], [For70]) and the VA based on the state transition diagram of the dual encoder can be used to decode channel outputs without any loss of information. It turns out that the dual encoders are always feed-forward and hence the decoder can be implemented on a set of processors connected according to the product graph of de Bruijn graphs of appropriate orders.

33.4   Concluding Remarks

In this short paper we have described some parallel architectures that are suitable for efficient implementations in VLSI and can execute the VA for decoding convolutional codes. They are shown to be related to well-known architectures for parallel processing such as shuffle-exchange and de Bruijn networks. More detailed accounts of the material presented here can be found in [RGMK87] and [RGK87].

References

[CK84]   J. B. Cain and R. A. Kriete. A VLSI r=1/2, k=7 Viterbi Decoder. *Proc. of NAECON*, May 1984.

[CY86]   C. Y. Chang and K. Yao. Systolic Array Processing of the Viterbi Algorithm. *Submitted to IEEE Transactions on Information Theory*, June 1986.

[For70]   G. D. Forney. Convolutional Codes I: Algebraic Structure. *IEEE Transactions On Information Theory*, IT-16, No. 6:720–738, Nov. 1970.

[For73a] G. D. Forney. Structural Analysis of Convolutional Codes via Dual Codes. *IEEE Transactions on Information Theory*, IT-19:512–518, July 1973.

[For73b] G. D. Forney. The Viterbi Algorithm. *Proceedings of The IEEE*, 61, No. 3:268–278, March 1973.

[GK]   P. G. Gulak and T. Kailath. Locally Connected VLSI Architectures for the Viterbi Algorithm. *IEEE Journal on Selected Areas in Communications*, to appear in 1988.

[Gol82]   S. W. Golomb. *Shift Register Sequences*. Aegan Park Press, 1982.

[GS86]   P. G. Gulak and E. Shwedyk. VLSI Structures for Viterbi Receivers: Part I- General Theory and Applications. *IEEE Journal on Selected Areas in Communications*, SAC-4:142–154, Jan. 1986.

[Gul84]   P. G. Gulak. *VLSI Structures For Digital Communications*. PhD thesis, University of Manitoba, Winnipeg, Canada, Dec. 1984.

[Kai80]   T. Kailath. *Linear Systems*. Prentice-Hall Inc., Englewood Cliifs, N.J., 1980.

[KLLM81] D. Kleitman, F. T. Leighton, M. Lepley, and G. L. Miller. New Layouts for the shuffle-exchange graph. *Proc. of the 13th ACM Symposium on Theory of Computation*, 278–292, May 1981.

[Lei81]   F. T. Leighton. *Layouts for the shuffle-exchange graph and lower bound techniques for VLSI*. PhD thesis, Department of Mathematics, Massachusetts Institute of Technology, 1981.

[Omu69]   J. K. Omura. On the Viterbi Algorithm. *IEEE Transactions Information Theory*, IT-15:171–179, Jan. 1969.

[PV81]   F. P. Preparata and Jean Vuillemin. The Cube-Connected Cycles: A Versatile Network for Parallel Computation. *Communications of the ACM*, 24:300–309, May 1981.

[RGK87]   V. P. Roychowdhury, P. G. Gulak, and T. Kailath. Optimal VLSI layouts of de Bruijn Graphs. *to be submitted to IEEE Trans. on Computers*, 1988.

[RGMK87] V. P. Roychowdhury, P. G. Gulak, A. Montalvo, and T. Kailath. Decoding of Rate k/n Convolutional Codes in VLSI. *to be submitted to IEEE Trans. on Info. Theory*, 1988.

[SP84]   M. R. Samatham and D. K. Pradhan. A Multiprocessor Network Suitable for Single-chip Implementation. *Proc. 11th Ann. Symp. on Computer Architecture*, 328–337, June 1984.

[Sto71]   Harold S. Stone. Parallel Processing with the Perfect Shuffle. *IEEE Transactions On Computers*, c-20, No. 2:153–161, Feb. 1971.

[Tho80]   C. D. Thompson. *A Complexity Theory for VLSI*. PhD thesis, Dept. of Comp. Science, Carnegie Mellon University, Pittsburgh,PA, 1980.

[Ull81]   J. D. Ullman. *Complexity of VLSI Design*. MIT Press and John Wiley Sons, Inc., 1981.

[Vit67]   A. J. Viterbi. Error Bounds for Convolutional Codes and an asymptotically optimum decoding algorithm. *IEEE Transactions Information Theory*, IT-13:260–269, Apr. 1967.

[VO79]   A. J. Viterbi and J. K. Omura. *Principles of Digital Communication and coding*. New York: McGraw Hill, 1979.

Chapter 34

IC* Supercomputing Environment

E. J. Cameron [1]
D. M. Cohen [1]
B. Gopinath [1]
W. M. Keese II [1]
L. Ness [1]
P. Uppaluru [1]
J. R. Vollaro [1]

34.1 Introduction

The IC* project is an effort to create an environment for the design, specification, and development of complex systems. Examples of such complex systems are communications and real time software and protocols, hardware, and physical and econometric models.

The basis of our work is the IC* model of computation. The IC* model can support many different user languages and one such has been used in Bell Communications Research for communications protocols since 1984 [1,2]. This model allows a complex system to be described in a highly parallel fashion and supports the specification of performance requirements and the easy reuse of existing specifications.

In the IC* model, a system is described by defining a set of state variables and a set of *invariants* that specify how the values of the state variables change in time. A *path* is a function which assigns a set of values for the state variables to every

---

[1] Bell Communications Research, Morristown, NJ

instant of time. The invariants specify the logical properties that the paths must satisfy. There are three types of invariants: *static invariants, differential invariants,* and *conditional invariants*. In analogy to the theory of differential equations, the static invariants specify a surface that the set of values for all the state variables must lie on, and the differential and conditional invariants specify how the paths move in time on this surface.

A supercomputer, called the I*C machine, is being built to execute IC* programs. It will have one gigabyte of semiconductor memory, will execute one gigafloating point instructions persecond, and will have many gigabytes of secondary laser disk storage. The prototype of the I*C machine is called the Y machine. A scaled down version of the Y machine will be used as a communications processor for direct implementations of communications protocols from their IC* specifications. One of the main goals in designing these machines is to provide a real time software engineering environment aided by a special purpose supercomputer.

The IC* environment provides tools for writing, documenting, analyzing, simulating, and implementing IC* specifications. IC* programs are observed through symbolic animation using audio and visual icons and controlled by various input devices such as joysticks and pressure gauges. A version of this environment, including a system to produce prototypes of communications protocols, currently runs on some commercially available personal, micro, and mini computers.

34.2　IC* Model of Computation

In the IC* Model of Computation, a system evolves over time, according to a set of *invariant expressions*, creating a particular *history* for the system. The set of invariant expressions determines how the *system state* and the set of *active invariants* change in time. For the purposes of this paper, time is assumed to be discrete and described by a succession of non-negative integers. Though the set of active invariants can be arbitrarily large, their *description* in terms of the invariant expressions is bounded and finite. A description need not determine a unique evolution; many different histories may satisfy it.

After a system has evolved for t time units, two sequences of sets describe the history of this particular evolution. One sequence, $St : S[0], \ldots, S[t]$, is the sequence of system states. The other sequence, $At : A[0], \ldots, A[t]$, is the sequence of sets of active invariants. The system state and the set of active invariants at time t represent the structure and behavior of the system at that time. We will call St the *system history*, and we will call At the *activation history*.

Given a particular history up to the current time t, the next elements in each of the two sequences are computed by identifying $P[t+1]$, the set of *involved predicates* at time $t+1$. The next activation state $A[t+1]$ is constructed from $A[t]$, using St, the

current system history; then, $P[t + 1]$, the set of involved predicates is constructed
from $A[t]$ and $A[t + 1]$, again using St; finally, the next system state, $S[t + 1]$, is
constructed from the current system state $S[t]$, using the new set $P[t+1]$. The next
system state, $S[t + 1]$, is one of a set of allowable next system states, from which it
is non-deterministically chosen.

There are three types of invariants: *differential invariants*, *static invariants*, and
conditional invariants. They are called invariants because they describe invariant
properties that the history must satisfy. In analogy to the theory of differential
equations, the differential and conditional invariants specify how the states move in
time, and the static invariants specify surfaces the state space must be contained
in.

A *differential invariant* is a cause-effect pair and is stated as

> *WHENEVER* < Cause Expression > :> < Effect Expression >.

A *static invariant* is a constraint, and is stated as

> *MAINTAIN* < Constraint Expression >.

A *conditional invariant* is stated as

> *WHENEVER* < Cause Expression > :> < Effect Expression >
>
> *ACTIVATE* {Child Invariants}
> *UNTIL* < Cause Expression > :> < Effect Expression >

The words *MAINTAIN, WHENEVER, ACTIVATE*, and *UNTIL* are suggestive of
the semantics of each of the invariants. The cause, effect, and constraint expressions
are predicate expressions.

A differential invariant specifies that whenever the system state satisfies the predicate defined by the cause expression, then it must satisfy the predicate defined
by the effect expression in the next instant of time. A static invariant states that
the system state transitions must maintain the predicate defined by the constraint
expression.

A conditional invariant has three parts; the *activator*, the *child invariants*, and the
deactivator. Both the activator and the deactivator are differential invariants that
control the activation and deactivation of the child invariants. A child invariant can
be a static invariant, a differential invariant, or a conditional invariant.

All invariants are rules for determining what choices and constraints are placed on system state transitions and invariant activations. A differential invariant specifies permissible transitions from certain states to other states. A static invariant may allow further transitions or disallow certain system states. Only conditional invariants can change the set of invariants in a future activation state.

Each type of invariant is stated using predicate expressions. Evaluation of a predicate expression at a particular time determines a predicate which may depend on the history of this particular evolution of the system.

A predicate is defined by a domain set and a solution set. [2] The solution set is a subset of the power set of the domain set. [3] Each element of the solution set is called a *solution*. A state satisfies a predicate if its intersection with the predicate's domain, is a solution.

Every system state, $S[t]$, is a subset of the union of the domains of the predicates defined by the specification. Thus, this union of domains, which we will denote by U, is the *universal state space* for the specification.

The predicate expressions may reference history. Such references in the cause and effect expressions are relative to the time at which the predicate expression is evaluated. References in the child invariants can also be relative to the time at which they are activated. When a predicate expression is evaluated, the historical references evaluate to constants in the resulting predicate.

At each time t, the cause expression in a differential invariant evaluates to a *cause predicate* on the current state, $S[t]$. The effect expression in a differential invariant evaluates to an *effect predicate* on the next state, $S[t+1]$. The constraint expression in a static invariant evaluates to a *constraint predicate* on the current state, $S[t]$.

$A[0]$ is the initial set of active invariants and $S[0]$ is the initial system state. We assume $S[0]$ satisfies all the constraint predicates defined by $A[0]$. For $t \geq 0$, the next activation state, $A[t+1]$, is computed by the following steps:

1. If the cause predicate of the deactivator of a conditional invariant in $A[t]$ is true on the current system state, $S[t]$, then all sets of descendant invariants of the conditional invariant, which are in $A[t]$, are removed from $A[t]$.

2. Any deactivator invariant whose cause predicate was true on $S[t-1]$ is removed from $A[t]$.

3. If the cause predicate of the activator of a conditional invariant in $A[t]$ is true on the current system state, $S[t]$, then its child invariants and deactivator are added to $A[t]$ to form $A[t+1]$.

---

[2] The domain set specifies all possible predicate arguments and the solution set specifies the subsets that satisfy the predicate.

[3] The power set of a set is the set of all its subsets.

Note that for a child invariant that is a conditional invariant, only the activator is added. Thus, for any time t, $A[t]$ contains only the differential and static invariants that are governing the current behavior of the system.

$P[t+1]$, the set of involved predicates at time $t+1$, is computed from $A[t]$, $S[t]$, and $A[t+1]$ by the following steps:

1. Any effect predicate of a differential invariant whose cause predicate is true on $S[t]$ and which is still included in $A[t+1]$, is an *involved* predicate, and is included in $P[t+1]$. We also refer to such effect predicates as *triggered*.

2. Any constraint predicate determined by a static invariant in $A[t+1]$ is involved, if its domain intersects the domain of an involved predicate, and is thus included in $P[t+1]$.

Note the computation of $P[t+1]$ is a transitive closure on the involvement relation.

The next system state, $S[t+1]$, can be computed from $S[t]$, by replacing a subset of $S[t]$, by another set. The set to be deleted, $D[t]$, and its replacement, $R[t]$, are determined by the predicates in $P[t+1]$. Let $U[t+1]$ be the union of the domains of each predicate in $P[t+1]$. Then, $D[t]$ is the intersection of $S[t]$ with $U[t+1]$. Any subset of $U[t+1]$, whose intersection with the domain of each predicate in $P[t+1]$ is a solution of that predicate, is a *possible replacement set* for $D[t]$. Once all the possible replacement sets have been computed, one, $R[t]$, is non-deterministically chosen.

$S[t+1]$ is obtained by deleting $D[t]$ from $S[t]$, then taking the union of the result with $R[t]$, and checking if any uninvolved constraint predicate is not satisfied by $S[t+1]$. All of the set operations occur in the universal state space, U.

Note that *every* choice of a possible replacement set results in a correct construction of the next system state, $S[t+1]$. Any particular system history is just one evolution in the set of all possible correct system histories. Hence evolution of the system is non-deterministic. Further, note that if the constraint expression of a static invariant contains historical references and the corresponding constraint predicate is uninvolved, it is possible that the predicate may not be satisfied by $S[t+1]$.

If no replacement set exists or if any uninvolved constraint predicate is not satisfied by $S[t+1]$, the system halts and $S[t+1]$ is rejected. Note that if C is a static invariant which is in two successive activation states, $A[t]$ and $A[t+1]$, but which is not involved in $A[t+1]$, then the solution of C in $S[t+1]$ is unchanged from its solution in $S[t]$, i.e. static invariants cause state transitions only if they are involved; however, they are checked every instant of time even if they are not involved. Thus, static invariants can be used as correctness criteria; the system will halt if the specification is not operating correctly by these criteria. A more detailed discussion, including examples, of the IC* model may be found in [1].

34.3   IC* Environment

Complex systems frequently possess many parts and there is inherent awkwardness in dealing with large objects or with any object that has many parts. The IC* environment is designed to help users deal with complex systems by providing tools to focus on parts and aspects of the system.

Tools have been created to help manage the complexity of naming, to observe and control running simulations, to analyze their behavior, and to produce documentation.

Because of the problem of state space explosion, simulation is often believed to be the best way to study a complex system [3], and so, to ease and encourage simulation, the IC* environment was designed to allow real-time interaction between the user and the simulation. The users should be able to get the "feel" of a system they are simulating. The basis of the IC* environment, illustrated in Figure 34.1, is provided by the I* system, the M* system, the N* system, and the D* system.

34.3.1   The I* System

One of the major problems in simulating and "playing" with a large system is interpreting the enormous amounts of information a simulation can produce. This information should be filtered so that the designer can determine how a part of the system affects its global behavior. The I* system, also called the Interactive Information Interpretation system, provides real-time filtered observation of a running simulation. A prototype of the I* system has been built and it runs on some commercially available personal and mini computers.

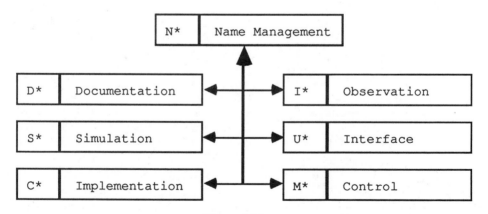

Figure 34.1

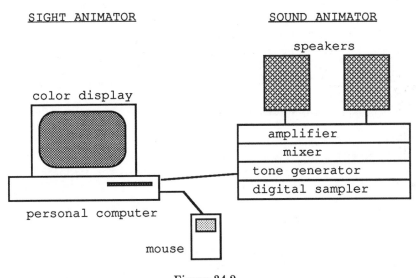

Figure 34.2

The I* system is the window through which the simulation is monitored. A simulation is observed through symbolic animation using both visual and audio icons as indicated in Figure 34.2. The user can see both text and high resolution animated color graphic displays, such as moving gauges and dials. The user can also hear speech, sound effects, and musical patterns whose changes are controlled by the simulation.

The I* system is used by attaching a set of state variables as inputs to a *filter* and a filter to a *gauge*. Each filter performs a transformation of its input values, such as scaling, averaging, or evaluating the maximum. The filters, and the connections between the filters and the state variables are specified in the IC* language. The system has libraries of visual and sound gauges. The user can select any gauge to see or hear the output of a filter. Several gauges can be connected to the same filter, so the user can find which gauge portrays the data in the most convenient form. Sound gauges can be very helpful if continued but not close attention to a particular output is desired.

The system has a graphics editor that allows the user to configure the gauges, e.g., change the color, position, or type of gauge, as well as add or delete gauges. A sound editor lets the user manipulate the pitch, volume, tempo, and various other audio parameters.

34.3.2 The M* system

The M* system, also called the Manual Model Manipulation system, provides for the run time input of control and data. The user can manipulate the parameters of the simulated system so that "tuning" or "optimizing" can be done while simulating with real-time feedback. The M* system consists of all the devices for entering information into the system such as keyboards, joysticks, digitizing tablets. In addition to entering specifications before the simulation starts, the user can enter, during execution, commands to control the simulation e.g., stop, start, change speed, or change a value of a state variable. There is a variety of input devices and the user can select those most applicable to the simulation.

Connecting an input device to one of the system's variables or parameters, permits its value to be changed by the user, in real-time. The user can see, via the I* system, what effect that change had on the rest of the system. This feedback mechanism can be very useful in optimizing or studying a system.

With the aid of the M* and I* systems, the user can observe the simulation, stop it at any point, make changes, restart it and see what effect the changes produced. These systems can be used with the history buffer produced during a simulation, for back-tracking. When the user observes an error or other undesirable behavior in a system during a simulation, backtracking can pinpoint the exact place where the system begins to misbehave.

34.3.3   The N* system

Real problems may involve many objects and a plethora of names, and the N* system, also called the Network for Naming and Notation, manages the names of objects, variables, invariants etc., in a way that mirrors the structure of the system being described and helps keep related information within "easy reach".

The N* system is both a library manager and an interactive user interface. It provides facilities for composing many IC* descriptions, some possibly from its libraries, into one system and answering questions about the result of such compositions. Notation can be introduced dynamically to describe and refer to various parts of the system. The N* system controls the compilation and execution of IC* simulations, the filtering of the results into user understandable terms either via the I* system or through the history file produced during a simulation, and the production of documentation.

34.3.4   The D* system

The D* system, also called the Direct Descriptive Documentation system, helps automate the generation of documentation by producing graphical representations of a system directly from its IC* description. In the current system, each object is represented by a large box which contains smaller boxes that represent its components. The boxes are joined by various types of lines which represent the type of relationship between the components. For example, static invariants that are equivalence relationships are represented by solid lines. Relationships produced by differential and conditional invariants can also be represented, for example, arrows are used to indicate the flow of data from one part to another. The user may draw the entire system, part of the system, or the relationships between parts.

34.3.5 Verifiable Direct Compilation System

Although having a clear and precise specification greatly eases implementation, incompatible and incorrect implementations are possible, if not likely. Producing implementations mechanically from the specification will increase confidence in the implementations, as the tools used to produce the implementations gain in accuracy. The C* system has been designed to produce a prototype of a communications protocol directly from its IC* specification. Test scripts and performance measurement descriptions can be written in the IC* language and included in the input to the compiler.

The protocol's environment in the target machine needs to be modeled in the IC* language. State variables that can be changed by the environment are flagged. The interface to the environment is handled by differential invariants that use functions from a library specific to that interface. For example, in a data communications protocol, the state variable **port** would be flagged and opening the port would be done by a differential invariant with the effect **port'** = **open(port)** where **open** was an appropriate library functions.

We have used the system to produce a prototype of a new data communications protocol, and a description of the work may be found elsewhere [2]. The direct compiler currently works for some commercially available personal, micro and mini computers.

34.4   Y Machine

The Y machine is a prototype implementation of the I*C architecture. It has been especially designed to execute IC* programs. The execution model is illustrated in Figure 34.3. The Y machine is composed of several processors interconnected by multiple busses, as shown in Figures 34.4 and 34.5. The function of the processors is to maintain the invariants. Several mechanisms to help maintain the invariants have been implemented directly in the hardware, resulting in a substantial reduction in software overhead.

A set of processors maintains the invariants of the program and another set is assigned to the observation and control of the program. Each processor knows the system time which can be referenced in an IC* program as a state variable. Since the observation and control is also specified in IC*, identical hardware is used for all the processors. The observation and control system is connected to a bank of micro computers which run the symbolic audio and visual animation and the input devices. An optical mass storage system stores the history of the program and is used for backtracking. A large semiconductor memory acts as a cache between the processors and the mass storage system.

Each processor has a name selector, a name translator, and a synchronizer. The name selector speeds up the identification of the triggered causes. The name translator speeds up the detection of involved static invariants. The synchronizer resolves changes in system time within the instruction cycle time of the processor's cpu and this speeds up the synchronization needed to keep the evolution of the system consistent.

At every instant in time, each processor is responsible for a set of invariants and a set of state variables. If an invariant assigned to a particular processor can change a state variable, then that processor is also assigned all static invariants that use that state variable. Thus any static invariant may be assigned to several processors. Each processor is assigned any state variable that appears in a cause or effect predicate of one of its assigned differential invariants, or in one of its static invariants.

At every change in system time, the cause expressions of the active differential invariants are checked to find which invariants are to be triggered. Solutions for the triggered effect predicates are computed and then a set of new values for the state variables is found which satisfies all of the triggered effect predicates and all the involved static invariants. If a triggered invariant is the deactivator of a conditional invariant, then the children of that invariant are immediately made inactive, and the deactivator is made inactive in next unit of time. If a triggered invariant is the activator of a conditional invariant, then the child and deactivator invariants are made active. After the state variables have been assigned their new values and the activations have been done, the system time is allowed to advance. The synchronizer hardware is used to insure that the different processors do not get out of step during this cycle.

Some of the calculations needed to compute the solution sets of the effect predicates can be done before the invariant is triggered. For example, many deactivators have effect predicates whose solution sets can be computed at the time the invariant was made active. These calculations can be performed in advance, while a processor is finished with its work for the next unit of time and is waiting for the system time to advance.

If none of the state variables in a cause predicate has changed from one instant of time to the next, then the cause predicate does not have to be reevaluated. The name selector hardware identifies, for the software, which state variables have changed.

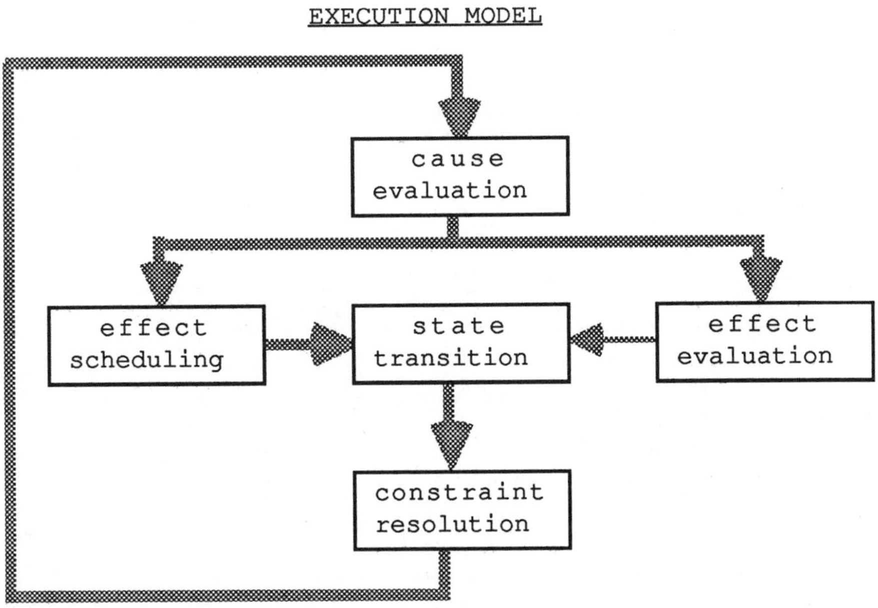

EXECUTION MODEL

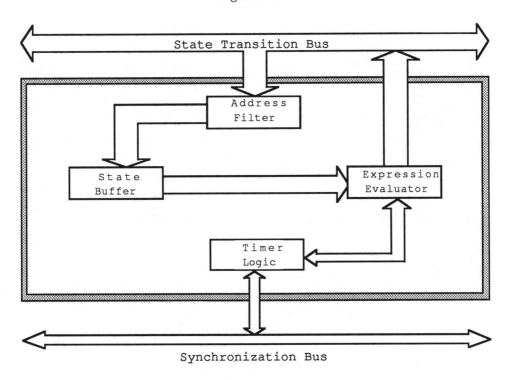

Figure 34.3

Figure 34.4

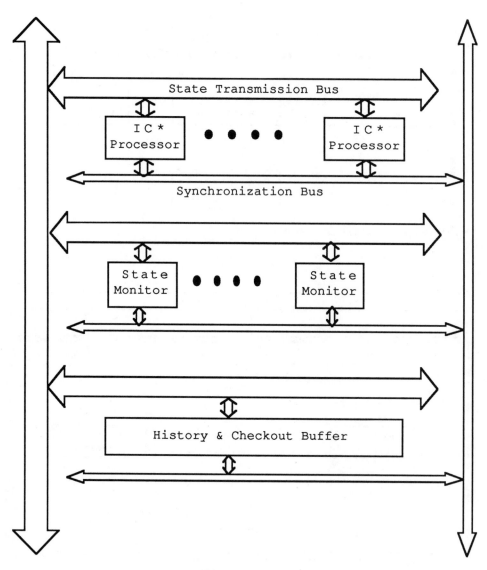

Figure 34.5

Many cause predicates involve state variables that are related to the system time. Often this relationship is satisfiable only when the system time reaches a value that can be computed at the time the invariant is activated. The exact triggering time of the deactivator is then known at the activation time. To avoid having to check these cause predicates at every unit of time, the synchronizer has a timer mechanism that will notify the processor the next time one of them can be true.

At any change in system time, if every variable in a cause predicate is either linked to system time or else has the same value as before, the cause may not have to be checked. If none of the cause predicates assigned to a processor has to be checked, then the processor will be screened by the synchronizer and name translator hardware, and will be allowed to calculate effect predicates for a future time without being interrupted to calculate cause predicates.

34.5   Conclusions

Although the IC* model has similarities to classical rewriting systems [4], and their extensions [5], a better analogy is the theory of differential equations on a surface because there is an explicit notion of time and many invariants can simultaneously determine the next state of the system. The model gives a precise meaning to parallel coordinated computation. The model's inherent fine grained parallelism has been very useful in writing specifications of communications systems. A complex system can be specified incrementally by using references and static invariants to compose descriptions of its component parts or of its different aspects. The partial, but still precise, description of a system can be meaningfully discussed, documented, analyzed, and even simulated or implemented. Having an explicit notion of state and time allows clear reasoning about the paths of the system. The model's mathematical foundation supports tools for the production of documentation and analysis directly from the IC* specification and for the automated translation of an IC* specification into a simulation or prototype implementation. The simulation environment supports real-time interaction between a user and a running simulation. This research is aimed at understanding the interplay between parallel programming and parallel architectures with special emphasis on design and implementation of complex systems.

References

[1] D. M. Cohen, B. Gopinath, M. L. Honig, W. M. Keese, P. Levin, J. Myers, U. Premkumar, D. Slepian, and J. R. Vollaro, "I*C: An Environment for Specifying Complex Systems", IEEE Global Telecommunications Conference, Houston, Texas, Dec 1 - 4, 1986.

[2] D. M. Cohen and E. J. Isganitis, "Automatic Generation of a Prototype of a New Protocol From its Specification," IEEE Global Telecommunications Conference, Houston, Texas, Dec 1 - 4, 1986.

[3] C. H. West, "Protocol Validation by Random State Exploration", Protocol Specification, Testing, and Verification, VI, North Holland, 1986.

[4] A. Thue, "Ueber Unendliche Zeichenreihen", Skrifter utgit av Videnskapsselskapet i Kristiania. I, 1906, pp 1-22.

[5] A. Salomaa, "Computation and Automata", Encyclopedia of Mathematics and its Applications, Volume 25, Cambridge University Press, 1985.

Chapter 35

The Distributed Macro Controller for GSIMD Machines

W. Holsztynski [1]
R. Raghavan [2]

Abstract

The Distributed Macro Controller (DMC) is a computer architecture which has been implemented as a controller for Geometric Single Instruction Multiple Data (GSIMD) machines. This note contains a presentation of the major characteristic features of the DMC, such as the modularity, macroprogramming (including the Cartesian factorization of the macros), memory management, usage of stacks versus randomly accessed variables. Also, we discuss some research questions that the DMC suggests. One of the themes of the DMC is that understanding and compression are nearly synonymous terms.

35.1 Characteristic Features of the Distributed Macro Controller (DMC)

We first discuss the philosophy of the DMC, invented by W. Holsztynski, and some research issues in information processing that the DMC addresses. The next section contains a description of an embodiment of the DMC that we have constructed, and may be referred to for concrete details. One of the themes of the DMC is that understanding and compression are nearly synonymous terms.

---

[1] Consultant, Lockheed R&DD, Palo Alto, CA
[2] Lockheed R&DD, Palo Alto, CA

Existing digital computational devices may be regarded as finite state automata which at each instant of time, or clock, compute (i) a transition function to a new internal state and (ii) perhaps produce an output. These computations are based on the present state and the input. It would be natural if the logical hardware of the machine were also divided into two components performing these two tasks. Thus these two components form a simple distributed system. The strength of the classic von Neumann architecture is also its weakness. Since the same hardware performs both functions, speed is sacrificed for hardware savings. The division just mentioned may also be regarded as the division into (an electronic) strategist and tactician. If delays are to be avoided completely, then each step of implementation by the tactical component must be long enough for the strategist to be able to devise the next plan. By a similar token, the strategic plan cannot take too long, or be too detailed. The DMC embodies this strategy; the lower bound on the duration of macros is three cycles, allowing penalty free execution of subroutine calls, macro calls, loops, memory management and other high level tasks.

The DMC is a configuration of Flow Control Units (FCU's) and Macro Generator Units (MGU's) serving as the "strategists" and "tacticians" respectively (Figure 35.1). The organization, though invoving multiple processors, may be considered as an evolution of the von Neumann model, rather than a radical departure from it. The language of the DMC is not a high level language, by the usual standards. However, it has many features, including the ability to overlap macros in the memory, that allow for considerable code compactness.

An important difference between the DMC and other controllers is the flexibility of address generation. Most existing controllers stress the generation of op-codes, and leave the difficult problem of generation of addresses to pre-computation. Also, the DMC is unique in its memory management features implemented in hardware, which is (primarily) the allocation and the freeing of memory without loss of machine cycles (always with the proviso that macros are at least three cycles long). Thus unpredictable program paths are allowed for in the presence of limited memory per processor. Memory management considerations, where memory accesses may depend on the input and cannot be computed in advance, suggest the following research questions. What is the mean time for which the main memory of a processor is sufficient ? (assuming ideal, instantaneous memory management). The same question may be asked for a particular scheme of memory management. The ratio of these times would be a measure of the efficiency of the management scheme.

Another interesting issue for study is the following. While the assumption of three cycle macros is sufficient for programming without loss of speed in almost all applications, the difficulty of generating such optimal code means that many programs are conveniently written only by sacrificing some speed. Thus, there is a trade-off between the compactness of the final machine code (i.e. the speed of program execution) and the modularization of the program, allowing easy debugging. In all cases, we have studied this trade-off before making a decision, and in about 90% of

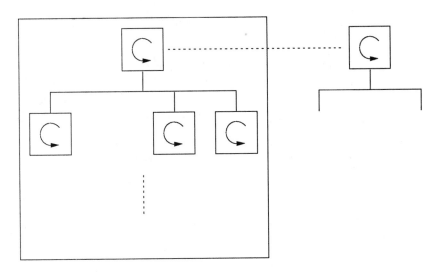

Figure 35.1: Hierarchical Control Systems. An evolution of the von Neumann architecture. Each box is a von Neumann Type machine: i.e. self-referential as indicated by the curved arrow. The multiple processor systems follows a human management-type hierarchy.

the cases, we do provide the ideal DMC code.

The DMC uses many features, including the use of stacks, to promote flexibility and compression. Stacks are useful for variables that are accessed rather locally in time ("short term memory") while random access memory is more useful when variables are required in disconnected or separated intervals in time ("long term memory"). A research issue suggested by the use of stacks is a form of program and data compression. Namely, by allowing frequently used instructions to stay near the top of the stack, and studying the time for access of the little used instructions, residing perhaps in an external memory, one can achieve program compression that learns from its recent history. The same philosophy can be applied to study forms of data compression.

The DMC considers macros in a new way, namely as a subset of a Cartesian product of macro components. It uses multiple macro generators allowing one, in principle, to construct many combinations from a few macro components. The particular form that we have used for the GSIMD array using GAPP (Geometric-Arithmetic Parallel Procesor) chips uses two macro generator units for the addresses and the op-codes respectively (described in more detail below). Thus one op-code component may be synchronized with various address components and vice-versa. Apart from advantages of compression, constructions common to one type of processing can be interpreted and adapted for the other. The amount of compression that such a factorization achieves remains a matter for research. Development of these ideas may also lead to new forms of program construction, hardware etc.

Re-interpretation is another construction which is useful and enforces a better understanding of macros. Thus for a GSIMD machine it is natural to store shifts along only one direction, the other directions then being generated by re-interpretation. Many geometric, arithmetic (addition versus subtraction) and pattern recognition (template matching) symmetries are obvious and can be taken advantage of.

In brief, the DMC has many features allowing optimization of code for parallel machines, and suggests directions of research in computer science.

35.2   The Parallel Processor System

We have constructed a prototype, which we now describe, of our parallel system, schematically shown in Figure 35.2. In this description, we shall focus primarily on the DMC implementation that is fully functional. Processing within the parallel processor system is completely self-contained so that once started by the host, program execution can proceed independently. Our present system loads data via either a DMA (Direct Memory Access) channel in the host computer, or in real time through a video RAM (Random Access Memory) linked to a fast disk. The results can be unloaded similarly to the host or to a real time video display unit. A software console program was developed for use in the host computer to control the parallel processor system interactively.

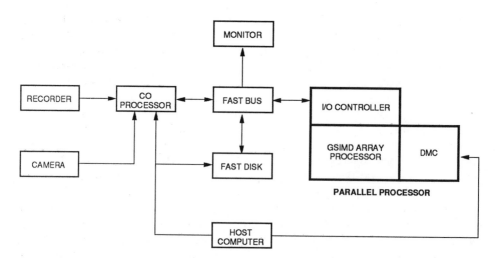

Figure 35.2: Block diagram of parallel processing system. Peripheral equipment including data sources for real time applications are also indicated.

35.2.1 Processor Array

The processor array presently consists of several boards of parallel processor chips. We are currently using the GAPP [1]. Up to three arrays of chips can be separately controlled to make a MIMD (Multiple Instruction Multiple Data) system, or in simpler applications a SIMD machine. We are presently using a configuration of 22,000 processing elements, easily expandable to greater numbers.

35.2.2 The Distributed Macro Controller (DMC)

The DMC is meant to allow ready implementation of adaptive programming decisions made by the host; that is, without loss of machine cycles. The top level architectural innovation is that the controller is a MIMD machine that processes several different instructions streams simultaneously, as shown in Figure 35.3. A Flow Control Unit (FCU) feeds several (here two) Macro Generator Units (MGU). The instruction streams from the MGU's are combined to feed the control lines of the GAPP array. Each of these units would fit on a single VLSI chip when implemented as integrated circuits. All MGU's are identical. (The general construction of the DMC is, naturally, not limited to one FCU or two MGU's).

The FCU supervises program flow within DMC while the MGU's produce output instructions for the parallel processor array. The MIMD architecture is hierarchical; i.e., the Flow Control Unit directs the production of the programs from the Macro Generator Units. The controller offers a very high degree of program compression. Existing sequencers have wide microcode words but little program optimization or compression.

The Flow Control Unit allows nested subroutines and nested loops. While loops increment, the loop counts of interior loops can be changed. Subroutine calls and returns are performed in 3 clock cycles. With the provision that subroutines are at least three instructions long, this allows penalty-free macroprogramming. External inputs may be tested for conditional operatings (branching, looping, calling, and returning). The Flow Control Unit is designed to be a single chip, and can function as a sequencer although its primary function in the present controller is to direct the internal flow (within the DMC) of the program.

The Macro Generator Units, which are physically identical and each designed to be a single chip, have several features:

1. Callable macro and address routines,

2. Automatic memory management,

3. Static and dynamic reinterpretation logic, and

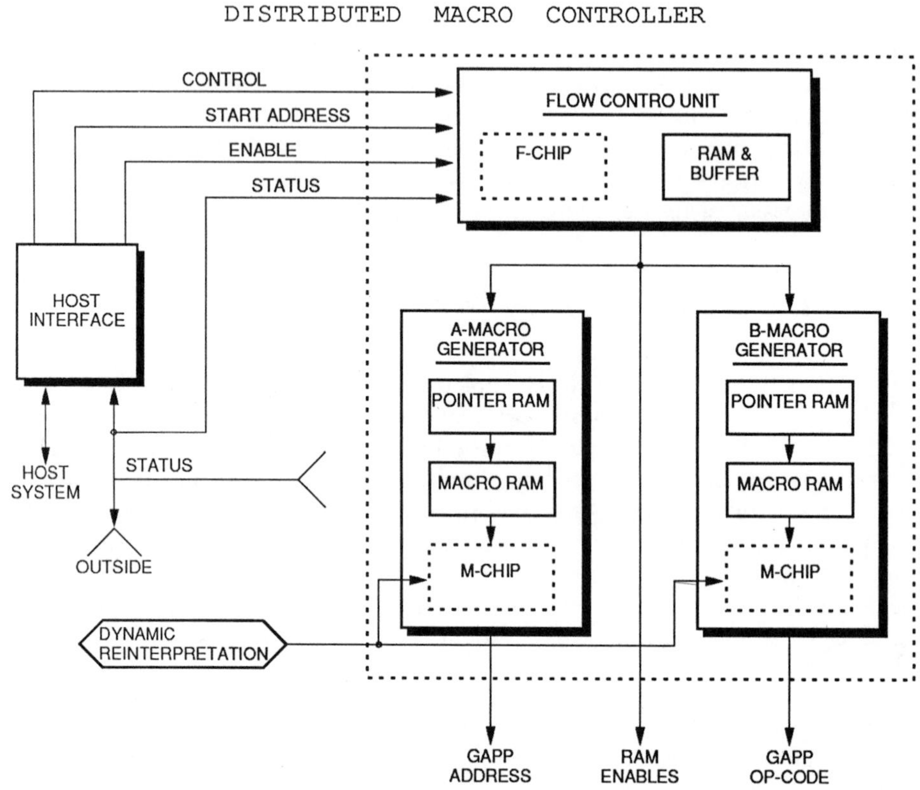

DISTRIBUTED   MACRO   CONTROLLER

Figure 35.3: Block diagram of DMC for GSIMD array of GAPP chips.

4. Several stack operations.

Feature (1) is for program compression. Pre-loaded instruction streams can be called by specifying a pointer and length. Typically, these instructions cause the Macro Generator Units to produce machine code indirectly.

Memory management calculates physical addresses given logical ones. Thus all of the memory addressing is indirect and penalty free. A linked list of memory segments with "occupied" and "free" areas is maintained. This handles allocation of memory and makes the task of the programmer much easier. Also, if these functions were to be performed in software, the processing system would not be able to operate at full speed. More importantly, pre-computation of addresses may be impossible in many real-time situations. Thus, the DMC allows an essentially new capability to processors.

Reinterpretation logic is useful both for program compression and for program speed, and is useful from both an op-code and memory point of view. The usefulness of reinterpretation as a program compression feature is that applications natural to GSIMD machines tend to be highly patterned. Addition, subtraction, and template matching to either zero or one differ only in the selection of CARRY and BORROW; loops proceed by alternately selecting one stack or another; filters and multiplication of an array by a constant differ only in shifts of the bit pattern denoting a number, and subsequent addition or subtraction, etc. A study of the class of transformations natural to GSIMD machines reveals the frequent occurrence of such patterns allowing switching between instructions with the use of reinterpretation bits. Reinterpretation of address bits allows symmetric operations within address space. This feature directly leads to program speedup: the if-then-else construction becomes available to parallel processors without penalty.

Finally, a rich set of stack operators is also provided by the controller. One can operate simultaneously on two stacks holding address pointers to the processor memory. Stacks offer a way of changing the instruction sequence in nonconsecutive or nonlinear ways.

35.3 Programming the DMC

We have previously described the language of the DMC [2,3,4], as well as provided details on our hardware. Here, we only mention the most characteristic features of the machine language of the DMC. As mentioned, the FCU, is the strategist, and its instructions include: specification of macros in the MGU, subroutine calls within the FCU, loop calls and memory management operations for banks of memory. The MGU specifies memory management for bits, as well as generating the machine code for macros specified from the FCU. Other operations include re-interpretation logic specified by a string of bits either from a memory location or from external (real-time) input.

In conclusion, the realization of the DMC for our parallel system, allows for considerable code compactness without sacrificing the speed of the parallel processor. Just as a caricature of a conventional sequencer may view it as all memory and little logic, a corresponding one for the DMC would look at it as a lot of logic and little memory. Thus, a conventional sequencer can run all programs of upto a given length and no longer ones. Conversely, the DMC can run only a small fraction of possible machine instructions of any length, but can in principle run programs of infinite length without halting. The detailed statistics proving the significance, or otherwise, of this philosophy of programming parallel processors remains to be delineated

References

[1] W. Holsztynski, Canadian Patent 1,201,208 Issued: Feb 25, 1986, *Geometric Arithmetic Processor.*

[2] *Geometric Single Multiple Data Processing,* W. Holsztynski, R. Raghavan, H. T. Nguyen and C. H. Ting, Parallel Architectures Group Report No. 4, Lockheed R&DD, September 1986.

[3] *High Density Parallel Processing,* H. T. Nguyen, R. Raghavan, C. H. Ting and H. S. Truong, Proceedings of 7th Rochester Forth Conference on Advanced Architectures, June 1987.

[4] *Compiler for the Distributed Macro Controller,* Parallel Architectures Group Report No. 19, Lockheed R&DD, August 1987.

Chapter 36

The Linda Machine

Venkatesh Krishnaswamy [1]
Sudhir Ahuja [2]
Nicholas Carriero [1]
David Gelernter [1]

Abstract

The Linda Machine is a parallel computer that has been designed to support the
Linda parallel programming environment in hardware. Programs in Linda communicate through a logically shared, associative memory called tuple space. Physically-
shared memory seems, however, to be a more complicated and less scalable basis for
a multi-computer architecture than distributed memory, and the goal of the Linda
Machine project is accordingly to implement Linda's high-level shared-memory abstraction efficiently on a non-shared-memory architecture. We describe the machine's special-purpose communication network and its associated protocols, the
design of the Linda coprocessor and the way its interaction with the network supports global access to tuple space. The Linda Machine has been designed and is
the process of fabrication. We discuss the machine's projected performance and
compare this with software versions of Linda.

[1] Yale University, Department of Computer Science, New Haven, Connecticut
[2] AT&T Bell Labs, Holmdel, New Jersey

36.1   Introduction

The Linda Machine is a scalable, MIMD parallel processing system that supports
the Linda parallel programming environment in hardware. It consists of a collection of *Linda nodes* interconnected by a grid of buses as shown in Figure 36.1. Each
Linda node consists of a computing element and a *Linda engine*. The Linda engine,
in co-operation with the customized interconnection network, execute the communication operations on behalf of the computing element. Linda supports asynchronous
communication between multiple independently executing strands of a parallel program. An experimental prototype of the Linda machine is under construction at
AT&T Bell Laboratories. The configuration can support upto 80 single-board nodes
each with a MC68020 as the computing element. In this paper, we describe the design issues and decisions involved in deriving an architecture for a Linda machine
and discuss some specifics of our implementation. We also discuss the projected
performance of the machine, comparing it to existing software versions of Linda.

Considerable research on architectural support for higher-level parallel computation
models has been reported in the literature. A wide variety of parallel computation
models have been studied including shared memory architectures [G82], message
based architectures [W85], object oriented architectures [D87], dataflow machines
[GKI85] [AGP78], systolic architectures [F83], concurrent prolog machines [W87],
and data-parallel architectures [H85]. The Linda Machine project addresses a set
of implementation challenges that makes it unique and distinct from each of these.
These challenges stem from the fact that Linda is intended to make the task of the
parallel programmer easier by providing him a simple, high-level abstract machine
model. The machine model shields the programmer from the realities of the underlying machine architecture and is designed to enable him to express the parallelism

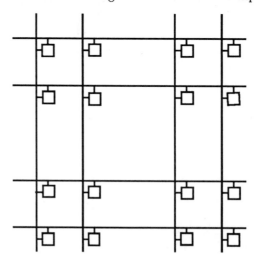

Figure 36.1: The Linda Machine Grid

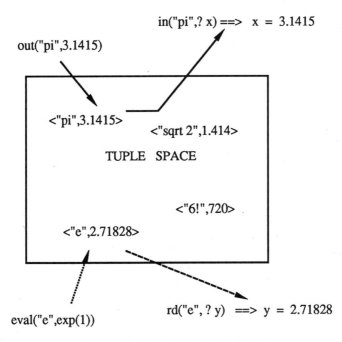

Figure 36.2: The Linda Model

inherent in the application well. The performance of Linda programs is, therefore, critically dependent on its system implementation. We explain these challenges by first describing the Linda model and then discussing the set of requirements for its efficient implementation.

Linda processes communicate through a globally shared associative memory mechanism called *Tuple Space*. The unit of communication is a *tuple* - a self-contained object containing an arbitrary number of data elements. Processes communicate by adding, reading and withdrawing tuples from tuple space through the operations out(), rd() and in() respectively. All tuples are accessible to every process. Tuples, however, do not have addresses; they are identified by content. Hence a process that wishes to read or withdraw a tuple must present a template that matches that tuple. Note that an in() or a rd() request will be satisfied by *any* tuple that matches the corresponding template. If at the time of the request no matching tuple is found, the request blocks until one is deposited in tuple space by an out(). Templates may contain either actual parameters(data values that must be matched exactly) or formals, which specify a data type. The formals in the template are filled in from the corresponding data fields in the matching tuple. Figure 36.2 illustrates the Linda model.

From the description of the model we stipulate the following as requirements for a machine architecture for Linda:

- It should provide uniform access and a consistent view of memory from all the processors in the system.

- It should store variable sized and multi-element objects compactly and be capable of delivering these to requesting processors with minimal latency.

- It should incorporate an efficient search mechanism to locate and retrieve objects requested by processors.

In addition to these, we impose the conditions that it should scale to hundreds or even thousands of nodes and be implementable in conventional technology. The challenges are immediately apparent: How to build a shared memory system that scales to the numbers we mention? Clearly, a single bus shared memory system will not suffice – how, then, do we emulate shared memory on a network with distributed memory and how do we ensure that it provides a coherent view to all the processors? Linda encourages programs that completely disregard the machine architecture; how do we, therefore, optimize for highly arbitrary communication patterns including multicasts and broadcasts? To complicate matters, communication in Linda programs is asynchronous – how do we deal with memory requests that may have to wait a while before they can be resolved? How to represent tuples compactly and build hardware that stores and accesses these efficiently? How do we build an associative matching system that scales to very large memory sizes and arbitrarily complex patterns? We provide, in this paper, an empirical solution to the questions posed by describing the architecture of a machine that incorporates innovations in network design and node architecture and in compiler technology within an integrated framework.

We stress that this paper does not attempt to defend Linda – many previous papers have argued its case. Linda and its software implementations have been described extensively in the literature and it has been shown to be an expressive tool for parallel programming [ACG86] [CGL86] [BCG87] [Ca87]. It is particularly appropriate in domains where (a) a highly asynchronous or "uncoupled" style of parallelism is useful, or (b) a programming language that differs (in a syntactic sense) only slightly from a sequential base language like C or Fortran is desirable, or (c) a parallel language that shields the programmer from the details of the machine architecture is useful.

This rest of the paper is organized as follows: in Section 36.2 we give an overview of Linda and compare the model with other models of parallel computation. Section 36.3 describes the architecture of the Linda machine and the design decisions involved therein. In Section 36.4, the performance of the machine is discussed and Section 36.5 concludes the paper with a summary and status of the project.

36.2 The Linda Model

Tuple space holds two kinds of tuples. Process tuples are under active evaluation; data tuples are passive. To build a Linda program, we ordinarily drop one process tuple into tuple space; it creates other process tuples. The process tuples (which are all executing simultaneously) exchange data by generating, reading and consuming data tuples. A process tuple that is finished executing turns into a data tuple, indistinguishable from other data tuples.

Tuple space is accessible via four basic operations: out(), in(), rd() and eval(). out(*t*) causes tuple *t* to be added to tuple space; the executing process continues immediately. in(*s*) causes some tuple *t* that matches template *s* to be withdrawn from tuple space; the values of the actuals in *t* are assigned to the formals in *s*, and the executing process continues. If no matching *t* is available when in(*s*) executes, the executing process suspends until one is, then proceeds as before. If many matching *t*'s are available, one is chosen arbitrarily. rd(*s*) is the same as in(*s*), with actuals assigned to formals as before, except that the matched tuple remains in tuple space. eval(*t*) is the same as out(*t*), except that *t* is evaluated after rather than before it enters tuple space; eval implicitly forks a new process to perform the evaluation.

Tuple space is an associative memory. Tuples have no addresses; they are selected by in or rd on the basis of any combination of their field values. Thus the five-element tuple (A, B, C, D, E) may be referenced as "the five element tuple whose first element is A," or as "the five-element tuple whose second element is B and fifth is E" or by any other combination of element values. To read a tuple using the first description, we would write

rd(A, ?w, ?x, ?y, ?z)

(this makes A an actual parameter – it must be matched against – and w through z formals, whose values will be filled in from the matched tuple). To read using the second description we write

rd(?v, B, ?x, ?y, E)

and so on. Associative matching is in fact more general than this: formal parameters (or "wild cards") may appear in tuples as well as match-templates, and matching is sensitive to the types as well as the values of tuple fields.

The Linda programming paradigm we've found most useful so far involves *distributed data structures* and a bunch of identical worker processes (or several bunches of several different kinds of process) crawling over the data structures simultaneously. We use the term *distributed data structure* [CGL86] to refer to a data structure that is directly accessible to many processes simultaneously. Any datum sitting in a Linda tuple space meets this criterion: it is directly accessible to any process that currently occupies the same tuple space. A single tuple constitutes a simple distributed data structure. We can build more complicated multi-tuple structures (arrays or queues, for example) as well.

Tuple space has been implemented in software on a broad range of parallel machines, including AT&T Bell Labs' S/Net, the Encore Multimax and Sequent Balance shared-memory multiprocessors and the Intel iPSC hypercube. These implementations have been used to program real applications and have been shown to perform well [Ca87]. Why, then, build a hardware Linda Machine? These software versions run on machines that have tens of processors. Linda is implemented as a layer on top of a more basic (and often very primitive) communication mechanism. In the absence of a communication mechanism that is more directly suited to the Linda model, software implementations may not scale well to hundreds or thousands of processors. For example Lucco [Lu86] reports that, in the Linda implementation on the Bell Labs Hypercube, up to 80% of time spent in the Linda kernel represents software overhead for communication. This overhead represents a potentially serious bottleneck for machines with a large number of nodes. Many approaches are evolving to deal with the software latency bottleneck in parallel computation; almost all of them recognize the need for a co-processor dedicated to interprocessor communication. Each node in the Linda Machine has such a co-processor that largely replaces the software layer that executed tuple space operations in previous incarnations of Linda. The Linda Machine, however, is more than an attempt to add a few hardware features to a conventional architecture in order to grease the wheels of software Linda. It represents an integrated approach to designing a new kind of parallel computer. Although our prototype won't grow larger than roughly a hundred nodes, the architecture we describe forms the framework for a machine with hundreds and potentially thousands of powerful computing nodes.

The Linda Machine supports tuple space on a foundation consisting entirely of local, per-node memories - no memory is physically shared among processors. It is designed as a grid of processor nodes: broadcst buses form the rows and columns and a processor inhabits every cross point. What the program sees as a tuple object, the machine realises as a *tuple beam* that spans an entire row of processors: the tuple's image has been broadcast and is now stored by each node in the row. When a program needs to find a matching tuple, the machine flashes an inverse beam along a column. When an inverse beam intersects a matching tuple beam, both beams disappear and a tuple is delivered to a tuple requestor as is shown in Figure 36.3. (The dashed box in the figure indicates the node on which the request broadcast along the second column from the left is resolved. The matching tuple was broadcast along the second row from the bottom.) The Linda Machine's tuple space, then, doesn't occupy any particular box. It is distributed, instead, throughout the parallel computer and distributed in a strong sense: not only is tuple space distributed but each individual tuple inhabits many nodes simultaneously.

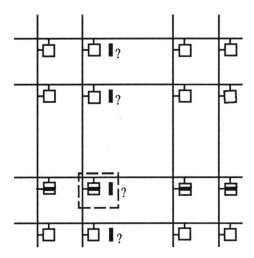

Figure 36.3: Intersection of tuple beams.

Before describing the Linda Machine's design, we take up a basic question: why is Linda a good choice for a parallel machine language? We explain, comparing it with some competing parallel programming models.

Linda as the Unconnection Machine. The field of parallel-programming models is too broad to survey here, but it's useful to contrast Linda with two influential and interesting systems that represent very different approaches. Occam is a parallel language based on Hoare's CSP; it is one of the few languages that has (like Linda) inspired a multi-computer architecture (the Inmos Transputer) designed in top-down fashion to support it. The Connection Machine project seemingly proceeded in the more conventional direction: an architectural model inspired the construction of a new kind of machine, and new programming languages were subsequently cut to fit.

Occam and the Connection Machine's "data parallel" languages are widely different; but they also have something significant in common, which sets them both apart from Linda. They both tend to bind the concurrent elements of a parallel program tightly together, whereas in Linda the opposite is true. The elements of a Linda program are as *unconnected* as possible.

Why is this desirable? In Occam, for example, processes are tightly connected both spatially and temporally: spatially, insofar as output data must be sent to a specific location (a particular channel); temporally, insofar as a "send" operation blocks the sender until a data transfer is complete. But this kind of tight coupling tends (we believe) to force programmers to think in simultaneities. As far as possible, we would like programmers to be able to develop the code for any given process without having to envision other simultaneous execution loci. Specifically, we want it to be possible for processes to develop output data and release it into the system without knowing or caring which other processes will accept this data as input: that way, we're free to change communication patterns transparently, or to let them develop and change dynamically at runtime. Nor is it acceptable for a data-producer's forward progress to depend on the consumer's. In Occam, the producer stops until the consumer takes delivery; in Linda, the consumer process might not even exist when the producer runs. The consumer might take delivery three weeks later: the producer doesn't care.

The Connection Machine eliminates Occam's tight-coupling by doing away with explicit process interaction altogether. One operation is applied to an entire data structure in parallel; processors execute the same instruction stream in lockstep. The result, however, is tighter synchronization then some problems seem to need. For example: we wrote a C-Linda version of the factorization step of the Dongarra Linpack benchmark. This program shows good speedup on the Encore Multimax as we add worker processes, up to ten workers (the maximum we tested). The program repeatedly reduces the columns of a shrinking sub-matrix against a pivot column. One reason it performs so well is that, under some circumstances, a given worker can plunge on into the next iteration before other workers have finished the last one. Linda's uncoupled character encourages asynchronous solutions of this sort, solutions in which individual processes are allowed to charge ahead with a minimum of supervision from the home office. It also encourages solutions involving processes of various unrelated types, and, of course, Linda programs are candidates for execution on a variety of multicomputers – which may even include inhomogeneous nodes.

The loose coupling of Linda's processes also enables efficient use of processor resources. It is easy to float tuples representing work-assignments from different sub-problems, or even unrelated program runs, in one tuple space. This in turn makes it possible for evaluator processes to choose sub-problems to work on dynamically, based on supply and demand. This in turn makes it possible for the system to partition its processor resources dynamically among competing tasks – a goal which seems harder to accomplish in Occam or on the Connection Machine.

There's a final respect in which Linda differs both from the Occam machine and from the Connection machine. A tuple space is an object with an independent existence. It may be stored in a file system; it may be reactivated or operated-on any number of times. It may even overflow the boundaries of its multicomputer of origin to take up residence on another machine, or on many other machines simultaneously.

36.3 Linda Machine Architecture

Each node in the Linda machine executes one or more processes of the user program. Tuple space is implemented over the network formed by the Linda coprocessors in-

terconnected by the grid of buses. It is physically distributed across the nodes of
the network and is stored in the local memories of the Linda coprocessors. Processes executing on a node present their out(),in() and rd() requests to the Linda
coprocessor which executes the corresponding tuple space operation[3]. Each of the
Linda coprocessors stores a segment of tuple space in its local memory; to retrieve
a tuple not in its segment, a Linda coprocessor accesses the network using protocols
built into the microcoded hardware. These access protocols reflect the particular
scheme used for distributing tuple space.

36.3.1 Designing a Distributed Tuple Space

Linda is inherently a shared memory model. However, for reasons of scalablity of a
machine architecture, we decided against a machine with a physically shared memory and concentrated instead on mapping tuple space on to a distributed memory
machine. Supporting this is a problem with no single obvious solution. Broadly,
however, there are two approaches - hashing and uniform distribution. Under the
first, tuples are mapped onto individual nodes by hashing on the tuple format or on
one or more fields. Retrieving a tuple is then a matter of locating a specific node
in the network on which a matching tuple is likely to reside and searching among
the tuples there to find an exact match. The uniform distribution schemes attempt
to spread the tuples uniformly through the network. It is not possible, therefore,
to predict which specific node will house a particular tuple; finding a tuple involves
searching on several nodes.

Bjornson et al. [BCG87] and Lucco [Lu86] report experiences with hashing schemes
in software Linda implementations on hypercubes. These experiments have been
fairly successful; however, their scalablity is in question. In particular, Lucco reports
that extensive runtime heuristics are required to fine-tune the performance of these
schemes. The problem is that it is not easy to distribute the communication load
evenly under these schemes. This assumes greater importance as the size of the
network increases – tuple traffic increases and hot-spots caused by non-uniform
loading can become a major bottleneck. Uniform distribution offers better promise
of balancing the communication load in a scalable way; hence we have focussed on
this in deriving a communication architecture for the Linda Machine. We show that
this choice enables us to exploit parallelism among separate tuple-space operations
and also *within* each operation, and is particularly well-suited to strong hardware
and low-level protocol support.

---

[3]Linda considers processes to be tuples also and provides the primitive eval() for evaluating
them. In the Linda Machine, however, the tuple space network is designed not for computation but
to move data around. Hence we distinguish between processes and tuples at an implementation
level, although there is no semantic distinction in Linda.

The central idea in uniform distribution is to partition the nodes of the computer into 'out-sets' and 'in-sets' such that the intersection of each 'in-set' with each 'out-set' contains at least one node. An out() operation causes a copy of the tuple to be installed in each node of its pre-determined out-set. To find a tuple, through an in() or a rd() operation, the template is broadcast to all nodes of an orthogonal, pre-determined 'in-set'. The in-set must contain at least one copy of each tuple in the system for this scheme to implement Linda correctly – the intersection condition guarantees this. There is a spectrum of uniform distribution schemes. At one extreme, each tuple is stored only on the node in which it is generated. The in-set must consist, therefore, of all the nodes in the system. At the other extreme, each tuple is broadcast to all the nodes in the network; each node then forms a complete in-set. Carriero and Gelernter [CG86] describe the implementation of these schemes in the software Linda kernel on the S/Net [Ah83] multiprocessor.

There is a third scheme, however, that is precisely intermediate between these two. Suppose there are N nodes in the network. To implement out(t), we broadcast t to $\sqrt{N}$ nodes (the out-set); to implement in(s) or rd(s), we broadcast s to a different $\sqrt{N}$ nodes (the in-set). Out-sets and in-sets must be designed in such a way that each in-set includes at least one member of each out-set.

This intermediate scheme has several advantages. Unlike the first two possibilities, it doesn't require network-wide broadcasts, which – particularly on large networks – can be expensive. The number of nodes required to participate in a given out(t)-in(s) transaction are minimal in the intermediate case. The first two schemes each require that N nodes participate in every such transaction (because they each require a network-wide broadcast, by the node responsible for the out in the first case, by the in-performing node in the second). The intermediate scheme requires only $2\sqrt{N}$ node-participations: $\sqrt{N}$ nodes when an out is performed, a second $\sqrt{N}$ nodes on in. (By this criterion, in fact, the intermediate case is provably optimal even if we consider the entire spectrum of uniform-distribution schemes [Gel84].)

Having decided that the intermediate scheme is a good choice, we notice that there is an easy and convenient way to implement it on a network of processors. If we build a $\sqrt{N} \times \sqrt{N}$ grid of processors, each processor has exactly $\sqrt{N}$ nodes (itself included) in its row, and $\sqrt{N}$ nodes in its column. If we define each node's out-set as the nodes in its row, and its in-set as the nodes in its column, then out-sets and in-sets have the overlapping quality they are required to have: over the entire network, each in-set contains exactly one member of each out-set. Finally, we use broadcast buses to implement the rows and columns. Broadcasting a tuple to every node in an out-set now requires a single operation (a broadcast over a row-bus); tuple-request broadcasting works likewise, over the column bus. We have arrived at the Linda Machine's communication architecture.

It should be noted, though, that the story doesn't end with our prototype Linda Machine. The tuple-space implementation strategy outlined above will work on any network that can support the embedding of a grid. If, for example, we build a network in the shape of a 2^{20}-node binary hypercube, we can define in-sets and out-sets as consisting, for every node, of two 2^{10}-node sub-hypercubes that intersect at that node. These sub-hypercubes have the same characteristics as the rows and columns of the Linda Machine grid: each includes $\sqrt{N}$ nodes, and they can be chosen in such a way that each in-set includes exactly one member of each out-set.

36.3.2 Implementing tuple beams

A tuple that is **out**'ed is broadcast to every node in the row containing the **out**-performing node; a tuple-request generated by an **in** or **rd** is broadcast to every node in a column. Each node, then, maintains a local database consisting of some tuples (ones that were **out**'ed on this row) and some tuple-requests (ones that were **in**'ed or read on this column). Whenever a node finds itself holding a tuple and a tuple-request that match, it dispatches the matched tuple to the node that wanted it. In the case of an **in**, the matched tuple must first be successfully deleted from tuple space: if other processes want the same tuple, the network protocol must ensure that only one process gets it.

Maintaining a partially replicated database of tuples while allowing many processors to operate on it concurrently and independently requires the use of appropriate network protocols. These must ensure that each operation is performed correctly irrespective of other concurrent operations, and that the database is maintained in a consistent state.

To simplify the design of these network protocols and to minimize the transfer of administrative information between nodes, we made the following design choice: the database is stored identically across the tuple memory address spaces of the nodes in any given **out** set. In other words, a tuple is located at the same physical address in the tuple memories of all nodes in the set.

Specifically, the network protocols address the following problems:

1. Two or more processes in the same out-set simultaneously wish to install a tuple. The conflict arises because a tuple has to be written at the same address in all node memories.

2. Two or more nodes in the same in-set simultaneously find a match to the same request. Only one of these must be allowed to delete the matching tuple and satisfy the request.

3. Two or more nodes in different in-sets resolve requests to the same tuple and simultaneously wish to delete it. Only one node must be allowed to delete the tuple; the rest continue their search.

In each of these situations, distributed arbitration is required to resolve the conflict. The network protocols make use of the fact that processors have to contend for use of the broadcast bus. Hence the out() conflict is resolved simply by allowing the conflicting processors to contend for the out bus; the processor that wins does the first tuple installation.

The protocol for in() and rd() is a little more complicated. Not only must the Linda engine at a node process requests from its local CPU, it also has to respond to requests from all the other nodes in its column. Since all in() requests generated among the nodes of a column are broadcast to all the other nodes in the column, each node maintains a copy of a column-wide *in queue*. queue. To process a local request, then, the Linda engine first contends for the network in bus (so that it has exclusive access to the in queue) and then, when it has won control, broadcasts the request. Eventually every in processor in the column will get to work on this request. The first node to find a match sends a message over the in bus to all the other nodes instructing them to suspend their searching. If the request is a rd(), then the matching tuple is sent to the requesting node. If the request is an in() the matching tuple has to be deleted. To do this, the node on which the match was found must contend for its out bus. If the tuple is still there when the node wins the contention, it deletes the tuple and returns the match to the requesting node; otherwise it instructs all nodes in the column to reactivate the search.

To maintain the replicated tuple database along a row in a consistent state at all times, all update operations along the row bus are performed synchronously. In other words every update is seen at exactly the same time (or close enough) by all nodes. The single-access out bus implicitly resolves conflicts that may occur in memory allocation for a new tuple that is being installed. out()s generated simultaneously within one set must therefore be performed in sequence. Nodes in an in set, on the other hand, need only synchronize *after* resolving a request. Hence the search process may be performed concurrently across all nodes in a set. This property can be exploited to realize the full 'search bandwidth' of the network.

Linda's memory, like other shared memory systems, requires that global consistency of data be maintained at all times. The scheme we have described represents one solution to the problem of data consistency. The problem is similar to the cache coherency problem in multicache shared memory systems which has been addressed by many researchers [BD86]. On the Linda machine, coherency must be maintained at the level of individual tuples rather than at the level of memory pages. Our problem is simpler because the Linda model addresses the problem of synchronizing access to shared data at the program level. It disallows *in situ* modification of data in the shared memory; rather it forces a process to declare 'ownership' of a datum by physically hauling it out of shared memory. The coherency protocols, accordingly, need only address the issue of atomicity of data additions and deletions, not data updates. This atomicity is implicitly achieved by virtue of the fact that a processor wishing to perform a destructive memory operation must first claim exclusive use to the appropriate network bus.

The question then arises: can the row bus become a bottleneck? The bus can support the installation of a 32-byte tuple every 4μsecs. On a bus with 15 nodes, this means that each node can generate an out() or a delete every 60μsecs under maximum bus loading conditions(roughly 200 instructions on a high performance microprocessor). By itself, this indicates that the synchronization overhead due to bus contention is not very significant. In addition, it must be pointed out that each out() in the program has a corresponding in() (and possibly many rd()'s as well). These operations require tuple matching which does not incur any overhead on the row bus but which does cost node processing time. In complex programs which generate large numbers of tuples, the search time on the nodes can be expected to overshadow the communication time. Under this scenario, it is extremely unlikely that the row buses will be under sustained high load during the lifetime of a program. A similar argument holds for the column buses as well. Our estimate is that the Linda Machine configuration will scale (using the above protocol) to about a thousand nodes without the buses being a limiting factor.

36.3.3 Operation of the Linda Machine Node

As we mentioned above, the Linda node has two components, the computing engine and the Linda engine. The computing engine runs Linda processes; any conventional processor may be used for this purpose. (The Linda Machine prototype uses MC-68020's.) It is the operation of the Linda engine that is of interest here.

The Linda engine consists of a tuple memory and a working store, the operations controller and the interfaces to the buses and the processor. A memory controller regulates access to the tuple memory and the working store. Figure 36.4 shows a schematic.

The Linda engine processes in() and out() requests generated by the computation processor and co-operates in the processing of requests from other nodes in the in and out sets. It maintains two operations queues in the working store. The in queue holds pending in() requests received from the computation processor or from any other node in the in set. The out queue holds pending out() requests from the computation processor. The operations controller consists of two independent units, the in and the out processors that work through the in and the out queues respectively.

The computation processor initiates a request by writing it to a statically assigned area in the Linda engine's working store. Associated with this request is a *status word* that the computation processor may poll to determine the completion status of the request. The appropriate unit of the operations controller updates this word when it is working on the request.

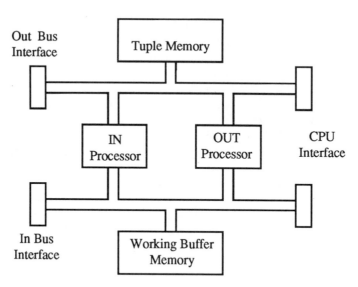

Figure 36.4: The Linda Engine

In servicing an out() request, the operations controller first contends for the out bus. When it wins control, it allocates space for the new tuple in tuple memory, and then writes the tuple into the memories of all nodes in the out-set simultaneously. Recall that the tuple is written at identical memory locations across all the memories. By gaining control of the out bus before it allocates space, the operations controller ensures that there is no memory conflict across the nodes in the out-set. As a consequence, all nodes in a row store the same tuple memory state at all times.

In servicing an in() request, the controller's first action is to ship the request off to all nodes in the in-set, causing the request to be appended to the in queues of all these nodes. Hence, all nodes in an in-set (i.e. in the same column) store an identical in queue. The controller then matches the new template against the tuples in a series of specific tuple sub-groups. (The C-Linda compiler partitions tuples into classes of similar type and "matchability". Hence we are able to restrict runtime tuple-matching based on compile-time tuple analysis. This system is already in use in the Encore and Sequent implementations [Ca87].) Meanwhile, all other nodes in the in-set eventually get to work on the same request. When a match is found on any of these nodes, the operations controller on that node resolves any conflicts that may arise according to the protocol described in the previous section. If no match is found immediately, the tuple request is kept pending on the in queue. When a match is eventually resolved, the matching tuple is returned to the requesting node (again at a pre-determined area of the working store), and the status word updated accordingly. The computation processor picks up the tuple from here.

36.3.4 Tuple Storage and Retrieval

Tuples are stored in tuple memory as self-describing data structures. Storage is allocated for tuples by the coprocessor in units of 128-bit words. Each word is sub-divided into four 32-bit fields. Words are linked together by pointers in a data structure that has sufficient space to store all the data and control information needed to process the tuple. The backbone of this data structure is a *tuple hanger*. Figure 36.5 illustrates the representation of the 5-element tuple [10, fml int,9.2,"acceleration",fml float]. The first field in the hanger is a header that contains the following information: the number of data elements in the tuple, single bit tags to identify each element as formal or actual, and single bit tags to identify each element as immediate or indirect. All formals and actual 32 bit integers are identified as immediate data; all other actual data types are identified as indirect. Actual indirect data is stored in linked lists of memory words. The hanger field for an actual indirect element contains a pointer to a linked list storing the element. In the example, the first two elements are immediate – the first an actual and the second a formal. The third and fourth elements are indirect actual fields. The first sub-word of the indirect element stores its type and size. The fifth element in the example tuple is a formal also; the field stores the type.

Note that the header seemingly allows for no more than 8 elements in a tuple. Tuples with greater than 8 elements are stored in *tuple hieirarchies*. Any field of the representation of a tuple may contain a pointer to the representation of another tuple. Hence in the representation of a 10-element tuple, it may be broken into two tuple objects, a 8-element tuple one of whose elements is a tuple containing the remaining 3 element. The extra field in the 8-element tuple is tagged to indicate that it points to another tuple.

Why the choice of 128 bits as the word size for memory? In the Linda programs we've seen so far, tuples most commonly contain between two and five elements. Non-integer elements are typically floating point numbers or short strings. This format stores the commonly occurring tuples with little overhead for pointers; one or two words are required for the hanger and most non-integer elements can be represented in one or two words. The tagged representation of data is reminiscent of architectures for Lisp [M85] and object-oriented languages [BSUH87]. The tagging scheme used for tuples is simple and compact and fits in with a standard 32-bit data format that makes the coprocessor compatible with commercially available microprocessors.

The Linda engine allocates storage out of a heap of free words. At program initialization time, all the available words in memory are linked together into a free list. Words are returned to this list on being deallocated. The algorithm for reclaiming words depends upon the distribution protocol; in a scheme with no tuple replication, storage may be reclaimed when the tuple is withdrawn from tuple space. Garbage

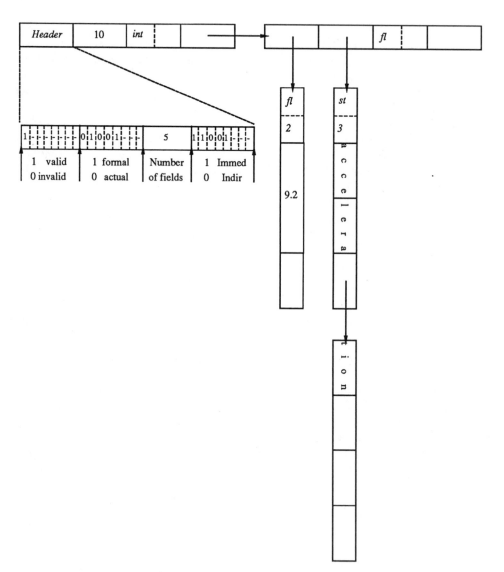

Figure 36.5: Tuple Representation

collection is more complicated under the schemes that involve tuple replication. We do not discuss this further here.

36.3.5   Linda Machine Hardware

The prototype Linda Machine now under construction uses single board Linda nodes hooked up through backplane buses along one dimension and ribbon cables along the other. All backplane buses are connected to a host computer which oversees the initial distribution of code and handles I/O. (It is our eventual plan to have an I/O system to attach to each backplane bus). Each Linda Machine board contains both the computing processor and the Linda engine and uses off-the-shelf MSI and LSI components. The decision to use off-the-shelf components was made so that we could have a working prototype as soon as possible. The architecture of the Linda engine was, however, designed with a view to our ultimate goal of a VLSI implementation.

The **in** and **out** processors have a simple datapath architecture controlled by horizontally microcoded instructions. Each processor has six address registers (three of these can address tuple memory and the other three can address the working store) and 3 special purpose registers. As shown in figure 36.3, there are two separate 32-bit datapaths connecting the processors with the tuple memory and the working store. Memory is implemented using high performance CMOS static RAMs. The instruction sequences for executing the tuple handling protocols are stored in PROMs and instructions are issued once every 60ns cycle. Each memory access takes two cycles; the instructions are sequenced such that there are two to four cycles of processing for each memory access (involving one or both of the memories). Thus available memory bandwidth is used efficiently to accomodate requests from the **in** and **out** processors, the bus interfaces and the CPU. Each node in the prototype has 256Kbytes of memory to store tuples and 64Kbytes for the working store.

The **in** and **out** bus interfaces provide a one-word buffer for bus messages. The bus interfaces operate as slaves of the **in** or the **out** processor when either needs to send a message. The processor reads the message out of memory and ships it off directly on the bus via the one-word buffer. When an interface is receiving messages from the bus, though, it operates independently of the **in** and **out** processors. It tranfers each word as it arrives directly into memory, stealing cycles on the appropriate internal bus. Thus the **in** and the **out** processors deal with the network only when they have to. The **in** and the **out** buses can transfer a 32 bit word in 240 ns. Each bus should support up to 30-35 nodes without serious performance degradation; hence the current configuration of the Linda Machine should support roughly a thousand nodes.

36.4   Performance estimates

The Linda Machine has been designed and is in the process of fabrication; we are
not yet in a position to quote performance figures. Our reference points are the
various software Linda implementations; on S/Net, hypercubes and shared memory machines like the Encore Multimax and the Sequent Balance. Our hardware
implementation improves upon these in the following respects:

1. Communication functions are offloaded from the computing processor. In the
   S/Net implementation, for example, every **out()** (or every **in()** depending
   on the tuple distribution protocol) causes a message broadcast that interrupts
   every processor on the bus. In the Linda Machine, the computing processor
   involves itself only with its own transactions. In the hashing schemes on
   the hypercubes, the computing processor at a node is required to suspend
   its computation in order to respond to **in()** and **out()** requests from other
   nodes.

2. Tuple storage management is offloaded from the computing processor.

3. Tuple search operations are highly optimized through fine tuned microcode
   that runs on a custom architecture.

4. Communication performance is improved beacuse, one, the network architecture reduces the size of the broadcast sets and therefore the distance each
   message has to travel. Secondly, the microarchitecture is tailored to the semantics of tuples. Hence the overhead involved in 'packetising' tuples on more
   general communication hardware (the so-called 'software latency' of messages)
   is minimized.

Our estimates based on the characteristics of the hardware indicate an impressive
improvement over current software versions of Linda. The following exercise gives
an informal idea of this improvement. Consider the following programs running on
two processors in the same out-set.

Prog1: **in("ping");**

Prog2: **out("ping");**

The tuple involved is a minimal (16byte) tuple and hence this program estimates the
bare communication cost involved in moving a tuple from one processor to another.
Prog2 should execute in about 13 μsecs from the time the CPU initiates the out.
Of these, 6 μsecs is the estimated time for the CPU to communicate the request to
the out processor and 7 μsecs is the estimated time for the installation of the tuple
in the out-set. Prog1 should execute in about 21 μsecs after the tuple is received.
(Match resolution takes about 15 μsecs, and the CPU picks up the matching tuple

in about 6 μsecs). This gives a total of 34 μsecs for the maximally simple in-out pair. A comparable figure for Linda on the 18 processor Encore Multimax shared memory machine is approximately 200 μsecs; on the 10-processor S/Net, about 1.3ms [Ca87].[4] It must be stressed here that the numbers quoted for the software implementations (in particular on the shared memory machines) will *not* scale to machines much larger than their present size. The figures quoted for the Linda Machine, on the other hand, will hold for upto a thousand nodes. Also, the Linda Machine, with its optimized search mechanism, will scale better as tuple spaces grow larger and more complex. Quite soon we expect to report performance studies backing these claims.

36.5 Conclusions

We've omitted significant details in describing the Linda Machine architecture, but the discussion above is sufficient, we hope, to convey this machine's fundamental character. We start with a programming abstraction that we find powerful and attractive: a swarm of tuples, some passive and some active, grows, shrinks and maintains internal coordination by generating and consuming more tuples. Every active tuple (every process) must have unimpeded access to the whole of whatever tuple space it inhabits. It must be able to see and grab any tuple it can describe. This illusion of a global bag of tuples is conjured up by the intercrossing tuple-beams of the Linda Machine network. A tuple-beam in turn is a bus-wide broadcast that is trapped in local store and maintained by the appropriate synchronization and matching routines.

The Linda machine we are building is a framework for large parallel engines—on the order, ultimately, of thousands of powerful computing nodes. The purpose of such a machine is to serve as the core engine in an integrated environment. Local workstations (themselves parallel machines) are outcroppings of the core. Such an environment offers the user the opportunity to develop his own parallel applications, and the use of a large collection of parallel utilities. These should provide (for example) basic numerical routines, graphics and design tools, fast parallel compilers and intelligent inferencing from huge, constantly-changing databases. The resulting environment should be extremely powerful and a lot of fun.

References

[Ah83] S.Ahuja, "S/Net: A high speed interconnect for multiple computers," *IEEE Selected Areas in Communication*, pp751-756, Nov. 1983.

---

[4] S/Net Linda's performance seems roughly comparable to the performance of highly-optimized LAN communication systems like the V kernel [CG86].

[ACG86] S.Ahuja, N.Carriero, D.Gelernter, "Linda and Friends," *IEEE Computer*, pp110-129, May 1986.

[ACGK87] S.R. Ahuja, N. Carriero, D. Gelernter and V. Krishnaswamy, "The Linda Machine", *Yale University Tech. Report*, 1987.

[AGP78] Arvind, K.P. Gostelow and W. Plouffe, "An Asynchronous Language and Computing Machine," *TR114a University of California, Irvine*, January 1978.

[BD86] P. Bitar and A. Despain, "Multiprocessor Cache Synchronization – Issues, Innovations, Evolution," *Proceedings, 13th Annual Intl. Symposium on Computer Architecture*, pp 424-433, 1986.

[BCG87] R. Bjornson, N. Carriero and D. Gelernter, "Linda on Distributed Memory Sytems," *Proc. 1988 Workshop on Hypercube Multiprocessors*, to appear.

[BSUH87] W.R. Bush, A.D. Samples, D. Ungar and P.N. Hillfinger, " Compiling Smalltalk-80 on to a RISC," *Proceedings, 2nd Intl. Conf. on Architectural Support for Prog. Languages and Operating Systems*, pp112-116, October 1987.

[Ca87] N. Carriero, "Implementing Tuple Space Machines," *PhD thesis, Yale University.*

[CG86] N.Carriero, D.Gelernter, "The S/Net's Linda kernel," *ACM Trans. on Computer Systems*, pp110-129, May 1986.

[CGL86] N.Carriero, D.Gelernter, J.Leichter "Distributed Data Structures in Linda," *Proc. ACM Symp. on Principles of Prog. Lang.*, Jan 1986.

[D87] W.J. Dally, L. Chao, A. Chien, S. Hassoun, W. Horwat, J. Kaplan, P. Song, B. Totty and S. Wills, "Architecture of a Message Driven Processor," *Proceedings, 14th Annual Intl. Symposium on Computer Architecture*, pp 189-196, 1987.

[F83] A.L. Fisher, H.T. Kung, L.M. Monier, H. Walker and Y. Dohi, "Design of the PSC: A Programmable Systolic Chip," *Proceedings of the Third Caltech Conference on Very Large Scale Integration*, March 1983.

[H85] D.Hillis, "The Connection Machine," *The ACM Distinguished Dissertation Series, MIT Press*, 1985.

[Gel84] D.Gelernter, "Dynamic global name spaces on network computers," *Proc. Intl. Conf. on Parallel Processing*, pp25-31, Aug. 1984.

[G82] A. Gottlieb, R. Grishman, C.P. Kruskal, K.P. McAuliffe, L. Rudolph and M.Snir, "The NYU Ultrcomputer - Designing an MIMD Shared Memory Parallel Computer," *Proceedings, 9th Annual Intl. Symposium on Computer Architecture*, pp 27-42, 1982.

[GKI85] J.R. Gurd, C.C. Krikham, I.Watson, "The Manchester Prototype Dataflow Computer," *Communications of the ACM*, vol. 28, no. 1, January 1985.

[M85] D.A. Moon, "Architecture of the Symbolics 3600," *Proceedings, 12th Annual Intl. Symposium on Computer Architecture*, pp 76-83, 1985.

[Lu86] S.Lucco, "A heuristic Linda kernel for hypercube multiprocessors," *Proc. 1986 Workshop on Hypercube Multiprocessors*, Sept. 1986.

[W85] C. Whitby-Strevens, "The Transputer," *Proceedings, 12th Annual Intl. Symposium on Computer Architecture*, pp 292-300, 1985.

[W87] M.J. Wise "Prolog Multiprocessors," *Prentice-Hall, Australia*, (in press).

INDEX

Absolute addressing, 382
Acceptance probability, 517
Active-data model, 353
Adaptive programming, 693
ADM(Augmented Data Manipulator) network, 541, 542
Algebraic straightline program, 172
Alignment, 81
All-Prefix-Sums problem, 141, 144
Allocation (array), 610
Alternating Turing machine, 163, 170
Alternative paths, 79
Amorphous silicon, 38
Analog computation, 33, 636, 640
Anti-delay, 252
Applicative task, 476
ARBITRARY model, 120
Arithmetic expression, 191
Array (amplifier), 36
Array processor, 76, 100, 269, 630
Artificial intelligence, 159
ASAD (Application-Specific Architecture Design), 326, 327
ASEN (Augmented Shuffle-Exchange Network), 495, 497
Associative memory, 36, 385, 699, 701
Associative processing, 646
Associative retrieval, 636
Asynchronous messages, 609
Asynchronous parallel tree contraction, 154
ATM (see Alternating Tree Machine)
Atomic actions, 476
Atomic systolic computation, 276
Augmentation technique, 559

Augmented networks, 495, 497, 541
Availability, 567

Balanced network, 85, 189, 199
Bandwidth, 78, 326, 529, 600, 608
Belvedere, 403
Bilinear form, 172
Binary n-cube, 206, 209, 212
Binary tree, 103, 210
Bit serial, 604, 646
Blocking, 82, 497, 529
Boltzmann machine, 642
Boolean circuit, 25, 184, 185
Bounded Reduction, 140, 146
Bridge, 141
Broadcast, 226, 290, 359, 363, 603, 611, 700, 706
Buffering, 380
Built-in-test, 425, 439
Bus contention, 82, 460
Butterfly, 210, 641

Cache coherence, 708
Capacity function, 130
Case statement, 233
Causal consistency, 224
Causal dependence, 228
CCRC, 592
CEDAR multiprocessor, 478
Cellular architecture, 645
Cellular automata, 3, 14
Cellular logic, 25
Census function, 161
C-equivalent, 141
Channel object, 225

Character recognition, 39
Checkpointing, 476
CHiP computer, 265, 393, 400, 436
Circuit switching, 498, 517
Circuit value problem, 168
Class NC, 117, 125, 158, 160
Class P, 120, 158
Classes (complexity),101, 157
Classes (objects), 219
Clock buffer, 87
Clock distribution, 51, 54, 87
Clock period, 90
Clock skew, 87, 88
Coarse grain, 646
Code optimization, 117
COLLAPSE operation, 153
Collective modes, 6
Column broadcast 616
Combination tag, 545
COMMON model, 119
Communicating Sequential Processes
        (Hoare), 264
Communication
    bandwidth, 5, 30, 78, 367
    bottleneck, 204
    constraints, 23
    cost, 213, 248
    delay, 51, 70
    overhead, 100, 702
    processor, 330
    protocols, 675
    rate, 75
    scheme, 187
    space complexity, 216
    space cost, 216
    stream, 291
    time complexity, 215
    time cost, 215
Compilation (architecture), 327, 331
Complexity
    algorithm, 341, 344
    class, 157
    iteration, 341
COMPRESS operation, 121, 141,
        149, 192

Computation
    cost, 204
    cycles, 291
    specification, 327
    time, 205
Computational origami, 25
Concurrency, 228
Concurrency (functional), 8
Concurrency (true), 218
Concurrent diagnosis, 436
Concurrent error detection, 592
Concurrent testing, 439
Conditional execution, 612
Conditional expressions, 233
Conditional invariants, 676
Conflict, 228, 233
Conjugate switch, 504
Connection Machine, 106, 703
Connectionist model, 33, 34
Connectivity, 5, 529
Consistency checks, 425
Contention, 365, 460, 606
CONTRACT operation, 121, 149, 192
Convolution codes, 660
Correlation distance, 640
Cost
    communication, 204, 248
    computation, 204
    physical space, 215
    physical time, 215
Coupling (intercomponent), 8
CRCW (see PRAM)
CREW (see PRAM)
Critical loop, 314
Critical path, 304, 307, 341
Crossbar network, 496, 526
Cube-Connected-Cycles,   100,   160,
        190, 209, 212, 641
Cyclo-stationary, 303, 319

DAG (directed acyclic graph), 168,
        310
Dahbura-Mason algorithm, 416
Data assertion, 226
Data dependence graph, 620

Data-driven, 363
Data flow, 326, 328, 476, 698
Data reduction, 610
DCG (directed cyclic graph), 310
De Bruijn graph, 210, 661
Deadlock, 82, 355
Debugging (parallel), 402
Decidable problem, 157
Defect density, 60
Delay fluctuation, 71
Delay optimal, 316
Delay variation, 71
Delta Network, 503
Dependability (system), 450
Depth first search, 195
Derandomization, 196
Destination distribution, 73
Deterministic algorithm, 183, 188
Deterministic coin tossing, 198
Deterministic scheduling, 303
Deterministic traffic flow, 348
Diagnosis, 293, 413, 418
Differential invariants, 676
Digit exchange graph, 189
Digit routing, 512
Dilation, 102
Diogenes approach, 437
Distributed buffering, 94,95
Distributed computing 65
Distributed data structures, 701
Distributed fault effect, 579
Distributed queue, 361
Distributed routing, 79,496
Distributed scheduling, 346
Distributed (tasks), 33
Divide-and-conquer, 99, 108
DMC (Distributed Macro Controller), 689
DRONE processor, 440
Duality, 252
Dynamic expression evaluation, 191
Dynamic programming, 649, 660, 663
Dynamic reconfiguration, 19, 437, 501
Dynamic tree expansion, 157, 158, 163, 164

Edge-symmetric graph, 210
Efficiency (programming), 600, 604
Elastic path, 78
Electro-optic switches, 16
Embedded edges, 205
Embedding, 99, 102, 265, 276
Equational reasoning, 249
Equivalence (observable), 237
Equivalence (on hardware), 237
EREW (see PRAM)
Error
    1-sided, 2-sided, 186
    checking, 16
    confinement, 482
    correcting codes, 424
    detection, 15, 271, 424, 486
    fixed, 5
    propagation, 481
    transient, 5
ESC (Extra-state cube network), 533
Euler tree tour, 145
Execution time, 68
Expansion, 102
Extra Stage Cube Network (ESC), 533
Extra stage MADM network, 559

Failure, 567
Failure (server), 453, 455
Failure (task), 453, 454
Fast Fourier Transform, 4, 105, 329, 614
Fast parallel algorithms, 118
Fault, 100, 412, 567, 571
    avoidance, 101, 565
    containment, 475, 483, 484
    coverage, 484
    detection 350, 422, 586
    hard, 5
    management, 8
    recovery, 350
    soft, 5
    tolerance, 15, 60, 76, 349, 422, 436, 449, 463, 475, 497, 511, 542, 554, 559, 567, 602, 615

Fault-free, 447
Feature extraction, 39
Fine grain, 54, 65, 70, 304, 353, 375,
        402, 476, 645, 646, 687
Finite event, 217, 220
Fixed connection network, 184, 186
FLASHSORT algorithm, 187, 190
Floating-point array 329
Flow control, 380, 693
Flow graph, 313
Flow network, 130
Fluid-flow model, 325, 332
FPS T-Series computer, 607
Functional programming, 245, 246,
        475, 476

GaAs, 10, 47, 48, 77
Gamma network, 547
Gantt chart, 307
GAPP (Geometric Arithmetic Paral-
        lel Processor), 691, 693
GCN (Gated Connection Network),
        646, 647
GCP (Generic Code Processor), 396
Givens transformation, 626
Global
    address, 378
    clock, 51
    consistency, 708
Graceful degradation, 526
Granularity, 375
    variation, 68
Graph embedding, 99
Guards, 224

HADM (Half Augmented Data Ma-
        nipulator) network, 541, 553
Hard-core circuit, 438
Hardware maintenance, 425
Hashing, 705
Heap segment, 480
HEARTS, 265, 269
Heterostructures, 9
Hexagonal array, 587

Hierarchical, 646
    control, 691
    design, 95
    model, 327, 450
Hoare's CSP model, 264, 267
Hopfield model, 636
Hopfield/Tank model, 569, 636
H-tree network, 93
Hybrid wafer-scale circuits, 48, 75
Hypercube, 99, 100, 106, 159, 160,
        265, 478, 699, 603, 606, 608,
        616, 641, 705

IADM (Inverse ADM) network, 541,
        542
IC* model, 675
INTEGERSORT algorithm, 190
Image processing, 637, 645
Initialization (array) 293
Interconnection bandwidth, 23
I/O bottleneck, 6
ISOLATE operation, 150
ISOLATE-COMPRESS   operation,
        150
Isolation (technique), 149
Iteration count, 341
Iteration period, 309, 314
Iterative problems, 303

J-Machine, 376
Josephson junction, 82

Kalman filter, 620

Labelled transition system, 234
Labelled event structures, 228
Large grain, 70, 476
Las Vegas algorithm, 186
Latch, 252
Latency, 50, 308, 314, 601, 606, 616
Layout (efficient), 671
LINDA machine, 697, 698
Linear parallel algorithm, 191
Link-fault model, 501

List ranking, 121, 141, 144
Load balancing, 183, 358, 360, 600, 602, 615
Load-factor, 102
Local communication, 7, 72, 73
Loop unrolling, 303
Lowest cost path, 341
LU-decomposition, 585

MADM (Minimum ADM) network, 541, 556
Maintainability, 567
Majority voting, 15
Malfunction, 567
Mapping, 7, 66, 75, 100, 101, 331, 355, 398, 619
Matrix vector product, 272, 279
Maximum flow, 130, 131
M-contraction, 144
M-critical, 141
MDP (Message-Driven Processor), 376
Memory
    contention, 606
    latency, 606
    management, 690, 694
Memory-processor bottleneck, 603
MESFET, 50
Mesh array, 75, 77, 105, 280, 603, 646
Mesh-of-Trees, 101
Message based architecture, 270, 698
Message passing, 294, 375, 384, 477, 607
MIADM (Minimum Inverse ADM) network, 541
Micro-cables, 80
Microtransmission line, 62
MIN (Multistage Interconnection Network), 96, 478, 495, 526, 541, 542
Minimum cost path, 649
Minimum cut, 130
Minimum model, 120
Minimum parallelism, 315
Minimum vertex cover set, 416

Module binding, 331
Module (software), 328
Modules (physical), 329
Monotone circuit, 168
Monte Carlo algorithms, 186
MOSAIC, 376
MPP (Massively Parallel Processor), 600, 604
MTTF (Mean time to failure), 466, 515
Multicast, 700
Multiple-commodity fluid-flow, 332
Multistage network (see MIN)
Multiterminal reliability, 515
Multivalued logic, 15

Nanoelectronics, 3, 9
NC class, 117, 125, 158, 160
N-Cube, 106, 188, 372, 376, 606
NDF (Network Design Frame), 376
Nearest neighbor, 37, 264
Network delay (access), 78
Network
    flow, 345, 418
    queue, 358
    saturation, 74
Neural network, 4, 33, 567, 620
NMR (N- modular redundancy), 423, 443
Non-deterministism
    communication, 325
    distribution, 361
    routing, 364
Non-local communication, 72
Non-preemptive scheduling, 303, 304
Non-recursive problem, 157
NP-complete problem, 307
N-version programming, 423

Object oriented, 363, 698
Object-based paradigms, 219
Objects (named), 224
Oblivious algorithm, 189
Occam, 264, 703
Omega network, 661

Optical computing, 23, 27, 568, 640
Optical interconnections, 16, 23, 54, 77, 81
Optical logic etalon (OLE') -28
Optical modulator, 16
Optimum parallel algorithms, 191
Optimum speedup, 161
Optoelectronic device, 77
Overhead (message reception), 378, 382

Packet routing, 361, 364, 609
Parallel algorithm design, 158
Parallel architecture (multipurpose), 103
Parallel comparison tree model, 184, 185
Parallel hardware, 119
Parallel hashing, 192
Parallel prefix, 161
Parallel time, 119
Parallel tree contraction, 139, 149, 152, 191
Parallel tree evaluation, 161
Parallelizability of algorithms, 205
Partially augmented data manipulator, 545
Partitioning, 25, 376
P-class, 120, 158
Perfect shuffle, 24
Performability, 567
Permanent fault, 464
Permutation (data) 611
Piggyback array, 51
Pipeline algorithms, 264
Pipelined (array), 88
Pipelining, 8, 17, 47, 161
Planar circuit, 168
PMC model (Preparata, Metze, Chien), 411, 412, 439
Point event, 220
Point task, 67
POKER programming, 265, 391, 392
Polylog time, 119
Polylogarithmic algorithm, 159
Polylogarithmic time, 158, 160

Polynomial algorithm, 159
Polynomial product, 280
Polynomial size, 160
Polynomial time, 118, 158
PRAM (Parallel Random Access Machine), 159
    CRCS, 118, 119
    CRCW, 119, 139, 140, 184
    CREW, 118, 119, 184
    EREW, 118, 119, 139, 184
Precedence, 220
Prefix sum, 121
Prep-P Mapping, 391, 392
Principle of superposition, 6
Processor
    bound, 316
    optimal,316
    304, 319schedule, 330
Programmability, 110
Programmable interconnections, 39
Programming environment, 265, 599
Promotion (of FP laws), 256
Proof tree, 158, 164
PRUDENT COMPRESS operation, 153
Pseudo-dynamic traffic model, 348
Pseudo-random
    generator, 197
    sequence, 197

Quantum effects
    cellular automaton, 15
    coupled device, 3, 9, 11, 15
    dots, 10, 11
    size effects, 8, 10
    well, 11, 16, 27
QUEST (Quantum-well envelope state transition), 28
Queue, 3346, 358, 367, 382

RAKE operation, 121, 128, 149, 192
RAM (Random Access Machine), 118, 159, 203
Random mating lemma, 195
Randomization, 182
Randomized Boolean circuit, 185

Randomized algorithm, 181, 182, 186, 196

Randomized sorting algorithm, 187

Rate optimal, 316

Reconfigurable register, 606

Reconfiguration, 8, 17, 100, 349, 365, 436

Recovery, 456, 475, 487, 488

Rectangular grid, 105, 280

Recursive problem, 157, 230, 303

Reducible flow graph, 117, 123

Reduction, 146, 359, 363, 610

Redundancy, 423, 424, 436, 486, 497, 526, 533

Redundancy graph, 502

Regular array, 76

Regular control flow, 159

Regular data structure, 159

Regular interconnection, 24, 402

Regular layout, 372

Regular structures, 25

Relational database, 393

Relative addressing, 382

Reliability, 297, 436, 463, 464, 495, 498, 597

Rendezvous, 347

Repair, 62, 470

Resonant tunneling, 9, 11

Reversible logic, 15

RNC (Random NC) class, 121

Robustness (array), 100, 212

Routing, 80, 183, 187, 358, 380, 503, 512
    Upfal algorithm, 189
    Valiant algorithm, 189

Row broadcast, 616

RP3, 376

RPA (Reconfigurable Processor Array) computer, 353

Sanity message, 425

Scalability, 7, 393

Scaling
    devices, 7, 76, 88
    limits, 4, 5, 76

Schedule, 291, 303, 304, 319, 325, 331, 338, 378, 453, 609, 621

Schreiber design (matrix product), 281

SDEF, 263

Self-checking circuits, 424

Self-electrooptic effect device (SEED), 27

Self-implicating systems, 418

Self-organizing, 19

Self-reconfiguration, 435, 440

Self-repairing, 19

Self-scheduling, 349

Self-testing, 435

Self-timed clocking, 83

Semantics (of composition operators), 219

Separator set, 192

Shared memory, 118, 160, 184, 384, 477, 526, 606, 698, 705

SHARPE, 449, 471

Shortest path problem 542, 545

SHRINK operation, 128

Shuffle permutation, 614

Shuffle-exchange, 210, 503, 660, 667

SIGNAL, 328

Signal flow graph, 621

Silicon circuit board, 48

Simple Simon, 391, 392, 396

Simulated annealing, 640

Simulation, 95, 680

Simultaneity, 220

Skew, 51

Skin effect, 51

Small grain parallelism, 330

Sorting, 161, 183, 190

Space-time, 205, 271, 290

Spatial projection, 273

Spatial properties, 66

Special purpose VLSI, 34

Speed-of-light, 82

Speedup, 453

SR-redundancy, 440

SSIMD (Skewed single instruction multiple data), 321

ST-redundancy, 440

Stacks, 367
Statelessness, 257
Static communication, 291
Static invariants, 676
Static reconfiguration, 437, 501
Store-and-forward, 608
Stream communication, 291
String pattern matching, 280
Stuck-at fault, 575
Superconductivity, 81
Switch complexity, 498
Switch-fault model, 501
Switching node design, 330, 346
Symbolic substitution, 25
Symmetric communications network, 205
Symmetric complementation games, 194
Synchronization, 219, 226, 304, 376, 684
Synchronous communication, 51, 87, 294, 586, 609
System dependability, 450
System-level fault diagnosis, 411, 439, 440
Systolic array, 4, 17, 48, 159, 263, 265, 445, 585, 620, 698

Tags, 385, 545, 548
Target architecture, 327
Task
    descriptor, 479
    execution time, 68, 450
    switching time, 601
    token, 480
Temporal assertions, 236
Temporal logics, 221
Temporal path, 68
Temporal projection, 27
Terminal connectivity, 502
Test scheduling, 427
Testing, 271, 413, 438
Three dimensional systems, 17, 51, 75
Time complexity, 184
Time reversal, 16
Timelessness, 254

TMR (Triple modular redundancy), 476, 566
Traffic (segmentation), 343
Transient fault, 465, 586
Transitive closure, 120, 121
Translation domain dependences, 277
Transputer, 264, 293, 295, 353, 376, 608, 703
Traveling salesman problem, 569
Tree contraction, 121, 128, 150
Tuple (in LINDA), 699, 701
Turing machine, 204
Two-dimensional array, 320

Ultracomputer, 160
Unbounded reduction, 140
Uniform data dependences, 277
Unsolvable problem, 157
Upfal's routing algorithm, 189

Valiant's routing algorithm, 188
Validation (programmability), 101
Validity checks, 425
Vector computer, 601
Virtual address, 378
Virtual processors, 356
Virtual switch, 610
Vision processing, 637, 646
Viterbi algorithm, 659, 663
VLSI model (of computation), 665
Von Neumann bottleneck, 203

W2, 265, 267
Wafer-scale hybrid packaging, 48, 75
Wafer-scale integration, 48, 75, 427, 436, 500, 526, 565, 603
Warp computer, 264, 294
Warp (time), 609
Watchdog, 425, 485
Wavefront array, 620
Wire length/space, 213
Wormhole routing, 380

Y Machine, 683
Yield, 5, 436